Lecture Notes in Networks and Systems

Volume 357

The series "Lecture Notes in Networks and Systems" publishes the latest developments in Networks and Systems—quickly, informally and with high quality. Original research reported in proceedings and post-proceedings represents the core of LNNS.

Volumes published in LNNS embrace all aspects and subfields of, as well as new challenges in, Networks and Systems.

The series contains proceedings and edited volumes in systems and networks, spanning the areas of Cyber-Physical Systems, Autonomous Systems, Sensor Networks, Control Systems, Energy Systems, Automotive Systems, Biological Systems, Vehicular Networking and Connected Vehicles, Aerospace Systems, Automation, Manufacturing, Smart Grids, Nonlinear Systems, Power Systems, Robotics, Social Systems, Economic Systems and other. Of particular value to both the contributors and the readership are the short publication timeframe and the world-wide distribution and exposure which enable both a wide and rapid dissemination of research output.

The series covers the theory, applications, and perspectives on the state of the art and future developments relevant to systems and networks, decision making, control, complex processes and related areas, as embedded in the fields of interdisciplinary and applied sciences, engineering, computer science, physics, economics, social, and life sciences, as well as the paradigms and methodologies behind them.

Indexed by SCOPUS, INSPEC, WTI Frankfurt eG, zbMATH, SCImago.

All books published in the series are submitted for consideration in Web of Science.

More information about this series at https://link.springer.com/bookseries/15179

Yassine Maleh · Mamoun Alazab ·
Noreddine Gherabi · Lo'ai Tawalbeh ·
Ahmed A. Abd El-Latif
Editors

Advances in Information, Communication and Cybersecurity

Proceedings of ICI2C'21

Springer

Editors
Yassine Maleh
Sultan Moulay Slimane University
Beni Mellal, Morocco

Noreddine Gherabi
Sultan Moulay Slimane University
Béni Mellal, Morocco

Ahmed A. Abd El-Latif
Gamal Abd El-Nasir
Menoufia University
Menofia Governorate, Egypt

Mamoun Alazab
Charles Darwin University
Darwin, NT, Australia

Lo'ai Tawalbeh
San Antonio One University Way
Texas A&M University
San Antonio, TX, USA

ISSN 2367-3370 ISSN 2367-3389 (electronic)
Lecture Notes in Networks and Systems
ISBN 978-3-030-91737-1 ISBN 978-3-030-91738-8 (eBook)
https://doi.org/10.1007/978-3-030-91738-8

This Springer imprint is published by the registered company Springer Nature Switzerland AG
The registered company address is: Gewerbestrasse 11, 6330 Cham, Switzerland

Preface

The recent advances in machine intelligence and cloud computing have profoundly changed people's daily lives. Implementing such techniques addresses various fields and makes things "smart" in different domains, such as big data and cloud computing, AI and machine learning, Internet of things, cybersecurity, elearning, transportation systems, environment, and industrial systems.

In the context of COVID-19 pandemic, the obligation for structures is to accelerate their digital transformation. Their priority is to protect employees and ensure the continuity of the company's activities. It is imperative to continue adapting its IT system, which is necessary for employees to work from home and promote collaborative digital tools. It is essential to strengthening transformation efforts capable of increasing agility, flexibility, security, and resilience in cybersecurity, tools around data analysis, planning, and even value supply chain optimization.

This volume constitutes the refereed proceedings of the International Conference Information, Communication, and Cybersecurity ICI2C 2021, held in Khouribga, Morocco, on November 10–11, 2021. ICI2C2021 was jointly co-organized by the National School of Applied Sciences of Sultan Moulay Slimane University, Morocco, and Charles Darwin University, Australia.

The 54 full papers were carefully reviewed and selected from 159 submissions. The papers presented in the volume are organized in topical sections on synergies between machine intelligence and information retrieval; smart systems and networks; online learning; smart healthcare; cybersecurity, and information assurance. All contributions were subject to a double-blind review. The review process was highly competitive. We had to review near to 159 submissions. A team of over 80 program committee members and reviewers did this terrific job. Our special thanks go to all of them.

ICI2C2021 offered again an exciting technical program as well as networking opportunities. Outstanding scientists and industry leaders accepted the invitation for keynote speeches:

ICI2C2021 offered again an exciting technical program as well as networking opportunities. Distinguished scientists accepted the invitation for keynote speeches:

- **Vincenzo Piuri**
  University of Milan, Italy
- **Helen Karatza**
  Professor Emeritus, Aristotle University of Thessaloniki, Greece
- **Imed Romdhani**
  University of Edinburgh Napier, UK
- **Yves Duthen**
  Toulouse 1 Capitole University, France
- **Al-Sakib Khan Pathan**
  Independent University, Bangladesh
- **Mamoun Alazab**
  University Charles Darwin, Australia
- **Nabil Benamar**
  Moulay Ismail University, Morocco
- **Philippe Roose**
  University of Pau, France

We want to take this opportunity and express our thanks to the contributing authors, the members of the program committee, the reviewers, the sponsors (OCP, EMSI, ARCIC, CNRST, and IEEE CN Morocco), and the organizing committee for their hard and precious work. Thanks for your help—ICI2C 2021 would not exist without your contribution.

Yassine Maleh
Mamoun Alazab
Noreddine Gherabi
Lo'ai Tawalbeh
Ahmed A. Abd El-Latif

Organization

Committee

Honorary Committee

Nabil Hmina	President of Sultan Moulay Slimane University, Beni Mellal, Morocco
Mohammed Sajieddine	Director of ENSA Khouribga, Beni Mellal, Morocco
Hafida Hanine	Vice President, Sultan Moulay Slimane University, Beni Mellal, Morocco

General Chairs

Yassine Maleh	SMIEEE, National School of Applied Sciences, Khouribga, Morocco
Noreddine Gherabi	National School of Applied Sciences, Khouribga, Morocco

Conference Co-chairs

Mohamed Moughit	National School of Applied Sciences, Khouribga, Morocco
FATANI Imade Fahd Eddine	National School of Applied Sciences, Khouribga, Morocco
Mamoun Alazab	Charles Darwin University, Australia

Technical Program Chairs

Imed Romdhani	Edinburgh Napier University, Edinburgh, UK
Mustapha Belaissaoui	Hassan 1st University, Morocco

Mohammad Shojafar	SMIEEE, University of Surrey, UK
Abdelkrim Haqiq	SMIEEE, Hassan 1st University, Morocco

Publication Chairs

Yassine Maleh	SMIEEE, National School of Applied Sciences, Khouribga, Morocco
Mamoun Alazab	Charles Darwin University, Australia
Noreddine Gherabi	National School of Applied Sciences, Khouribga, Morocco
Lo'ai A. Tawalbeh	SMIEEE, UTSA, USA
Ahmed A. Abd El-Latif	Menoufia University, Egypt

Steering Committee

Aboul Ella Hassanien	Cairo University, Egypt
Christoph M. Friedrich	University of Applied Sciences and Arts, Dortmund, Germany
Arpan Kumar Kar	Indian Institute of Technology Delhi, New Delhi, India
Patrick Siarry	Universit de Paris 12, France

Sponsoring Chair

Brahim EL BHIRI	EMSI, Morocco

Special Sessions Chairs

Makhrout Mohammed	ENCG, Hassan 1st University, Morocco
Mohamed Amnai	Ibn Tofaïl University, Kenitra, Morocco
Mounim Kabouri	ENCGS, Morocco

Publicity Chairs

Alireza Souri	Halic University, Istanbul, Turkey
Aswani Kumar Cherukuri	Vellore Institute of Technology, India
Ferda Özdemir Sönmez	Imperial College London, UK
Saad Motahhir	Sidi Mohamed Ben Abdellah University, ENSA, Fez, Morocco
Youssef Baddi	University Chouaib Doukkali, El Jadida, Morocco

Technical Program Committee

Francesco Flammin	Linnaeus University, Sweden
Kim-Kwang Raymond Choo	UTSA, USA
Imed Romdhani	Edinburgh Napier University, Edinburgh, UK
Mohammad Shojafar	SMIEEE, University of Surrey, UK
Hossain Shahriar	Kennesaw State University, USA
Patrick Siarry	Universit de Paris 12, France
Aswani Kumar Cherukuri	Vellore Institute of Technology, India
Xuping Huang	Meiji University, Japan
Sang-Bing Tsai	University of Electronic Science and Technology, China
Matthieu Perrin	University of Nantes, France
Khalifa Mansouri	University Hassan II Casablanca, Morocco
K. P. Chow	University of Hong Kong, Hong Kong
Christoph M. Friedrich	University of Applied Sciences and Arts, Dortmund, Germany
Tzung-Pei Hong	National University of Kaohsiung, Taiwan
Wenjuan Li	Hong Kong Polytechnic University, Hong Kong
Said El Kafhali	Hassan 1st University, Morocco
Yudi Prayudi	Center for Digital Forensics Studies, Indonesia
Neetesh Saxena	Cardiff University, UK
Youssef Qasmaoui	Hassan 1st University, Morocco
Kun Hao	Northeastern University, USA
Antonio J. Jara	University of Applied Sciences Western Switzerland, Switzerland
Rafigul Islam	Charles Sturt University, Australia
Chck Yoo	Korea University, South Korea
Khizar Hameed	University of Tasmania, Australia
Ahmad Sghaier Omar	University of Waterloo, Canada
Lo'ai A. Tawalbeh	Florida A&M University, USA
Sheraz Anwar	Xiamen University, China
Kevin Curran	Ulster University, UK
Youssef Baddi	University Chouaib Doukkali, El Jadida, Morocco
Shareeful Islam	University of East London, UK
Mustapha Belaissaoui	Hassan 1st University, Morocco
Rajgopal Kannan	Louisiana State University, Las Angeles, USA
Chang-Tsun Li	University of Warwick, UK
Rajgopal Kannan	Louisiana State University, Las Angeles, USA
Mohammed Amin Almaiah	King Faisal University, Saudi Arabia
Nilesh Uke	Pimpri Chinchwad College of Engineering, India
Ayman Alkhalifah	La Trobe University, Australia
Abderrahim Bajit	ENSAK, Ibn Tofail University, Morocco
Abdelwahab Boualouache	University of Luxembourg, Luxembourg

Khalid El Makkaoui	University Mohamed 1, Oujda, Morocco
Franklin	Kennesaw State University, Marietta Campus, USA
Xuping Huang	Advanced Institute of Industrial Technology, Japan
Halim Khelalfa	University of Wollongong in Dubai, United Arab Emirates
Ramgopal Kashyap	Amity University, India
Hridoy Mahanta	Assam Don Bosco University, India
Yusuf Albayram	Central Connecticut State University, USA
Mauro Pappaterra	Uppsala University, Sweden
Mounia Zaidi	Hassan 1st University, Morocco
Ferda Özdemir Sönmez	Middle East Technical University, Turkey
Alireza Souri	Halic University, Istanbul, Turkey
Pedro Antonio Martin Cervantes	University of Almeria, Spain
Homero Toral Cruz	University of Quintana Roo, México
Sumaiya Thaseen	VIT University, India
Sang-Bing Tsai	University of Electronic Science and Technology of China, China
Xiaohong Yuan	North Carolina A&T State University, USA
Mohamed Amine Ferrag	Guelma University, Algeria
Abdelkebir Sahid	University Hassan 1st, Morocco
Abdelaziz Amara korba	Badji Mokhtar-Annaba University, Algeria
Jawad Awan	University of Sindh, Pakistan
Franklin	Kennesaw State University, Marietta Campus, USA
Xuping Huang	Advanced Institute of Industrial Technology, Japan
Yassine Sadqi	FP, Beni Mellal, Morocco
Giuseppe Ciaburro	Università della Campania Luigi Vanvitelli, Italy
Hridoy Mahanta	Assam Don Bosco University, India
Youssef Gahi	Ibn Tofail University, Kenitra, Morocco
Ranesh Kumar Naha	University of Tasmania, Australia
Della Krachai Mohamed	University of Science and technologies of Oran, Algeria
Sheraz Anwar	Xiamen University, China
Khizar Hameed	University of Tasmania, Australia
Saad Motahhir	Sidi Mohamed Ben Abdellah University, ENSA, Fez, Morocco
Moez Esseghir	Technology University of Troyes, France
Samuelson W. Hong	SMIEEE, Oriental Institute of Technology, Taiwan
Gayatri Sunilkumar Pandi	LJ Institute of Engineering and Technology, Ahmedabad, India

Organizing Committee

Yassine Maleh	SMIEEE, National School of Applied Sciences, Khouribga, Morocco
Noreddine Gherabi	National School of Applied Sciences, Khouribga, Morocco
Mohamed Moughit	National School of Applied Sciences, Khouribga, Morocco
FATANI Imade Fahd Eddine	National School of Applied Sciences, Khouribga, Morocco
Soufyane Mounir	National School of Applied Sciences, Khouribga, Morocco
Khalid El Gholami	National School of Applied Sciences, Khouribga, Morocco
Ouafaa Ibrihich	National School of Applied Sciences, Khouribga, Morocco
Mostafa Saadi	National School of Applied Sciences, Khouribga, Morocco
Mohamed Nasri	National School of Applied Sciences, Khouribga, Morocco
Aziza Abid	National School of Applied Sciences, Khouribga, Morocco
Omar El Bennay	National School of Applied Sciences, Khouribga, Morocco

Contents

Advances in Machine Intelligence and Information Retrieval

Adaptive Learning Algorithms and Platforms: A Concise Overview

Hammadi Mezin, Saad Yasser Kharrou, and Ayoub Ait Lahcen(✉)

Engineering Sciences Laboratory, National School of Applied Sciences, Ibn Tofail University, Kenitra, Morocco
ayoub.aitlahcen@uit.ac.ma

Abstract. The distance education has become an indispensable teaching method, especially after the situation of Covid-19. For this reason, educational institutions are looking for e-learning platforms that offer better course management, ease of administration, user friendly, and achieving the learning objectives. Adaptive learning is considered an active research area, it enables to detect learning style of learners based on their behaviors and learning purposes in order to recommend relevant course materials. The objective of this article is to present an overview of personalization in the traditional learning system and the new developed systems as well as the approaches used to understand the learner's individual needs. Furthermore, this work analyzes the problems in these systems and presents the prospect of development.

Keywords: Adaptive E-learning · Intelligent tutoring systems · Learning models · Knowledge tracing

1 Introduction

Online learning has become an essential element in the educational strategy adopted by the institutions and universities. The instructional strategies are transitioned from face-to-face learning to distance education. However, the current e-learning systems face several challenges that retard the utilization of such systems, we mention the changed management issues, the technical problems, and the financial support issues [1].

In fact, the lack of customization and adaptivity is one of the main reasons of e-learning failures [2]. Therefore, learners find difficulties to take courses individually following the current learning models. This is due to organizational factors (e.g. inadequate platforms) and psychological reasons (e.g. feelings of isolation). Also, the learners need to interact with the system and its adapted content to acquire new skills and competences.

According to the literature [3,4], different categories of personalization have been proposed in e-learning systems in order to enhance the way of teaching and learning. These include the development of new adaptive systems, the introduction of web-based online platforms with the adaptation of learning style,

Y. Maleh et al. (Eds.): ICI2C 2021, LNNS 357, pp. 3–12, 2022.
https://doi.org/10.1007/978-3-030-91738-8_1

and plugins added to learning management system (LMS). In this work, we are interested in the adaptation of pedagogical sequences and the commonly used techniques for recommendation of learning materials. The adaptive systems recommend learning resources to learners according to their goals, their mastery of knowledge, and behaviors. Knowledge tracing models can be used to estimate the state of mastery of a given concept.

The paper is organized as follows. Section 2 presents the modules of adaptive e-learning systems. Section 3 describes the approaches used to detect the learning style and trace the knowledge state of students. Section 4 examines the customization of learning based on the learners' needs in the selected educational systems and tools. Section 5 discusses open issues and expectations to enhance the quality of learning. Finally, Sect. 6 concludes the paper and remarks.

2 Background

The Adaptive E-learning Systems (AES) are introduced to make online education more accessible, interactive and to select the shortest learning path. These systems track intelligently the progress of the learner on a specific learning goal, also they provide intelligent assistance when the learner faces regularly problems with particular knowledge.

The AES consist of three principal modules which are: learner model, domain model, and adaptation model [5].

Learner Model: The learner model represents targeted and pertinent information about the student, characteristics and problem-solving performance [6]. It is considered the core of every adaptive e-learning system [7]. This model contains all the actions and skills related to the learner's profile. Moreover, it updates learner's profile attributes based on his activities and deliver personalized content for each learner [8]. The authors [9] provided an extensive study of learner modeling in adaptive educational hypermedia.

Domain Model: The domain model, also called expert model, describes the expert knowledge in a field of expertise [10]. It consists in structuring and organizing the knowledge of a particular domain [11].

The domain model can be represented using the frame-based which contains attributes describing learning components, network-based which composed of nodes and edges to represent the concepts and the relation between them, and logic-based which represent the concepts based on rules and conditions [12].

Adaptation Model: The adaptation model matches the characteristics of the learner model with the concepts of the domain model in order to find suitable learning items for the student. It selects the teaching choices based on the student preferences, performance, and his progress in a course [10].

The adaptation mechanisms have been studied since the nineties, the article [13] is considered among the first works treating adaptive systems. Later, the author [14] mentioned the most known forms of adaptation: the first based on presentation and the other focused on adaptive navigation.

3 Adaptive Learning Algorithms

In this section, we present the knowledge tracing models employed to predict the student performance, the learning style detection approaches, and adaptive test methods.

3.1 Item Response Theory (IRT)

The IRT [15] is a test theory that models the mastery of concepts by the learner based on the relationship between the characteristics of question and the answers on this question. An Item Characteristic Curve (ICC) is used to estimate the probability of student answering an item correctly [16]. Indeed, the IRT estimates the ability of an examinee based on his responses (true/false) to past test questions. The ICC can be modeled according to the number of parameters in logistic function: 1-parameter logistic model (1PL), 2-parameter logistic model, and 3-parameter logistic model (3PL) [17].

3.2 FSLSM (Felder-Silverman Learning Style Model)

The authors [18] created the FSLSM model that combines the expertise in educational psychology and the experience in engineering education. The FSLSM describes the preferences that the learner chooses to understand a learning concept. For example, some learners prefer to learn a course by means of videos and discussing with other students, while others prefer textual presentation and self-reflection after the end of the course. The model represents the learning styles of the student in four dimensions: Sensing/Intuitive Learners, Visual/Auditory Learners, Inductive/Deductive Learners, and Active/Reflective Learners.

3.3 Bayesian Knowledge Tracing (BKT)

The BKT approach is initially described in the early 1995s [19]. It represents the knowledge state in two binary variables: the concept is acquired or not by the student. This model uses two-state Hidden Markov Model (HMM) in order to update the probabilities of the hidden variables based on the student's answer to the previous question (correct or incorrect). Afterward, the approach has undergone several improvements: introducing item difficulty to the model [20], individualized BKT [21], and estimating when learners acquire concepts heuristically [22].

The authors [23] developed a python implementation of BKT named pyBKT. This library estimates learner cognitive mastery based on problem solving exercises.

3.4 Performance Factors Analysis (PFA)

The PFA [24] is based on the reconfiguration of the Learning Factor Analysis (LFA) data mining technique [25]. This algorithm takes the form of a standard logistic regression model with student performance as the dependent variable [26]. With a simple change in the LFA method, the PFA ignores the learner's competence. Instead of taking the total number of repetitions of the KC (knowledge component), this model takes into account the number of successes and failures obtained while examining this KC.

3.5 Computerized Adaptive Test (CAT)

In Computerized Adaptive Test, the tests are designed according to the abilities of each learner in a flexible manner [27]. With the emergence of the Web, the adaptive tests are widely used in online learning during the assessment [28], and medical field [29]. These new tests replace traditional measurement (fixed tests content). The adaptive assessment followed the CAT approach and mostly applied the IRT [16] to provide the most relevant items for students based on their individual performance.

3.6 Deep Knowledge Tracing (DKT)

The DKT model predicts learners' responses to next question based on their previous history [30]. This approach uses Recurrent Neural Networks (RNNs) architecture. The RNNs is characterised by its capacity to exploit input data from much later in time, Long Short Term Memory (LSTM) is the most known type of the RNN that demonstrates its performance.

Giving an input sequence of vectors $x_1, ..., x_T$, the objective is to predict an output sequence of vectors $y_1, ..., y_T$ which corresponds to the probabilities of answering the question correctly. The hidden state function is defined by the equation:

$$h_t = \tanh(W_{hx}x_t + W_{hh}h_{t-1} + b_h) \quad (1)$$

Where h_{t-1} the vector of the previous state, W_{hx} the weight matrix, W_{hh} the recurrent weight matrix, and b_h the bias.

The outputs y_t are calculated using the Sigmoid function:

$$y_t = \sigma(W_{yh}h_t + b_y) \quad (2)$$

4 Adaptive Learning Platforms

Learning platforms can apply content personalization in different learning activities. The previous studies provided an idea on how the customization can be integrated into the process of learning. The Table 1 summarizes the educational systems reviewed in this article.

ALOSI (Adaptive Learning Open Source Initiative): ALOSI is an open source adaptive learning framework developed by Harvard university and Microsoft in partnership with edX [31]. The proposed solution is composed of an adaptive engine, bridge for adaptivity, learning management system, and content source. This framework uses BKT to model students' knowledge state. In

Table 1. Systems summaries provided personalized learning

System	Type	Personalization strategies	Approaches used	Main contribution
ALOSI [31,32]	Adaptive engine	Offering different learning paths	Bayesian Knowledge Tracing (BKT)	Increase learning gains
Moodle [33,34]	Learning Management System	Offering different learning paths	FSLSM questionnaire	The personalization affects students' motivation and performance
Canvas LMS	Learning Management System	Offering different learning paths	Manual customization	Achieve mastery learning goals in a course
Protus [3,8]	Programming tutoring system	Suggesting different learning activities	FSLSM questionnaire	Learners have gained more concepts in less time
Concerto [28]	Adaptive testing	Providing adaptive assessments	Computer Adaptive Tests (CAT), Item Response Theory (IRT)	Reduce the length of the tests, enhance the precision, improve the user experience
Mod-Knowledge agent [35]	Intelligent agent	Recommending different learning activities	Dynamic Key-Value Memory Net (DKVMN)	Improve the accuracy of recommendation
MathBot [36]	Chatbot	Providing practice questions with personalized feedback	Natural Language Processing (NLP)	Produce learning gains
SKOPE-IT [37]	Intelligent tutoring system	Offering self-explanation tutoring dialogs	Natural Language Processing (NLP)	Increase knowledge level

the study [31], the system is tested with Microsoft Massive Open Online Course (MOOC) on edX including 35 knowledge components and 400 assessment associated with those knowledge. Three experimental groups are constructed, two groups have different level of adaptivity (remediation, continuity), the third group has no adaptive algorithm. The results indicated that the adaptive groups had access to more questions in the learning, furthermore the number of learners left is lower than the control group.

In another study [32], the authors implemented the adaptivity on multiple choice assessment problems.

Moodle: The authors in [33] proposed an approach to detect the learning style of learners using data mining techniques. The proposed system is based on a hybrid model that uses automatic detection of learning style and Index of Learning Styles (ILS) questionnaire for better precision. The contribution consists in adding new extensions for adaptation instead of modifying the Moodle core. The authors formed two groups of learners, the first group includes the adaptive mechanism in Moodle course and the second without any adaptivity. The results showed that the adaptive mechanism leverages the performance of students.

Another work in [34] extended Moodle platform (plugins) in order to provide adaptive lessons, exercises and subject practices. In the proposed approach, the recommendation of the learning path is based on learning style and previous knowledge. The authors used FSLSM model to describe the learning style of each student.

Canvas LMS: Canvas is an open source learning management system developed by the educational technology company Instructure. This platform provides many features to design courses and deliver instructions according to the teaching models. Further, Canvas offers MasteryPaths feature [38] that allows instructors to create differentiated assignments, these assignments direct automatically the student to learning path based on their achieved scores. After the grading of the finished module of the course, conditional assignments are applied to the obtained score in order to redirect the student to the next learning path.

Protus: In [3,8], the authors presented the architecture and the implementation of an intelligent web-based programming tutoring system Protus (PROgramming TUtoring System). Protus provides two kinds of personalization, the first based on adaptive hypermedia and the second based on the recommender systems techniques. Adaptive hypermedia guides the student in the course through suggesting the pages that meet the learner's requirement or adding automatically explanation in the current page. This system includes the learner's interactions, assignment grades, ontology-based recommender system, and user interface customization.

Concerto: In [28], the presented research focuses on new adaptive system named *Concerto*. This system is an open source platform developed by the university of Cambridge Psychometrics Centre using for adaptive assessments. It is the most known platform dedicated to provide web-based application of the CAT approach cited above. Concerto employs R statistical computing language to create the personalized test, it is designed to meet the abilities of each learner profile through specific tests.

Mod-Knowledge agent: In the study [35], the authors presented a multi-agent system that provides the possibility to change the algorithm used by each agent, depending on the external environment. The system includes an agent called "Mod- Knowledge" that traces the knowledge state taking into account the learner external navigation. In this agent, the Dynamic Key-Value Memory Net (DKVMN) algorithm is modified according to the context of e-learning platform: generalizing the semantic correlation to avoid manual labeling of question tags, considering user questions and his answers on the forum instead of restricting only to the questions proposed by the system. The DKVMN model is an improvement of DKT model [39]. Furthermore, the algorithm can detect struggling students, who encounter advancement difficulties in the learning.

Educational Chatbots for Adaptive Learning: The authors in [40] built and evaluated a dialogue-based adaptive learning system named QuizBot that assists students to study and memorize factual knowledge. The developed chat-based agent provides answers to the students based on their inputs. Afterwards, another similar system was developed by [41] for argumentation skills based on Natural Language Processing (NLP) and machine learning techniques. Recently, the authors [36] implemented MathBot tutoring system, a personalized conversational agent that teaches math content. The MathBot uses the interaction between the learner and the system to learn math concepts, instead of browsing online learning resources. This system applies fuzzy matching to examine the student responses and direct them to the correct path.

SKOPE-IT: The Shareable Knowledge Objects as Portable Intelligent Tutors (SKOPE-IT) system [37] provides a conversational tutoring that can be embedded into existing learning system. The developed method allowed to annotate HTML pages with natural language-tutoring dialogs that explain and discuss the content in each block of the page. The data used was collected from ALEKS commercial adaptive system, system logs, student preferences, student abilities, and questionnaire test based on mathematical concepts.

5 Analysis and Discussion

After the analysis of the above-mentioned systems, we found that each system or tools focuses on a specific mechanism for understanding the learner and neglects other factors. The ALOSI provides personalized learning based on mastery. Whereas, the learner behaviors and career interests are, however, not included in this adaptive engine.

The majority of customization solutions implemented in Moodle consist in developing new extensions that adapt content to learners' needs. This choice allows to avoid the modification in moodle core and its main functionalities, as well these plugins are unaffected by moodle updates. Despite the finding that participants achieved, its validity depends on the subjectivity of student responses when answering the FSLSM questionnaire.

Although, adaptive models have been proposed, there is still a need for a clear model that represents the learner and his learning goals. The learner attributes are not fully exploited in these systems, consequently the most appropriate resources and activities are not recommended for the target student. Several factors are not considered such as the question response time taken by the student, the type of question and the interdependence of such elements of the knowledge.

Additionally, the test of acquired knowledge is important in the educational process. The integration of adaptive assessment algorithms in comprehensive learning system will improve the user experience, for example the exploitation of CAT models. Nevertheless, the implementation of the CAT and the IRT methods in educational assessment activities face several limitations [29].

Several knowledge tracing models have been developed to estimate the learner's performance. Yet these approaches suffer from several problems and limitations. The binary representation of the knowledge state in the BKT model cannot accurately determines the student's learning state due to the possible states of understanding. According to the authors [20], the IRT model has better prediction performance than BKT. Afterwards, the DKT model has been proposed that worked without the need for domain expert annotations. Despite its performance, the DKT is more practical when a large amount of data is available in order to train the model. The BKT and the DKT models establish the knowledge state for each concept separately. For this reason, the DKVMN model has been introduced that employ the relationship between the nearest neighbour concepts. Experiments in [39] showed that the DKVMN outperforms the previous knowledge tracing models.

6 Conclusion

Adaptive learning systems assist the learner to acquire the maximum knowledge in the course. These systems use advanced techniques to detect the learning style for each user as well as his learning needs. Despite the methods developed in the previous solutions, the huge number of resources available remains a limiting factor because this situation leaves the learner lost, confused, and frustrated.

In this paper, we have analyzed the technique and the approaches used in the current adaptive e-learning systems. Further, we have detailed the process of personlization employed in the educational systems: first the system detects the learning style of each student through questionnaires, next it estimates the mastery of knowledge by applying knowledge tracing models, and then recommend personalized suitable content. We have discussed the challenges faced when implementing adaptive learning systems and the opportunities that will make the learners more satisfied.

Acknowledgments. This work was supported by the Al-Khawarizmi Program funding by Morocco's Ministry of Education, Ministry of Industry and the Digital Development Agency (ADD) under Project No. 451/2020 (Smart Learning).

References

1. Almaiah, M.A., Al-Khasawneh, A., Althunibat, A.: Exploring the critical challenges and factors influencing the E-learning system usage during COVID-19 pandemic. Educ. Inf. Technol. **25**, 5261–5280 (2020)
2. Ueda, H., Furukawa, M., Yamaji, K., Nakamura, M.: SCORMAdaptiveQuiz: implementation of adaptive e-learning for moodle. Proc. Comput. Sci. **126**, 2261–2270 (2018)
3. Klašnja-Milićević, A., Vesin, B., Ivanović, M., Budimac, Z.: E-learning personalization based on hybrid recommendation strategy and learning style identification. Comput. Educ. **56**(3), 885–899 (2011)
4. Vesin, B., Mangaroska, K., Giannakos, M.: Learning in smart environments: user-centered design and analytics of an adaptive learning system. Smart Learn. Environ. **5**(1), 1–21 (2018). https://doi.org/10.1186/s40561-018-0071-0
5. Knutov, E., De Bra, P., Pechenizkiy, M.: Ah 12 years later: a comprehensive survey of adaptive hypermedia methods and techniques. New Rev. Hypermed. Multimed. **15**(1), 5–38 (2009)
6. Self, J., et al.: The defining characteristics of intelligent tutoring systems research: ITSs care, precisely. Int. J. Artif. Intell. Educ. **10**(3–4), 350–364 (1999)

7. Brusilovsky, P., Somyürek, S., Guerra, J., Hosseini, R., Zadorozhny, V., Durlach, P.J.: Open social student modeling for personalized learning. IEEE Trans. Emerg. Top. Comput. **4**(3), 450–461 (2015)
8. Vesin, B., Budimac, Z., et al.: Applying recommender systems and adaptive hypermedia for e-learning personalization. Comput. Inform. **32**(3), 629–659 (2013)
9. Brusilovsky, P., Millán, E.: User models for adaptive hypermedia and adaptive educational systems. In: Brusilovsky, P., Kobsa, A., Nejdl, W. (eds.) The Adaptive Web. LNCS, vol. 4321, pp. 3–53. Springer, Heidelberg (2007). https://doi.org/10.1007/978-3-540-72079-9_1
10. Buche, C., Querrec, R., De Loor, P., Chevaillier, P.: MASCARET: a pedagogical multi-agent system for virtual environments for training. Int. J. Dist. Educ. Technol. (IJDET) **2**(4), 41–61 (2004)
11. Clark, M., et al.: Automatically structuring domain knowledge from text: an overview of current research. Inf. Process. Manag. **48**(3), 552–568 (2012)
12. Alshammari, M., Anane, R., Hendley, R.J.: Adaptivity in e-learning systems. In: 2014 Eighth International Conference on Complex, Intelligent and Software Intensive Systems, pp. 79–86. IEEE (2014)
13. Brusilovsky, P.: Methods and techniques of adaptive hypermedia. In: Kobsa, A., Vassileva, J. (eds.) Adaptive Hypertext and Hypermedia, pp. 1–43. Springer, Dordrecht (1998). https://doi.org/10.1007/978-94-017-0617-9_1
14. Brusilovsky, P.: Adaptive hypermedia. User Model. User-Adap. Inter. **11**(1), 87–110 (2001)
15. Hambleton, R.K., Shavelson, R.J., Webb, N.M., Swaminathan, H., Rogers, H.J.: Fundamentals of Item Response Theory, vol. 2. Sage (1991)
16. Hambleton, R.K., Swaminathan, H.: Item Response Theory: Principles and Applications. Springer, Heidelberg (2013)
17. Wang, F.H.: Application of componential IRT model for diagnostic test in a standard-conformant eLearning system. In: Sixth IEEE International Conference on Advanced Learning Technologies (ICALT 2006), pp. 237–241. IEEE (2006)
18. Felder, R.M., Silverman, L.K., et al.: Learning and teaching styles in engineering education. Eng. Educ. **78**(7), 674–681 (1988)
19. Corbett, A.T., Anderson, J.R.: Knowledge tracing: modeling the acquisition of procedural knowledge. User Model. User-Adap. Inter. **4**(4), 253–278 (1994)
20. Pardos, Z.A., Heffernan, N.T.: KT-IDEM: introducing item difficulty to the knowledge tracing model. In: Konstan, J.A., Conejo, R., Marzo, J.L., Oliver, N. (eds.) UMAP 2011. LNCS, vol. 6787, pp. 243–254. Springer, Heidelberg (2011). https://doi.org/10.1007/978-3-642-22362-4_21
21. Yudelson, M.V., Koedinger, K.R., Gordon, G.J.: Individualized Bayesian knowledge tracing models. In: Lane, H.C., Yacef, K., Mostow, J., Pavlik, P. (eds.) AIED 2013. LNCS (LNAI), vol. 7926, pp. 171–180. Springer, Heidelberg (2013). https://doi.org/10.1007/978-3-642-39112-5_18
22. Hawkins, W.J., Heffernan, N.T., Baker, R.S.J.D.: Learning Bayesian knowledge tracing parameters with a knowledge heuristic and empirical probabilities. In: Trausan-Matu, S., Boyer, K.E., Crosby, M., Panourgia, K. (eds.) ITS 2014. LNCS, vol. 8474, pp. 150–155. Springer, Cham (2014). https://doi.org/10.1007/978-3-319-07221-0_18
23. Badrinath, A., Wang, F., Pardos, Z.: pyBKT: an accessible python library of Bayesian knowledge tracing models. arXiv preprint arXiv:2105.00385 (2021)
24. Pavlik Jr, P.I., Cen, H., Koedinger, K.R.: Performance factors analysis–a new alternative to knowledge tracing. Online Submission (2009)

25. Cen, H., Koedinger, K., Junker, B.: Learning factors analysis – a general method for cognitive model evaluation and improvement. In: Ikeda, M., Ashley, K.D., Chan, T.-W. (eds.) ITS 2006. LNCS, vol. 4053, pp. 164–175. Springer, Heidelberg (2006). https://doi.org/10.1007/11774303_17
26. Gong, Y., Beck, J.E., Heffernan, N.T.: Comparing knowledge tracing and performance factor analysis by using multiple model fitting procedures. In: Aleven, V., Kay, J., Mostow, J. (eds.) ITS 2010. LNCS, vol. 6094, pp. 35–44. Springer, Heidelberg (2010). https://doi.org/10.1007/978-3-642-13388-6_8
27. Özyurt, Ö., Özyurt, H., Baki, A.: Design and development of an innovative individualized adaptive and intelligent e-learning system for teaching-learning of probability unit: details of UZWEBMAT. ESWA **40**(8), 2914–2940 (2013)
28. Scalise, K., Allen, D.D.: Use of open-source software for adaptive measurement: Concerto as an R-based computer adaptive development and delivery platform. Br. J. Math. Stat. Psychol. 478–496 (2015)
29. Oppl, S., Reisinger, F., Eckmaier, A., Helm, C.: A flexible online platform for computerized adaptive testing. Int. J. Educ. Technol. High. Educ. **14**(1), 1–21 (2017). https://doi.org/10.1186/s41239-017-0039-0
30. Piech, C., et al.: Deep knowledge tracing. arXiv preprint arXiv:1506.05908 (2015)
31. Rosen, Y., et al.: Adaptive learning open source initiative for MOOC experimentation. In: Penstein Rosé, C., et al. (eds.) AIED 2018. LNCS (LNAI), vol. 10948, pp. 307–311. Springer, Cham (2018). https://doi.org/10.1007/978-3-319-93846-2_57
32. Rosen, Y., et al.: The effects of adaptive learning in a massive open online course on learners' skill development. In: Proceedings of the Fifth Annual ACM Conference on Learning at Scale, pp. 1–8 (2018)
33. Karagiannis, I., Satratzemi, M.: An adaptive mechanism for Moodle based on automatic detection of learning styles. Educ. Inf. Technol. **23**(3), 1331–1357 (2018)
34. Arsovic, B., Stefanovic, N.: E-learning based on the adaptive learning model: case study in Serbia. Sādhanā **45**(1), 1–13 (2020)
35. Trifa, A., Hedhili, A., Chaari, W.L.: Knowledge tracing with an intelligent agent, in an e-learning platform. Educ. Inf. Technol. **24**(1), 711–741 (2018). https://doi.org/10.1007/s10639-018-9792-5
36. Cai, W., et al.: Bandit algorithms to personalize educational chatbots. Mach. Learn. **110**(9), 2389–2418 (2021). https://doi.org/10.1007/s10994-021-05983-y
37. Nye, B.D., Pavlik, P.I., Windsor, A., Olney, A.M., Hajeer, M., Hu, X.: SKOPE-IT: overlaying natural language tutoring on an adaptive learning system for mathematics. Int. J. STEM Educ. **5**(1), 12 (2018)
38. MasteryPaths in canvas LMS. https://community.canvaslms.com. Accessed 13 June 2021
39. Zhang, J., Shi, X., King, I., Yeung, D.Y.: Dynamic key-value memory networks for knowledge tracing. In: Proceedings of the 26th International Conference on World Wide Web, pp. 765–774 (2017)
40. Ruan, S., et al.: QuizBot: a dialogue-based adaptive learning system for factual knowledge. In: Proceedings of the 2019 CHI Conference on Human Factors in Computing Systems, pp. 1–13 (2019)
41. Wambsganss, T., Kueng, T., Soellner, M., Leimeister, J.M.: ArgueTutor: an adaptive dialog-based learning system for argumentation skills. In: Proceedings of the 2021 CHI Conference on Human Factors in Computing Systems, pp. 1–13 (2021)

A Minimum and Maximum of Regional Information Method to Improve the Sobel Edge Detector

Brahim Bouda(✉) and Ahmed Alifdal

Department of Computer Science, Faculty of Sciences and Techniques, Moulay Ismail University of Meknes, P.O. Box 509, Boutalamine, 52000 Errachidia, Morocco

Abstract. Edge detection is the most common approach for detecting meaningful discontinuities in grayscale images. In this paper we develop an improved edge detector based on Sobel method which produces promising results. We analyze the effect of considering the minimum and maximum of the regional information around a given pixel in the stage of the gradient computation. We treat each image as a set of dark or bright regions depending on a predetermined threshold. Then, we create a pixel set by linearly increasing or decreasing the range of the image intensities. To finish, the edge map and direction map are generated. We experimentally demonstrate that the method is competitive with some conventional methods such as Sobel, Prewitt and Canny.

Keywords: Edge detection · Image processing · Regional information · Sobel edge detector

1 Introduction

Edge is a collection of the pixels where we have abrupt change of gray level. It represents the most basic feature of the image. Therefore, edge detection is a very important process in many applications of computer vision such as pattern recognition, robot vision, image retrieval, motion tracking, medical image analysis and a many other applications.

During the early years, a series of edge detection approaches are elaborated. They can be grouped in two classic methods: first-order differential methods [1–4] and second-order differential methods [5–9]. Basically, they utilize masks to do convolution on the image to detect the edges based on the abrupt change of the gray level. The most famous ones are Prewitt [3], Sobel [2] and Laplacian [7,9]. Subsequently, these works were followed by both Marr and Hildreth [10] and Canny [11] who developed one of the most elaborated filters based on analytical considerations.

Recently, a variety of edge detection approaches have been developed, such as the statistical methods [12–15], the morphological methods [16–18], the wavelet

Y. Maleh et al. (Eds.): ICI2C 2021, LNNS 357, pp. 13–22, 2022.
https://doi.org/10.1007/978-3-030-91738-8_2

methods [19–21], the neural methods [22,23] and the fuzzy pattern matching methods [24,25].

The majority of the above methods encounter a set of frequent problems such as thick, unclear and broken edges. Consequently, it is still a challenging issue to develop a simple method which can remedy such problems.

The aim of this paper is to develop a new and simple method, which can extract thin and clear edges in a grayscale image. The method is based on the famous edge detector of Sobel [2]. We generalize this method by considering a regional information around a given pixel using an aggregation of the minimum and maximum functions and the algebraic sum and difference operators.

The outline of this paper is as follows. The preliminaries of some basic concepts are presented in Sect. 2. The Sobel method will be presented in Sect. 3. Our proposal will be described in Sect. 4. The evaluation of the proposed method using a number of real images is illustrated in Sect. 5. Finally, Sect. 6 summarizes the main conclusions.

2 Basic Concepts

In this paper we propose an usage of the regional information to improve the Sobel edge detector. We define such a reginal information (RI) as any information that can be obtained by considering the central pixel (C) and its neighborhood (N). In order to determine an estimated region ER from an original region R of an image I, we propose the usage of two statistical functions: the minimum of the reginal information ($Min(C, R)$) and maximum of the reginal information ($Max(C, R)$). These functions will be used in conjunction with the algebraic sum and difference operators to expressed some edge detection responses as treated again in Sect. 4.

3 Sobel Method

In the Sobel method, originally introduced by Irwin Sobel and Gary Feldman [2], the authors proposed an isotropic 3×3 image gradient operator. This operator is composed of two directionally masks Sx and Sy as shown in Fig. 1. After the convolution process with Sobel masks on the grayscale image, the edge intensities Ex and Ey in the x and y directions are obtained, respectively. The edge intensity of the mask center is defined as $|Ex| + |Ey|$. By setting an appropriait threshold, a pixel location is declared as an edge point if the value of the edge intensity exceeds the threshold.

However, in the Sobel method [2], the edge lines detected are usually thicker and some broken edges appear in the result. This problem is general and often arises when using many other differential operators [1–8] which depend only on the intensity change. This is not desirable in many precise edge-based applications and further signaled in the Canny's criteria "only one response to a single edge" [11]. Concretely, it is agreed that for a good edge detection, the edge line should be thin and with few broken edges. Canny [11] modified Sobel method

−1	0	+1
−α	0	+α
−1	0	+1

Sx

+1	+α	+1
0	0	0
−1	−α	−1

Sy

Fig. 1. The convolution masks in Sobel method ($\alpha = 2$).

and used the non-maxima suppression (NMS) to produce thinner edges. Unfortunately, the NMS cannot be applied in the stage of gradient computation. An illustrative example is the one in Fig. 2, where the edges of the classical 301 $\times$ 263 Shapessm image have been provided with Sobel and Canny methods. Concretely, the two detectors produce double and thick edges. This is not very souhaitable and can affect the result of many precise pattern recognition systems.

(a) (b) (c)

Fig. 2. Gradient magnitudes for Shapessm image: (a) Original image; (b) Sobel; (c) Canny.

4 Our Proposal

The Sobel method is based on the density difference and distance to neighbor and does not consider any other information about edge detection such as the central pixel and its neighborhood. Consequently, a percentage of information about edge detection is going to be missed. In this Section, such vitally information comes in the shape of a vector.

On the other hand, when using the Sobel method or any differential operator, both brighter and darker regions tend to be assigned the same relevance. So, the method does not take into account the specificities of each image region. This problem is also tackled in this work.

Our idea is to determine at first the dark and bright region from every original region R. Then, treat each one of them differently. To do it, we need a threshold value. Considering the interval [0, 1] as the range of intensities of a grayscale image I, which contains different intensity regions. Therefore, it is reasonable to consider the center of the image intensities as a threshold value i.e., $T_h = 0.5$. Consequently, we consider that a region R is a dark region if all intensities of its pixels are lower than T_h and inversely, a region R is a bright region if at least one of the intensities of its pixels is upper than T_h.

Once the dark and bright regions are determined, we propose to empower dark regions by adding to them the value of the terms $Min(C, R)$ and $Max(C, R)$ and to disempower (disqualify) bright regions by subtracting from them the value of the terms $Min(C, R)$ and $Max(C, R)$. Therefore, darker regions will be assigned more relevance, which conduct to larger gradients and brighter regions will be assigned less relevance, which conduct to smaller gradients.

The first version of the estimated region with the minimum of regional information method ($ERMin$) is defined as

$$ERMin = \begin{cases} R + Min(C, R) & if\ R \leq T_h \\ R - Min(C, R) & otherwise \end{cases} \tag{1}$$

We define the second version of the method as: the estimated region with the maximum of regional information ($ERMax$). It can be expressed by the following equation

$$ERMax = \begin{cases} R + Max(C, R) & if\ R \leq T_h \\ R - Max(C, R) & otherwise \end{cases} \tag{2}$$

In order to leave doubts about the usage of the proposed method, we have repeated the experiments in Fig. 2 for Shapessm image. We have performed the convolution process with Sobel and Canny methods on each estimated region ER by Eq. 1 and Eq. 2. For comparison, we have also performed the convolution process with Sobel and Canny methods on the grayscale image (the traditional Sobel and the traditional Canny). Results are illustrated in Fig. 3. Concretely, the positive consequence of considering the regional information on the detectors is clearly visible where, the edges become sharpener and finer.

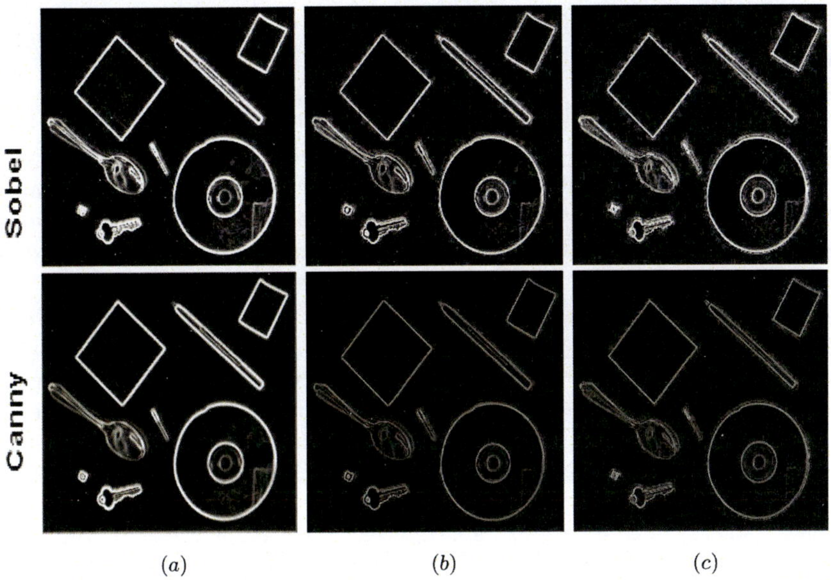

Fig. 3. Gradient magnitudes for Shapessm image: (a) Traditional; (b) ERMin; (c) ERMax.

5 Experimental Results

The set of images which we will be used for a qualitative comparison include four real images which are commonly used as benchmarks for edge detection are illustrated in Fig. 4. These images include the 256 × 256 classical Cameraman image, the 481 × 321 253027 image taken from the Berkeley Segmentation Dataset (BSDS) [26] and two other medical images of specific interest in this work which are the 225 × 225 MRI1 and the 221 × 228 MRI2. These images contain several illuminated areas and some specific textures in which, most of edge detection algorithms cannot perform better.

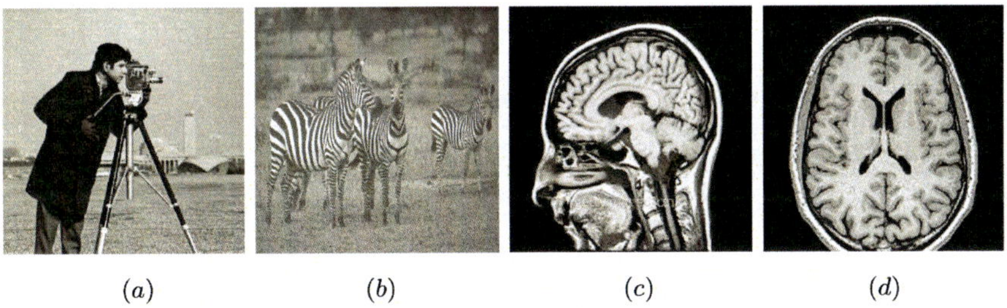

Fig. 4. The images for comparison. (a) Cameraman; (b) 253027; (c) MRI1; (d) MRI2.

To evaluate the proposed method, we have applied it to a large set of natural images taken from the BSDS [26], University of Cordoba (UCO) [27] and many other natural images. The obtained results are always promising. However, only the images of Fig. 4 will be presented into the final version of this work. On the other hand, we restrict here to present only the $ERMin$ results because the two methods $ERMin$ and $ERMax$ are going to give very close results and the difference is marginal. Furthermore, we have performed the convolution process with the above cited operators on the estimated region ER by using the $ERMin$ method. For comparison, we have also applied the original operators of Sobel [2], prewitt [3] and Canny [11] on the above test images. Figures 5, 6, 7 and 8 illustrate the obtained results.

By analyzing the different results of Figs. 5, 6, 7 and 8, we can see that the traditional methods of Sobel, Prewitt and Canny produce thicker edge lines and some broken edges. The proposed method shows to be superior to different traditional operators on which it is implemented on and usually produces sharp and thin edges. This result is very souhaitable by the scientific community especially, in the case of the medical images. However, if we want to compare the three versions of the detector between them depending on the traditional operator on which they are implemented on, we can say that, the $ERMin$ with the Canny gives thinner edges and reduces broken edges. Moreover, the $ERMin$ method with the Sobel and Prewitt give very close results. However, they also thin relatively edge lines and reduces a little broken edges.

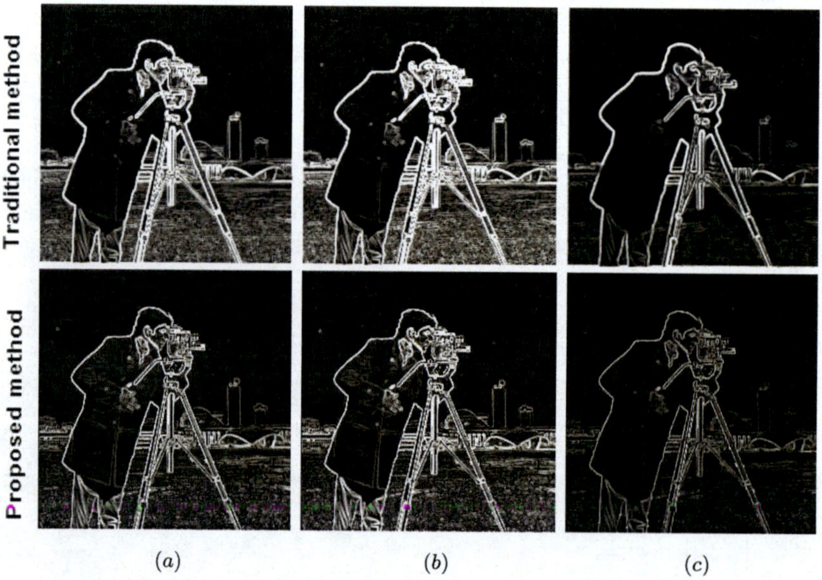

Fig. 5. Gradient magnitudes for Cameraman image obtained with different traditional mask-based methods and with the proposed method. (a) Prewitt; (b) Sobel; (c) Canny.

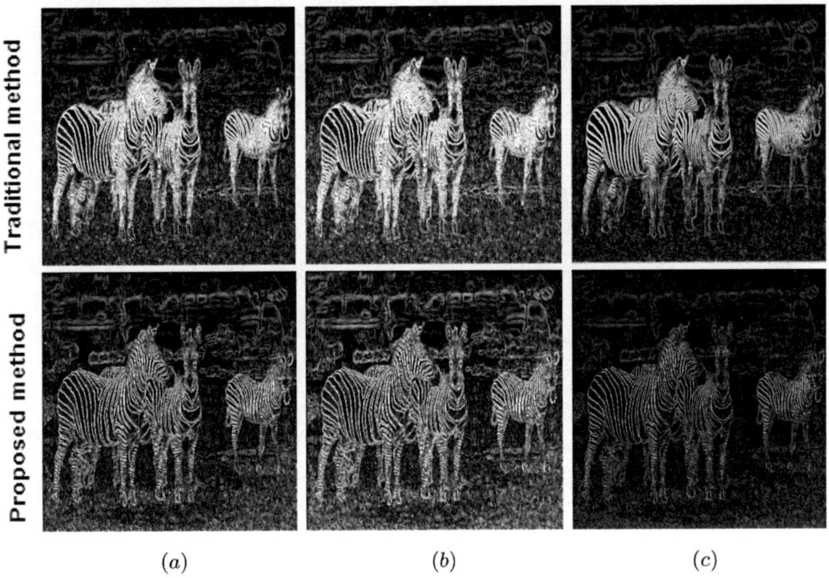

Fig. 6. Gradient magnitudes for 253027 image obtained with different traditional mask-based methods and with the proposed method. (a) Prewitt; (b) Sobel; (c) Canny.

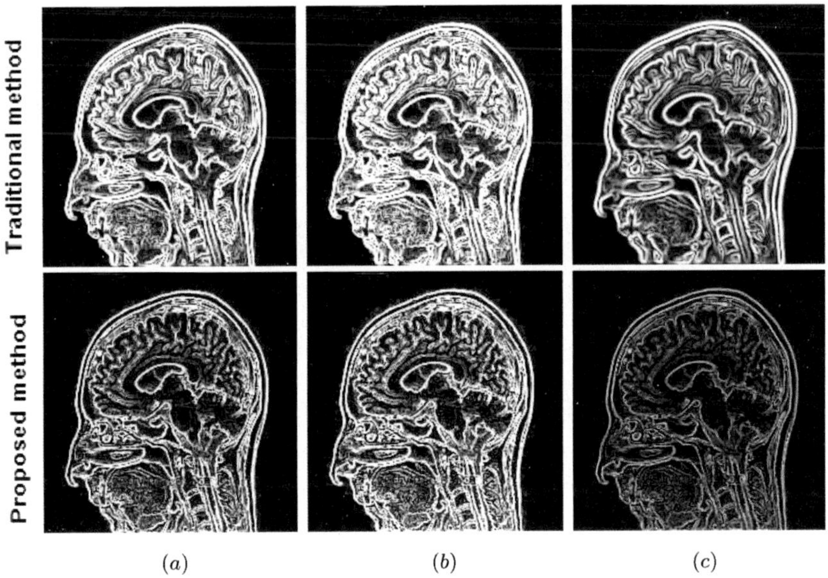

Fig. 7. Gradient magnitudes for MRI1 image obtained with different traditional mask-based methods and with the proposed method. (a) Prewitt; (b) Sobel; (c) Canny.

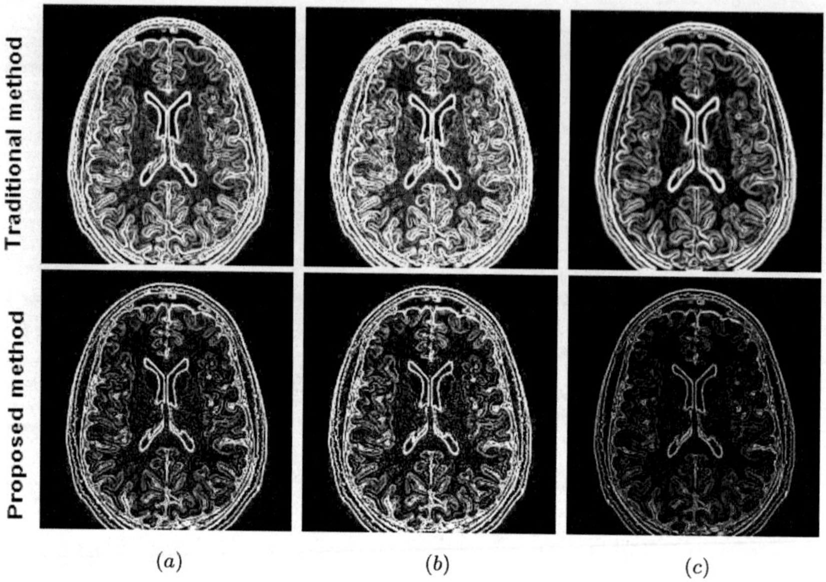

Fig. 8. Gradient magnitudes for MRI2 image obtained with different traditional mask-based methods and with the proposed method. (a) Prewitt; (b) Sobel; (c) Canny.

6 Conclusion

Based on regional information, a new edge detection method has been proposed in this paper to improve the well known Sobel edge detector. It is simple and easy in implementation. It is still a masks-based method but, gives competitive results on a series of natural images and in some concrete situations which require an special accuracy in comparison with the common edge detection methods. The computer experiments prove its effectiveness. The proposed method can be applied to more general fields like medicine and treatment which require more precision.

Acknowledgment. The authors are very thankful to anonymous Editors and Reviewers for their careful reading and valuable comments and suggestions.

References

1. Roberts, L.: Machine perception of three-dimensional solids. In: Optical and Electro-Optical Information Processing, pp. 159–197. MIT Press, Cambridge (1965)
2. Sobel, I., Feldman, G.: A 3 × 3 isotropic gradient operator for image processing. Presented at a Talk at the Stanford Artificial Project (1968)
3. Prewitt, J.G.: Object enhancement and extraction. In: Lipkin, B., Rosenfeld, A. (eds.) Picture Processing and Psychopictorics, pp. 75–149. Academic, New York (1970)

4. Kirsh, R.: Computer determination of the constituent structure of biomedical images. Comput. Biomed. Res. **4**(3), 315–328 (1971)
5. Clarrk, J.J.: Authenticating edges produced by zero crossing algorithms. IEEE Trans. Pattern. Anal. Mach. Intell. **11**(1), 43–57 (1989)
6. Torre, V., Poggio, T.: On edge detection. IEEE Trans. Patt. Anal. Mach. Intell. **8**(2), 147–163 (1986)
7. Lee, J., Haralick, R., Shapiro, L.: Morphologic edge detection. IEEE J. Robot. Autom. **3**(2), 142–156 (1987)
8. Korn, A.F.: Toward a symbolic representation of intensity changes in images. IEEE Trans. Pattern Anal. Mach. Intell. **10**(5), 610–625 (1988)
9. Chidiac, H., Ziou, D.: Classification of image edge. In: Vision Interface 1999, Troise-Rivieres, Canada, pp. 17–24 (1999)
10. Marr, D., Hildreth, E.: Theory of edge detection. Proc. Roy. Soc. London **B207**, 187–217 (1980)
11. Canny, J.: A computational approach to edge detection. IEEE Trans. Pattern Anal. Mach. Intell. **8**(6), 679–698 (1986)
12. Gregson, P.H.: Using angular dispersion of gradient direction for detecting edge ribbons. IEEE Trans. Pattern Anal. Mach. Intell. **15**(7), 682–696 (1993)
13. Ando, S.: Image field categorization and edge/corner detection from gradient covariance. IEEE Trans. Pattern Anal. Mach. Intell. **22**(2), 179–190 (2000)
14. Malik, J., Belongie, S., Leung, T.K., Shi, J.: Contour and texture analysis for image segmentation. Int. J. Comput. Vis. **43**(1), 7–27 (2001)
15. Lin, W.C., Wang, J.W.: Edge detection in medical images with quasi high-pass filter based on local statistics. Biomed. Signal Process. Control **39**, 294–302 (2018)
16. Rivest, J.F., Soille, P., Beucher, S.: Morphological gradients. J. Electron. Imaging **2**(4), 326–336 (1993)
17. Zhu, S.Y., Plataniotis, K.N., Venetsanopoulos, A.N.: Comprehensive analysis of edge detection in color image processing. Opt. Eng. **38**(4), 612–625 (1999)
18. Yu, X., Lin, X., Dai, Y., Zhu, K.: Image edge detection based tool condition monitoring with morphological component analysis. ISA Trans. **69**, 315–322 (2017)
19. Zhang, L., Bao, P.: Edge detection by scale multiplication in wavelet domain. Pattern Recogn. Lett. **23**(14), 1771–1784 (2002)
20. Ducottet, C., Fournel, T., Barat, C.: Scale-adaptive detection and local characterization of edges based on wavelet transform. Signal Process. **84**(11), 2115–2137 (2004)
21. Azeroual, A., Afdel, K.: Fast image edge detection based on Faber Schauder Wavelet and Otsu threshold. Heliyon **3**, e00485 (2017)
22. Lu, S., Wang, Z., Shen, J.: Neuro-fuzzy synergism to the intelligent system for edge detection and enhancement. Pattern Recogn. **36**(10), 2395–2409 (2003)
23. Vardhana, M., Arunkumar, N., Abdulhay, E.: Convolutional neural network for biomedical image segmentation with hardware acceleration. Cognit. Syst. Res. **50**, 10–14 (2018)
24. Kim, D.-S., Lee, W.-H., Kweon, I.-S.: Automatic edge detection using 3 × 3 ideal binary pixel patterns and fuzzy-based edge thresholding. Pattern Recogn. Lett. **25**(1), 101–106 (2004)
25. Orujov, F., Maskeliunas, R., Damaevicius, R., Wei, W.: Fuzzy based image edge detection algorithm for blood vessel detection in retinal images. Appl. Soft Comput. **94**, 106452 (2020)

26. Fernandez-Garcia, N.L., Carmona-Poyato, A., Medina-Carnicer, R., Madrid-Cuevas, F.J.: Automatic generation of consensus ground truth for comparison of edge detection techniques. Image Vis. Comput. **26**(4), 496–511 (2008). https://www2.eecs.berkeley.edu/Research/Projects/CS/vision/grouping/segbench/BSDS300/html/dataset/images.html
27. Fernandez-Garcia, N.L., Carmona-Poyato, A., Medina-Carnicer, R., Madrid-Cuevas, F.J.: Images from automatic generation of consensus ground truth for comparison of edge detection techniques (2008). http://www.uco.es/~ma1fegan/Comunes/investigacion/imagenes/ground-truth.html

Hybrid Mammogram Segmentation Using Watershed and Region Growing

Mohammed Rmili(✉), Abdelmajid El Moutaouakkil, and Mousatapha M. Saleck

LAROSERI Laboratory, Computer Sciences Department, Faculty of Sciences, University of Chouaïb Doukkali, EL Jadida, Morocco

Abstract. The present paper introduces a new hybrid approach for mass segmenting in mammography images. In this work, special attention is paid to breast mass as they are the most common sign of cancer in mammograms. The human radiology experts analyze the mammograms manually to identify the malignancy of a mass. However, such a task takes time and depends on the decision of an expert, which can be affected by many factors. Therefore, the computer-aided diagnosis systems (CAD) in mammography are increasingly used to help radiologists in their decisions. Segmentation is an essential step in (CAD) systems. The proposed method, aiming to come up with a better mass segmentation, uses both watershed transformation technique and region growing algorithm. The watershed transformation was applied in the initial segmentation of the pretreated image while the region growing algorithm was applied in the final segmentation of the mass. We have tested our approach on mammogram images taken from mini-MIAS database. The suggested method can be put to the test through a quantative evaluation based on sensitivity, specificity, accuracy and overlap. The results demonstrate that the proposed approach could produce a precise mass edge.

Keywords: Mammography · Mass segmentation · Watershed transformation · Rcgion growing

1 Introduction

Obviously, breast cancer is a very serious health issue all over the world [1]. It is by far the most widely spread type of cancer among women and the most deadly one. Early detection is the main way to reduce breast cancer deaths. It is largely agreed that mammography is an outstandingly effective method to detect breast cancer [2,3]. However, studies demonstrated that a rate ranging from 10% to 30% of breast cancer cases are misdiagnosed [4,5]. Computer-aided diagnosis systems (CAD) were developed to help radiologists provide a much more accurate mammogram diagnosis.

The mass is the most recurrent type of cancer. (CAD) systems largely depend on mass segmentation. Radiologists invest the mass edge and shape to decide if

Y. Maleh et al. (Eds.): ICI2C 2021, LNNS 357, pp. 23–32, 2022.
https://doi.org/10.1007/978-3-030-91738-8_3

this mass is malignant or benign. Many methods were introduced with the aim of overcoming the issue of mass segmentation. For instance, Li et al. [6], suggested a deep learning method that brings together U-Net and attention Gates (AGs). Another method suggested by Patil et al. [7] depends on the region growing technique and the optimized fuzzy classifier technique.

This academic work introduces a brand new approach in the field of mass segmentation. This new mass segmentation approach uses the watershed transform algorithm. It is true that this algorithm is known for its over-segmentation errors, namely in mammogram images; however, we intend to overcome this issue by integrating in our approach the region growing algorithm. The two algorithms will work together; the watershed transform algorithm will split the initial image into segments while the region growing algorithm will merge the relevant segments to produce a clear mass shape and edge.

The next four sections of this work shall be as follows: Sect. 2 below will review the watershed transform algorithm. Section 3 will present the new mass segmentation approach. Section 4 will evaluate this approach experimentally and quantitatively. Finally, Sect. 5 will lay out conclusions and future work.

2 Watershed Transform

Vincent and Soille's watershed transform algorithm [8] is a widely used segmentation method; it is applied to the gradient images. Put in plain words, gradient images are seen as a 3D topographic surface. The gray level of each pixel represents a certain altitude and the contours are made up of crest lines. Water floods the topography via regional minima to fill the catchment basins. The moment the waters of two nearby catchment basins are about to meet, a dam preventing the merging rises as in Fig. 1. The flooding continues until the water reaches the highest point of the topographic relief. Finally, the dams represent the contours of the segmented image. The application of this method in the segmentation of mammographic images or their gradients often gives inaccurate results like over-segmentation. This can be due to potential noise or the large number of minima in a given mammographic image. The problem of over-segmentation has been widely discussed in the literature and many solutions/tools have been

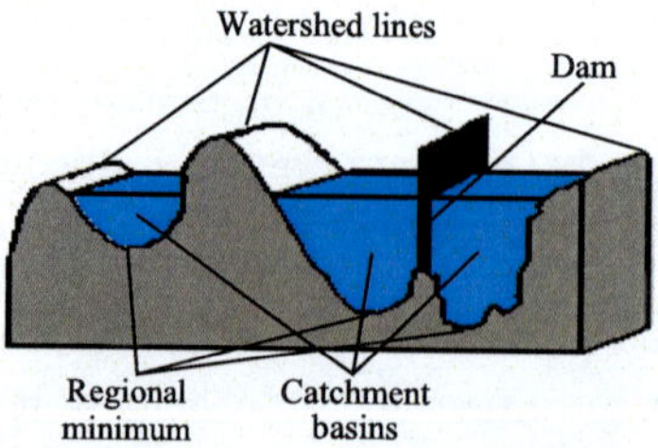

Fig. 1. Topographic view of watershed transform.

proposed to overcome this problem. Xu et al. [9] invested this method in getting the mass boundary with the help of internal and external markers. They claim that watershed transform provides much better results compared to some other methods like dynamic programming boundary tracing and dynamic programming and plane fitting. Arnab et al. [10] has coined his proper technique of breast cancer segmentation depending on watershed technique. Their study introduces a novel marker controlling watershed algorithm to finally spot all the suspicious regions distinctly. It is a morphological operation that relies on extracting watershed lines from the topographic surface of the original image. To minimize over-segmentation in mammogram images, Varsha et al. [11] applied watershed algorithm with a morphological marker controlled method. Shavi et al. [12] proposed a new technique of breast cancer detection based on extracting catchments basins from the topographic image. To avoid the over-segmentation of the watershed algorithm, they have decided to preprocess the mammogram image under study as a first step. Next, they have identified the distance transform of the preprocessed image. Finally, they have applied watershed algorithm on this image.

3 The Proposed Technique

Our paper suggests a mass segmentation approach using the regions of interest (ROI) in the mammogram. A ROI image contains useful information about a suspicious and very specific area of the image. The method is based on watershed transform algorithm and region growing (RG) algorithm. Our method consists of the following steps: image acquisition, pretreatment, and mass segmentation. Figure 2 shows an illustration of the method.

3.1 The Image Acquisition

Mammogram databases are available to scientific research and contribute to the development of cancer detection techniques [13]. Most databases are not accessible to the public. The most easily accessible and most commonly used database is Mammographic Image Analysis Society (MIAS) which is available for researchers [14]. MIAS provide valuable information about the lesion area like type, location, severity, central coordinate and radius. Each image is 1024×1024 pixels. To decrease computation costs, we propose using ROI to instead of the original mammogram images.

3.2 Preprocessing

Preprocessing helps in building much clearer images and therefore the possibility of further image segmentation. It is very important because the mammography images have low contrast and contain more noise. The median filter [15], a non-linear spatial filter, suggests itself as very effective noise removal technique that we will adopt in our present study. In the median filter, the value of one pixel

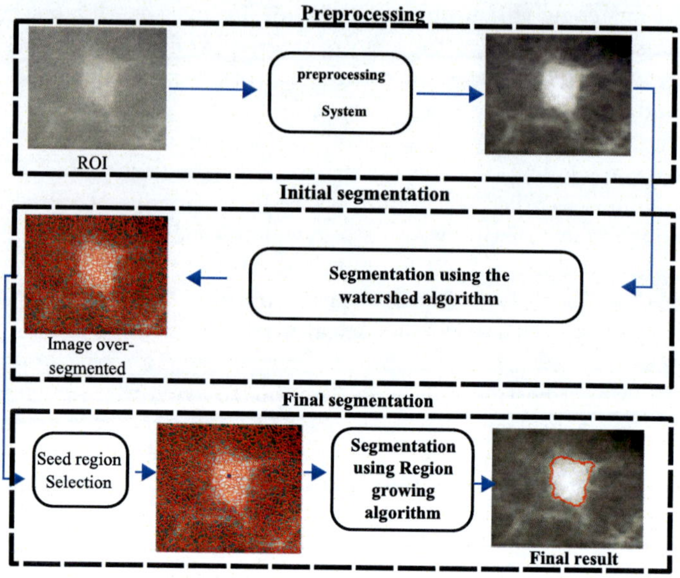

Fig. 2. Block diagram of mass segmentation methodology.

is substituted by the middle ones of the nearby pixels. It sorts pixels by gray level and then selects intermediate values as output pixel values. This filtering technique makes it possible to remove noises and, in the same time, keep the best image quality and the maximum of information. To enhance the contrast of mammograms, we propose Contrast-Limited Adaptive Histogram Equalization (CLAHE) technique. This technique is very practical in image enhancement thanks to its simplicity and great performance with almost all sorts of images compared to other methods [16].

3.3 Segmentation

Initial Segmentation: The initial segmentation stage is about initially segmenting the original region of interest (ROI) by means of the watershed transform technique. This technique is applied on the gradient of images and gives an over-segmented image (Fig. 3(a)). The outcome is a partition image used in the construction of the region adjacency graph (RAG) that will work as an input in the region merging procedure. In a RAG, Each graph node represents a certain region. Between each pair of adjacent regions, "X" and "Y" for example, exists an edge labeled (X,Y). Regions are neighbors if they share at least two neighboring pixels. Figure 3(b) illustrates such a situation: regions A and B are neighbors, but C and D are not.

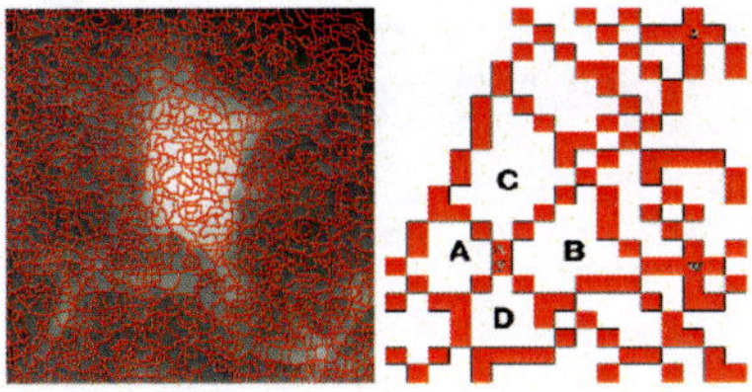

(a) (b)

Fig. 3. Initial segmentation: (a) represents the result obtained using watershed; (b) is zoomed part of (a).

Final Segmentation: In this step, we propose the region growing algorithm as an effective solution to solve the over-segmentation problem. It stands as the most frequently used and productive method in the field of segmentation applications. The algorithm focuses on similarity measure between the object of interest and its neighbor pixels [17,18]. Nevertheless, this technique is known for its unsatisfactory segmentation results in irregular intensity images and also known for its long computation time. Selecting the appropriate value of the threshold has been considered as an important step. But, the threshold optimal value identification through trial and error method or histogram analysis is not only fault prone but also time consuming. In our study, we first start with a seed region that the radiologist in charge selects manually. This seed region which grows iteratively agglomerates with its similar neighbor regions. The growing process stops when there is no candidate region for agglomeration. Finally, the contour with the maximum value of average gradient represents the accurate mass contour. Average gradient is represented in the following equation.

$$A_g = \frac{1}{n}\sum_{i=1}^{n} g(v_i) \tag{1}$$

$g(v_i)$ denotes the gradient value for the contour element v_i. As an example, Fig. 4 illustrates the outcome of applying proposed segmentation approach.

4 Experimental Results

4.1 Qualitative Evaluation

The suggested mass segmentation approach was tested on 20 mammogram images. The performance was assessed by comparing the drawn contours to the outlines a breast radiologist has drawn. As we can see in Fig. 5, the black contours are drawn by a breast radiologist while the yellow contours are drawn by the algorithm. Figure 5 proves that we can have much better results thanks to this approach. We observe that the results (yellow contours) are much closer to the ground truth even in the blurred regions. However, the result in the

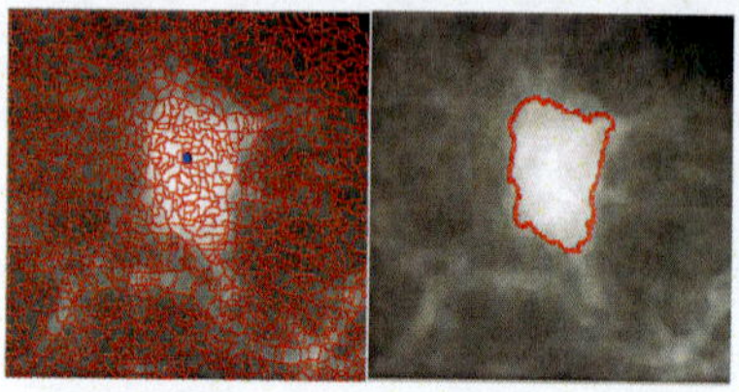

(a) (b)

Fig. 4. Final segmentation: (a) Seed region selection, (b) The final result

image c_1 has a false positive, the anomalies of this approach reflect that uses only the intensity property and the gradient of the image, but the specifications are ignored: like shape and size. Therefore, this work requires a continuity of addition of other geometric criteria such as the shape.

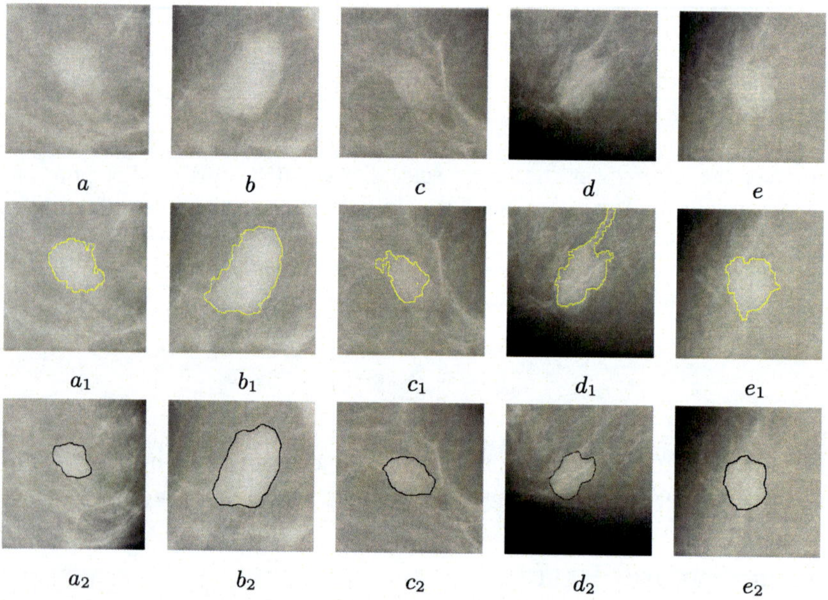

Fig. 5. Experimental results of five mammograms: (a)-(e) are the original ROI. (a_1)-(e_1) are the results. (a_2)-(e_2) The ground truth

4.2 Quantitative Evaluation

Besides the visual evaluation of segmentation performance, we apply the four quantitative evaluation measures specificity, sensitivity, accuracy and overlap which can be calculated as follows.

$$Sensitivity = \frac{TP}{TP+FN} \times 100 \tag{2}$$

$$Specificity = 1 - \frac{FP}{FP+TN} \times 100 \tag{3}$$

$$Accuracy = \frac{TP+TN}{TP+FP+FN+TN} \times 100 \tag{4}$$

$$Overlap = \frac{TP}{TP+FP+FN} \times 100 \tag{5}$$

TP, TN, FP and FN stand for True Positive, True Negative, and False Positive and False Negative numbers (see Fig. 6). According to the proposed method, the average sensitivity is 91%, the average specificity is 94,35%, the accuracy is 91.15% and the overlap is 80,3%. Table 1 represents our quantitative analysis results which prove the effectiveness of the proposed algorithm. Comparing these quantitative results to other results obtained via the latest methods proves the high performance of our method (Table 2). These results demonstrate that the proposed approach obtained satisfactory results that are close to the best results achieved in the four metrics by the latest methods. Results from Table 2 demonstrates that the proposed approach obtained better results for most of the metrics. Indeed, the proposed method allows to obtain fine and continuous contours, and their execution does not require any parameterization by the expert. However, this present method has some limitations. For instance, the object to be segmented is already part of the ROI images and has been previously selected in complete mammograms.

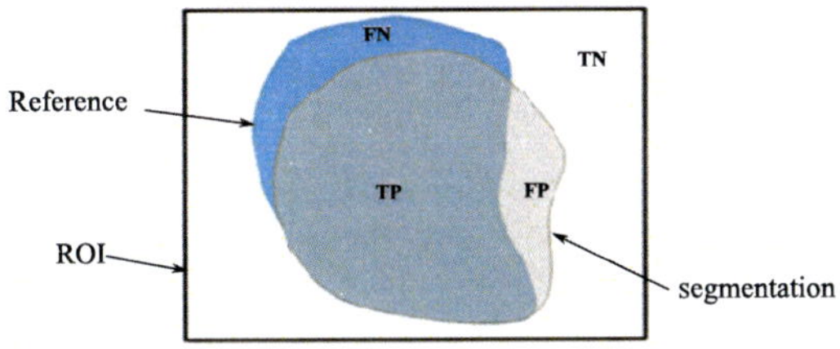

Fig. 6. An illustration of TP, FP, TN and FN.

Table 1. The performance criterion of the proposed method evaluated on 20 images of the mini-MIAS dataset

Image	Overlap(%)	Sensitivity(%)	Specificity(%)	Accuracy(%)
mdb010	84	92	95	94
mdb005	80	90	93	92
mdb012	79	90	92	91
mdb015	80	98	93	92
mdb021	91	92	89	97
mdb023	81	90	92	93
mdb025	80	93	97	96
mdb028	89	89	91	93
mdb069	79	71	92	89
mdb092	70	98	96	82
mdb097	72	91	100	84
mdb132	89	90	94	97
mdb132	81	98	98	92
mdb134	84	97	100	93
mdb142	77	93	92	95
mdb144	67	87	91	86
mdb202	82	91	97	93
mdb267	74	80	98	89
mdb271	81	92	94	80
mdb184	86	98	93	95
Average	**80,3**	**91**	**94,35**	**91.15**

Table 2. Performance comparison

Ref	Year	Technique	Performance				Dataset
			Overlap	Sensitivity	Specificity	Accuracy	
[19]	2018	Fuzzy C-means, GLCM	82.02%	98.23%	74.20%	96.28%	MIAS
[20]	2018	Region Growing	79%	83%	97%	91%	MIAS
[6]	2019	Deep learning	–	–	–	98%	DDSM
[21]	2019	Clustering K-means, FCM	–	–	–	91.18%	MIAS
[7]	2020	Improved region growing	–	–	–	78.83%	MIAS
Proposed method	**2021**	**Watershed and RG**	**80,30%**	**91%**	**94,35%**	**91.15%**	**MIAS**

5 Conclusion and Future Work

The present study proposed a hybrid method of mass boundary detection in mammograms. The new method uses both watershed transformation and region growing. We applied the watershed transform on preprocessed mammography images. Then, we showed that the application of watershed transform on a mammographic images produce an over-segmentation problem. To solve this issue, we applied the region growing algorithm. The data is taken from mini-MIAS dataset. The results proved the high performance of our method.

The next step of this academic contribution is to analyze the segmented mass to get more descriptive and informative information to say if the mass is malignant or benign.

References

1. Sharma, R.: Breast cancer incidence, mortality and mortality-to-incidence ratio (MIR) are associated with human development, 1990–2016: evidence from global burden of disease study 2016. Breast Cancer **26**(4), 428–445 (2019)
2. Tang, J., et al.: Computer-aided detection and diagnosis of breast cancer with mammography: recent advances. IEEE Trans. Inf Technol. Biomed. **13**(2), 236–251 (2009)
3. Elter, M., Horsch, A.: CADx of mammographic masses and clustered microcalcifications: a review. Med. Phys. **36**(6Part1), 2052–2068 (2009)
4. Kerlikowske, K., et al.: Performance of screening mammography among women with and without a first-degree relative with breast cancer. Ann. Intern. Med. **133**(11), 855–863 (2000)
5. Giger, M.L.: Computer-aided diagnosis in radiology. Acad. Radiol. **9**(1), 1–3 (2002)
6. Li, S., et al.: Attention dense-u-net for automatic breast mass segmentation in digital mammogram. IEEE Access **7**, 59037–59047 (2019)
7. Patil, R.S., Biradar, N.: Improved region growing segmentation for breast cancer detection: progression of optimized fuzzy classifier. Int. J. Intell. Comput. Cybern. (2020)
8. Vincent, L., Soille, P.: Watersheds in digital spaces: an efficient algorithm based on immersion simulations. IEEE Trans. Pattern Anal. Mach. Intell. **13**, 583–598 (1991)
9. Xu, S., Liu, H., Song, E.: Marker-controlled watershed for lesion segmentation in mammograms. J. Digit. Imaging **24**, 754–763 (2011)
10. Chattaraj, A., Das, A., Bhattacharya, M.: Mammographic image segmentation by marker controlled watershed algorithm. In: 2017 IEEE International Conference on Bioinformatics and Biomedicine (BIBM), pp. 1000–1003 (2017)
11. Gaikwad, V.J.: Marker-controlled watershed transform in digital mammogram segmentation. Int. J. Res. Appl. Sci. Eng. **3**, 18–21 (2015)
12. Gupta, S., et al.: Semi-automatic segmentation of breast cancer for mammograms based on watershed segmentation. In: Proceedings of the 5th National Conference for Nation Development (2011)
13. Smaoui, N., Hlima, A.: Designing a new approach for the segmentation of the cancerous breast mass. In: 2016 13th International Multi-Conference on Systems, Signals & Devices (SSD), pp. 313–317 (2016)
14. Suckling, J.P.: The mammographic image analysis society digital mammogram database. Digit. Mammo 375–386 (1994)
15. Verma, K., Singh, B.K., Thoke, A.: An enhancement in adaptive median filter for edge preservation. Procedia Comput. Sci. **48**, 29–36 (2015)

16. Kaur, M., Kaur, J., Kaur, J.: Survey of contrast enhancement techniques based on histogram equalization. Int. J. Adv. Comput. Sci. Appl. **2** (2011)
17. Adams, R., Bischof, L.: Seeded region growing. IEEE Trans. Pattern Anal. Mach. Intell. **16**, 641–647 (1994)
18. Zucker, S.W.: Region growing: childhood and adolescence. Comput. Graph. Image Process. **5**, 382–399 (1976)
19. Saleck, M.M., et al.: Breast mass segmentation using a semi-automatic procedure based on fuzzy c-means clustering. Telkomnika **16**, 665–672 (2018)
20. Rmili, M., et al.: A new approach to the detection of mammogram boundary. Int. J. Electr. Comput. Eng. **8**, 3587 (2018)
21. Kamil, M.Y., Salih, A.M.: mammography images segmentation via fuzzy c-mean and k-mean. Int. J. Intell. Eng. Syst. **12**, 22–29 (2019)

Human Activity Recognition Using Stacked LSTM

Zakaria Benhaili(✉), Ihsane Kabbaj, Youssef Balouki, and Lahcen Moumoun

Faculty of Sciences and Techniques, Mathematics, Computer Science and Engineering Sciences Laboratory (MISI), Hassan First University of Settat, 26000 Settat, Morocco
z.benhaili@uhp.ac.ma

Abstract. Recently, the rapid growth of artificial intelligence has led to the deployment of Deep learning techniques for Human Activity Recognition (HAR) issues. Lately, this application has received a lot of attention because of its effectiveness in gaining knowledge from signals generated by sensors. It can use nonlinear transformations to extract features from raw data in a hierarchical way, unlike traditional methods that depend on feature engineering. This paper proposed a human activity recognition model that can deal directly with data sensors without any pre-processing. It uses stacked LSTM layers to recognize several daily life activities. Experience on two datasets shows that we get accuracy comparable or better than the existing state-of-the-art methods.

Keyword: Human activity recognition · LSTM · Deep learning · Sensors

1 Introduction

Human activity recognition is grounded on the gathering of data from different sensing devices. We distinguish between two types: vision-based HAR, which requires cameras, and sensor-based HAR based on sensors like accelerometers, gyroscopes, GPS, and others. These sensors are embedded in smartphones and other smart objects.

The efforts made in the last years in the electronics industry have allowed to reduce sensors' size and produce others with smaller sizes, which has motivated the development of several context-aware applications in various fields such as healthcare [1], supervision [2], security [3], and other daily life applications primarily the domain of internet of things [4].

Vision-based HAR suffers from problems related to privacy, restrictions related to mobility and power consumption, and difficulties in obtaining good images in some climatic conditions. These complications directly impact recognition accuracy and push researchers to adopt sensor-based HAR, which is more accurate, flexible, and straightforward.

Sensors take advantage of being ubiquitous. They can now be embedded into wearable devices like phones, watches, bracelets, and non-wearable items like automobiles, doors, and furniture, thanks to the growth of smart gadgets and the Internet of

Y. Maleh et al. (Eds.): ICI2C 2021, LNNS 357, pp. 33–42, 2022.
https://doi.org/10.1007/978-3-030-91738-8_4

Things. Sensors are broadly present in our environment, logging people's motions in a non-intrusive and continual manner.

Today, everybody has a smart device that contains at least one sensor. These devices are an excellent tool for collecting data and monitoring human activities. Currently, there is a shift, especially towards smartphones, to realize this recognition.

For good recognition, traditional Machine Learning algorithms necessitate domain knowledge to pre-process data and to select features. Recent achievements of deep learning in computer vision, speech recognition, and natural language processing attracted researchers to investigate its effectiveness in recognizing human activities. Deep learning can extract features from the sensor's raw data automatically, and thus substituting its manual selection and engineering. This capacity of learning permits to save time and to design more general models.

The key contributions of this paper: We propose a deep neural network architecture capable of extracting features from sensor data directly. Our method can recognize a wide range of basic activities that frequently occur in our daily lives. The optimal parameters in the model were finally selected after analyzing its impact on performance. We put our approach to the test on two commonly used datasets. Our model improves prior findings, as evidenced by the results. We constructed our article as follows: The second section examines related works in human activity recognition. In part four, we provide our HAR model. The experimental setup is presented in section five. The results are presented and examined in section six. Finally, and last we give our conclusion.

2 Related Works

Deep learning models have greater learning ability, according to current reviewed research in HAR. Convolutional neural networks have been employed as feature extractors, either alone or combined with recurrent neural networks and its variants.

Oluwalade et al. [5], have investigated the difference in data generated by two types of devices that embed the same sensors (a watch and a phone). They used four models: Long short-term memory (LSTM), Bidirectional Long short-term memory (LSTM), Convolutional Neural Network (CNN), and Convolutional LSTM (ConvLSTM) to classify fifteen hand and non-hand-oriented activities. They also used GRU to forecast the last 30 s of data generated from the watch accelerometer. They obtained an average classification accuracy of more than 91%. In [6] Agarwal et al. proposed an LSTM-CNN Architecture for Human Activity Recognition learning model for HAR. The model requires little computational power, and it was developed by combining a shallow RNN and LSTM algorithm. It can achieve an accuracy of 95.78% on the WISDM dataset. In [7], the authors proposed a training methodology based on several stages by performing several transformations on the time series to facilitate feature space exploration. After this comes a deep CNN network that performs an extraction of the features of each space. The result is combined to better performance on three databases UCI HAR, USC HAR, and SKODA. In [8] The authors proposed a CNN-LSTM network capable of classifying the activities of the HAPT open dataset with an accuracy of 85.87%. First, they transformed the times series data to an image-like two-dimensional array. Then they deployed three Convolutional neural network layers for the extraction of the motion features. Finally, the LSTM layer establishes a relation between the actions and the sequences. Authors in [9] followed the same logic combined four convolutional layers, two recurrent LSTM layers and an ELM classifier to take advantage of its generalization and learning speed capabilities for activity recognition. Cao et al. [10] used smartphones to collect data, and adapted a classic CNN architecture called LeNET [11], after training on six elemental activities, the generated model is transferred back to the android phone.

3 Background

3.1 Recurrent Neural Network

RNN is a type of neural network initially created in the 1980s to anticipate what's coming straightaway in a highly exact way. This ability makes them the most suitable for sequential data like time series prediction, speech and audio recognition, financial data, weather forecasting and much more. RNN possesses cyclic connections that allow it to learn the temporal dynamics of sequential inputs. Each node in the hidden layer has a function that input the current hidden layer state h_t and generate the output y_t based on its current input x_t and the previous hidden state h_{t-1} according to the following equations:

$$h_t = F(W_h h_{t-1} + U_h x_t + b_h) \quad (1)$$

$$y_t = F\left(W_y h_t + b_y\right) \quad (2)$$

W_h, U_h, and W_y are the weight for the hidden-to-hidden recurrent connection, input-to-hidden connection, and hidden-to-output connection, respectively. b_h And b_y are bias terms for the hidden and output states, respectively. Each node also has an activation function F connected with it [12]. This element-wise non-linearity function is commonly selected from various existing functions, such as the sigmoid, hyperbolic tangent, or rectified linear unit (ReLU) [12].

3.2 Long Short Term Memory

The vanishing and exploding gradient are serious problems that delay the gradient to backpropagate through long-range time intervals, making the training of regular RNNs a very challenging Task. LSTM is designed to solve the problems of RNN. This method is based on changing the structure of hidden units from "sigmoid" or "tanh" to memory cells, with gates controlling their inputs and outputs [13]. These gates regulate the flow of information to hidden neurons and preserve extracted features from earlier time steps [14].

4 The Proposed Architecture

4.1 Methodology

We designed a deep learning model that can recognize basic activities systemically based on the raw signals from sensors, fed directly into the neural network, with no pre-processing or feature engineering. Compared to traditional machine learning methods, this ability of deep learning offers more simplicity and eliminates the need for domain knowledge. For this, we used three consecutive LSTM layers to capture the temporal dependencies from sensor data, which is typically characterized as sequential and heterogeneous. Figure 1 presents a schema of our approach based on stacked LSTM layers. We used LSTM to exploit the temporal dependencies within the movements. This network solves traditional RNN gradient problems. Furthermore, it is capable of capturing long-term dependencies.

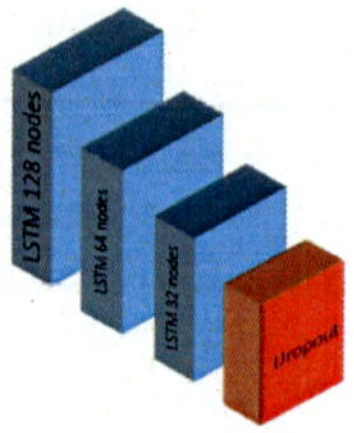

Fig. 1. Deep stacked LSTM network

Table 1. The parameters of our network

Parameters	Value
LSTM layer_1	128
LSTM layer 2	64
LSTM layer 3	32
Dropout	90%
Flatten	–
Dense layer_3	Output numbers = 6 (softmax)

A dropout follows all the LSTM layers after the third layer and before the final dense layer to reduce overfitting. Before we arrived at the final architecture in Table 1, we investigated the most impacting hyper-parameters.

4.2 Hyper-parameter Tuning

We built several variants of our network to verify the impact of the architecture and the number of nodes on the total performance. The evaluation and the choice of the best model are made based on the number of parameters and the accuracy of the UCI dataset.

Table 2. Details and parameters of the tested structures

Structure	Accuracy %	Parameters	Time/epoch
A: lstm(32)-lstm(32)-lstm(32)	94.19	46,598	2 s
B: lstm(64)- lstm(64)- lstm(64)	95.82	134,150	2 s
C: lstm(128)- lstm(128)- lstm(128)	95.14	432,134	3 s
D: lstm(128)-lstm(64)-lstm(32)	96.098	157,062	2 s
E: cnn(6) –lstm(128)-lstm(64)-lstm(32)	95.35	155,034	3 s

The number of nodes is indicated between brackets. It can be seen in Table 2 that structure D, provided the best performance while maintaining deepness and a limited number of parameters without slowing down computation speed. In what follows, we adopted this structure and we continued to tune the hyper-parameters.

Effect of Dropout. Dropout was applied after the third layer of the network with the probability of retaining the unit being p = (0.9, 0.8, 0.7, 0.6, 0.5, 0.4, 0.3, 0.25, 0.2, and 0.1) Fig. 2 shows the resulting accuracies. We can see that 10% is a little dropout value for all those different values, yet it performs well. However, although 70% and 90% are large, they enhanced the performance to 0.9623 and 0.9687, respectively.

Effect of Batch Size. To identify the optimal batch size, we tested 192, 128, 64, and 32. As stated in [15], the higher batch size will speed up the network's computation and reduce the number of updates required for the network to reach convergence. Table 3 shows that the best score is obtained when the batch size is set to 32.

Effect of the Optimizer. Table 4 shows that, compared to the Adadelta, Nadam, and Sgd optimizers, Adam optimizer produced the best results with the fastest calculation using its default values (Learning rate = 0.001). Rmsprop also provided us with a good level of accuracy.

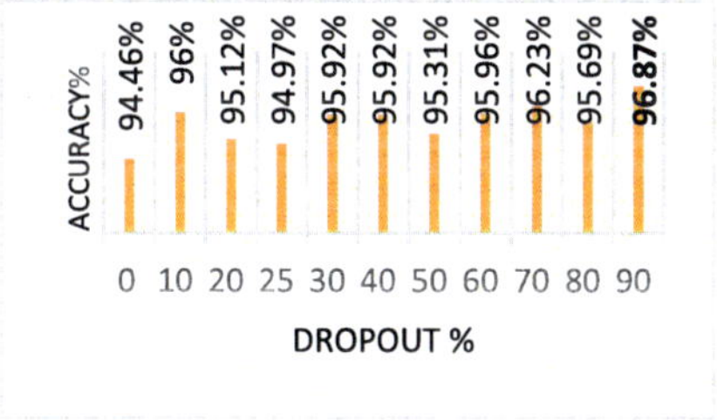

Fig. 2. Impact of Dropout

Table 3. Impact of batch size on accuracy for UCI

Batch size	Accuracy	Time per epoch
32	97.15%	23 s
64	96.10%	12 s
128	94.70%	3 s
192	96.064%	4 s

Table 4. Impact of the optimizer on accuracy UCI HAR

Optimizer	Adadelta	Nadam	SGD	Adam	RMSProp
Accuracy %	50,39	96,06	90,74	97,15	96,47

5 Experimentation

5.1 Datasets

We considered two public benchmark datasets for HAR to evaluate the performance of our network. Here is a brief description of each:

UCI HAR. [16] is a dataset collected by 30 users of different ages (from 19 to 48 years) wearing Samsung Galaxy SII smartphones on the waist and performing six activities. The accelerometer and gyroscope sensors are used to collect data. The available dataset contains 10,299 samples which are already separated into a training set and a test set.

WISDM. v1.1 [17], provided by the wireless sensor data mining (WISDM) Lab, is a dataset collected using only one smartphone accelerometer in the front pants leg pocket. Sampled at 20Hz, the chosen activities were selected depending on their performance regularity in daily life. Walking, Jogging, Upstairs, Downstairs, Sitting, and Standing were performed by 36 volunteer participants.

5.2 Performance Measure

We utilized four widely used evaluation metrics to compare the model's performance: Accuracy, Recall, Precision, F-Measure, defined as follows (TP: True Positive, TN: True Negative, FP: False Positive and FN: False Negative):

$$\mathrm{Accuracy} = \frac{\mathrm{TP} + \mathrm{TN}}{\mathrm{TP} + \mathrm{TN} + \mathrm{FP} + \mathrm{FN}} \tag{3}$$

$$\mathrm{Precision} = \frac{\mathrm{TP}}{\mathrm{TP} + \mathrm{FP}} \tag{4}$$

$$\mathrm{Recall} = \frac{\mathrm{TP}}{\mathrm{TP} + \mathrm{FN}} \tag{5}$$

$$\mathrm{F - Measure} = 2 \times \frac{\mathrm{Precision} \times \mathrm{Recall}}{\mathrm{Precision} + \mathrm{Recall}} \tag{6}$$

5.3 Training

UCI HAR dataset is devised to 70% for training and 30% for testing. As a result, subjects 2, 4, 9, 10, 12, 13, 18, 20, and 24 are left unseen by the model. This splitting will allow a fair comparison with previous works using the same evaluation technique. We used Adam optimization algorithm to minimize the cost function. The softmax layer exists in the final layer, and the gradient back propagates from this layer to the LSTM layers of the input. The dropout is 90%. During training, we tested different batch sizes. For over 200 epochs, we selected the size of 32. The signals in this dataset have been pre-processed using a low pass filter with 50% overlap. The sliding window has a size of 2.56 s.

To verify the generalization capability of our network, we tested it on the WISDM dataset on raw data with no feature selection. We just applied a sliding window, with a width of 180 and a time step of 100.

6 Results and Discussion

To compare our results, we draw Table 5 which contains the test accuracy of various approaches. The score achieved surpasses other studies based on feature selection, feature engineering, or methods based on traditional machine learning classification algorithms.

Table 5. Comparison with other works on UCI HAR dataset

Approach	[18]	[19]	Ours	[20]	[21]	[22]	[22]	[23]
Accuracy %	91.76	93.18	**97.15**	95.49	93.6	95.75	94.79	92.16

Table 6. Comparison with other works on WISDM

Approach	Catal et al. [24]	Alsheikh et al. [25]	Ours	Ravi et al. [26]
Accuracy %	94.3	98.2	**99**	98.6

To analyze the behavior of the detection of each class, we draw the confusion matrix presented in Fig. 3. it's clear that making the distinction between the sitting and standing classes was a bit of a challenge for the model. We can understand this misclassification if we remember the localization of the sensor on the waist. This position tends to make the gathered signals from those activities very similar. For the WISDM dataset, we adopted two evaluation methods, using a split of 70% training-30% testing and using 5-fold cross-validation. Although those two techniques are different, they have led to approximately the same accuracy score. We compare in Table 6 the results with the current state-of-the-art work. For the Per-user classification, Ignatov et al. [27] chose the first 26 subjects for their training set and then tested the results on the remaining ten users, scoring 93.32%. We adopted the same splitting and we got 99% accuracy again. Figure 4 shows there is a slight difficulty in the detection of upstairs and downstairs activities. To get a comprehensive vision of which activities our model can detect the most, we've included in Table 7 the classification results for each class.

WISDM and UCI har have nearly identical activities, except jogging and laying, which are distinct. We can divide them into two categories: static and dynamic. The former includes laying and standing, while the latter includes walking downstairs, jogging, upstairs, and walking. They also present the vast bulk of the data in WISDM and almost half of the UCI HAR dataset. A recall of 1 for activity C signifies that every item belonging to this class, was labelled as belonging to class C. As shown in Table 7, we had a good recall for all of the classes in both datasets except sitting, where data was gathered using an accelerometer and a gyroscope, and classification was a little more complex. It is misclassified withstanding. Downstairs has a lower recall of 0.96 than the others, indicating that recognizing the down direction with only one accelerometer is challenging. The remaining activities have all a good recall reaching 1. The presented

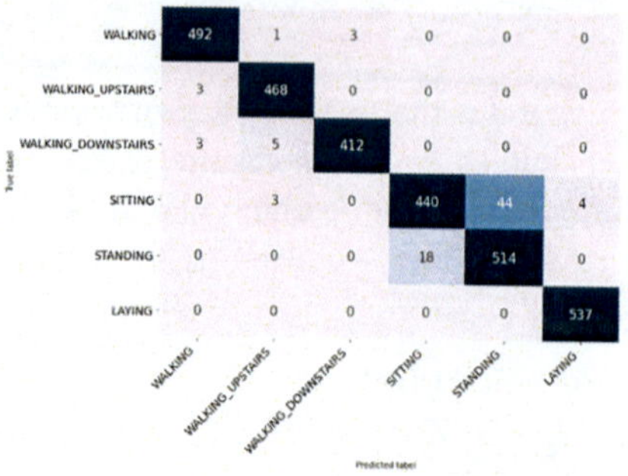

Fig. 3. Confusion matrix for UCI HAR dataset

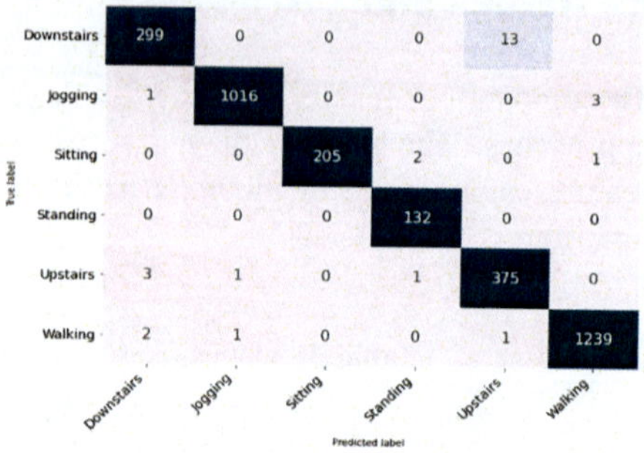

Fig. 4. Confusion matrix for WISDM dataset

results show that our model can detect those basic behaviors even when the position, number, and kind of sensors are changed.

Table 7. Per activity classification of our model

	WISDM v1		UCI HAR	
	Precision	Recall	Precision	Recall
Downstairs	0.98	0.96	0.99	0.98
Jogging	1	1	–	–
Sitting	1	0.99	0.96	0.90
Standing	0.98	1	0.92	0.97
Upstairs	0.96	0.99	0.98	0.99
Walking	1	1	0.99	0.99
Laying	–	–	0.99	1.00
F1-score	0.99		0.97	

7 Conclusion

In this work, artificial intelligence and the IoT are exploited to design a system capable of recognizing human activities. Our approach is based on using stacked LSTM layers in an end-to-end fashion without any data pre-processing. It eliminates feature engineering and provides better performance than traditional methods based on feature design. We plan to look into other datasets with more complex actions in the future and energy-efficient implementation of this model in smart devices.

References

1. Zhou, X., et al.: Deep-learning-enhanced human activity recognition for internet of healthcare things. IEEE Internet Things J. **7**(7), 6429–6438 (2020). https://doi.org/10.1109/JIOT.2020.2985082

2. Htike, K.K., Khalifa, O.O., Ramli, H.A.M., Abushariah, M.A.: Human activity recognition for video surveillance using sequences of postures. In: The Third International Conference on e-Technologies and Networks for Development (ICeND2014), pp. 79–82 (2014). https://doi.org/10.1109/ICeND.2014.6991357
3. Ke, S.-R., Thuc, H., Lee, Y.-J., Hwang, J.-N., Yoo, J.-H., Choi, K.-H.: a review on video-based human activity recognition. Computers **2**(2), 88–131 (2013). https://doi.org/10.3390/computers2020088
4. Lu, W., Fan, F., Chu, J., Jing, P., Yuting, S.: Wearable computing for internet of things: a discriminant approach for human activity recognition. IEEE Internet Things J. **6**(2), 2749–2759 (2019). https://doi.org/10.1109/JIOT.2018.2873594
5. Oluwalade, B., Neela, S., Wawira, J., Adejumo, T., Purkayastha, S.: Human activity recognition using deep learning models on smartphones and smartwatches sensor data (2021). http://arxiv.org/abs/2103.03836. Accessed 27 June 2021
6. Agarwal, P., Alam, M.: A lightweight deep learning model for human activity recognition on edge devices. Procedia Comput. Sci. **167**, 2364–2373 (2020). https://doi.org/10.1016/j.procs.2020.03.289
7. Tanvir Mahmud, A.Q.M., Sayyed, S., Fattah, S., Kung, S.-Y.: A novel multi-stage training approach for human activity recognition from multimodal wearable sensor data using deep neural network. IEEE Sens. J. **21**(2), 1715–1726 (2021). https://doi.org/10.1109/JSEN.2020.3015781
8. Wang, H., et al.: Wearable sensor-based human activity recognition using hybrid deep learning techniques. Secur. Commun. Netw. **2020**, 2138–2138 (2020). https://doi.org/10.1155/2020/2132138
9. Sun, J., Yongling, F., Li, S., He, J., Cheng, X., Tan, L.: Sequential human activity recognition based on deep convolutional network and extreme learning machine using wearable sensors. J. Sens. **2018**, 1–10 (2018). https://doi.org/10.1155/2018/8580959
10. Cao, J., Lin, M., Wang, H., Fang, J., Yueshen, X.: Towards activity recognition through multidimensional mobile data fusion with a smartphone and deep learning. Mobile Inf. Syst. **2021**, 1–11 (2021). https://doi.org/10.1155/2021/6615695
11. Lecun, Y., Bottou, L., Bengio, Y., Haffner, P.: Gradient-based learning applied to document recognition. Proc. IEEE **86**(11), 2278–2324 (1998). https://doi.org/10.1109/5.726791
12. Murad, A., Pyun, Jae-Young.: Deep recurrent neural networks for human activity recognition. Sensors **17**(11), 2556 (2017). https://doi.org/10.3390/s17112556
13. Cong, L., Tang, K., Wang, J., Zhang, Y.: AlphaPortfolio for investment and economically interpretable AI . SSRN Electron. J. (2020). https://doi.org/10.2139/ssrn.3554486
14. Jiuxiang, G., et al.: Recent advances in convolutional neural networks. Pattern Recognit. **77**, 354–377 (2018). https://doi.org/10.1016/j.patcog.2017.10.013
15. Bengio, Y.: Practical recommendations for gradient-based training of deep architectures. In: Montavon, G., Orr, G.B., Müller, K.-R. (eds.) Neural Networks: Tricks of the Trade. LNCS, vol. 7700, pp. 437–478. Springer, Heidelberg (2012). https://doi.org/10.1007/978-3-642-35289-8_26
16. Anguita, D., Ghio, A., Oneto, L., Parra Perez, X., Reyes Ortiz, J.L.: A public domain dataset for human activity recognition using smartphones. In: ESANN, Bruges, Belgium, pp. 437–442 (2013)
17. Kwapisz, J., Weiss, G., Moore, S.: Activity recognition using cell phone accelerometers. ACM SIGKDD Explor. Newsletter **12**(2), 74–82 (2011). https://doi.org/10.1145/1964897.1964918
18. Ronao, C.A., Cho, S.B.: Human activity recognition using smartphone sensors with two-stage continuous hidden Markov models. In: 2014 10th International Conference on Natural Computation, ICNC 2014, pp. 681–686 (2014). https://doi.org/10.1109/ICNC.2014.6975918

19. Ronao, C., Cho, Sung-Bae.: Recognizing human activities from smartphone sensors using hierarchical continuous hidden Markov models. Int. J. Distrib. Sens. Netw. **13**(1), 155014771668368 (2017). https://doi.org/10.1177/1550147716683687
20. Lin, Y., Jianning, W.: A novel multichannel dilated convolution neural network for human activity recognition. Math. Prob. Eng. **2020**, 1–10 (2020). https://doi.org/10.1155/2020/5426532
21. Zhao, Y., Yang, R., Chevalier, G., Ximeng, X., Zhang, Z.: Deep residual bidir-LSTM for human activity recognition using wearable sensors. Math. Prob. Eng. **2018**, 1–13 (2018). https://doi.org/10.1155/2018/7316954
22. Ronao, C., Cho, Sung-Bae.: Human activity recognition with smartphone sensors using deep learning neural networks. Expert Syst. Appl. **59**, 235–244 (2016). https://doi.org/10.1016/j.eswa.2016.04.032
23. Li, Y., Shi, D., Ding, B., Liu, D.: Unsupervised feature learning for human activity recognition using smartphone sensors. In: Prasath, R., O'Reilly, P., Kathirvalavakumar, T. (eds.) MIKE 2014. LNCS (LNAI), vol. 8891, pp. 99–107. Springer, Cham (2014). https://doi.org/10.1007/978-3-319-13817-6_11
24. Catal, C., Tufekci, S., Pirmit, E., Kocabag, G.: On the use of ensemble of classifiers for accelerometer-based activity recognition. Appl. Soft Comput. **37**, 1018–1022 (2015). https://doi.org/10.1016/j.asoc.2015.01.025
25. Alsheikh, M.A., Selim, A., Niyato, D., Doyle, L., Lin, S., Tan, H.P.: Deep activity recognition models with triaxial accelerometers (2016). http://arxiv.org/abs/1511.04664
26. Ravi, D., Wong, C., Lo, B., Yang, Guang-Zhong.: A deep learning approach to on-node sensor data analytics for mobile or wearable devices. IEEE J. Biomed. Health Inform. **21**(1), 56–64 (2017). https://doi.org/10.1109/JBHI.2016.2633287
27. Ignatov, A.: Real-time human activity recognition from accelerometer data using convolutional neural networks. Appl. Soft Comput. **62**, 915–922 (2018). https://doi.org/10.1016/j.asoc.2017.09.027

Optimized Influencers Profiling from Social Media Based on Machine Learning

Bahaa Eddine Elbaghazaoui(✉), Mohamed Amnai, and Youssef Fakhri

Laboratory of Computer Sciences, Faculty of Sciences Kenitra, IbnTofail University, Kenitra, Morocco
{bahaaeddine.elbaghazaoui,mohamed.amnai,fakhri}@uit.ac.ma

Abstract. The massive amount of data and information posted by social networks is attracting growing interest because of the several applications fields it can be utilized. In fact, publishing allows users to express opinions about products and to influence other users. Therefore, identifying influencers on online social networks with their products perception and preferences is crucial to enable marketers to apply effective techniques of viral marketing and recommendation. In this paper, our main purpose is to identify the optimal machine learning model for profiling influencers from social media networks and identifying the correlation between users. However, to achieve this goal, our solution illustrates around determining influence nodes over social network graph using the Pagerank algorithm. Then, through machine learning, the results of our experiment have shown that the correlation is always different when adding new parameter to our models, also to determine the suitable for our needs. In any case, the results we have obtained are very important and with very high precision.

Keywords: User profiling · Machine learning · Influencer · Information diffusion · Social media · PageRank

1 Introduction

Many social network platforms are gaining impressive popularity because writing posts allow users to express opinions, sentiments, and preferences about different topics and products in a very concise form [1]. The availability of large amounts of information released by users has attracted growing interest from research groups and commercial companies. In fact, keeping abreast of the motivations of the theme or the product's popularity and the possibility of people's views on user content can be a valuable help in designing more effective promotional activities. In fact, with opinions expressed by people, we have the possibility to understand the motivations of the popularity about topics or products, also can be a valuable help to design more effective promotional campaigns.

Now-a-days, social media content has been used by various brands for content promotion and dissemination. As we all know, it is beneficial in all areas of

Y. Maleh et al. (Eds.): ICI2C 2021, LNNS 357, pp. 43–56, 2022.
https://doi.org/10.1007/978-3-030-91738-8_5

business and management, including social commerce [2], e-government [3], political marketing [4] and digital marketing [5]. Moreover, the large amount of data is changing the landscape of digital marketing and posing huge challenges. It is necessary to use analytical modeling and management techniques to transform brand marketing data into business insights.

Websites use social marketing techniques to recommend their products to customers. In this case, marketers need to better identify users who have the ability to influence other users' choices in order to authorize product recommendation technology. Influencing users on social media can be judged by their participation in the posted content [7]. Therefore, a crucial research activity is to profile Influencers and their product perception on online social networks [6].

In this article, our objective is to profiling and identify influencers using machine learning models from online social media networks, however, our system starts with the collection of necessary data and clean up the ambiguities and errors from our database, then we used the PageRank algorithm for profiling influencer. Finally and depending on the data that we have extracted, we used machine learning models to train our system to identify influencers quickly and in real-time.

Many methods have been proposed to automatically identify influencers from social networks [18,19]. However, the performance of these methods largely depends on the parameters that are collected by each user like the number of followers, followings, likes, etc. In our approach, we focused on profiling influencers based on the content of their posts. Then train our system to speed up our approach. For this, our system will help to carry out several applications such as:

- Detect the "trending topics".
- Find new hot topics reported in the social media.
- Detect new and interesting stocks based on a set of expert articles.
- Organize scientific texts around specific themes.
- Find the different aspects of a product to be criticized in order to improve it.

By summarizing all previous thoughts, the structure of this article is as follows. In the first part, we first outline the related work and the main issues of our approach. Second part, give the main idea of our solution. In the third part, we introduce our implementation. Finally, we gave a global conclusion and proposed future work.

2 Related Works

This section describes the most important previous work related to finding influencers in social networks. The basic definitions, algorithms and techniques of graph theory and network analysis have been extensively studied in the works of Kleinberg and Easley [20] or Cook and Holder [21]. Many new studies have focused on the use of platform-specific functions [22,23]. These functions include the number of follower users, the number of followed users, the verification status, the number of users added lists, etc.

Generally, on Twitter, influencing people is measured by the number of retweets received by their tweets [23,24]. The Twitter API does not provide information about who reposts whom. Therefore, we cannot find out who is spreading information on Twitter. Therefore, it seems that the most appropriate approach is to use platform-specific features to identify the individuals who have the greatest impact on platforms such as Twitter.

Zengin Alp and Gunduz Oguducu [23] proposed a method called "Personalized PageRank", which integrates information obtained from the Twitter network and user activities. The proposed method aims to identify influencers on a particular topic, who are considered experts on the selected topic. Cataldi and Aufaure [25] proposed a method that analyzes the multiple paths that information follows on the network and provides a way to estimate the impact between users by evaluating the relationship between users. They use directed graphs to model the Twitter network on selected topics, where directed edges represent forwarding actions between nodes (users). Kwak et al. [26] Compare different impact metrics based on parameters (such as the number of reposts, the number of followers). They also proposed a "retweet tree", but did not use this feature as part of the impact measurement. Cha et al. [27] puts forward an in-depth comparison of three influence levels, namely degree, retweet and mention.

Our approach would focus on profiling influencers from twitter data using PageRank algorithm. Through this analyze, we will be able to predict influencers and tweet spreading due to machine learning models.

3 Problem Formulation

Many studies have focused on extracting metadata from social media platforms to fully understand the behavior of users. Understanding user behaviors can help correct the quality of various products and services [8], then present their products according to user needs. One way to easily present a product for multi-users or spreading information is to use influencers. In fact, influential marketing is one of the fastest growing areas of marketing and a key of marketing research. Sometimes it is difficult to determine the right influencer. Currently, there are many approaches proposed to solve the problem of extracting influencers from social networks [18]. The recommendations suggested are based on profiling influencers depends on parameters (number of followers, following, likes ...) and not on the content of their publications. Therefore, our objective is to find the most appropriate model for our problematic depend on the user's post content . On this paper, we focus on making a comparison between several machine learning models as well as taking the publication content into consideration.

4 Contribution

Influential marketing is one of the fastest growing areas of marketing and a key area of marketing research [9]. On social media, influential people share their opinions on products, services, and brands every day. These influential people can be regarded as opinion leaders, they communicate with the huge social network of people who follow them [10]. Some publications of these influencers spread at a very significant rate. We have noticed that this is a trend or a tendancy, these trends makes it possible to form public opinion and define several social standards. However, predicting the possibility of such an aspect may increase the impact of product advertising, form positive or negative public opinion, or detect and prevent such manipulation [11].

Recently, many algorithms have appeared and created for automating the process of identifying and profiling influencers in social networks. However, the evolution of the solution of the influence maximization (IM) problem is started and also based on the greedy algorithm. Indeed, the greedy algorithm has mainly solved the problem of finding top-k influencers in social networks [13]. Greedy algorithm is also suitable for Dijkstra's shortest path, Prim's MST and other problems, but it will also fail in some NP Hard problems.

In this paper, our workflow approach present as shown in the following Fig. 1. In the first step (1), we need to collect the necessary social network data about many users. The second step (2), we determined the profile that has a great impact and influence on others. Indeed, in this part, we modeled the problem in mathematical and graphical form to facilitate the solution (user as node and its influence as arc). Finally, determine the nodes that have an impact on neighboring nodes. In the third part (3), we applied some known and most powerful machine learning models on our data, then compare them to find the appropriate model with the best results for our needs.

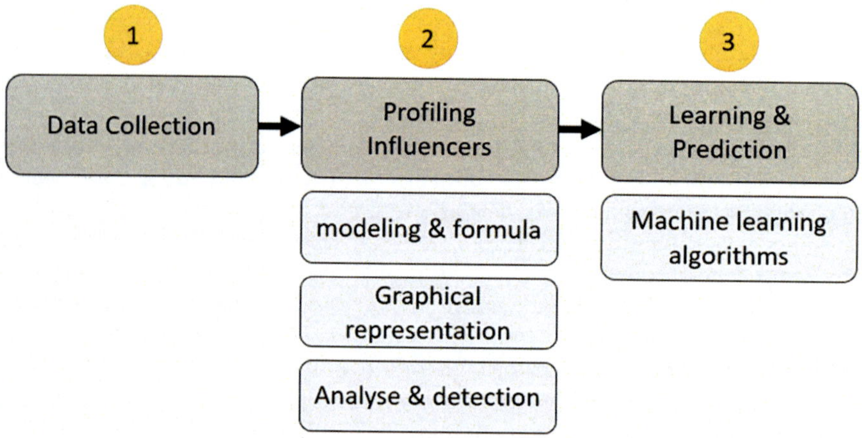

Fig. 1. Workflow process approach

In the first step is to extrat and profile data [28]. In fact ,the process of data collection in our approach is to collect and measure accurate insights data from a social network. Social media promises complete real-time recording of "natural" user activity. Issues related to validity and representativeness are often discussed [12].

The second step of our methodology is the identification of influential users by analyzing the data. Recall that in the influence maximization problem, we are trying to determine the seed set of k users in the social network, which will lead to the maximum spread of information that is considered a specific topic or problem to the rest of the social network. This problem is NP-hard, and there is a greedy approximation algorithm with an approximation ratio of 63% [14], which can solve this problem. This greedy algorithm inserts vertix into the seed set iteratively by adding vertices that try to maximize the influence spread or have the largest center value in each iteration.

In the graph, the centrality of a vertex defines its importance in the network. Some of the most common centralities are degree centrality, feature vector centrality, centrality between centers, and PageRank centrality. In this article, we use a method based on PageRank centrality to detect the seed set of influencers. The PageRank centrality illustrates the importance of the vertices connected to the vertex [15].

PageRank's first patent was filed on September 1, 1998, and became the original algorithm Google used to calculate the importance of web pages and rank these web pages. In short, Google is actually based on the idea of Sergey Brin [16], which can rank information on the Internet according to the link popularity of the page. The more links to the page, the higher the ranking.

Influential nodes are those nodes that are followed by the most number of other users/nodes. In addition, if the followers of a node are influential themselves, then the followed node becomes even more influential. For example, if X has 20 followers and Y has 10 followers, then X is supposed to be more influential than Y. If Z also has 10 followers, then Y and Z are equally influential. However, if the followers of Z are more influential than the followers of X then Z becomes more influential than X (simply because it is followed by more influential people).

Use the PageRank algorithm to calculate the impact of each node. The algorithm repeatedly tries to find the score of each node. Ideally, when the score does not change between subsequent iterations, the algorithm will stop (converge). The mathematical formula 1 used to calculate PageRank.

$$Rank[V_x] = \frac{1-d}{N} + d(\sum_{i=1}^{n} \frac{Rank[V_i]}{C[V_i]}) \tag{1}$$

- V_x: is vertex x.
- d: is a damping factor, generally 0.85.
- N: total numbers of nodes in the graph.
- V_i: is one of the n neighboring vertices of V_x.
- $C[V_i]$: is the count of neighbors of Vertex V_i.
- n: number of nodes in the graph.

Each node starts with an initial score of 1/n, where n is the number of nodes in the graph. For a node without any output vertex, its weight will be evenly distributed to all nodes in the graph (just like the virtual drawing edge from this

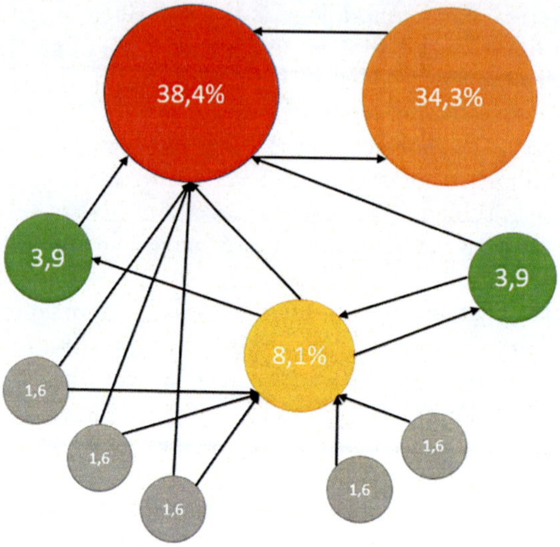

Fig. 2. The PageRank algorithm

node to all nodes in the network). As illustrat in the following Fig. 2, the score of a node is some weighted average of the scores of its immediate neighbors.

To generalize, the pagerank algorithm work as following 1:

Algorithm 1: PageRank Algorithm

```
G : graph
d : damping factor
iter : maximum iteration
PR : list of number (result pagerank)
Input  : G, d, iter
Output: PR
A = List[]
N = graph.nodes
for node in N do
| A[node] = 1/N
end
for node in N do
| PR[node] = 1-d
| for neighbor connect to node do
| | O = number of outgoing edges of neighbor
| | PR[node] = PR[node] + d * A[neighbor] / O
| | for node in N do
| | | A[i] = PR[i]
| | end
| end
end
```

In the last step of our methodology, we have worked with machine learning techniques to identify the most influential nodes in the network, study different machine learning models to determine the most suitable model for the network, and understand how to apply information cascade technology [17]. However, during this phase ,we have trained our machine learning models to accelerate influencer profiling from social media networks. Indeed, we have applied this procedure in two ways. The first, we recorded only the parameters of the users in the form of a matrix, then we divided it into two parts, one of the trainings and the other for the test. After, we concluded the suitable learning model in our case. The second, we respected the same previous procedure, but we also took into consideration the content of users' tweets.

5 Approach Implementation

Following the workflow that we have presented in our contribution. As the first step, our case is to collect a group of related user's data. Twitter is the target of our data processing and extraction, mainly for the following three reasons:

- Twitter is an open social network. Compared with other social networks, Twitter's data access is very easy to manage.
- Twitter pays more attention to the textual aspects of social network image to text range. In terms of vision, vision plays a role, but it is insignificant compared with other alternative methods.
- The retweet mechanism is unique to Twitter, allowing you to follow the posting along the posting path.

To start our process, we need to have access to the twitter application programming interface. We used anaconda ecosystem to run Tweepy which is a module executable in python in order to collect data from twitter, the use of Tweepy gave us the ability to extract data by accessing the twitter API.

The twitter API returns an object that represents a tweet in the form of "JSON", which will be used as a dictionary and has a large number of properties (keys) as showing in the following Fig. 3. Indeed, we can query the tweet object to see the keys of its dictionary. Among these keys we notice that some are also dictionaries, for example the "user" attribute has quite a few keys.

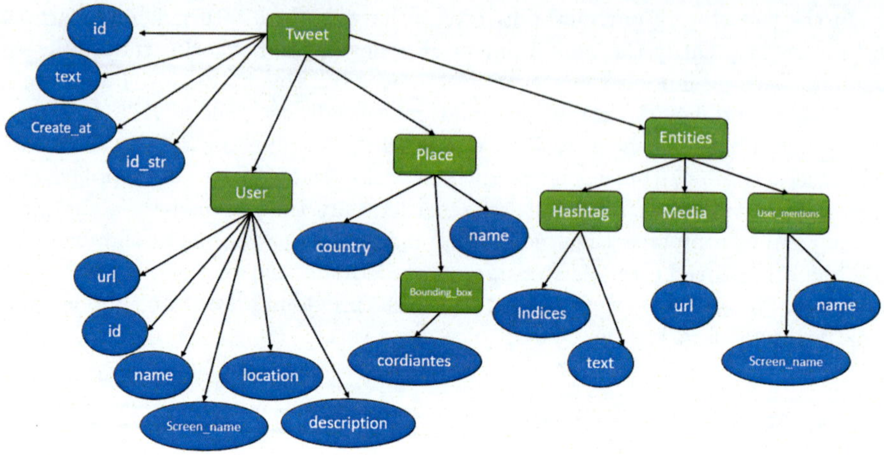

Fig. 3. Representation of tweet object

After extracting the data from Twitter. Now we have to identify these influencers among hundreds of thousands of users, we go from one chaotic form to another ordered matrix which can be well manipulated and navigated using mathematical formulas and apply the computer tool to this set of data going through several stages before arriving at a better form which is the graphical form.

According to the user data, we obtain a matrix with a coefficient of 0 or 1 (0: not follow & 1: follow). Moreover, we obtain an adjacency list or a list of neighbors of each vertex. In our case, it is a directed graph, so the head is a user X and the queue Y is the follower of the X. In another way, Y is an element of the adjacency list of the X.

These interconnections between people gave us a social graph. Based on the "NetworkX" library on Python, we provided a graph G with the set of nodes and the list of the arcs between these nodes which gave us an important base for our social graph. As a result, we can see in the following Fig. 4 that we have 178819 nodes and 218091 arcs.

So far, we have extracted all the users from our database and their interconnections. To properly classify users by their impact on others. We need to apply the PageRank algorithm. However, After performing the PageRank on our social graph, we identify each node by a PageRank centrality indicator.

After ordering users by their pagerank indicator. We end up having an initial result for our objective of detecting influencers that we call potential influencers. Indeed, we filter these nodes and remove the numerical indicators since our searches will henceforth only be indexed by the account id or by its username (screen_name), this set of potential influencers is not in real life. as a list of all users who are ordered according to their PageRank score.

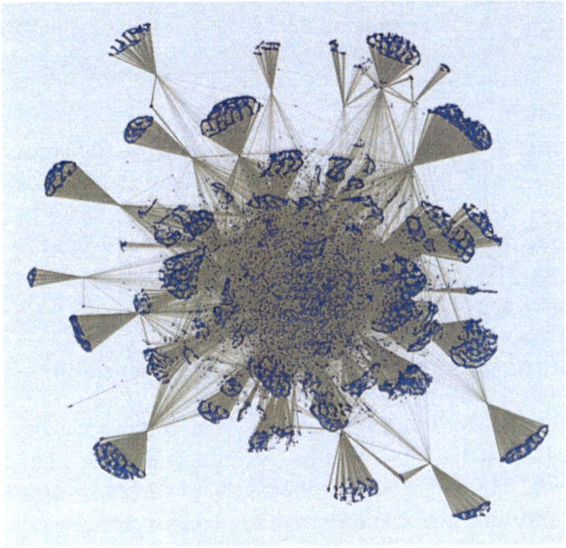

Fig. 4. Social graph

Users who do not meet at least a number of retweet greater than 13 (the number we used to determine a tweet in trend).In our case, we only took each tweet that was not retweeted 20 times, or a low number of links or that it is a private account will all be removed from the list of influencers. However, to achieve more concrete results, a numerical indicator "r" has been introduced, defined in the Eq. 2 as follows:

$$r = \frac{1}{20}\sum_{i=1}^{20}\frac{C_i}{T} \tag{2}$$

- C_i: the number of retweet of a T_i tweet.
- T: the number of followers of this user.

Going through these stages we arrive at a set of 78 users that we can rate influencers, these users represent a minority compared to the number of starting users, on the rating we sort these indicators to have our influencers from the most impactful to the Less impactful, the remark that we noticed to our surprise is, most of the influencers in our data set end up with accounts whose content in English.

No, we have to move on to the machine learning step. In our case, we concluded by taking 8 fields that we put in boxes to represent them in a numerical form within a matrix of integers. Indeed, the first 7 boxes will enter our model learning algorithms, and the last would be the result of finding the relationship with these data. Then, we note that this dataset has been repartitioned with a ratio of 60% for the training elements and 40% for the test. These fields represent as following:

- **Id**: user identifier.
- **Number of Followers**: the number of followers of the creator of this tweet.
- **Friends count**: The number of other users that the creator of this tweet is following.
- **Verified**: if the account is verified we put a 1 if not we put a 0.
- **Tweet length**: the length of the tweet without integrating the links.
- **Entities**: set to 1 if the tweet has a link if not 0.
- **Likes**: all "likes" received by the tweet.
- **Retweet count**: the trend arbiter set to 0 or 1 depending on its spread.

Our first step of analysis and process is to go through user data and regardless of the content of the tweet itself, only the information of the creator and the form of that tweet. The approach we have seen is to use several models to feed, compare them, and find out if the results are consistent.

So we consider 5 models which are as follows:

- Decision tree
- Random Forest
- Neural Networks
- Linear regression
- Logistic regression

These algorithms were all treated in the same way, providing the same data set to avoid any form of bias in our results.

There are two kinds of models among these. Linear Regression and Logistic Regression their precision would be calculated based on the variance and the other 3 (Decision Tree, Random Forest and Neural Networks) are calculated from a predictive nature where the result is discrete. A measurement of precision would be made by comparison between the result obtained and the actual existing result.

We have summarized the results of the models tested (without content) in the following Table 1.

Table 1. Comparison between machine learning models (without content)

Model	Precision (%)	Variance
Decision tree	92.66	0.29
Random Forest	93.15	0.23
Neural Networks	75.27	0.23
Linear regression	—	0.11
Logistic regression	92.23	0.25

After having implemented these models we evaluated the performance of each model in order to determine the most suitable one for our dataset, this would be done through the error metrics(MAE: Mean Absolute Error, RMSE: Root Mean Squared Error & MSE: Mean Squared Error). Throught these metrics, we got the results as presented in Table 2.

Table 2. Performance metrics (without content)

Model	MAE	RMSE	MSE
Decision tree	0.0815	0.0815	0.2855
Random Forest	0.0776	0.0776	0.2786
Neural Networks	0.1689	0.1689	0.4110
Linear regression	0.2277	0.1197	0.3461
Logistic regression	0.1032	0.1032	0.3213

Note that all models except linear regression generally have precision between 92% and 94%, which shows a strong correlation between the user and the tendency of a tweet without going through its content. The variances are all acceptable and very close to each other which shows minimal dispersion, therefore a prediction whose noise does not have a great effect. To conclude this task, we consider that the random forest is the best choice for our given database.

In the second step of our treatment, we must consider the content of each tweet. However, we start the preprocessing by removing all noise from the text (links, emoticons, special characters, etc.), we limit the acceptable characters to the 26 letters of the alphabet by removing the numerical forms and the links (Http/Https).

Afterwards, we transformed each text into Tokens "Tokenization". For example, the sentence "the earth is round" will undergo a partitioning in relation to the words one by one, which produces the following form ("the", "earth" , "is", "round"), where each element of this list is a Token of this sentence.

This is not sufficient since the terms "the" and "is" do not serve to reflect the subject of our sentence but above all to add a free complexity to our model, we opted for a simple solution. Indeed, we must remove all the empty words "STOPWORDS", consequently the same sentence becomes: ("earth", "round").

So each tweet text is then transformed into a list of tokens. This method allows us to classify and extract the theme or the cluster of each tweet. Indeed, according to the method of Latent Dirichlet Allocation (LDA), we can obtain the most used words in a topic where the human being must always identify.

By using the id of each tweet, we also added the topic and the trend to our database. In addition, we train the same previous models by the new data and we obtain the results illustrated in the following table:

Model	Precision (%)
Random Forest	84
Neural Networks	84

We notice that :

- The neural network surprises us that it is more efficient at this task than the previous one.
- The extraction of a topic is not very reliable, we notice that in each topic we find at least one intruding word. For example, we found the word "COVID" in most of the topics since the current word situation.
- The percentage that a tweet from our database is trending is 20%.

By highlighting these points, we can deduce that the content contributes to the tendency of a tweet but in a less important way than the profile of the user itself.

The prediction that a tweet is trending and a spot where the amount of data comes first, the limitations imposed by the Twitter API limit us to have better results specifically for content analysis, the machine limitation is another problem. Just to save the binary matrix we needed more than 120 GB of RAM.

6 Conclusion and Future Works

The main background of this work is to combine four fundamental tasks in big data. The collection and storage of data, the analysis and graphic representation, the learning and the prediction and finally the manipulation of textual data. However, throughout this project, we profiled the influencers from social media network (Twitter), then using machine learning, we made a comparison between several models to identify the best model suitable to our needs. Finally, we consider the random forest is the best choice for our database regardless on the content of the tweets with a precision of 93.15%. But, if we consider the content of each tweet, we notice that the best models are random forest and neural network with a precision of 84%.

In future works, we plan to add more data to our models (like images, voice messages, etc.). Also, classify users by groups, this will help us to make adaptable suggestions for each user.

References

1. Ahmad, A.: Social network sites and its popularity. Int. J. Res. Rev. Comput. Sci. Kohat **2**(2), 522–526 (2011)
2. Chen, A., Lu, Y., Gupta, S.: Enhancing the decision quality through learning from the social commerce components. J. Glob. Inf. Manag. (JGIM) **25**(1), 66–91 (2017)
3. Vakeel, K.A., Panigrahi, P.K.: Social media usage in e-government: mediating role of government participation. J. Glob. Inf. Manag. **26**(1), 1–19 (2018)

4. Grover, P., Kar, A.K., Dwivedi, Y.K., Janssen, M.: Polarization and acculturation in US Election 2016 outcomes-can twitter analytics predict changes in voting preferences. Technol. Forecast. Soc. Change (2018)
5. Parsons, A.L., Lepkowska-White, E.: Social media marketing management: a conceptual framework. J. Internet Commer. 1–15 (2018)
6. Oro, E., Pizzuti, C., Ruffolo, M.: A methodology for identifying influencers and their products perception on Twitter. In: National Research Council of Italy (CNR), Institute for High Performance Computing and Networking (ICAR), Via Pietro Bucci, 4-11C, 87036 Rende (CS), Italy (2018)
7. Romero, D.M., Galuba, W., Asur, S., Huberman, B.A.: Influence and passivity in social media. In: Gunopulos, D., Hofmann, T., Malerba, D., Vazirgiannis, M. (eds.) ECML PKDD 2011. LNCS (LNAI), vol. 6913, pp. 18–33. Springer, Heidelberg (2011). https://doi.org/10.1007/978-3-642-23808-6_2
8. Giannakis, M., Dubey, R., Yan, S., Spanaki, K., Papadopoulos, T.: Social media and sensemaking patterns in new product development: demystifying the customer sentiment. Ann. Oper. Res. 1–31 (2020). https://doi.org/10.1007/s10479-020-03775-6
9. Boerman, S.: The effects of the standardized Instagram disclosure for micro- and meso-influencers. Comput. Hum. Behav. **103**, 199–207 (2020)
10. De Veirman, M., Cauberghe, V., Hudders, L.: Marketing through Instagram influencers: the impact of number of followers and product divergence on brand attitude. Int. J. Advert. **36**(5), 798–828 (2017)
11. Goswami, A., Kumar, A.: A survey of event detection techniques in online social networks. Soc. Netw. Anal. Min. (2016)
12. Diaz, F., Gamon, M., Hofman, J. M., Kiciman, E., Rothschild, D.: Online and social media data as an imperfect continuous panel survey. PLoS One **11**(1) (2016). http://dx.doi.org/10.1371/journal.pone.0145406
13. Manaskasemsak, B., Dejkajonwuth, N., Rungsawang, A.: Community centrality-based greedy approach for identifying top-k influencers in social networks. In: Vinh, P.C., Alagar, V. (eds.) ICCASA 2015. LNICST, vol. 165, pp. 141–150. Springer, Cham (2016). https://doi.org/10.1007/978-3-319-29236-6_15
14. Kempe, D., Kleinberg, J.M., Tardos, E.: Maximizing the spread of influence through a social network. Theory Comput. **11**, 105–147 (2015)
15. Zafarani, R., Abbasi, M.A., Liu, H.: Social Media Mining: An Introduction. Cambridge University Press, Cambridge (2014)
16. Page, L., Brin, S., Motwani, R., Winograd, T.: The PageRank citation ranking: bringing order to the web. Technical report, Stanford InfoLab (1999)
17. Makhija, R., Ali, S., Jaya Krishna, R.: Detecting influencers in social networks through machine learning techniques. In: Hassanien, A.E., Bhatnagar, R., Darwish, A. (eds.) AMLTA 2020. AISC, vol. 1141, pp. 255–266. Springer, Singapore (2021). https://doi.org/10.1007/978-981-15-3383-9_23
18. Harrigan, P., Daly, T.M., Coussement, K., Lee, J.A., Soutar, G.N., Evers, U.: Identifying influencers on social media. Int. J. Inf. Manag. (2021)
19. Huynh, T., Nguyen, H., Zelinka, I., Dinh, D., Pham, X.H.: Detecting the influencer on social networks using passion point and measures of information propagation. Sustainability (2020)
20. Easley, D., Kleinberg, J.: Networks, Crowds, and Markets: Reasoning About a Highly Connected World. Cambridge University Press, Cambridge (2010)
21. Cook, D.J., Holder, L.B. (eds.): Mining Graph Data. John Wiley & Sons, Hoboken (2006)

22. Cossu, J.V., Labatut, V., Dugue, N.: A review of features for the discrimination of twitter users: application to the prediction of offline influence. Soc. Netw. Anal. Min. (2016)
23. Alp, Z.Z., Oguducu, S.G.: Identifying topical influencers on twitter based on user behavior and network topology. Knowl.-Based Syst. (2018)
24. Pal, A., Counts, S.: Identifying topical authorities in microblogs. In: Proceedings of the fourth ACM International Conference Web Search Data Min (2011)
25. Cataldi, M., Aufaure, M.A.: The 10 million follower fallacy: audience size does not prove domain-influence on Twitter. Knowl. Inf. Syst. (2015)
26. Kwak, H., Lee, C., Park, H., Moon, S.: What is Twitter, a social network or a news media? In: Proceedings of the 19th International Conference on World Wide Web (2010)
27. Cha, M., Haddadi, H., Benevenuto, F., Gummadi, K.: Measuring user influence in Twitter: the million follower fallacy. In: International AAAI Conference on Weblogs and Social Media (2010)
28. Elbaghazaoui, B.E., Amnai, M., Semmouri, A.: Data profiling over big data area. In: Gherabi, N., Kacprzyk, J. (eds.) Intelligent Systems in Big Data, Semantic Web and Machine Learning. AISC, vol. 1344, pp. 111–123. Springer, Cham (2021). https://doi.org/10.1007/978-3-030-72588-4_8

Natural Language Processing Based Approach to Overcome Arabizi and Code Switching in Social Media Moroccan Dialect

Soufiane Hajbi[1(✉)], Younes Chihab[1], Rachid Ed-Dali[2], and Redouan Korchiyne[1]

[1] Laboratory of Computer Sciences, Ibn Tofail University, Kenitra, Morocco
{soufiane.hajbi,younes.chihab,redouan.korchiyne}@uit.ac.ma
[2] Faculty of Letters and Humanities, Cadi Ayyad University, Marrakesh, Morocco
rachid.eddali@uca.ac.ma

Abstract. Over the past decade, the use of social networks has become more common and people have found a convenient place to share information and express opinions. The massive volume of data generated provides a good opportunity to extract valuable knowledge to reveal people's needs and behaviours. For this purpose, Sentiment Analysis techniques are widely used. The results are very accurate when they are applied to common languages, namely English, Spanish or French. However, these techniques are still at development stage for Modern Standard Arabic (MSA) and derived dialects. In the case of Moroccan Dialect used in Social Media, the main challenge is the phenomena of Code Switching; two or more languages appear in the same sentence (Arabic, Tamazight, French, English or Spanish). The second is the Arabizi of words using Latin script combined with numbers instead of Arabic characters. As a consequence, the preprocessing became one of the important steps of data analysis. This paper proposes a new method based on Natural Language Processing (NLP) to address the challenges of preprocessing text that contains Arabizi and Code Switching forms. We aim to build a multilingual corpus that includes linguistic features and reflects the structure of text written in Social Media Moroccan Dialect.

Keywords: Moroccan Dialect · Code Switching · Arabizi · Natural Language Processing · Data Mining

1 Introduction

The poor quality of the collected data sets is often an obstacle to knowledge discovery. The presence of noise, missing values and inconsistent data can give rise to misleading results. Thus, data preprocessing is a major task and the most time-consuming step in the data mining process [1]. It provides several techniques to process such data before being analyzed, which can significantly improve the performance of any ML prediction model [2].

Y. Maleh et al. (Eds.): ICI2C 2021, LNNS 357, pp. 57–66, 2022.
https://doi.org/10.1007/978-3-030-91738-8_6

Social media are amongst the important sources of data that can be useful in knowledge discovery and help identifying society trends and needs. The majority of this data exists in textual forms which are highly unstructured and are written in different ways that vary from dialect to standard language. Compared with MSA, the Moroccan Dialect on social networks has special characteristics. The text is a mixture of symbols, numbers, repeated letters, non-linguistic expressions like emojis and emoticons, and Arabizi of Arabic words. Additionally, it contains spelling mistakes and words from different languages. Although ML and NLP provide many tools for MSA, they remain not adapted to process Moroccan Dialect [3]. The cleaning and preprocessing of data in such a dialect is a major challenge to research and data mining applications [4]. The purpose of this work is to present two significant contributions based on ML and NLP techniques. The first one is to overcome Code Switching by identifying the languages used in the collected data before processing the words according to the NLP rules applied to their original languages. The second is to detect the text written in Arabizi and non-standard Romanization of Arabic by using Latin script and numbers [5].

The remainder of this paper is structured as follows: Sect. 2 deals with the problem statement and motivation and a brief literature review is presented in Sect. 3. In Sect. 4, we introduce the NLP techniques used in this work. Section 5 sheds light on the proposed approach to tackle the difficulties caused by the usage of Arabizi and Code Switching in Moroccan Social Media Dialect. Lastly, Sect. 6 concludes the paper.

2 Problem Statement and Motivation

The Moroccan Dialect is a homogeneous and varied mixture of different languages that reflects the history of the country through several eras. Morocco has two official languages: Arabic and Tamazight. The latter has three varieties including Tarifit, Tamazight and Tachelhit. These varieties are generally used in oral communication but do not have a unified form. Administratively, MSA is the most used language. French and Spanish are also present, while English is starting to gain interests due to globalization and the ability to have access to information and communication technologies [6].

Recently, the Moroccan society has started to exhibit Diglossia [7]. This phenomenon is becoming more perceptible with the rise of social medial networks. In their publications, Moroccan users do not only use multiple languages: MSA, Moroccan Dialect, French and English, but also the Arabizi written form. This makes the automatic knowledge discovery a very challenging task. The following couple of examples of Social Medial Moroccan Dialect illustrate the previously mentioned phenomena:

- **Code Switching**
  - Original Sentence: واش الابناء الصغار معنيين ب ***test PCR***
  - English translation : *do children have to take a PCR test?*

  - Transliteration[1] : wA$ Al>bnA' AlSgAr mEnyyn b test pcr

- **Arabizi**
  - Original Sentence: ***kayan ba3d nass kay3al9o bla mayfahmo***
  - Moroccan Dialect: كاين بعض ناس كيعلقو بلا مايفهمو
  - English translation: *some people comment without understanding the context*
  - Transliteration (See footnote 1): kAyn bED nAs kyElqw blA mAyfhmw

In this paper, we propose an NLP based approach to overcome the Arabizi and Code Switching in Social Media Morrocan Dialect. Further, we aim to classify words by language and build a clean and well structured multilingual corpus.

3 Related Works

Several works on sentiment analysis for MSA and Arabic Dialects varieties have been conducted. In general, the scope was either to compare the accuracy of the common ML prediction models or to propose new data mining methods. In this section, we have selected some researches that shed light on the aforementioned scopes. Mainly, we discuss the aim of the related works and eventually mention the preprocessing methods applied.

Farha & Magdy [8] conducted a comparative study by testing the effectiveness of different ML approaches on popular benchmark Arabic data sets. The applied preprocessing techniques were Letter normalization, Elongation removal, Emoji replacement and data cleaning. They found that deep learning models are more effective compared to classical ML algorithms but due to the ambiguity of the data annotation and sarcasm sentiments in some data sets, the results are not quit high. Soufan [9] worked on sentiments analysis on Arabic Language using text extracted from Twitter and GoodReeds. The author used different models, such as Multinomial Naive Bayes, SVM classifiers as well as LSTMs, Word-Level CNN and Char-level. The standard data cleaning and normalization techniques were applied to enhance the analysis accuracy. The experiment shows that the basic ML models are suitable for small data sets while the accuracy of deep learning is better on large datasets. Furthermore, the main challenges faced are the lack of annotated data and the limited number of preprocessing tools built for MSA and Arabic Dialects.

Mallek et al. [10] have proposed a statistical machine translation model for Arabic tweets into English. They used the parallel corpus (MSA and English) from the United Nation (UN) as training data set. The results of the automatic translation show that the initial preprocessing step is very important, as it can improve the bilingual evaluation understudy (BLEU) scores by 1 to 4 points. However, the system was not able to translate words written in Arabic

[1] Arabic transliteration according to Buckwalter System. Retrieved June 07, 2021, from http://www.qamus.org/transliteration.htm.

Dialect. Hegazi et al. [4] presented a methodology that is completely focuses on the preprocessing of MSA text on social media aiming to make datasets more suitable for further processing and knowledge extraction. For this purpose, the authors provide, based on NLP techniques a set of algorithms for data cleaning, tokenization, stemming, root and morphological generation.

Abdouli et al. [11] proposed an approach based on a distributed system for analyzing the tweets of Moroccan users. The preprocessing was the principal task of the work because of the difficulties in dealing with the Moroccan Dialect. Besides the conventional preprocessing techniques, the authors deployed a manually developed dictionary that includes mapping between words written in Moroccan Dialect and their correspondence in Arabic or French so that the final corpus contains only words from standard languages. For the sentiment analysis of the Saudi dialect tweets, Ghaith [12] presents SaudiSentiPlus, a new lexicon built by collecting words from various sources. First, he applied the automatic translation of English sentiment lexicons to Arabic. Then he manually extracted the Saudi dialect words from Twitter datasets. Furthermore, 4431 words have been taken from SauDiSenti lexicon. In the preprocessing step, the author applied some techniques to clean and remove redundant letters in tweets. Additionally, he developed two algorithms for preprocessing lexicon words and handle the known prefixes and suffixes in Saudi dialect words which helped to achieve 81% of accuracy.

4 Natural Language Processing (NLP)

NLP is a cross-discipline of computer sciences and computational linguistics. It provides several techniques that help machine recognize and interpret human language, such as Sentence splitting, Part-of-speech tagging, Semantic role labelling, Syntactic parsing, etc. In this work, we are interested in exploring and applying these techniques that are relevant for preprocessing the MSA and Moroccan Dialect. To this end, we will use Natural Language Toolkit (NLTK), which provides basic classes for building programs in python [13]. First, we will use Tokenization to split the data into a collection of unique words. Then, we will apply the regular expression and the language detection techniques to extract the Arabizi words and detect Code Switching by classifying words according to their language.

Tokenization is an important part of text processing. It relates to segmenting an input sequence of orthographic elements into small units known as tokens [14]. The presence of Arabizi and Code Switching makes the analysis of social media Moroccan Dialect a very challenging task. In our approach, tokenization is performed on the collected data to prepare it for further processing that involves the extraction of all Arabizi occurrences and then the classification of words by language.

Language Detection over text that contains words from more than one language has been considered as an open research question [15]. Generally, the

NLP techniques are designed to process monolingual texts, so that the incorporation of additional standard and dialectal languages generally decreases the accuracy of data mining systems [16]. We choose the python version of Nakatani Shuyo's language-detection library [17] to perform the language identification. The library supports 49 languages (including Arabic) and uses a Naive Bayes algorithm with character n-grams features. Naive Bayes is widely used in NLP for classification, filter, and sentiment analysis tasks. The core concept relies on a probabilistic classifier that assigns a class label to instances that are represented by a set of features. This becomes feasible by the strong assumption of conditional independence among the features that characterize a class [18].

Regular Expression is the most fundamental tool for text processing. It is a formal language which offers a practical way of specifying various kinds of NLP patterns. We used this tool to extract the words in Arabizi form, where the Arabic script is transcribed into Latin script combined with Arabic numerals (*ba3d, corresponding Arabic word: بعض, English translation: some*). Additionally, we use language detection techniques to identify the second form, where only Latin script is used (*nass, corresponding Arabic word: الناس, English translation: people*).

5 Proposed Approach

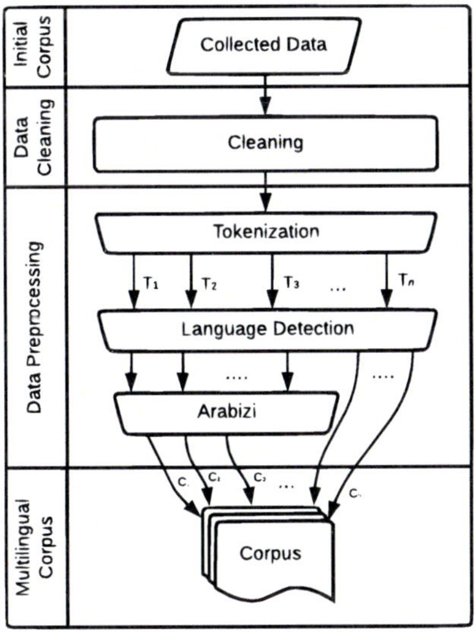

Fig. 1. Moroccan Dialect preprocessing method.

As we can see from the cited examples in Sect. 2, sentences may include words from different languages. Moreover, using numbers, Arabic and Latin characters form texts in the Moroccan Dialect, along with noise, missing values and inconsistent data pose more challenges to NLP techniques since they are mainly adapted to standard languages. Thus, Our main goal in this work is to transform the text from informal and unstructured way to a well-formed multilingual corpus where the words are classified by language (Moroccan Dialect, MSA, English, Spanish, French) and writing script (Arabic, Arabizi, Latin). Therefore, we will apply the NLP tokenization, language detection and regular expressions techniques to build this corpus (Fig. 1).

5.1 Data Collection

According to Alexa rank[2], Facebook is the most popular social media site in Morocco. Hence, we decided to choose the Facebook page of the online news website Hespress, which counts over 18M followers, to collect data. For the extraction, we choose Facepager [19]. Facepager is an open-source application for the automatic data extraction. It uses APIs for fetching openly available data from Facebook, Twitter and other websites. The data are stored in an SQLite database before being exported to a CSV file. Table 1 shows four chosen sentences, source texts (STs) and target texts (TTs) that demonstrate the common modes of transcription used by Moroccan users in social media:

Table 1. The transcription modes used by Moroccan users in social media

Source Text / Target Text	Arabizi	Code Switching	Languages
اللقاح هو الحل لكن هل من لقاح آمن وفعال!؟ / *The vaccine is the solution, but is there a safe and effective vaccine!?*	-	-	MSA
Étant vacciné ne signifie pas que tu ne peux pas attrapé le virus / *Being vaccinated doesn't mean you can't be infected with the virus*	-	-	Fr
Ndir **lhajr walakin** mayfrdoch 3lia hôtel / *I accept a quarantine but not a mandatory hotel stopover*	✓	✓	**MSA**,Dialect, Fr
لكونفينمون سالا ممكن نسافرو سيمانة الجاية / *lockdown has been ended, we can travel next week*	-	✓	**MSA**,Dialect, Fr,Es

---

[2] Alexa Ranking: Top social media sites in Morocco, http://www.alexa.com/topsites/countries/MA, visited Mai 07, 2021.

5.2 Data Cleaning

Data cleaning is the second step in our proposed approach. Generally, it is an important task in the data mining process. But when it comes to unstructured text, this step is even more important. Usually, the text in our case consists of missing data, duplicated data, misleading data, Arabic characters, Latin script and numbers, etc. Furthermore, the Arabizi and Code Switching forms in the Moroccan Dialect increase the complexity of this step. Therefore, the data cleaning in our approach consists of (1) removing punctuation, (2) removing emojis and emoticons, (3) removing urls and htmls, (4) removing language irrelevant characters that does not impact further analysis like: + / - — ( ) [ ], and (5) removing Elongation; some users tend to repeat a character to intensify the emotions expressed. This repetition will be removed and the word will be reduced to its original form. To do so, Python provides a rich selection of libraries. First, we use the pandas library to store the data in tabular form - data frame, which is essential for efficient data manipulation. Then we use the Regular Expression module (re) in combination with the string module to perform the above-cited steps 1 to 5.

5.3 Tokenization

Tokenization is an essential part of text processing. Every NLP task starts by cutting text into identifiable small units such as sentences or words to transform the input sequence into a machine-readable form [13]. It depends significantly on the structure and composition of each language. Usually, the default tokenization process includes a morphological analysis that aims to identify the linguistic units of a given language such as lemma, root, prefix, etc. [20]. But one of the major challenges of such word tokenizer is the processing of Out Of Vocabulary (OOV), words that do not exist in the vocabulary. In our case, all occurrences of Arabzi, Dialect and Code switching forms are OOV words. Therefore, we decided to use the regular expression techniques to extract the tokens independently of their languages or transcription modes. For this, NLTK tokenize package includes a Regular-Expression Tokenizers that splits text into tokens based on regular expression. This is the best adapted tokenizer for our purpose. We have applied the following pattern `\w+|\$[\d\.]{1}|\S+` that matches all word forms that

Table 2. RegExp tokenizer results

Original Text	RegExp Tokenizer Results
اللقاح هو الحل لكن هل من لقاح آمن وفعال!؟	اللقاح، هو، الحل، لكن، هل، من، لقاح، آمن، وفعال
Étant vacciné ne signifie pas que tu ne peux pas attrapé le virus	Étant, vacciné, ne, signifie, pas, que tu, ne, peux, pas, attrapé, le, virus
Ndir lhajr walakin mayfrdoch 3lia hôtel	Ndir, lhajr, walakin mayfrdoch, 3lia, hôtel
لكونفينمون سالا ممكن نسافرو سيمانة الجاية	لكونفينمون، سلا، ممكن، نسافرو، سيمانة، الجاية

occur in social media Moroccan Dialect. The tokenization results of the above-cited STs in Table 1 are shown in Table 2.

5.4 Arabizi and Language Detection

In this work, the automatic language detection is the main task. For this, we propose an algorithm based on the python version of Nakatani Shuyo's language-detection library [17]. This library evaluates the input text T and returns the corresponding language id Lid and a precision score A. The library was tested with texts in 49 standard languages with 99.8% average precision. Thus, we consider all tokens with a detection score below this precision average $A < 0.998$ as Moroccan Dialect Md words. First, we set the list of possible languages to MSA-Ar, Englisch-En, Spanisch-Es, and French-Fr. Then, we classify the detected Moroccan Dialect words according to their writing script-ws; *Arabic*, *Arabizi*, and *Latin*. Finally, we store the input Token T_i together with the detected language Lid and the writing script ws in the Corpus C. We have chosen the following notation: (1) The language-detection of Nakatani is denoted as $ld(T_i)$ with the result $Lid \in \{Ar, En, Es, Fr\}$ and precision A. (2) We set the result of the automatic language detection in our algorithm to $L_w(T_i)$ with T_i as input token.

$$ld(T) = \begin{bmatrix} Lid \\ A \end{bmatrix} \tag{1}$$

Where Lid is the result of the language detection and A the precision score.

$$L_w(T_i) = \begin{bmatrix} ld(T_i) & ,if\ Lid \in \{Ar, En, Es, Fr\}\ and\ A > 0.998 \\ Md & ,otherwise \end{bmatrix} \tag{2}$$

Where $ld(T_i)$ is Arabic, English, Spanish, or French if A has scored over 0.998. Otherwise, the language of token (T_i) is set to Moroccan Dialect Md.

In the Algorithm 1, we first set the precision score A to 0.998 and the possible language result to the interval L (Arabic, English, Spanish, and French). Then, for each token Ti, we check both: the language result $R[Lid_i]$ and the calculated precision Ai. If the detected language $R[Lid_i]$ belongs to the possible languages in L and the precision A is over 0.998, then Ti language is the result $R[Lid_i]$. Otherwise, the language is set to Moroccan Dialect Md. Additionally, a check of the used writing script ws_i is performed: ws_i is set to Latin when writing script of Ti is English, Spanish, or French and to Arabic if the Arabic alphabet is used. In the case of token written in Arabizi form, ws_i is set to Arabizi. Finally, all the tokens are stored in Corpus C with the appropriate language and writing form.

Algorithm 1. Arabizi and Language Detection

Input: $T = (T_1, T_2, T_3, ..., T_n)$

Output: $C = \left(\begin{bmatrix} T_1 \\ Lid_1 \\ ws_1 \end{bmatrix}, \begin{bmatrix} T_2 \\ Lid_2 \\ ws_2 \end{bmatrix}, \begin{bmatrix} T_3 \\ Lid_3 \\ ws_3 \end{bmatrix}, ..., \begin{bmatrix} T_n \\ Lid_n \\ TS_n \end{bmatrix} \right)$

Initialize: $A = 0.998$ $L = (Ar, En, Es, Fr)$

for T_i in T **do**
  $R = ld(T_i)$ ▷ Result of langdetect library
  if $R[Lid_i] \in L$ *and* $R[A_i] \geq A$ **then**
    ws_i = get writing script by Lid_i ▷ $Lid_i = ar$ then Arabic, otherwise *Latin*
  else
    $Lid_i = Md$ ▷ Moroccan Dialect token
    switch writing script of T_i **do** ▷ writing script detection with RegExp
      case *Arabic*
        $ws_i = Arabic$
      case *Arabizi*
        $ws_i = Arabizi$
      case *Latin*
        $ws_i = Latin$
  end if
  add $C_i = (T_i, Lid_i, ws_i)$ to C
end for
return C

6 Conclusion

The preprocessing of MSA and Dialectal Arabic is still at an early stage and needs more attention to improve the data quality. Usually, the nature of the texts in similar languages increases the difficulty of the preprocessing step. For this reason, most works use the human crafted approach to clean and prepare data before applying prediction models. In this paper, we present a new approach to tackle the challenges related to the preprocessing of Social Media Moroccan Dialect. It aims to overcome the Arabizi and Code-Switching and enhance the input data quality. Based on the NLP techniques Regular Expressions and Language detection, the above Algorithm 1 builds a multilingual corpus where each token is classified by its language and writing script. This Corpus gives us an insight into Moroccan Dialect and can be useful in further analysis of Social Media Moroccan Dialect. For instance, we can measure the influence of foreign languages on Moroccan Dialect or build a dictionary entirely based on Moroccan Dialect.

References

1. Alasadi, S.A., Bhaya, W.S.: Review of data preprocessing techniques in data mining. J. Eng. Appl. Sci. **12**(16), 4102–4107 (2017)

2. Alshdaifat, E., Alshdaifat, D., Alsarhan, A., Hussein, F., El-Salhi, S.M.F.S., et al.: The effect of preprocessing techniques, applied to numeric features, on classification algorithms' performance. Data **6**(2), 1–23 (2021)
3. Harrat, S., Meftouh, K., Smaili, K.: Machine translation for Arabic dialects (survey). Inf. Process. Manag. **56**(2), 262–273 (2019)
4. Hegazi, M.O., Al-Dossari, Y., Al-Yahy, A., Al-Sumari, A., Hilal, A.: Preprocessing Arabic text on social media. Heliyon **7**(2), e06191 (2021)
5. Talafha, B., Abuammar, A., Al-Ayyoub, M.: ATAR: Attention-based LSTM for Arabizi transliteration (2088–8708). Int. J. Electr. Comput. Eng. **11**(3), 2327–2334 (2021)
6. Chakrani, B.: Between profit and identity: analyzing the effect of language of instruction in predicting overt language attitudes in Morocco. Appl. Linguis. **38**(2), 215–233 (2017)
7. Ferguson, C.A.: Diglossia. Word **15**(2), 325–340 (1959)
8. Farha, I.A., Magdy, W.: A comparative study of effective approaches for Arabic sentiment analysis. Inf. Process. Manag. **58**(2), 102438 (2021)
9. Soufan, A.: Deep learning for sentiment analysis of Arabic text. In: Proceedings of the ArabWIC 6th Annual International Conference Research Track. ArabWIC 2019. Association for Computing Machinery (2019)
10. Mallek, F., Belainine, B., Sadat, F.: Arabic social media analysis and translation. Procedia Comput. Sci. **117**, 298–303 (2017). Arabic Computational Linguistics
11. El Abdouli, A., Hassouni, L., Anoun, H.: Mining tweets of Moroccan users using the framework Hadoop, NLP, k-means and basemap. In: 2017 Intelligent Systems and Computer Vision (ISCV), pp. 1–7. IEEE (2017)
12. Al-Ghaith, W.: Developing lexicon-based algorithms and sentiment lexicon for sentiment analysis of Saudi dialect tweets. Int. J. Adv. Comput. Sci. Appl. **10**(11), 83–88 (2019)
13. Bird, S., Klein, E., Loper, E.: Natural Language Processing with Python: Analyzing Text with the Natural Language Toolkit. O'Reilly Media Inc., Sebastopol (2009)
14. Habash, N., Rambow, O.: Arabic tokenization, part-of-speech tagging and morphological disambiguation in one fell swoop. In: Proceedings of the 43rd Annual Meeting of the Association for Computational Linguistics (ACL'05), pp. 573–580 (2005)
15. Hughes, B., Baldwin, T., Bird, S., Nicholson, J., MacKinlay, A.: Reconsidering language identification for written language resources. In: Proceedings of the Fifth International Conference on Language Resources and Evaluation (LREC'06). European Language Resources Association (2006)
16. Jauhiainen, T., Lui, M., Zampieri, M., Baldwin, T., Lindén, K.: Automatic language identification in texts: a survey. J. Artif. Intell. Res. **65**, 675–782 (2019)
17. Shuyo, N.: Language detection library for java (2010). http://code.google.com/p/language-detection/
18. Friedman, N., Geiger, D., Goldszmidt, M.: Bayesian network classifiers. Mach. Learn. **29**(2), 131–163 (1997). https://doi.org/10.1023/A:1007465528199
19. Jünger, J., Keyling, T.: Facepager. An application for automated data retrieval on the web. Facepager. An application for generic data retrieval through APIs. Source code and releases available (2019). https://github.com/strohne/Facepager
20. Chiny, M., Chihab, M., Chihab, Y., Bencharef, O.: LSTM, VADER and TF-IDF based hybrid sentiment analysis model (2021)

Towards a New Lexicon-Based Features Vector for Sentiment Analysis: Application to Moroccan Arabic Tweets

Moncef Garouani[1,2(✉)] and Jamal Kharroubi[1]

[1] LISIC Laboratory, Univ. Littoral Cote d'Opale Calais, Calais, France
moncef.garouani@etu.univ-littoral.fr
[2] LSIA Laboratory, Faculty of Sciences and Techniques Fez, USMBA, Fes, Morocco
jamal.kharroubi@usmba.ac.ma

Abstract. The emergence of the Web 2.0 technology generated a huge amount of raw data by enabling Internet users to post their opinions and reviews on the web. This data plays an important role in decision making for many peoples and organizations. An example of valuable insights that can be extracted from user's posts is their opinions and sentiments regarding topics, events, services, products, etc. The English language has been the subject of extensive research on sentiment analysis. The proposed solutions are largely dominated by the use of two main analysis approaches based on machine learning techniques and the lexical approach. This work focuses on the second one to analyze the sentiments expressed in Moroccan tweets written in Arabic language : Standard Arabic (SA) and Moroccan Dialect (MD), and proposes a new method for extracting characteristics and representing data. The main idea of this method is to represent the text as a weight vector of feelings. Due to the lack of resources (databases and lexicon dictionaries) for the Arabic language, especially for the Moroccan one, this work starts with the construction of a corpus of 18.000 valid tweets based on 36 114 collected tweets that are manually tagged and classified as MD or SA. Then describes the steps of the construction of the Moroccan Senti-lexicon, a dictionary of 30.000 words labeled as positive, negative or neutral. The results of this study prove to be superior to those obtained by other comparable state of the art approaches.

Keywords: Sentiment analysis · Opinion mining · Sentiment lexicon · Polarity classification · Sentiment vector · Moroccan tweets

1 Introduction

With the advent of the Web 2.0, Internet users became capable to share their feelings and opinions on particular topics (political, business or individual). Facebook, Twitter, Instagram and LinkedIn, these social platforms are now part of everyday life, allowing internet users to communicate, debate issues, and share

Y. Maleh et al. (Eds.): ICI2C 2021, LNNS 357, pp. 67–76, 2022.
https://doi.org/10.1007/978-3-030-91738-8_7

opinions. The data aspect of these social media, such as Twitter messages, generates a rich wealth of data about who is involved in communication.

This data plays an important role in decision making for many peoples and organizations. For example, it is important for a government to know the views and concerns of citizens and to reflect their moods and opinions in relation to the news [1]. Likewise, the example of companies wishing to know the reactions of consumers to their products. This could help them to change their strategies to improve the quality of their products and that of their services. For this reason, it is useful to analyze the content of social networks that have gained great popularity around the world [2]. Opinion Mining (OM) and Sentiment Analysis (SA) are interchangeable terms that refer to the process of automatic natural language processing methods that attempt to identify the polarity or sentiment orientation expressed in a text on a specific subject [2,3]. This subject can be a product, service, person, event, etc. In other words, the goal is to assess whether an opinion is positive or negative.

The benefits of performing SA are countless. As part of the modern society, sentiment analysis is vital to mining perceptions of a product or service by automatically analyzing customers feedback. The sentiment polarities of thousands of comments on a particular product or service can be determined with SA tools in a very short period [2,3]. By analyzing such comments, companies can better improve their services and products. In addition, governments can also use SA to determine public opinions and reactions to controversial issues as a quick and accurate alternative to public polls [3].

In this field, most of the researches have been carried out in European languages, notably English and French. Nevertheless, research carried out on the analysis of Arabic sentiment is considered very limited, especially the Moroccan dialect. This can be explained by two major factors. First of all, the lack of additional resources for this type of language [3]. Second, the complexity of the treatment of this language, namely the problem of syntactic interference between SA and MD, and the variety of dialectal Arabic spoken in Morocco depending on the regions [3,4].

The proposed solutions are largely dominated by the use of two analytical approaches. First, machine-learning techniques based on training corpora annotated with polarity information and, second, strategies based on lexicons polarity (lexical approach). Lexicon-based approaches are very popular in sentiment analysis and opinion mining, and they play a key role in all applications in this field. The main issue lexicon-based approaches face is that most polarity words are domain dependent because their subjectivity is very ambiguous. The same word can be provided with a subjective burden in a specific domain while it can refer to an objective information in another one.

A lexicon dictionary can be constructed in two ways: either manually or automatically from a seed of words. However, the manual construction requires significantly more time and affort, and suffers from many deficiencies, such as lack of accuracy and robustness [3]. The main aims of this study is to propose a novel method for representing data features for sentiment analysis, then automatically building polarity lexicons from corpora.

The remainder of the paper is organized as follows: In Sect. 2, we describe the related works and explore some challenges of using the Arabic language. Then, Sect. 3 describes the methods used to collect the dataset and to create the proposed lexicon and how to use it in the classification task. The Experiments are introduced in Sect. 4, where we also describe the evaluation and discuss the results. We draw conclusions in Sect. 5.

2 Background and Literature Review

A number of studies have been conducted on sentiment analysis and multiple approaches have been proposed and developed to deal with this problem. Typically, the majority these approaches are designed for the English language and are not applicable to more complex languages such as the Arabic. In the following, we present a comprehensive overview of recent works in this area and try to emphasize studies that are closely related to opinion mining and sentiment analysis of Arabic text.

2.1 Related Work

To construct lexicons, various methods have been implemented. Some researchers built their lexicon by hand [4], some expanded a small set with synonyms and antonyms [5,6] while others used methods based on statistics, such as PMI [7] and the links between terms [8].

Abdulla et al. [9] proposed a domain-based lexicon approach to deal with Arabic text (SA and colloquial Arabic). They created two lexicons for every domain (books, movies, society, politic, etc.), one for positive words and another for negative ones from a corpus of 1080 reviews compiled from different social networks. Their approach achieved an accuracy of 90%, however, they built the lexicon dictionary from the same test corpus used for the evaluation experiments.

An unsupervised technique for sentiment analysis of Arabic tweets was proposed by Al-Ayoub et al. [10]. The first step of their technique was collecting tweets and applying preprocessing methods (i.e. stemming and stop-word removal). Next, a sentiment lexicon was constructed with polarity scores between 0 and 100. Scores less than 40 indicated negative sentiment, between 40 and 60 corresponded to neutral, while scores from 60 to 100 indicate positive sentiment. Finally, all these scores were combined to compute the sentiment score of the text. This technique has achieved 86.89% in overall accuracy.

Itani et al. [11] conducted a comparative study between the corpus-based and the lexicon-based approach by using both SA and MD data. The experimental results show that the lexicon-based approach achieved an accuracy of 83.4% outperforming the corpus-based approach.

2.2 Challenges of Analyzing Arabic Language

Arabic language is one of six official languages used in the united nations [12] and it is the native spoken language for more than 300 million people worldwide [12,

13]. It is classified into three types: Classical, Modern Standard, and Colloquial Arabic (Dialect). The Arabic language is a morphologically rich language. Some circumstances are explained by the following points:

- Arabic has various diacritics, based on the presence or absence of them, the meaning of words can be totally reversed. For example, "كِتاب" (book) and "كُتّاب" (a sort of Sunday school where children learn about religion), both can be read as the same word "كتاب" when written without diacritics.
- A given word can have different meanings depending on its position in the sentence. For example, "شهر مُحَرَّمِ" that means the first month in the Islamic calendar and "هذا محرم" that means this is forbidden.
- Many users do not employ the SA but an Arabic that is proper to their countries ("أريد"="بغيت"). Moreover, dialect language can be divided into a group of subdialects according to regions of the country (accents).

3 Methodology

The proposed system is based on the NLP paradigm. Therefore, it contains all the necessary tools to handle the linguistic diversity that characterize the communication of Moroccan users on the social network Twitter. This section explains the followed steps as a general methodology, where the sentiment analysis process is divided into three main phases. Firstly, the process of gathering and cleaning corpora, followed by annotating them. Secondly, the lexicon construction, and finally, classification and evaluation.

3.1 Corpora Collection

Our data are exclusively extracted from the Twitter social network. This is justified by the high use of this social network by Moroccan Web users [3]. We have developed a module based on Twitter API[1] for the extraction of data. In order to target the right content, we used three methods of data collection provided by the Twitter API:

Streaming API: that allows getting the tweets in real time. We have selected tweets written in Arabic that meet the "Morocco" geolocation requirement.
User_timeline API: that returns a collection of the most recent Tweets posted by a given user. In this stage, we used SocialBakers[2] to determine the most active Moroccan Twitter accounts.
Search API: that returns a collection of relevant tweets matching a specific query. In this stage, we resorted to Trendsmap[3] to determine the most popular hash tags in Morocco and use them as queries.

---

[1] https://developer.twitter.com.
[2] https://socialbakers.com/statistics/twitter.
[3] https://www.trendsmap.com/local/morocco.

Dataset Properties. In total, we selected 18.000 tweets valid from 36.114 tweets collected. The collected data discuss various subjects: society, economy, politic, literature and arts, divers, etc. As aforementioned, all these tweets were assessed manually by defining their polarities (Positive, Negative, Neutral or Mixed). The collected data has been filtered and annotated by two native Moroccan speakers (ourselves). Table 1 and 2 show the distribution of the collected data according to their polarities, and some example entries form the annotated corpora respectively.

Table 1. Statistics about the collected corpora.

Type	Positive	Negative	Neutral	Mixed
SA	4993	3086	2214	747
MD	2085	2681	1906	288

Table 2. Some example entries form the annotated corpora.

Tweet	Type	Class
Ar: توقع الخير و افتح صباحک بالتفاؤل صباح النور **En**: Expect good things and start your day with optimism.	Positive	SA
Ar: من المؤسف ان هذا حالنا الذى نعيشه الأن **En**: Unfortunately, this is our current situation.	Negative	SA
Ar: تابعينى باش نقدر ندخلک **En**: Subscribe so that I can add you.	Neutral	MD
Ar: رغم الصعوبات والمشاكل .. كنحاول نضحک ونقول الحمد لله **En**: Despite the difficulties and problems,I try to be optimist.	Mixed	MD

3.2 Lexicon Construction

In sentiment analysis, *lexicons* are a synonym for dictionaries, except lexicons is sentiment analysis contain polarities along with the words instead of their definitions [14]. That is, every word has an associated sentiment orientation (e.i. positive, negative, or neutral).

In the literature, lexicons are generally assembled manually [3,4] or automatically from seeds of words [15]. The manual construction requires significantly more effort and time. It suffers from many flaws such as accuracy and robustness degradation [3,15]. However, through the pervasive domain diversity and the absence of domain prior knowledge, the automatic construction technique for sentiment lexicons has become a challenging research topic in the field of sentiment analysis and opinion mining [15].

The adopted lexicon in this study is created automatically from annotated training corpora under the supervision of native speakers (ourselves) of the considered language (Moroccan Arabic). It consists of about 30.000 Moroccan Arabic term, where each word is assigned a polarity (positive, negative or neutral). Table 3 presents the characteristics of the constructed sentiment lexicon, while Table 4 shows some example entries for each polarity category.

Table 3. Statistics about the constructed lexicon.

Type	Positive	Negative	Neutral
SA	2833	2057	13995
MD	1511	702	8902

Table 4. Some example entries form the constructed lexicon.

Type	Positive		Negative		Neutral	
	Ar	Eng	Ar	Eng	Ar	Eng
SA	نجاح	Success	صراع	conflict	حياة	Life
SA	جميل	Nice	حزن	Sadness	يوم	Day
MD	زوين	Beautiful	خايب	So bad	شوية	Few
MD	توحشتك	Miss you	عيان	terrible	شحال	How many

3.3 Tool Design and Functionalities

Preprocessing Task. The text-preprocessing is a crucial part of any SA application, especially when using the lexicon-based approach. The first to be discussed is the text representation. The simplest method is to represent the text as a bag-of-words, wich extracts and treats each word as an individual token. The following give more details about each step.

Repeated Letters Elimination. Repeated characters are eliminated after the tokenization step. For instance, a social media text may include words such as "جميييييل" which means "very good". This kind of expressions is used it to intensify the meaning (which became popular in online comments), and could negatively affect any NLP tool. Therefore, removing such repeated characters, either manually or automatically, is imperative.

Normalization. The second step is the normalization of characters. Arabic language is known as a morphologically rich language. Letters can take different forms depending on their position within words. For example, the letter "A" in Arabic "Alif" has four forms "آ إ أ ا" that even native speakers may make mistakes regarding which form to use. Hence, normalizing some letters such as "ة" to "ه" and "إ أ آ" to "ا" becomes an essential processing stage to prevent the classification performance degradation.

Stemming. In this study, light stemming was employed as a third stage of preprocessing. For the Arabic language, two morphological analysis methods are proposed: light stemming and root-based. In the Root-based stemming, all affixes (prefixes and suffixes) are removed from each Arabic word and each root pattern is restored. Light stemming, on the other hand, removes only the common affixes from words without altering the word's origin (root).

The light stemming algorithm ISRI is applied in this study. The main reason for using the light stemming is that there are many words whose roots share the same meaning, but convey completely different meanings or emotions. For instance, the following words “اللاعبون” and “يتلاعب” which mean “the players” and “falsifies” respectively have the same root “لعب” which means “play” but semantically as sentiment-wise, they are very different. In contrast, if we only apply light stemming, the affixes will be simply removed and the result will be “لاعب” and “تلاعب”” which means “player” and “falsify.” Therefore, the stemmed word maintains the correct connotation.

Sentiment Analysis Tool. One noteworthy aspect of the proposed tool is that it is capable of handling both SA and MD negations, which most Arabic language tools in this field fail to do [3]. Usually, in language, negation refers to words that reverse the polarity (connotation) of the ones that come afterwards. The following is the list of the most common negation words in SA: “لم، لا، لن،”. For example, “لم أعجب به” (which means “I didn't like it”), the word “أعجب” (like) is a positive word in our lexicon with polarity *Positive*. Yet, due to the negation word “لم” (did not) that came before it, the polarity turns to *Negative*. Therefore, the polarity of any word preceding or following a negation term is switched completely to the opposite. In addition, the negation words list is further enriched by the addition of negation forms from the Moroccan dialect such as “ما \ش”. The negation in the Moroccan dialect is expressed by adding affixes to terms, for example, the term “بغيت” which means “I love” is switched to “I don't love” by adding the prefix “ما” and the suffix “ش”, to give “مابغيتش” which means “I do not like”.

Text Sentiment Vector. Multiple tools and methodologies have been developed to deal with sentiment analysis using the lexicons [7,13,15]. For example, the authors of [14] discussed the sentiment analysis for Jordanian tweets, and built a tool for extracting the polarity of unstructured text where a weight representing the polarity is assigned to each word in the lexicon (+1 and -1 for positive and negative words, respectively). In our research approach, we propose a tool for extracting the sentiment polarity of a Moroccan Arabic text, in which each word has a label for the polarity (positive, negative, or neutral). In order to perform features representation, we apply the Unigram method. That is, each word is considered as an individual token in the text regardless to the surrounding words or terms. Once each word's polarity has been extracted from the built lexicon, a vector sentiment is generated for all input text by aggregating the total weights of each class. The classes weight is calculated using the following equation:

$$Weight_{class} = \frac{Number\ of words\ of\ a\ class\ in\ the\ tweet}{Total\ number\ of\ words\ in\ the\ tweet} \tag{1}$$

The final values of the weights determine the polarity of the whole tweet, representing it as a vector:

$weight_{positive}$	$weight_{negative}$	$weight_{neutral}$

Below are the possible vectors for each class (Fig. 1):

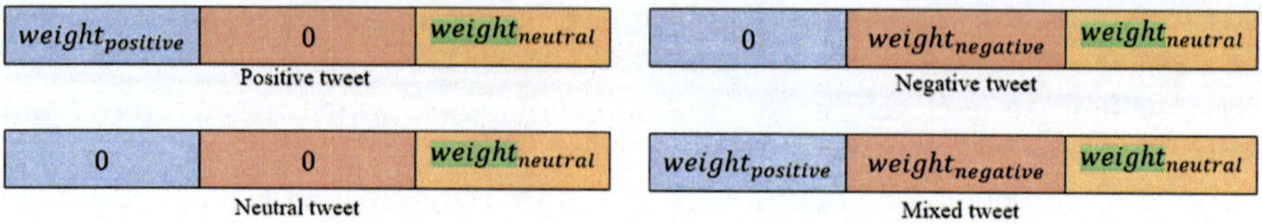

Fig. 1. Tweets polarity vectors.

3.4 Sentiment Classification

As our aim is to evaluate the efficiency of our features representation methodology, we train different classifiers by making use of simple lexicon-based features, namely: the weight of positive, negative and neutral terms in the document (tweet) forming the sentiment vector. Those features were used to train a Convolutional Neural Networks, Recurrent Neural Network, Support Vector Classifier and Logistic Regression classifier.

4 Experimentations and Results

This section presents the experimental results of the proposed approach. Three experiments were run based on our constructed corpora. The first experiment concerns the SA corpus. The second experiment concerns the MD corpus. Finally, the last experiment concerns the completely collected corpora (SA+MD). In all three experiments, we tested the influence of stop words on the classification by using the classic precision measure (accuracy).

Table 5 shows the evaluation results of the proposed vector representation applying the lexicon based approach using classical classifiers (SVM and LR) and deep learning classifiers (CNN and LSTM).

Table 5. Evaluation results of the proposed vector representation.

Model	SW filter	Accuracy		
		SA	MD	SA_MD
CNN	0	90.85	**85.42**	89.25
	1	90.80	85.30	89.14
LSTM	0	**90.88**	84.53	**89.62**
	1	90.63	84.02	88.66
SVM	0	82.04	74.14	78.11
	1	81.49	73.25	77.80
LR	0	81.08	71.77	77.96
	1	80.63	71.51	77.54

1. The CNN classifier obtained an accuracy of 90.85% by eliminating functional words, and 90.80% by keeping them in the first corpus (SA), and convincing results in the dialectal corpus with an accuracy of 85.42%, and 89.25% in the combination of the two corpora.
2. The LSTM classifier with elimination of functional words obtained the best result of 90.88% in the first corpus (SA), outperforming the results of the other classifiers.
3. The SVM classifier reaches a precision level of 82.04% in the first corpus by keeping the functional words and a precision of 78.11% in the third corpus (SA+MD). The overall predictive performances on the dialectal corpus are relatively inferior to those on the SA corpus. This fact is due to the reduced size of the corpus on the one hand and by the great variety of writing of the Moroccans.

The Mixed polarity contains tweets in which the author of the tweet has expressed both a positive and a negative feeling. While sometimes the positive side is the most dominant or the opposite. Hence, the system classify the tweet to the dominant type, which gives a precision degradation in the results. The following table shows the classification results for three classes (positive, negative, and neutral). As shown in the Table 6, the SVM classifier with 3 classes, the accuracy is of 84.56%, while with 4 classes (positive, negative, neutral and mixed) it is 82.04% (decrease of 2%) which demonstrates the affect of the mixed class on the performances of models.

Table 6. Evaluation results of the proposed vector representation

Model	SW filter	Accuracy		
		SA	MD	SA_MD
SVM	0	**84.56**	76.02	76.02
	1	83.89	**83.89**	83.89
LR	0	**83.78**	83.78	**86.78**
	1	73.12	75.61	77.89

5 Conclusion

In this paper, we addressed the sentiment analysis problem for Moroccan Arabic tweets. We studied the lexicon based approach since it is one of the most widely studied approaches to other languages. To this effect, we built a sentiment lexicon of about 30,000 Moroccan Arabic terms and built a SA tool that deals with some special cases of the used language. The proposed features representation approach showed great results, in terms of prediction (an accuracy of 90.85%) outperforming all state of the art representation approaches. As part of our future work, we plan to expand our lexicon to handle dialect-specifics and domain-specific issues, and the implication of other linguistic aspects such as the type of words (subject, verb, adjectives).

References

1. Bakshi, R.K., Kaur, N., Kaur, R., Kaur, G.: Opinion mining and sentiment analysis. In: 2016 3rd International Conference on Computing for Sustainable Global Development (INDIACom), pp. 452–455 (2016)
2. Cambria, E., Poria, S., Gelbukh, A., Thelwall, M.: Sentiment analysis is a big suitcase. IEEE Intell. Syst. **32**(06), 74–80 (2017). https://doi.org/10.1109/MIS.2017.4531228
3. Garouani, M., Chrita, H., Kharroubi, J.: Sentiment analysis of Moroccan tweets using text mining. In: Motahhir, S., Bossoufi, B. (eds.) ICDTA 2021. LNNS, vol. 211, pp. 597–608. Springer, Cham (2021). https://doi.org/10.1007/978-3-030-73882-2_54
4. Albraheem, L., Al-Khalifa, H.S.: Exploring the problems of sentiment analysis in informal Arabic. In: ACM International Conference Proceeding Series, pp. 415–418 (2012) . https://doi.org/10.1145/2428736.2428813
5. Abdulla, N.A., Ahmed, N.A., Shehab, M.A., Al-Ayyoub, M.: Arabic sentiment analysis: Lexicon-based and corpus-based. In: 2013 IEEE Jordan Conference on Applied Electrical Engineering and Computing Technologies (AEECT) (2013). https://doi.org/10.1109/AEECT.2013.6716448
6. Alsolamy, A.A., Siddiqui, M.A., Khan, I.H.: A corpus based approach to build Arabic sentiment lexicon. Int. J. Inf. Eng. Electron. Bus. **11**, 16–23 (2019). https://doi.org/10.5815/IJIEEB.2019.06.03
7. Mohammad, S., Salameh, M., Kiritchenko, S.: Sentiment lexicons for Arabic social media. In: Proceedings of LREC'16, pp. 33–37 (2016)
8. Eskander, R., Rambow, O.: SLSA: a sentiment lexicon for standard Arabic. In: Conference Proceedings - EMNLP 2015 Conference on Empirical Methods in Natural Language Processing, pp. 2545–2550 (2015). https://doi.org/10.18653/V1/D15-1304
9. Abdulla, N.A., Al-Ayyoub, M., Al-Kabi, M.N.: An extended analytical study of Arabic sentiments. Int. J. Big Data Intell. **1**, 103–113 (2014). https://doi.org/10.1504/IJBDI.2014.063845
10. Al-Ayyoub, M., Essa, S.B., Alsmadi, I.: Lexicon-based sentiment analysis of Arabic tweets. Int. J. Soc. Netw. Min. **2**, 101–114 (2015). https://doi.org/10.1504/IJBDI.2014.063845
11. Itani, M.M., Zantout, R.N., Hamandi, L., Elkabani, I.: Classifying sentiment in Arabic social networks: Naïve search versus Naïve Bayes. In: 2012 2nd International Conference on Advances in Computational Tools for Engineering Applications ACTEA 2012, pp. 192–197 (2012). https://doi.org/10.1109/ICTEA.2012.6462864
12. UNGEGN - United Nations Group of Experts on Geographical Names. https://unstats.un.org/unsd/geoinfo/ungegn/wg1.html. Accessed 02 July 2021
13. Abdulla, N.A., Ahmed, N.A., Shehab, M.A., Al-Ayyoub, M., Al-Kabi, M.N., Al-rifai, S.: Towards improving the lexicon-based approach for Arabic sentiment analysis. IJITWE **9**(3), 55–71 (2014). https://doi.org/10.4018/ijitwe.2014070104
14. El Abdouli, A., Hassouni, L., Anoun, H.: A new practical approach to automatically generate the trending topics in Morroccan society using the social network twitter. Revue Méditerranéenne des Télécommunications (2017)
15. Almatarneh, S., Gamallo, P.: Automatic construction of domain-specific sentiment lexicons for polarity classification. In: De la Prieta, F., et al. (eds.) PAAMS 2017. AISC, vol. 619, pp. 175–182. Springer, Cham (2018). https://doi.org/10.1007/978-3-319-61578-3_17

Enhanced Word Embeddings with Sentiment Contextualized Vectors for Sentiment Analysis

Mohammed Kasri[1(✉)], Marouane Birjali[1], Anas El Ansari[2], and Abderrahim Beni-Hssane[1]

[1] Computer Science Department, Faculty of Sciences, Chouaib Doukkali University, El Jadida, Morocco
kasri.m@ucd.ac.ma

[2] MASI Laboratory, Mohammed First University, FP Nador, Nador, Morocco

Abstract. Deep Learning models generally use pre-trained word embeddings as features to deal with Natural Language Processing problems. However, because most of these word embeddings are context-based, they cannot learn the complex characteristics of words, which may lead to words with similar vector representations expressing different meanings. For sentiment analysis, this represents a problem. Since these word embeddings fail to capture words' sentimental information, two words with similar vectors can have opposite sentiment polarities, thus degrading the performance of sentiment analysis. This paper addresses this problem and proposes a novel model named Continuous Sentiment Contextualized Vectors (CSCV). CSCV is designed to learn sentiment embeddings based on the Continuous Bag-of-Words (CBOW) model and sentiment lexicons for a given the word from its surrounding context words. The sentiment vectors obtained from this model are combined with existing pre-trained vectors. Experimental results show that: (1) CSCV can be combined with any pre-trained word-embeddings; (2) the nearest neighbors derived from the generated embeddings are more reasonable than other word embeddings because they are both semantically and sentimentally similar; (3) by using this approach, the accuracy can be improved on sentiment classification.

Keywords: Sentiment embeddings · Sentiment analysis · Word embeddings · Sentiment lexicon · Deep learning

1 Introduction

Word embedding (also called distributed representation) is a low-dimensional vector representation of a qualitative concept. In these representations, the information about a concept is dispersed all along the vector [1, 2]. These vectors are obtained using neural networks or vector space models, trained over a large corpus. Therefore, using word embeddings along with deep learning (DL) models has significantly improved the performance of most natural language processing (NLP) tasks such as text classification and sentiment analysis (SA) [3–6]. In these kinds of tasks, distributed representations automatically help deep learning models extract high-level features from text data.

Y. Maleh et al. (Eds.): ICI2C 2021, LNNS 357, pp. 77–86, 2022.
https://doi.org/10.1007/978-3-030-91738-8_8

However, the characteristics of word embeddings models make these features more related to semantic representation [7, 8].

Although the effectiveness of word embedding methods, they suffer from several weaknesses. Most of them need a large text corpus for training [9, 10]. In this situation, the computed word vectors will contain an important amount of semantic and syntactic regularities in the language [11], because all the characteristics of words are extracted from context words. As a result, words that appear in a common context, usually vectors, are located close to each other in the vector space, indicating that these words have similar linguistic features. This represents a problem for sentiment analysis because the word embedding techniques ignore sentiment information of words [12, 13]. Therefore, two words can have the closest vectors in the latent space since they have similar contexts, while their sentiment polarities are opposite. For example, the two words "good" and "bad" are considered 71% similar in word2vec, whereas they have extreme opposite sentiments. According to [14], in the CBOW model, about 30% of the top 10 similar words obtained using cosine similarity include at least one word with opposite polarity. This may hurt the sentiment analysis performance and reduce the efficiency of word embeddings for such tasks. The goal of this study is, therefore, to provide high-quality word embeddings for sentiment analysis.

The main contribution of this work is to introduce a sentiment embedding model called CSCV that learns sentiment embeddings using the CBOW model and a lexical resource. To do that, the model predicts the sentiment of a target word based on its surrounding context words. However, this approach aims to improve deep learning models that use word embeddings as features for text-based sentiment classification. Several experiments have been conducted using deep learning models on benchmark datasets to evaluate the effectiveness of the obtained sentiment embeddings. The results show that the proposed approach makes significant improvements over the existing state-of-the-art word embedding models in sentiment classification.

The rest of this paper is structured as follows: Sect. 2 discusses related work and the background. Section 3 describes the proposed approach for generating sentiment embeddings using the modified CBOW model. Section 4 discusses and reports the experimental results of the proposed approaches, and Sect. 5 concludes the work.

2 Related Work

To capture sentimental information, (e.g., sentiment polarity) of words, two methods have been proposed in the literature: (1) learn sentiment embeddings through word polarities (e.g., positive, negative) labels within a labeled corpus or (2) use sentiment lexicons to refine pre-trained word embeddings. The two approaches have shown a considerable improvement in the sentiment analysis task. In regard to the first approach, a semi-supervised approach has been introduced to learn semantics alongside sentiment information [15]. This method is based on two models where the first model is designed as a probabilistic topic model to capture semantic proprieties from unlabeled corpora. The objective of the second model is to learn sentiment embeddings from labeled corpus through a sentiment classification task using a logistic regression model. They have maximized the sum of their objective function to unify the two models. In addition, they introduced the Internet Movie Database (IMDB), a dataset of movie reviews.

Several studies have used deep neural network techniques to capture sentiment information of words from tweets during Twitter sentiment classification. The authors in [16] proposed a model called Sentiment Specific Word Embedding (SSWE). This model seeks to integrate sentiment information in word representation vectors. The authors used a multitask neural network model like the one proposed by [17], but to unify the two models, they used a weighted sum of individual loss functions. Similarly, [18] developed a neural network model based on [16, 17] methods to learn topic-enriched multi-prototype word embeddings.

The problem of the first approach is the need for labeled data, which is not easy to collect. Furthermore, the complexity of such models requires more parameters to tune. Therefore, other studies tackled the problem of sentiment embeddings by making use of existing sentiment lexicons to refine existing word representations. [14] proposed a method to refine word embeddings using the intensity score obtained from E-ANEW (Extended Affective Norms of English Words) lexicon [19]. [20] proposed Sentiment-augmented Convolutional Neural Networks (SCNN). This model has two inputs and lookup layers. While the first layer deals with word embeddings using the pre-trained word2vec, the second one provides sentiment embeddings based on SentiWordNet (SWN) sentiment lexicon [21]. (Naderalvojoud & Sezer, 2020) proposed a refining approach that uses a neural network model to predict the polarity score of a word based on its pre-trained vector and E-ANEW lexicon.

3 Learning Sentiment Embeddings

3.1 Sentiment Lexicon

We constructed pairs of (context window, target sentiment) from sentences to predict a sentiment from surrounding context words. Its sentiment category class changes the target word in the CBOW model (*Negative, Strong Negative, Neutral, Positive,* and *Strong Positive*) in our model. A combination of three lexicons was used to determine the sentiment category class of a given word. The three lexicons are (1) SentiWordNet [21], (2) SenticNet [22], and (3) VADER [23].

SentiWordNet is a lexical resource that provides positive, negative, and objective (neutral) scores in the range of 0.0 to 1.0 for each synonym set (syensets) of WordNet [24]. It contains about 117,658 words grouped in syensets based on their Part-of-Speech tags. The two other lexicons were employed to enrich the first one. SenticNet is a resource created for concept-level sentiment analysis in the English language. About 15,000 words can be found in this lexicon, with their quantitative polarity scores ranging between -1 and $+1$ (where -1 means extreme negativity and $+1$ extreme positivity).

Similarly, VADER provides polarity scores for 7,517 words. It is a lexicon and rule-based tool explicitly developed for sentiment analysis in social media. The score values of SenticNet and VADER, used as they are to classify words into sentiment categories. For SentiWordNet, the score of a word is obtained using the following formula:

$$Score = \frac{\sum_{i=1}^{k} Synset_{p(i)}}{k} - \frac{\sum_{i=1}^{k} Synset_{n(i)}}{k} \quad (1)$$

Where k represents the number of synsets that the word appears in, $Synset_{p(i)}$ denotes the positive score of synset i and $Synset_{n(i)}$ represents the negative score of synset i. The priority is given to the lexicon with a high number of words. If the word does not exist in the first lexicon, we look for it in the second one that certainly has fewer words than the previous lexicon. Based on which range the score obtained belongs, a sentiment class is assigned to the word.

3.2 CSCV Model

Contrary to CBOW, the main idea of this model is to replace the output word with its sentiment orientation to learn sentiment embeddings. Miller [24] shows the architecture of CSCV. It consists of a simple neural network with three layers. The words of local context $w_{t-c}, \ldots, w_{t+c}$ encoded to one-hot encoding before mapping to their unique vectors $v_{t-c}, \ldots, v_{t+c}$ in the projection layer, respectively. w_t is excluded from the local context because it represents the target word for a context window of size c. These vectors are averaged to construct the input h_t. Then h_t will be passed to the output layer to predict the sentiment polarity (negative, strong negative, neutral, positive, and strong positive) S_{w_t} of the word w_t (Fig. 1).

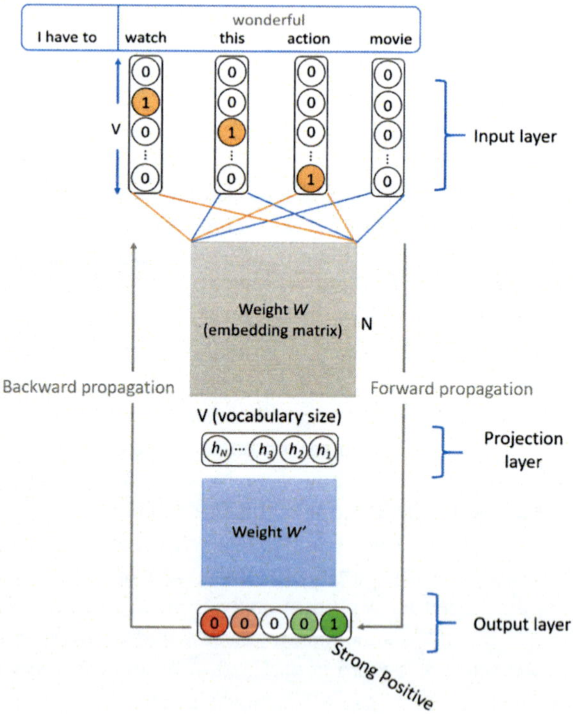

Fig. 1. The architecture of CSCV with an example of a strong positive word.

The input h_t is calculated using the following equation:

$$h_t = \frac{1}{C} W^T . \sum_{i=1}^{C} w_i \tag{2}$$

Where C is the context size (in our case, it is equal to 2, meaning that we take two words before and after the center word), w_i is the one-hot encoded vector of word i in the context and $\mathbf{W} \in \mathbb{R}^{V \times N}$ represents the embedding matrix (lookup table) of size V (vocabulary size) $\times$ N embedding size. $\mathbf{W}' \in \mathbb{R}^{N \times SC}$ is a weight matrix between the projection and the output layer. SC denotes the number of sentiment polarity (classes) in the output layer. Thus, we can compute the conditional probability for each node in the output layer using the following equation:

$$u_j = v_j^T h_t \tag{3}$$

where v_j^T is j^{th} column of matrix W′. These values are then passed through a SoftMax activation function to predict the most acceptable sentiment polarity of the target word. The objective of the CSCV model is to maximize Eq. 4 where S_{w_t} refers to the target word polarity providing its context words.

$$\frac{1}{SC} \sum_{1}^{SC} \log p(S_{w_t} | w_{t-c}, \ldots, w_{t+c}) \tag{4}$$

3.3 The CSCV Model Training

The CSCV model was trained using one of the well-known publicly available datasets for sentiment analysis, the IMDB dataset [15]. It contains 50,000 movie reviews labeled as positive or negative. However, because of the unsupervised manner of the CBOW model, we need only text for our proposed model and the pairs of (context window, target sentiment). This dataset has been chosen to partially alleviate the out of vocabulary (OOV) problem during the evaluation of the obtained vectors. After training, the sentiment embeddings are extracted then from the lookup table. Then, we combine these embeddings with pre-trained word vectors using Principal Component Analysis (PCA) technique:

$$EVs = \text{PCA}(\text{CSCV} \oplus \text{PV}) \tag{5}$$

where, *EVs*, CSCV, PV refer to the result vectors called *Enhanced Vectors*, the sentiment embeddings obtained from our proposed model, and context-based pre-trained vectors, respectively. PCA is a dimensionality reduction method used to decrease large data size to a smaller set of variables preserving all the necessary information. The obtained vectors *EVs* have the same size as both CSCV and PV. However, they contain all the words' important information for sentiment analysis, including semantic and sentiment. The training of CSCV is much faster than CBOW because we calculate the gradient for only five outputs instead of the whole vocabulary. This means it is easy to adopt such an approach for SA.

4 Experiments

To evaluate the performance of the proposed sentiment embedding models, we adopted sentence-level sentiment classification and carried out some experiments on various benchmarking datasets. This section will be consumed for describing the experimental setup and the baseline methods compared to our approach, in addition to the discussion of results.

4.1 Experimental Settings

Datasets. Movie Review[1] (MR) [25], and Stanford Sentiments Treebank[2] (SST) [26] are two common datasets used to demonstrate the effectiveness of our proposed approach. MR and SST are a set of sentences reviewing movies as good or not good, which means the sentence refers to a positive or negative review. SST-1 is a fine-grained dataset with five labels: very positive, positive, neutral, negative, and very negative. SST-2, on the other hand, is like SST-1, but without neutral reviews. The remaining reviews, in this case would take positive label if the previous labels were positive or very positive, and vice versa. Chen et al. [28] includes summary statistics of all datasets used for evaluation. CV (cross-validation) means the dataset does not have a standard train, dev, and test split. Therefore 10-fold CV was used to test the performance of the trained classifier (Table 1).

Table 1. Summary statistics of the datasets used for evaluation. Cs, SL, and VS denote the number of classes, sentence length, and vocabulary size, respectively.

Datasets	Cs	Max SL	VS	Size	Train	Dev	Test
MR	2	56	18765	10662	9596	–	CV
SST-1	5	53	17836	11855	8544	1101	2210
SST-2	2	53	16188	9613	6920	872	1821

Classifiers. Among the classifiers that achieved a good performance in sentiment analysis, we found: CNN (Convolutional Neural Network) [27], LSTM (Long Short Term Memory) [28], and BiLSTM (Bidirectional LSTM). We followed the same principles as these studies to conduct our experiments.

[1] http://www.cs.cornell.edu/home/llee/papers/pang-lee-stars.home.html.
[2] http://nlp.stanford.edu/sentiment/.

4.2 Sentiment Classification

Sentiment classification is the process of classifying given instances into coarse-grained classes such as positive, negative, and neutral, or fine-grained classes using the machine learning algorithms [29]. Chen et al. [30] show the performance obtained from EVs with word2vec pre-trained word embeddings for all datasets. As can be seen for MR, our approach outperformed word2vec in most cases. The accuracy of the CNN and BiLSTM models has improved by 1.3% and 0.3%, respectively. Similarly, EVs provide absolute accuracy improvement for SST, especially for fine-grained. Successively, the accuracy increased by 0.2%, 1.6%, and 2.3% for CNN, LSTM, and BiLSTM (Fig. 2).

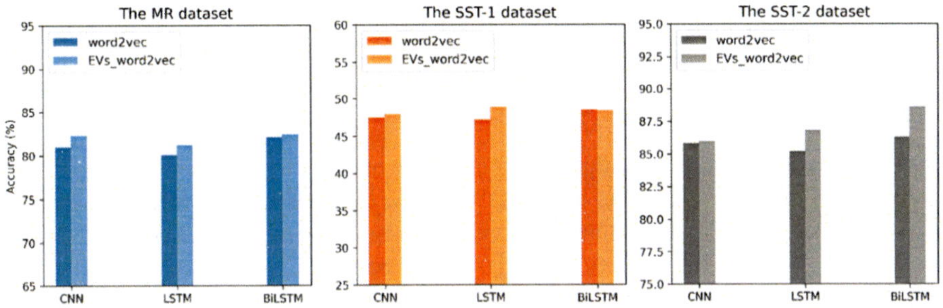

Fig. 2. The accuracy comparison of word2vec and refined word embeddings over MR, SST-1, and SST-2 using CNN, LSTM, and BiLSTM models.

4.3 Result and Discussion

Refined embeddings with CSCV approach achieved good performance since the combination of pre-trained word embeddings and sentiment contextualized vectors can remove noisy words from the top nearest neighbors, as shown in [30]. This means that the proposed approach can partially remove words with semantic similarity if they are sentimentally dissimilar from the top 10 similar words. The proposed refining method can be applied to any pre-trained word embedding model. The goal is to alleviate similar words with opposite sentiment polarity for sentiment analysis with word embeddings. Experimental results show that the refined word embeddings outperform some complex methods and achieve the state-of-the-art using the BiLSTM model. The obtained word embeddings retain valuable information for the sentiment analysis task, the sentiment orientation of words. More information can be obtained as in the study of [30] to improve the performance of NLP tasks by increasing the quality of word embeddings (Fig. 3).

word2vec

Word	Polarity	Score
good	SP	0.56
great	P	0.25
bad	N	-0.59
terrific	P	0.04
decent	P	0.39
nice	P	0.47
excellent	SP	1.00
fantastic	P	0.10
better	SP	0.52
solid	P	0.03
lousy	SN	-0.70

word2vec + CSCS

Word	Polarity	Score
good	SP	0.56
nice	P	0.47
terrific	P	0.04
fantastic	P	0.10
decent	P	0.39
perfect	P	0.37
fabulous	P	0.16
wise	P	0.16
awesome	SP	0.75
marvelous	P	0.25
unbelievable	SP	0.50

word2vec

Word	Polarity	Score
bad	SN	-0.59
good	SP	0.56
terrible	SN	-0.62
horrible	SN	-0.62
Bad	SN	-0.60
lousy	SN	-0.70
crummy	SN	-0.75
horrid	SN	-0.88
awful	N	-0.21
dreadful	SN	-0.70
horrendous	SN	-0.62

word2vec + CSCS

Word	Polarity	Score
bad	SN	-0.59
lousy	SN	-0.70
terrible	SN	-0.62
horrible	SN	-0.62
nasty	SN	-0.78
crummy	SN	-0.75
worse	SN	-0.54
dumb	N	-0.22
cruddy	SN	-0.62
stupid	SN	-0.34
scary	SN	-0.75

Fig. 3. Examples of nearest neighbors of the words "good" and "bad" from word2vec and its refined word embeddings. SP, P, N, and SN denote strong positive, positive, negative, and strong negative.

5 Conclusion

This paper presents a novel model to create sentiment embeddings from local contexts. It follows the same principle as the CBOW model to predict the sentiment (positive, strongly positive, neutral, strong negative, and negative) of a target word from its surrounding context. The sentiment of this word is obtained using the score value from a combination of well-known sentiment lexicons. Therefore, generated sentiment embeddings can be combined with any existing word representations applying the PCA technique. This combination aims to provide high-quality word embeddings that integrate both sentimental and semantic words' characteristics. Finding such features in a single word vector is beneficial for many NLP tasks, especially text-based sentiment analysis and emotion detection. According to the experimental results, this approach can improve sentiment analysis performance using deep learning models. Moreover, it achieved high accuracy compared to state-of-the-art methods over MR and SST-2 datasets. This reveals how our approach is effective in encoding words' sentimental information into word embeddings.

A future challenge in sentiment analysis is to exploit word embeddings to build a large sentiment lexicon. Since word embeddings effectively capture semantic and syntactic relations of words, it is possible to make use of such features to identify and generate a list of synonymous words. Thus, combining the obtained lists to create a sentiment lexicon.

References

1. Kasri, M., Birjali, M., Beni-Hssane, A.: Word2Sent: a new learning sentiment-embedding model with low dimension for sentence level sentiment classification. Concurr. Comput. Pract. Exp. **33**, 1–12 (2021). https://doi.org/10.1002/cpe.6149
2. Birjali, M., Kasri, M., Beni-Hssane, A.: A comprehensive survey on sentiment analysis: approaches, challenges and trends. Knowl.-Based Syst. **226**, 1–26 (2021). https://doi.org/10.1016/j.knosys.2021.107134

3. Atzeni, M., Reforgiato Recupero, D.: Multi-domain sentiment analysis with mimicked and polarized word embeddings for human–robot interaction. Future Gener. Comput. Syst. **110**, 984–999 (2020). https://doi.org/10.1016/j.future.2019.10.012
4. Ali, F., et al.: Transportation sentiment analysis using word embedding and ontology-based topic modeling. Knowl.-Based Syst. **174**, 27–42 (2019). https://doi.org/10.1016/j.knosys.2019.02.033
5. Dessí, D., Dragoni, M., Fenu, G., Marras, M., Reforgiato Recupero, D.: Deep learning adaptation with word embeddings for sentiment analysis on online course reviews. Presented at the (2020)
6. Kaibi, I., Nfaoui, E.H., Satori, H.: Sentiment analysis approach based on combination of word embedding techniques. Presented at the (2020)
7. Kasri, M., Birjali, M., Beni-Hssane, A.: A comparison of features extraction methods for Arabic sentiment analysis. In: Proceedings of the 4th International Conference on Big Data and Internet of Things, pp. 1–6. ACM, New York (2019)
8. Birjali, M., Beni-Hssane, A., Erritali, M.: Machine learning and semantic sentiment analysis based algorithms for suicide sentiment prediction in social networks. Proc. Comput. Sci. **113**, 65–72 (2017). https://doi.org/10.1016/j.procs.2017.08.290
9. Araque, O., Corcuera-Platas, I., Sánchez-Rada, J.F., Iglesias, C.A.: Enhancing deep learning sentiment analysis with ensemble techniques in social applications. Expert Syst. Appl. **77**, 236–246 (2017). https://doi.org/10.1016/j.eswa.2017.02.002
10. Giatsoglou, M., Vozalis, M.G., Diamantaras, K., Vakali, A., Sarigiannidis, G., Chatzisavvas, K.C.: Sentiment analysis leveraging emotions and word embeddings. Expert Syst. Appl. **69**, 214–224 (2017). https://doi.org/10.1016/j.eswa.2016.10.043
11. Mikolov, T., Yih, W., Zweig, G.: Linguistic regularities in continuous space word representations. In: Proceedings of the 2013 Conference of the North American Chapter of the Association for Computational Linguistics: Human Language Technologies, pp. 746–751 (2013)
12. Naderalvojoud, B., Sezer, E.A.: Sentiment aware word embeddings using refinement and senti-contextualized learning approach. Neurocomputing **405**, 149–160 (2020). https://doi.org/10.1016/j.neucom.2020.03.094
13. Rezaeinia, S.M., Rahmani, R., Ghodsi, A., Veisi, H.: Sentiment analysis based on improved pre-trained word embeddings. Expert Syst. Appl. **117**, 139–147 (2019). https://doi.org/10.1016/j.eswa.2018.08.044
14. Yu, L.-C., Wang, J., Lai, K.R., Zhang, X.: Refining word embeddings using intensity scores for sentiment analysis. IEEE/ACM Trans. Audio Speech Lang. Process. **26**, 671–681 (2018). https://doi.org/10.1109/TASLP.2017.2788182
15. Maas, A.L., Daly, R.E., Pham, P.T., Huang, D., Ng, A.Y., Potts, C.: Learning word vectors for sentiment analysis. In: Proceedings of the 49th Annual Meeting of the Association for Computational Linguistics: Human Language Technologies, pp. 142–150. Association for Computational Linguistics, Portland (2011)
16. Tang, D., Wei, F., Yang, N., Zhou, M., Liu, T., Qin, B.: Learning sentiment-specific word embedding for twitter sentiment classification. In: Proceedings of the 52nd Annual Meeting of the Association for Computational Linguistics (Volume 1: Long Papers), pp. 1555–1565. Association for Computational Linguistics, Stroudsburg (2014)
17. Collobert, R., Weston, J.: A unified architecture for natural language processing. In: Proceedings of the 25th international conference on Machine learning - ICML 2008, pp. 160–167. ACM Press, New York (2008)
18. Ren, Y., Zhang, Y., Zhang, M., Ji, D.: Improving Twitter sentiment classification using topic-enriched multi-prototype word embeddings. In: Proceedings of the Thirtieth AAAI Conference on Artificial Intelligence, pp. 3038–3044. AAAI Press (2016)

19. Warriner, A.B., Kuperman, V., Brysbaert, M.: Norms of valence, arousal, and dominance for 13,915 English lemmas. Behav. Res. Methods **45**(4), 1191–1207 (2013). https://doi.org/10.3758/s13428-012-0314-x
20. Yin, R., Li, P., Wang, B.: Sentiment lexical-augmented convolutional neural networks for sentiment analysis. In: 2017 IEEE Second International Conference on Data Science in Cyberspace (DSC), pp. 630–635. IEEE (2017)
21. Baccianella, S., Esuli, A., Sebastiani, F.: SentiWordNet 3.0: an enhanced lexical resource for sentiment analysis and opinion mining. In: Calzolari, N., et al. (eds.) Proceedings of the International Conference on Language Resources and Evaluation, {LREC} 2010, 17–23 May 2010. European Language Resources Association, Valletta (2010)
22. Cambria, E., Poria, S., Hazarika, D., Kwok, K.: SenticNet 5: discovering conceptual primitives for sentiment analysis by means of context embeddings. In: Thirty-Second AAAI Conference on Artificial Intelligence (2018)
23. Hutto, C.J., Gilbert, E.: VADER: a parsimonious rule-based model for sentiment analysis of social media text. In: Proceedings of the 8th International Conference on Weblogs and Social Media, ICWSM 2014 (2015)
24. Miller, G.A.: WordNet: a lexical database for English. Commun. ACM. **38**, 39–41 (1995). https://doi.org/10.1145/219717.219748
25. Pang, B., Lee, L.: Seeing stars: exploiting class relationships for sentiment categorization with respect to rating scales (2005)
26. Socher, R., et al.: Recursive deep models for semantic compositionality over a sentiment treebank. In: Proceedings of the 2013 Conference on Empirical Methods in Natural Language Processing, pp. 1631–1642. Association for Computational Linguistics, Seattle (2013)
27. Kim, Y.: Convolutional neural networks for sentence classification (2014)
28. Chen, T., Xu, R., He, Y., Wang, X.: Improving sentiment analysis via sentence type classification using BiLSTM-CRF and CNN. Expert Syst. Appl. **72**, 221–230 (2017). https://doi.org/10.1016/j.eswa.2016.10.065
29. Chen, X., Rao, Y., Xie, H., Wang, F.L., Zhao, Y., Yin, J.: Sentiment classification using negative and intensive sentiment supplement information. Data Sci. Eng. **4**(2), 109–118 (2019). https://doi.org/10.1007/s41019-019-0094-8
30. Hu, B., Tang, B., Chen, Q., Kang, L.: A novel word embedding learning model using the dissociation between nouns and verbs. Neurocomputing **171**, 1108–1117 (2016). https://doi.org/10.1016/j.neucom.2015.07.046

Arabic Topic Modeling-Based Sentiment Analysis on COVID-19 Feedback Comments

Mustapha Hankar[1(✉)], Marouane Birjali[1], Anas El-Ansari[2], and Abderrahim Beni-Hssane[1]

[1] LAROSERI Laboratory, Computer Science Department, Faculty of Sciences, University of Chouaib Doukkali, El Jadida, Morocco
[2] MASI Laboratory, Computer Science Department, University Mohammed First of Oujda, FP Nador, Nador, Morocco

Abstract. In March 2020, World Health Organization (WHO) declared COVID-19 disease a pandemic. People worldwide face stressful and overwhelming challenges and express emotions of loneliness, fear, and anxiety during quarantine periods. The pandemic strongly hits labor markets. As a result, many employees lost their jobs, and others are living at a high risk of losing theirs the next day. In this paper, we designed a hybrid approach that combines topic modeling and sentiment analysis methods to mine the impact of the COVID-19 pandemic on Moroccan citizens. The dataset used in this study contains collected comments from Hespress, an online news website. Our approach implements Latent Dirichlet Allocation (LDA) to extract topics from the comments. Consequently, we apply a topic-based sentiment analysis using a pre-trained model to mine people's feedback regarding the extracted topics. The results showed that expressed sentiments towards all extracted topics are more negative than neutral and positive topic-based sentiments.

Keywords: COVID-19 · Sentiment analysis · Topic modeling · CAMeL · Hespress · Feedback · Anxiety · Fear · Quarantine · LDA · Selenium

1 Introduction

Last December 2019, the novel coronavirus appeared in Wuhan. Back then, health officials thought that its origin might be from a seafood market in Wuhan [1]. The primary way of transmission [2] is through direct contact with respiratory droplets of an infected person through coughing, sneezing, or touching contaminated surfaces. Months later, WHO declared the disease a pandemic. A Conducted study [3] stated that people experienced psychological effects of being isolated and socially distanced from each other. It threatened to lose its jobs as the governments were shutting down the economy. These impacts include traumatic stress symptoms, confusion, anger, long quarantine suppress symptoms, infection fears, frustration, boredom, lack of supplies, and financial loss.

In this paper, we are designing a hybrid approach combining topic modeling and sentiment analysis to analyze COVID-19 effects on Moroccan people during the quarantine period. Our work is a subject of applying a topic modeling-based sentiment

Y. Maleh et al. (Eds.): ICI2C 2021, LNNS 357, pp. 87–95, 2022.
https://doi.org/10.1007/978-3-030-91738-8_9

analysis system to mine people's opinions about the COVID-19 pandemic regarding many topics. The system receives the preprocessed comments, extracts the topics from the dataset, and applies a topic-based sentiment analysis using the pretrained CaMel model to assign polarities to aspect-categorized texts. The dataset used for this study contains collected texts from Hespress written in both modern standard Arabic (MSA) and Moroccan dialect (MD). The system combines many techniques such as topic modeling for topic extraction, dimensionality reduction for word embeddings, and sentiment analysis to detect polarities on subjective opinions expressed about a particular topic related to COVID-19. For example, we want to know COVID-19 effects on various domains like people's mental health, the state economy, the labor market, or government measures to control the situation.

The remainder of this paper is organized as follows: in the second section, we will cite and discuss other related works to our study. In the third section, we describe and explore our dataset passed through cleaning and preprocessing tasks. We mention some issues in the Arabic language. Both the fourth and fifth sections are dedicated to citing the various methods used in our work, such as LDA for topic modeling, t-SNE for reducing dimensions and word embeddings, and sentiment analysis. In the sixth section, we present our findings for discussion and further perspectives.

2 Related Works

Natural language processing has been of considerable interest to the research community [4] to advance processing techniques to understand human languages. Arabic sentiment analysis (ASA) is still facing many challenges [5] related to specific characteristics of the language. This section will cite and discuss some associated works tackling sentiment analysis and topic modeling subjects.

Melville et al. [6] proposed a supervised (Naïve Bayes classifier) lexicon-based sentiment analysis approach to build a model based on labeled documents. Abdul-Mageed et al. [7], designed a system (SAMAR) for Subjectivity and Sentiment Analysis (SSA) to identify whether the text is objective or subjective before detecting its polarity. The system implements an SVMlight algorithm to perform the classification task. Zarra and al. [8] proposed a semi-supervised approach that combines topic extraction and sentiment analysis. The model aims to assign for each topic a specific sentiment. They implemented a supervised approach to extract the sentiments, and an unsupervised approach to extract the topic from a dataset that contains Facebook comments written in colloquial Maghrebi Arabic. Shelke et al. [9] proposed an approach for domain-independent aspect-oriented sentiment analysis for product reviews. Their approach uses the SentiWordNet lexicon to determine the polarity of identified features. Akhtar et al. [10] introduced an aspect-oriented sentiment analysis approach that combines aspect term extraction and sentiment classification. They used "three classifiers, namely Maximum Entropy, Conditional Random Field and Support Vector Machine. Experiments for aspect term extraction and sentiment analysis on two different kinds of domains show the effectiveness of our proposed approach". S.Asharaf et al. [11] proposed an aspect-specific sentiment analysis of product reviews using LDA. They extracted topic words, and then mapped with various aspects of an entity to perform the aspect-specific sentiment analysis on product reviews. Experiments with synthetic and real dataset show promising results compared to existing methods of sentiment analysis. Piryani et al. [12] reported a rule-based linguistic approach for aspect-level sentiment analysis for movie reviews. Their method identifies the aspects from movie reviews, locates opinions about that aspect, and computes the sentiment polarity of that opinion using linguistic approaches. Li et al. [13] proposed a client perception-based product feature mining algorithm that uses sentiment orientation. Product feature and sentiment orientation are sorted by user interest weights that help customers access review orientation more effectively.

2.1 Data Description

The dataset provided for this study was collected from Hespress (news platform) during the quarantine period. The collection task has taken more than three months (March, April, and May) to scrap more than 12000 comments written in Arabic (MSA or MD) and other texts written in French, Spanish, English, and other languages. We coded a python script using selenium library to crawl all comments from all articles and news related to the COVID-19 pandemic in various domains such as economy, society, health, politics, etc. Our interest is only on comments typed in Arabic; we filtered our data frame to select the Arabic language. The columns of our dataset contain the date on which the comment is published, the user's name, and the comment's text.

2.2 Data Cleaning and Preprocessing

The characteristics of the Arabic language make processing its text on the web a big challenge because of difficulties related to Arabic text morphology [14]. Another challenge is that Arabic web text is a mixture of classical Arabic (CA), modern standard Arabic (MSA), and Moroccan dialect (MD). The text may also contain non-Arabic words, notations, and orthographic features such as spelling mistakes, repeated letters, or express emotive words [15].

The structure of the Arabic language is different from other languages such as English which makes the processing task very challenging [16]. For example, the Arabic language is written from right to left. It contains 28 letters, including three vowels and diacritics called harakat (حركات) like fatha (فتحة), sukun (سكون), hamza (همزة), tanwin (تنوين), shadda (شدة), etc. Diacritics may affect the semantic and syntax, which makes the meaning of the terms more ambiguous. Cleaning Arabic text includes removing diacritics, punctuations, links, emojis, numbers, and stop words. For example, given an Arabic sentence like:
إِنْ الْقُرَّاْءَ يَقْرَؤُوْنَ الْقُرْآنَ قِرَاْءَةً جَمِيْلَــــةً. Cleaning the text will return an output text as follows: إن القراء يقرؤون القرآن قراءة جميلة. The next task in the data preparation pipeline after cleaning text data is to preprocess it. This includes splitting the texts into words, known as tokens, and the second step is to stem these obtained tokens by eliminating suffixes and prefixes. The third step is passing the extracted stems to a morphological analyzer that yields the root of the words (Table 1).

Table 1. Raw comments versus preprocessed comments

Raw comment	Preprocessed comment
الا بذكر الله تطمئن القلوب عزوزوا ايمانكم وفروا الى الله وقراءة القرآن فوالله لا علاج نفسي اقوى من هذا فجددوا ايمانكم تفوزوا بخير الدارين	ذكر, الله, تطمئن, قلوب, عزوزوا, ايمانكم, وفروا, الله, قراءة, قرآن, والله, علاج, نفسي, اقوى, جددوا, ايمانكم, تفوزوا, بخير, دارين
و هدوك اصحاب الرميد... واصحاب الرميد المنتهية صلاحيته و القطاع الغير المهيكل وهدوك كاين \n...لي غلطو او غلطو فيهم و و و و زيد و منهم مشدش حتى الدفعة الاولى	هدوك, اصحاب, رميد, اصحاب, رميد, منتهية, صلاحيته, قطاع, غير, مهيكل, هدوك, غلطو, غلطو, فيهم, زيد, كاين, منهم, مشدش, دفعة
هدشي اللي واقع فكازا والله حتى حرام 99 حالة فنهار بزاف والله حتى عيينا ومرضنا نفسيا وماديا.يا ربي تعفو علينا وتهدي الفيروس يضرب شي دورة بعيد	هدشي, اللي, واقع, فكازا, والله, حرام, حالة, نهار, الله,عيينا, مرضنا, نفسيا, ماديا, ربي, تعفو, علينا, تهدي, فيروس, يضرب, شي, دورة, بعيد
الحل هو اعطي الناس الفقراء والمحتاجين من المال ليكفل اسرته فلا تجد اي احد في الشارع	الحل, اعطي, ناس, فقراء, محتاجين, مال, يكفل, اسرته, تجد, شارع

3 LDA for Topic Extraction and Modeling

Topic modeling is an unsupervised classification of documents that finds some natural groups of topics in a given corpus. In a topic modeling process, we intend to discover the hidden topics in the collection, and classify the documents into these discovered topics [17]. LDA is a probabilistic model often used to extract topics from documents of a corpus. Each document of the corpus is a mixture of different topics, and each topic is an infinite mixture over an underlying set of topic probabilities. The topic probabilities provide an explicit representation of a document in the context of topic modeling [18]. In LDA terminology, a word is the basic unit of discrete data, defined to be an item from a vocabulary indexed by {1…V}.

A document is a sequence of N words denoted by W = (w_1,… w_n), where w_n is the nth word in the sequence, and a corpus is a collection of M documents denoted by D = W_1 … W_m. The generative process for each document "w" in a corpus "D" is represented as follows [18]:

1. Choose $N \sim \text{Poisson}(\xi)$
2. Choose $\theta \sim \text{Dir}(\alpha)$
3. For each of the N words w_n:
   - Choose a topic $z_n \sim \text{Multinomial}(\theta)$
   - Choose a word w_n from $p(w_n \mid z_n, \beta)$, a multinomial probability conditioned on the topic z_n.

A k - dimensional Dirichlet random variable θ can take values in the $(k-1)$ simplex (a k -vector θ lies in the $(k-1)$ simplex if $\theta_i \geq 0, \sum_{i=1}^{k}\theta_i = 1$), and has the following probability density on this simplex:

$$p(\theta|\alpha) = \frac{\Gamma\left(\sum_{i=1}^{k}\alpha_i\right)}{\prod_{i=1}^{k}\Gamma(\alpha_i)}\theta_1^{\alpha_1-1}\cdots\theta_k^{\alpha_k-1} \tag{1}$$

Where the parameter α is a k-vector with components $\alpha_i > 0$, and where $\Gamma(x)$ is the Gamma function. Given the parameters α and β, the joint distribution of a topic mixture θ, a set of N topics $\mathbf{z}$, and a set of N words $\mathbf{w}$ is given by:

$$p(\theta, \mathbf{z}, \mathbf{w}|\alpha, \beta) = p(\theta|\alpha)\prod_{n=1}^{N}p(z_n \mid \theta)p(w_n \mid z_n, \beta) \tag{2}$$

Where $p(z_n \mid \theta)$ is simply θ_i for the unique i such that $z_n^i = 1$. Integrating over θ and summing over z, we obtain the marginal distribution of a document:

$$p(\mathbf{w}|\alpha, \beta) = \int p(\theta|\alpha)\left(\prod_{n=1}^{N}\sum_{z_n}p(z_n|\theta)p(w_n|z_n, \beta)\right)d\theta \tag{3}$$

Finally, taking the product of the marginal probabilities of single documents, we obtain the probability of a corpus:

$$p(D|\alpha, \beta) = \prod_{d=1}^{M}\int p(\theta_d \mid \alpha)\left(\prod_{n=1}^{N_d}\sum_{d_{dn}}p(z_{dn} \mid \theta_d)p(w_{dn} \mid z_{dn}, \beta)\right)d\theta_d \tag{4}$$

4 Topic Modeling-Based Sentiment Analysis

Baseline Arabic sentiment analysis (ASA) methods were used to compute the opinion polarity ignoring the topic level in the text [19]. A subjective text expresses a mixture of various topics. Thus, our work intends to retrieve people's feedback towards various topics related to the COVID-19 pandemic. In the first phase, our framework aims to extract topics within the dataset using the LDA method to uncover the hidden topics in the collection and then reconstruct these topics based on the scoring probability of words belonging to a specific topic. In the second phase, we perform a full scan through the dataset and search for comments expressing the extracted topic as an input. This task yields a data frame of text comments that match the topic. We repeat this process for other extracted topics. We map these topics to corresponding aspects based on the word probability, which are formulated using a logical operator. Then we perform topic-level sentiment analysis to detect polarities using CAMeL pre-trained model [20] to classify sentiments as negative, positive, or neutral. The overall workflow of our approach is illustrated in Fig. 1 below.

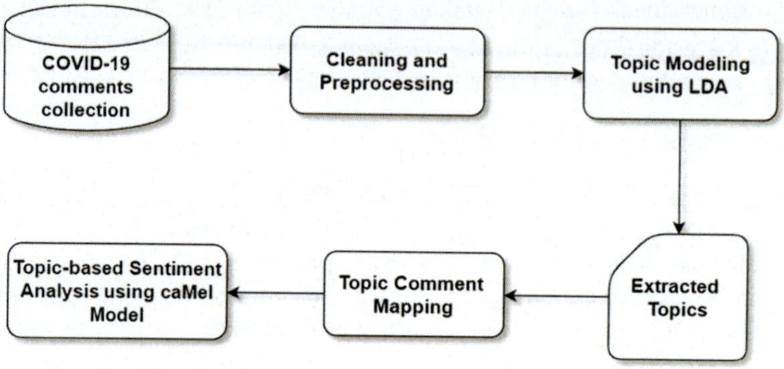

Fig. 1. Topic modeling-based sentiment analysis workflow

5 Results and Discussion

In this section, we present the obtained results of our work. Table 2 shows negative feedback towards the various extracted topics, their frequency per topic, and the rate of the polarity regarding the topic (the rate is obtained by dividing the negative sentiment frequency by the sum of all other polarities for a specific topic). In Tables 3 and 4, the same results for neutral and positive sentiments are given, and we notice that rates are very low compared to negative rates.

The obtained results support our anticipated hypothesis that the COVID-19 pandemic affected all ways of normal life calamitously. The findings in Table 3 show that expressed sentiments are highly negative concerning all the aspects. For example, we observe that most people felt deficient in how the pandemic affected the economy. They also expressed fear, anxiety, the discomfort of losing their jobs. People who lost their jobs were not satisfied with the governmental support plan during the quarantine period. This is obvious in Table 2 and Fig. 2 below, because 78% of expressed sentiments, regarding COVID-19 & support topic, were negative. We also mention that the highest frequency of negative comments expressed fear and anxiety towards the new coronavirus, which is normal, because all the comments were collected in the first months of the pandemic where the virus still poses an unknown threat to people's lives.

Table 2. Topic-based negative sentiment results

Sentiment	Extracted topic	Frequency	Rate (%)
Negative	Covid19 & Economy	197	77.56
	Covid19 & E-Learning	30	71.43
	Covid19 & Fear	836	55.18
	Covid19 & Government	341	75.44
	Covid19 & Health	153	57.30
	Covid19 & support	69	78.41

Table 3. Topic-based neutral sentiment results

Sentiment	Extracted topic	Frequency	Rate (%)
Neutral	Covid19 & Economy	40	15.75
	Covid19 & E-Learning	9	21.43
	Covid19 & Fear	200	13.20
	Covid19 & Government	64	14.16
	Covid19 & Health	102	38.20
	Covid19 & support	16	18.18

Table 4. Topic-based positive sentiment results

Sentiment	Extracted topic	Frequency	Rate (%)
Positive	Covid19 & Economy	17	6.69
	Covid19 & E-Learning	3	7.14
	Covid19 & Fear	479	31.62
	Covid19 & Government	47	10.40
	Covid19 & Health	12	4.49
	Covid19 & support	3	3.41

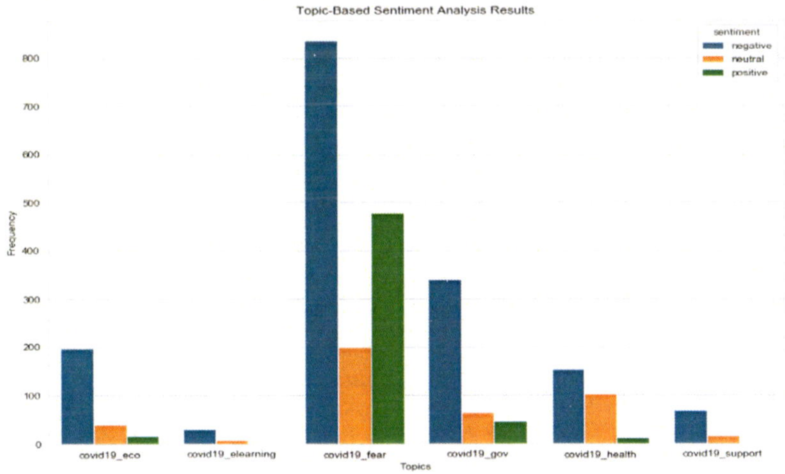

Fig. 2. Topic modeling-based sentiment analysis results

Our primary goal in conducting this study was to mine people's opinions on the web about COVID-19 impact on several domains like education, economy, mental health, etc. The achieved results were meant to answer how the critical situation affected people's normal lives. We assume that this study needs to be applied on a large scale in terms of data and aspects. As shown in the tables above, it is obvious that

COVID-19 has affected education dramatically and made it unreachable for the majority of a student, especially in the rural regions. As future work, we propose conducting a study on the challenges of eLearning in Morocco and finding solutions to make education possible and fruitful in situations like these.

6 Conclusion

This paper proposed a hybrid approach for topic-based Arabic sentiment analysis that combines the topic modeling method (LDA) with a sentiment analysis supervised model. The system was applied to collect data about the COVID-19 pandemic to perform an opinion mining task on a topic-based level. The obtained results showed that all expressed feedback regarding all extracted topics was highly negative.

References

1. Zhu, N., et al.: A novel coronavirus from patients with pneumonia in China, 2019. N. Engl. J. Med. (2020). https://doi.org/10.1056/nejmoa2001017
2. Karia, R., Gupta, I., Khandait, H., Yadav, A., Yadav, A.: COVID-19 and its Modes of Transmission. SN Compr. Clin. Med. **2**(10), 1798–1801 (2020). https://doi.org/10.1007/s42399-020-00498-4
3. Brooks, S., et al.: The psychological impact of quarantine and how to reduce it: rapid review of the evidence. Lancet **395** (2020). https://doi.org/10.1016/S0140-6736(20)30460-8
4. El-Ansari, A., Beni-Hssane, A., Saadi, M.: A multiple ontologies-based system for answering natural language questions. In: Rocha, Á., Serrhini, M., Felgueiras, C. (eds.) Europe and MENA Cooperation Advances in Information and Communication Technologies. Advances in Intelligent Systems and Computing, vol. 520, pp. 177–186. Springer, Cham (2017). https://doi.org/10.1007/978-3-319-46568-5_18
5. Alwakid, G., Osman, T., Hughes-Roberts, T.: Challenges in sentiment analysis for Arabic social networks. Proc. Comput. Sci. **117**, 89–100 (2017). https://doi.org/10.1016/j.procs.2017.10.097
6. Melville, P., Gryc, W., Lawrence, R. D.: Sentiment analysis of blogs by combining lexical knowledge with text classification. In: Proceedings of the 15th ACM SIGKDD international conference on Knowledge discovery and data mining, pp. 1275–1284. ACM (2009)
7. Abdul-Mageed, M., Kübler, S., Diab, M.: Samar: a system for subjectivity and sentiment analysis of arabic social media. In: Proceedings of the 3rd Workshop in Computational Approaches to Subjectivity and Sentiment Analysis, pp. 19–28. Association for Computational Linguistics (2012)
8. Zarra, T., Chiheb, R., Moumen, R., Faizi, R., El Afia, A.: Topic and sentiment model applied to the colloquial Arabic: a case study of Maghrebi Arabic (2017). https://doi.org/10.1145/3128128.3128155
9. Shelke, N., Deshpande, S., Thakare, V.: Domain independent approach for aspect-oriented sentiment analysis for product reviews. In: Satapathy, S., Bhateja, V., Udgata, S., Pattnaik, P. (eds.) Proceedings of the 5th International Conference on Frontiers in Intelligent Computing: Theory and Applications. Advances in Intelligent Systems and Computing, vol. 516, pp. 651–659, Springer, Bhubaneswar (2017). https://doi.org/10.1007/978-981-10-3156-4_69

10. Akhtar, M.S., Gupta, D., Ekbal, A., Bhattacharyya, P.: Feature selection and ensemble construction: a two-step method for aspect-based sentiment analysis. Knowl. Based Syst. **125**, 116–135 (2017)
11. Anoop, V.S., Asharaf, S.: Aspect-oriented sentiment analysis: a topic modeling-powered approach. J. Intell. Syst. **29**, 1166–1178 (2018). https://doi.org/10.1515/jisys-2018-0299
12. Piryani, R., Gupta, V., Singh, V.K., Ghose, U.: A linguistic rule-based approach for aspect-level sentiment analysis of movie reviews. In: Bhatia, S.K., Mishra, K.K., Tiwari, S., Singh, V.K. (eds.) Advances in Computer and Computational Sciences. AISC, vol. 553, pp. 201–209. Springer, Singapore (2017). https://doi.org/10.1007/978-981-10-3770-2_19
13. Li, J.J., Yang, H., Tang, H.: Feature mining and sentiment orientation analysis on product. In: Management Information and Optoelectronic Engineering: Proceedings of the 2016 International Conference on Management, Information and Communication (ICMIC2016) and the 2016 International Conference on Optics and Electronics Engineering (ICOEE2016), p. 79. World Scientific, Guilin (2017)
14. Guellil, I., Saadane, H., Azouaou, F., Gueni, B., Nouvel, D.: Arabic natural language processing: an overview. J. King Saud Univ. Comput. Inf. Sci. **33**, 497–507 (2019). https://doi.org/10.1016/j.jksuci.2019.02.006
15. Hegazi, M.O., Al-Dossari, Y., Al-Yahy, A., Al-Sumari, A., Hilal, A.: Preprocessing Arabic text on social media. Heliyon. **7**, e06191 (2021). https://doi.org/10.1016/j.heliyon.2021.e06191
16. Farghaly, A., Shaalan, K.: Arabic natural language processing: challenges and solutions. ACM Trans. Asian Lang. Inf. Process. (TALIP). **8**, 1–22 (2009)
17. Liu, L., Tang, L., Dong, W., Yao, S., Zhou, W.: An overview of topic modeling and its current applications in bioinformatics. Springerplus **5**(1), 1–22 (2016). https://doi.org/10.1186/s40064-016-3252-8
18. Blei, D., Ng, A., Jordan, M.: Latent Dirichlet allocation. J. Mach. Learn. Res. **3**, 993–1022 (2003). https://doi.org/10.1162/jmlr.2003.3.4-5.993
19. Al-Moslmi, T., Albared, M., Al-Shabi, A., Omar, N., Abdullah, S.: Arabic senti-lexicon: constructing publicly available language resources for Arabic sentiment analysis. J. Inf. Sci. **44**, 016555151668390 (2017). https://doi.org/10.1177/0165551516683908
20. Obeid, O., et al.: CAMeL tools: an open-source Python toolkit for Arabic natural language processing (2020)

Parallel Feature Selection Approaches for High Dimensional Data: A Survey

Naoual El Aboudi(✉) and Laila Benhlima

Mohammadia School of Engineering, Mohammed V University in Rabat, Rabat, Morocco
benhlima@emi.ac.ma

Abstract. The feature selection task represents an essential phase in machine learning that is time-consuming and requires significant computational resources, especially when considering current increasing trends in terms of data volume. To overcome these constraints, parallel computing models are adopted to reduce the computational time of these approaches while achieving promising results in classification performance. Although several studies reviewed feature selection methods, none of them provided a detailed classification of feature selection methods and viewed existing parallel and distributed feature selection methods. In this paper, we survey different categories of feature selection approaches. Then we present parallel computing paradigm, its models, its libraries, frameworks, and parallel feature selection methods. We also provide an in-depth analysis of their experimental results according to sequential versions of these algorithms. Finally, future issues and direction research are exposed.

Keywords: Parallelization models · Feature selection · Population-based algorithms · Big data

1 Introduction

Feature selection becomes an indispensable component of the learning process. Its principle consists of choosing relevant features from the entire feature set while dropping redundant and irrelevant ones to improve the classification performance [1]. The feature selection is an NP-hard optimization problem (Non-deterministic Polynomial-time hard) as the search space includes $2^n - 1$ feature subsets, with n number of features [2].

Since the feature selection task is considered a crucial step in data preprocessing, several research works have been proposed to handle the feature selection problem. In this context, a review of feature selection approaches was presented. In this study, different categories of feature selection based on evaluation criterion, search strategy and learning methods were exposed. Furthermore, several applications which adopt feature selection were presented [10].

Miao proposed a survey of existing feature selection methods, particularly unsupervised feature selection approaches that are hard since they don't use labeled data [21].

Y. Maleh et al. (Eds.): ICI2C 2021, LNNS 357, pp. 96–109, 2022.
https://doi.org/10.1007/978-3-030-91738-8_10

Recently, Cai conducted an interesting survey on feature selection methods [5] where he focused on supervised and unsupervised techniques highlighting their key aspects and presented some remained challenges concerning high dimensional datasets which can't be handled through traditional feature selection methods, ensemble feature selection which is the ensemble of base classifiers trained by different feature spaces obtained by feature selection can perform feature selection on high dimensional datasets.

Furthermore, Liu presented a survey on automatic feature selection and unsupervised feature learning for high-dimensional batch reinforcement learning [6]. In that study, the emphasis is on recent theoretical developments on applying statistical learning to establish finite-sample error bounds for batch RL algorithms based on weighted Lp norms. In this study, related works of feature selection in bioinformatics applications are presented.

According to the studies cited above, none of them provide a review of parallel and distributed feature selection approaches. In this study, different classes of feature selection methods are presented in Sect. 2. Centralized feature selection methods that are parallelized in the literature are exposed. Section 4 details some parallel feature selection models, methods, libraries and frameworks used for parallel solutions. Finally, direction research and challenges are depicted.

2 Feature Selection Classification

Feature selection can be classified according to supervision strategy, evaluation measures, search method, and data perspective. In the following section, we will present in detail different classes of feature selection methods cited above.

2.1 Feature Selection Based on Supervision Strategy

From supervision strategy, feature selection can be classified into supervised, unsupervised, and semi-supervised methods.

- *Unsupervised feature selection* is considered to solve clustering problems. Its computational time is reduced since it doesn't utilize la-bel information. To evaluate feature relevance, it considers all samples without using label information. It designs the cluster structures from all data instances according to the selected features. Based on the similarity measures, redundant features are discarded. If the features are similar to one or more features, then one of these features is eliminated. Variance Score and Spectral analysis-based feature selection are among popular unsupervised feature selection methods [3]. In Semi-supervised feature selection, the data set includes unlabeled and labeled data.

  On the one hand, these algorithms are time-consuming since they construct graphs in their processing through similarity matrix such as the Laplacian matrix. Therefore, they don't efficiently analyze high-dimensional data sets. On the other hand, these methods don't consider the correlation between features in the feature selection process [4].

- *Supervised feature selection methods* are adopted to handle classification problems and evaluate the relevance or correlation between features and the class label [5].

2.2 Feature Selection Based on the Evaluation Function

Feature selection approaches based on the evaluation function encompass three classes: filters, wrapper, and hybrid techniques [6].

- *Filter methods* perform without using a learning algorithm. They evaluate a candidate feature subset by ranking each feature by statistical methods and then comparing them with a predefined threshold to delete those with scores below the threshold [7, 8].
- *Wrapper methods* evaluate every feature subset through a learning algorithm by computing its classification performance [9].
- *Hybrid approaches* are the third class of feature selection approaches based on evaluation function. They combine both filter and wrappers methods (Table 1).

Table 1. Feature selection approaches based on evaluation

FS methods	Benefits	Shortcomings
Filters	- They are faster than wrappers - They consider feature dependencies - They don't use a classification algorithm	- They achieve worse results in terms of classification performance compared to wrappers
Wrappers	- They achieve better results in terms of classification performance than filters - They adopt a classification algorithm	- Their computational time is lower than filters - They don't consider feature dependencies
Embedded	- They consider feature dependencies - Their computational time is lower than wrappers - They use a classification algorithm	- Classifier dependent selection

2.3 Feature Selection from a Data Perspective

Feature selection approaches can be categorized into feature selection on conventional data, feature selection on structured data, feature selection on heterogeneous data, and feature selection with streaming data [10].

Feature selection based on conventional data methods encompasses four groups: similarity-based, information theoretical-based, sparse learning-based, and statistical-based methods.

- *Similarity-based feature selection methods:* don't use any learning algorithm and evaluate the importance of each feature through similarity measure by computing affinity such as measure related to label class for supervised methods and distance measure for unsupervised approaches. These methods can't perform feature redundancy analysis for all available features.
- Compared to similarity methods information, *theoretical-based feature selection algorithms* are based on filter approaches (independent for learning algorithm) and perform feature relevance feature redundancy analysis such as Minimum Redundancy Maximum Relevance and mutual information.
- Unlike the methods mentioned above, *sparse Learning-based Methods* like efficient and Robust Feature Selection minimize the fitting errors with sparse regularization terms. They incorporate the feature selection process into a learning algorithm such as linear regression or support vector machine. They spend high computational time in their execution.
- The fourth class of feature selection methods from a data perspective is called *statistical-based methods* (Table 2) .

Table 2. Feature selection from a data perspective

FS classes	Methods
Conventional data-based feature selection	• Similarity-based feature selection • Theoretical based feature selection • Sparse Learning-based Methods • Statistical based methods
Structured data-based feature selection	• Feature structure special • Temporal structure, • Groups, graphs or trees
Feature selection with heterogeneous data	• Linked data • Multi-source feature selection • Multi-view feature selection
Feature selection on data streams	• Online feature selection for data streams • Streaming feature selection for feature streams

2.4 Feature Selection Based on Search Strategy

Feature selection methods can be classified into: Exponential methods, sequential selection methods and metaheuristics based methods.

- *Exponential methods* consist of exploring the entire search space to obtain the optimal feature subset. These approaches can't be implemented easily in the case of high dimensional feature space.

- *Sequential feature selection methods* [11], which are based on hill-climbing, encompass both Forward selection [12] and Backward selection [13].
- *Metaheuristics-based algorithms* represent the last category of feature selection methods based on search strategy. These methods are composed of two classes: population-based feature selection methods and local search-based feature selection methods (Table 3).

Table 3. Feature selection based on search strategy

FS categories	Methods
Exponential	• Exhaustive search
Sequential	• Sequential forward • Sequential backward • Forward-backward selection
Metaheuristics based methods	• Genetic algorithm • Particle swarm optimization • Ant colony optimization • Migrating Birds optimization • Simulated Annealing • Tabu search

Traditional feature selection methods can't process an important volume of data and obtain worse results in both classification performance and computational time. Consequently, parallel and distributed computing can be adopted to improve the performance of traditional methods by parallelizing them. It consists of splitting each method into multiple tasks that can be run simultaneously on different machines. In this next section, we first present in detail some of the centralized feature selection techniques parallelized in the literature, then parallel computing paradigm and their models are exposed; finally, different parallel feature selection approaches are described closely.

3 Centralized Feature Selection Techniques

In this section, some of the widely adopted feature selection handle features parallelized in the literature are presented. On the one hand, population-based techniques for feature selection are detailed, then sequential feature selection methods are exposed. Finally, filter approaches for feature selection approaches are explained.

3.1 Population-Based Feature Selection

Population-based techniques are inspired by nature. They generate a near-optimal solution while spending a reasonable period [14]. They begin their search by initializing a population of solutions. Then they will improve each solution by applying operations of the chosen algorithm. Each candidate solution is evaluated through a fitness function. The algorithm continues its execution until it reaches a stop criterion. Finally, the best solution among its current population is returned.

Genetic Algorithms (GA) and Particle Swarm Optimization (PSO) represent population-based techniques that are widely adopted in the context of feature selection.

- ***Genetic algorithm for feature selection***
  The genetic method begins by generating an initial population of solutions called chromosomes which constitute candidates feature subsets. Each feature subset is encoded through a binary vector with a length equal to the total of a number of features [15].
  In order to reach the optimal feature subset, solutions are modified through GA operators at each iteration in order to obtain new solutions for the next generation of solutions. The GA operators are exposed as following:
  Selection: this GA operator chooses solutions randomly and groups them in the form of couples.
  Crossover: This operator is imitated from the reproduction. It produces two childs from each couple which represents parents by sharing a part of its sequence with one of its parents. The resulting sequence is obtained by dividing the parent's sequences.
  Mutation: The resulting solutions (childs) is modified by changing their genes through a predefined probability.
- ***Particle swarm optimization for feature selection***
  Particle swarm optimization (PSO) is a very promising optimization technique; it is easily implemented in several domains due to its benefit mechanism. It was applied to handle the feature subset problem and reached satisfactory results [16]. It is formed by a population of solutions called particles, each of them has two characteristics: velocity and position. The position of each article is modified through to the value of its velocity, its best previous position and the position of the best solution in all populations of solutions.
  Each particle in PSO for feature selection is encoded in the same way as GA. Each feature subset is represented by a binary vector with a length equal to the total number of features. 1 means that the corresponding feature is selected and 0 means it is discarded.
- ***Migrating bird optimization for feature selection***
  The MBO algorithm is a population-based optimization algorithm. it consists of imitating the V formation flight of the migrating birds. It was presented by Duman in 2012 and has been applied for several NP hard optimization problems such as the Quadratic Assignment Problem (QAP).
  MBO algorithm incorporates a neighborhood structure that allows full exploration of the search domain, and it is composed of the following phases:
  Representation of solutions: Each solution in MBO is encoded as a binary vector X which length corresponds to the dimensionality d of the feature set.

3.2 Sequential Techniques for Feature Subset Problem

These methods perform in an iterative manner and are categorized into forwarding selection techniques and backward selection techniques.

- ***Forward selection for feature selection***
  Forward selection starts the search of the optimal solution by an empty feature set S, and then it adds at every iteration a single feature that improves the classification performance. The algorithm stops its execution when the classification performance can't be enhanced by adding remaining features to the set S.
- ***Backward Selection for feature selection***
  On the contrary, the backward feature selection starts the search with a feature set S including all the available features, and then it begins the search by discarding at each iteration a single feature from the set S so that the classification performance is enhanced. The algorithm stops its execution when the classification result can be improved. Since backward feature selection begins with the feature set S containing all the existing features, it spends more computational resources than forwarding feature selection.
- ***Forward-Backward Selection algorithm (FBS)***
  It is a sequential feature selection method formed of two phases: forward and backward. The first phase chooses the feature which enhances the classification performance by evaluating its p-value and stops its execution when no more features improve the performance. In the second phase backward, the feature that doesn't decrease the performance is eliminated.

3.3 Filter Feature Selection Methods

Several techniques belong to the family of filter feature selection approaches. In the next section, widely adopted filter approaches for feature subset problems will be presented [9].

- ***mRMR***
  it is a feature selection technique that chooses features that are strongly correlated with the class while being weakly correlated with each other. It performs by ranking features by the minimal-redundancy-maximal-relevance criterion based on mutual information.
- ***The Relief***
  It's a quick method to rank attributes. For each feature, relief calculates a weight which value is between −1 and 1.
- ***Correlation-based Feature Selection (CFS)***
  CFS is one of the widely adopted techniques to calculate the relevance of features by computing correlation between features and classes and between features and other features.

4 Parallel Feature Selection Methods

In the next section, the parallel computing paradigm is presented with their models, then parallel libraries and frameworks are exposed. Finally, parallel versions of centralized feature selection methods, presented in the previous selection, are described.

4.1 Parallel Computing Models

In general, there are three parallelization Models. The first one is the grid model. The second model is the island model, whilst the third model is the global parallelization model, which is also identified as the fitness evaluation model [17].

- *The globalization model:*
  In this model, different solutions are affected to different nodes of a cluster, then all computations are performed in a centralized manner, but only the evaluation of the fitness is performed in a parallel manner.
- *The island model*
  divides the population into many subpopulations and assign them to islands. For each island, the method is applied to solutions. This model has many characteristics:
  - Subpopulations explore the search space deeply.
  - The search is diversified thanks to the migration of solutions between subpopulations permits.
- *The grid model:*
  Each solution represents a grid and evolves independently from other solutions. It is assigned to a node of the cluster, and all computations are performed in a parallel manner.

The implementation of parallel algorithms uses one of the existing parallel systems, such as MPI and MapReduce.

4.2 Parallel and Distributed Programming Frameworks and Libraries

In this section, parallel systems will be exposed. On the one hand, the Message Passing Interface (MPI) will be presented, on the other hand, the MapReduce paradigm will be explained, and finally, apache-spark will be presented. A few parallel libraries to run parallel algorithms will be detailed.

- ***MPI***
  MPI (message passing interface) represents a standard library allowing message passing between concurrent processes on distributed systems. It can be executed on PCs, workstations.
- ***Mapreduce***
  Mapreduce is a programming model in parallel computing which has presented for the first time by Google in 2004 to analyze high dimensional data. It is one of the most widely in parallel computing applications. It processes large datasets through a cluster of machines. Mapreduce includes three phases: Split, Map, and Reduce. In the beginning, the data is analyzed and transferred to the input format from HDFS

(Hadoop Distributed File System) then divided into data blocks. In the map phase, the map function is executed in the same node where the data block is stored. Mapreduce transforms data into pairs of the following format: (key, value).

$$\text{Map: } (\text{K1}, \text{V1}) \rightarrow (\text{K2}, \text{V2})$$

Finally, the reduce module takes the results of the map phase to generate the final output of the MapReduce job.

$$\text{Reduce: } (\text{K2}, \text{list}(\text{V2})) \rightarrow \text{list}(\text{V2})$$

- ***Apache spark***
  Spark is one of the most promising parallel computing systems. It is developed in the UC Berkeley AMP Lab for large-scale data. Spark works with HDFS and has high-level libraries for stream processing and for machine learning and graph processing, such as MLlib.
- ***Data Mining EXpressions Library (dmexl)***
  It is a java library that implements parallel data mining algorithms, for instance, parallel feature selection methods.
- ***PGAPack software***
  It is a collection of libraries that execute all GA steps. PGAPack deploys a parallel master/slave single population GA using the message-passing interface (MPI). The master node stores the initial population and applies the GA operations. Slave nodes are responsible for evaluating the fitness function of every chromosome.
- ***The Java Evolutionary Computation Library (JECoLi) for parallel computing***
  The Java Evolutionary Computation Library is a reliable, adaptable, flexible, extensible, and modular software framework that allows the implementation of metaheuristic optimization algorithms using the Java programming language. JECoLi already includes a large set of EC methods: general-purpose Evolutionary/Genetic Algorithms, Simulated Annealing (SA), Differential Evolution, Genetic Programming, and Linear GP, Grammatical Evolution, Cellular Automata GAs, and Multi-objective Evolutionary Algorithms. The parallelization of JEColi encompasses three modules. The first layer is the original JECoLi; the second contains the parallelism models to use, which are the parallel island model and the parallel solution evaluation. The last one introduces the mapping of the parallel behavior to the execution platform.

4.3 Parallel and Distributed Feature Selection Techniques for High Dimensional Data

Currently, parallel and distributed feature selection is applied to perform feature selection for high-dimensional datasets.

- ***Distributed Feature Selection Methods***
  In most cases, real-time applications generate data from different sources. To perform feature selection tasks, Bolon-Canedo proposed in his research different scenarios of distributed feature selection methods [22]. On the one hand, data can be

collected in a high dataset then distributed to several processors, and finally, the same feature selection method is applied to each data subset to obtain the results by combining results of all data subsets. The data can be situated in different locations in different datasets. In this case, the same feature selection method can be executed on each subset, and the results are merged. When the volume of data isn't important, the data can be stored in a single machine. In this situation, different feature selection methods can be applied to the whole data then the results will be combined to obtain the final feature subset.

There are several parallel feature selection techniques to handle the feature subset problem. On the one hand, parallel filter feature selection techniques will be described, on the other hand, parallel sequential feature selection methods will be explained, and finally, parallel population-based feature selection approaches will be examined (Table 4).

Table 4. Distributed feature selection methods

Article	Reference
Distributed parallel feature selection method based on a variance preservation	[23]
A scalable version of the filter method named mRMR using apache spark	[24]
A distributed Relief a filter feature selection approach named DiRelief	[25]
Distributed Correlation-Based Feature Selection in Spark	[26]
Multithreaded and Spark parallelization of feature selection filters	[27]
An Information Theory-Based Feature Selection Framework for Big Data Under Apache Spark	[28]

- ***Parallel feature selection based on Filter approaches***
  Zhao presented a distributed parallel feature selection method based on a variance preservation criterion using the proprietary software SAS High-Performance Analytics [23]. This method is used for both supervised and unsupervised feature selection. This algorithm reached satisfactory results for both datasets with a large number of instances and a high number of features. In this context, Gallego et al. proposed a scalable version of the filter method named mRMR using apache spark [24]. The obtained results for large datasets in terms of computational time are promising regarding those achieved by mRMR. A distributed Relief, a filter feature selection approach named DiRelief was exposed to handle large datasets with a high number of features and instances [25]. Eiras-Franco et al. [25] adopted three distributed filter FS algorithms called Info Gain, ReliefF, and the CFS to handle feature selection problems [26]. DiCFS-vp and DICFShp, two parallel methods of CFS are developed using spark for feature subset problems. The first one divides the data by rows, and the second one splits the data by columns.
- ***Parallel Feature selection based on sequential approaches***
  Parallel Forward-backward with pruning (PFBP) is a parallel sequential approach for feature selection [29], based on Forward–Backward Selection algorithm, which

is described in the previous section. It adopted the Early Dropping method to discard features from subsequent iterations to speed up the algorithm.
FBS uses p-values to rank feature subsets to determine if the classification performance is enhanced or not when a single feature is added or discarded. The data is divided into partitions and stored into workers in a spark cluster. The computations performed on each partition of data are executed by Workers, and the master node performs global computations in a centralized manner.

- ***Parallel feature selection based on population-based approaches***
Population-based techniques are adapted to parallelization due to their inherent properties. They can be parallelized for feature selection through Mapreduce following one of the three models:

  1. Global Parallelization Model
  2. Fine-grain Parallelization Model
  3. Coarse-grained Parallelization Model

In the Global Parallelization Model, the computation of fitness function for each solution is the alone task performed in a parallel manner while the other tasks are centralized. On the other hand, in the grid Model, each solution evolves independently, and all operations of population-based techniques are parallelized. Finally, in the island model, the population is decomposed into islands, and the population-based operations run on each of them.

On the one hand, Genetic Algorithms (GAs) are powerful metaheuristic techniques used to address many software engineering problems belonging to population-based methods that are usually executed on single machines as sequential programs, so scalability issues prevent that they are effectively applied to real-world problems. However, these algorithms are naturally parallelizable, as it is possible to use more than one population and execute the operators in parallel.

GA was parallelized to handle feature selection problems through MapReduce. It starts by ranking features through a filter approach and selects the best between them, then a parallel GA based on master-slave architecture is adopted for feature selection, the master node produces an initial population of solutions and executes GA operations while the slave node computes the fitness function for each solution. The results reached by DWFS are satisfactory for biomedical problems. On the other hand, several works proposed parallel versions of particle swarm optimization to handle feature selection problems. For instance, Gracia-Nieto proposed an approach called parallel GPSO to handle feature selection problems in high-dimensional datasets. It consists of adopting the island model to parallelize PSO, then uses a migration operation to exchange solutions between them. ACO was also parallelized through the MapReduce paradigm by Meena [36] in order to perform feature selection tasks for long documents. This solution is compared to other algorithms and demonstrates its efficiency to reduce the running time.

From the results of studies previously cited, we noticed that parallel GA is the most efficient method compared to another parallel population-based feature selection since it reached the highest accuracy, parallel PSO achieved the same accuracy as the sequential version of PSO. Finally, parallel ACO obtained the worse accuracy compared to other population-based methods and spent a high computational time (Table 5).

Table 5. Parallel feature selection based on population methods

Article	Reference
Evolutionary feature selection for big data classification: a MapReduce approach	[31]
DWFS: a wrapper feature selection tool based on a parallel genetic algorithm	[32]
Parallel feature selection algorithm based on rough sets and particle swarm optimization	[35]

5 Challenges and Future Directions

Feature selection is a crucial phase in data preprocessing and represents an active area that is still to be explored. In this context, several research works propose novel feature selection methods to analyze high-dimensional data sets—for instance, online Feature selection methods and deep learning-based approaches that will be exposed in the following.

- **Online feature selection methods**
  Currently, several real-time applications don't dispose of the whole feature space at the beginning of the learning. Therefore, conventional feature selection methods can't be adopted in this case. In this context, where data is changed over time, feature selection methods handling with dynamic feature space are involved. This type of algorithm is called online feature selection. This category of algorithms represents a challenging issue since there are few works where online feature selection methods are proposed [10].
- **Deep learning-based feature selection methods**
  Deep learning has become one of the widely adopted concepts in machine learning problems and achieved successful results in particular natural language processing. It is also used for feature learning in order to learn new feature representations. Recently, several researchers designed feature selection methods based on deep learning. For instance, a deep feature selection method is proposed to select features as inputs of a deep neural network [37].

6 Conclusion

This study provides a detailed classification from different perspectives: supervision, evaluation function, search strategy, data perspective. Then, centralized feature selection methods that are parallelized in the literature are exposed. Parallel feature selection models, approaches, and libraries are described. Finally, challenges and future direction research are presented in the last section.

References

1. Tabakhi, S., Moradi, P.: Relevance-redundancy feature selection based on ant colony optimization. Pattern Recogn. **48**(9), 2798–2811 (2015)
2. Davies, S., Russell, S.: NP-completeness of searches for smallest possible feature sets. In: AAAI Symposium on Intelligent Relevance, pp. 37–39. AAAI Press (1994)
3. Wolf, L., Shashua, A.: Feature selection for unsupervised and supervised inference: the emergence of sparsity in a weight-based approach. J. Mach. Learn. Res. 1855–1887 (2005)
4. Zhao, Z., Liu, H.: Semi-supervised feature selection via spectral analysis. In: SDM, pp. 641–646. SIAM (2007)
5. Cai, D., Zhang, C., He, X.: Unsupervised feature selection for multi-cluster data. In: Proceedings of the 16th ACM SIGKDD International Conference on Knowledge Discovery and Data Mining, pp. 333–342. ACM (2010)
6. Liu, H., Yu, L.: Toward integrating feature selection algorithms for classification and clustering. IEEE Trans. Knowl. Data Eng. **17**(4), 491–502 (2005)
7. Ding, S.: Feature selection based f-score and aco algorithm in support vector machine. In: Second International Symposium on Knowledge Acquisition and Modeling, KAM 2009, vol. 1, pp. 19–23 (2009)
8. Lee, S., Park, Y.T., d'Auriol, B.J.: A novel feature selection method based on normalized mutual information. Appl. Intell. **37**(1), 100–120 (2012)
9. Khalid, S., Khalil, T., Nasreen, S.: A survey of feature selection and feature extraction techniques in machine learning. In: Science and Information Conference (SAI), pp. 372–378 (2014)
10. Li, J., et al.: Feature selection: a data perspective. Comput. Surv. (CSUR) **50**(6), 1–45 (2017)
11. Narendra, P.M., Fukunaga, K.: A branch and bound algorithm for feature subset selection. IEEE Trans. Comput. **C–26**(9), 917–922 (1977)
12. Whitney, A.W.: A direct method of nonparametric measurement selection. IEEE Trans. Comput. **C–20**(9), 1100–1103 (1971)
13. Marill, T., Green, D.: On the effectiveness of receptors in recognition systems. IEEE Trans. Inf. Theory **9**(1), 11–17 (1963)
14. Xue, B., Zhang, M.: A survey on evolutionary computation approaches to feature selection. IEEE Trans. Evol. Comput. **99** (2016)
15. Lanzi, P.L.: Fast feature selection with genetic algorithms: a filter approach. In: IEEE International Conference on Evolutionary, Computation, pp. 537–540 (1997)
16. Chuang, L.Y., Chang, H.W.: Improved binary pso for feature selection using gene expression data. Comput. Biol. Chem. **32**(1), 29–38 (2008)
17. Hanchuan, P., Fuhui, L., Chris, D.: Feature selection based on mutual information: criteria of max-dependency, max-relevance, and min-redundancy. IEEE Trans. Pattern Anal. Mach. Intell. **27**, 1226–1238 (2005)
18. Rendell, L.A., Kenji, K.: A practical approach to feature selection. In: Proceedings of the Ninth International Workshop on Machine Learning, pp. 249–256 (1992)
19. Farahat, A.K., Elgohary, A., Ghodsi, A., Kamel, S.: Distributed column subset selection on MapReduce. Proceedings of the IEEE 13th International Conference on Data Mining (2013)
20. Filomena, F., Kechadi, M.T., Salza, P., Sarro, F.: A framework for genetic algorithms based on Hadoop (2013)
21. Miao, J., Niu, L.: A survey on feature selection. Proc. Comput. Sci. **91**, 919–926 (2016)
22. Bolón-Canedo, V., Sánchez-Maroño, N., Alonso-Betanzos, A.: Distributed feature selection: an application to microarray data classification. Appl. Soft Comput. **30**, 136–150 (2015)

23. Zhao, Z., Zhang, R., Cox, J., Duling, D., Sarle, W.: Massively parallel feature selection: an approach based on variance preservation. Mach. Learn. **92**(1), 195–220 (2013). https://doi.org/10.1007/s10994-013-5373-4
24. Ramírez-Gallego, S., et al.: Fast-mRMR: fast minimum redundancy maximum relevance algorithm for high-dimensional big data. Int. J. Intell. Syst. **32**(2), 134–152 (2017)
25. Palma-Mendoza, J.R.: On the design of distributed and scalable feature selection algorithms. Doctoral dissertation, Univer-sidad de Alcalá (2019)
26. Palma-Mendoza, J.R., de-Marcos, L., Rodriguez, D., Alonso-Betanzos, A.: Distributed correlation-based feature selection in spark. Inf. Sci. **496**, 287–299 (2019)
27. Eiras-Franco, C., Bolón-Canedo, V., Ramos, S., González-Domínguez, J., Alonso-Betanzos, A., Tourino, J.: Multithreaded and Spark parallelization of feature selection filters. J. Comput. Sci. **17**, 609–619 (2016)
28. Ramírez-Gallego, S., et al.: An information theory-ba- sed feature selection framework for big data under apache spark. IEEE Trans. Syst. Man Cybern. Syst. **48**(9), 1441–1453 (2017)
29. Tsamardinos, I., Borboudakis, G., Katsogridakis, P., Pratikakis, P., Christophides, V.: A greedy feature selection algorithm for Big Data of high dimensionality. Mach. Learn. **108**(2), 149–202 (2019)
30. Filippas, J., Amin, S., Naguib, R., Bennett, M.K.: A parallel system for the classification of colonic tissue using a genetic algorithm (2003)
31. Peralta, D., Del Río, S., Ramírez-Gallego, S., Triguero. I/, Benitez, J.M., Herrera, F.: Evolutionary feature selection for big data classification: a mapreduce approach. Math. Probl. Eng. (2015)
32. Soufan, O., Kleftogiannis, D., Kalnis, P., Bajic, V.B.: DWFS: a wrapper feature selection tool based on a parallel genetic algorithm. PloS One **10**, e0117988 (2015)
33. Ling Chen, H., Yang, B., Jing Wang, S., Wang, G., Zhong Li, H., bin Liu, W.: Towards an optimal support vector machine classifier using a parallel particle swarm optimization strategy. Appl. Math. Comput. **239**, 180–197 (2014)
34. García-Nieto, J., Alba, E.: Parallel multi-swarm optimizer for gene selection in DNA microarrays. Appl. Intell. **37**(2), 255–266 (2012)
35. Adamczyk, M.: Parallel feature selection algorithm based on rough sets and particle swarm optimization. In: IEEE Federated Conference on Computer Science and Information Systems, pp. 43–50 (2014)
36. Janaki Meena, M., Chandran, K.R., Karthik, A., Vijay Samuel, A.: A parallel ACO algorithm to select terms to categorise longer documents. Int. J. Comput. Sci. Eng. **6**(4), 238–248 (2011)
37. Li, Y., Chen, C.Y., Wasserman, W.: Deep feature selection: theory and application to identify enhancers and promoters. J. Comput. Biol. **23**(5), 322–336 (2016)

Metadata Quality in the Era of Big Data and Unstructured Content

Widad Elouataoui[1(✉)], Imane El Alaoui[2], and Youssef Gahi[1]

[1] Laboratoire des Sciences de l'Ingénieur, Ecole Nationale des Sciences Appliquées, Ibn Tofail University, Kenitra, Morocco
{widad.elouataoui,gahi.youssef}@uit.ac.ma

[2] Laboratoire des Systèmes de Télécommunications et Ingénierie de la Décision, Ibn Tofail University, Kenitra, Morocco
Imane.el.alaoui@uit.ac.ma

Abstract. Nowadays, companies rely on data-driven strategies more than ever to improve their business and develop a sustainable competitive advantage. However, building a successful data-driven approach requires all data ecosystem components, including metadata management, as a crucial element of any data processing. Managing Metadata is highly needed for data analytics since it provides valuable insights on data and enables time and cost-effective data management. Managing Metadata has received more attention since big data tools use Metadata for data storage and processing with the emergence of big data. Moreover, other metadata use purposes have appeared, such as data identification and discovery, data quality assessment, content tracking and monitoring, etc.

Nevertheless, metadata management could not be effective without considering the quality aspect of metadata. Indeed, in big data environments, the gathered Metadata are unstructured, heterogeneous, and may contain inconsistent values. Such Metadata could not support traditional data management and may negatively affects data processing. Despite the high dependency between metadata management and metadata quality, these two notions are generally addressed separately. Thus, this work highlights how metadata quality could be improved throughout the different phases of Metadata in a big data environment. Moreover, we present in this paper how successful metadata governance could be achieved. Finally, we conclude with some possible future work directions.

Keywords: Metadata quality · Big data · Metadata value chain · Metadata quality dimensions · Metadata governance

1 Introduction

Nowadays, making the right business decisions based on data analytics has never been more critical. Data analytics have shown high capabilities to enhance organizations' business value and transform the raw data into valuable knowledge. However, data analytics could not be used effectively without the proper metadata structure in place. Metadata could be defined as "Data about data" [1]. Indeed, Metadata is as essential as data since it contains information needed to understand and effectively use data such as

Y. Maleh et al. (Eds.): ICI2C 2021, LNNS 357, pp. 110–121, 2022.
https://doi.org/10.1007/978-3-030-91738-8_11

content, structure, context, quality, etc. Thus, managing Metadata allows extracting valuable and relevant insights from data.

For this reason, metadata has and is still receiving increasing attention from both researchers and business practitioners [2, 3]. With the arrival of big data, new challenges related to metadata processing have been raised. Indeed, managing Metadata in a big data environment involves handling a large volume of unstructured and heterogeneous Metadata that traditional metadata management tools could not process within a reasonable time. These challenges are highly related to the particular characteristics of big data, also known as 7V's [4, 5] (Fig. 1):

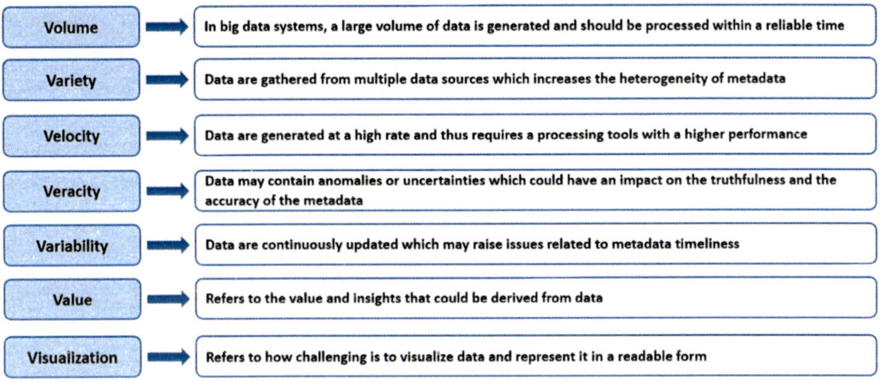

Fig. 1. Big data characteristics

In addition to the processing concerns [6, 7], the emergence of big data has raised new issues related to metadata quality. The collected Metadata in big data environments is usually not well structured, inconsistent, and of poor quality. Also, it may contain some errors such as empty fields, duplicated records, or conflicting values. Such Metadata could not support data analytics and are likely to lead to biased results. Therefore, enhancing metadata quality is highly required for effective and reliable data analysis. It is worth noting that metadata quality is positively related to the different phases that Metadata goes through since Metadata's quality loss and degradation can occur in any metadata processing phase. Even if metadata quality is highly related to metadata processing, these two notions are generally addressed separately. Therefore, this paper highlights how metadata quality could be improved throughout all metadata processing phases in big data environments. For this purpose, we present the metadata quality dimensions that could be impacted in each metadata processing phase. Furthermore, some solutions to overcome the presented quality issues will be suggested.

This paper is organized as follows: Sect. 2 reviews recent related works that have tackled metadata quality and metadata processing phases. Section 3 presents the most common metadata quality dimensions that were suggested in the literature. Section 4 describes the different steps that Metadata goes through in big data environments and highlights the metadata quality dimensions considered in each one. Section 5 suggests some good practices to ensure effective metadata governance. In the last section, conclusions are made, and future research directions are discussed.

2 Related Work

2.1 Metadata Management

Many metadata management frameworks for different metadata types have been suggested in the literature. For structured Metadata, Dong et al. in [8] have proposed a metadata management framework for big data that enables metadata integration throughout the whole information supply chain. The authors have also implemented their suggested framework in a data warehouse using the Metacube system and have presented the functional modules that should be considered for effective metadata management. The available modules of the Metacube include metadata generation, metadata analysis, metadata visualization, and metadata versioning. In [9], Xiao et al. have shown the importance of metadata management in data warehouses' construction process. Also, they have presented some guiding steps to follow for metadata construction, including determining the metadata management scope, generating and storing Metadata, selecting the appropriate metadata management tools, etc.

With the arrival of big data, new challenges related to extract, enrich, organize, and manage the semi-structured and unstructured metadata have been raised. Thus, recent studies have been performed to manage metadata in big data environments. Holom et al. have suggested in [3] an ontology-based approach for metadata representation using RDF technology to address big data challenges related to data and infrastructure management. The Metadata was represented according to the AVRO schema. Moreover, the suggested approach was implemented using Apache Hadoop for data storage and processing and Apache Spark for cluster computing. Golosova et al. in [10] have suggested a time-effective metadata integration system for big data environments for the same purpose. The recommended approach is based on Apache Kafka and was implemented in HEP experiments. Researchers in [8] have suggested a metadata management approach for textual documents using text mining techniques to address metadata management in data lakes. In this study, the authors have introduced three metadata types (inter-document Metadata, intra-document Metadata, and general Metadata), each requiring a particular approach to extract and store metadata. Other metadata management frameworks for big data context have been suggested in [11–13].

All studies mentioned above have provided significant insights into metadata management in big data environments. However, the suggested approaches focused on the functional metadata architecture and did not consider the Metadata's quality aspect.

2.2 Metadata Quality

Metadata quality has been introduced in the literature long before the emergence of big data. The first paper presented Metadata quality is [2], where Bruce and Hillman have defined seven metadata quality dimensions: accuracy, consistency, completeness, conformance to expectations, timeliness, accessibility, and provenance. In this paper, a detailed description of each metadata quality dimension has been provided. However, the authors have not given a formal definition or a metric to measure them. Later, extensive efforts have been performed to implement metadata quality dimensions. In [14], the authors have described the formulas to measure and assess five metadata quality metrics (completeness, weighted completeness, accuracy, the richness of information, and accessibility). The proposed metrics were implemented and applied to three public government data repositories. Similar studies aiming at implementing and measuring metadata quality dimensions have been suggested in [15, 16]. Starting from these studies, many research efforts have been conducted to offer new frameworks for metadata quality assessment and improvement in big data environments. In [17], researchers have suggested a novel framework to enhance the quality of data, source, and Metadata by defining the quality dimensions that should be considered for each element. Moreover, some guiding questions and good practices to follow in the different phases of the BDVC have been presented. Rousidis et al. in [18] have shown how poor metadata quality could bias data analytics. In this study, the authors have highlighted the data quality problems associated with metadata used in research data repositories and have suggested some solutions to overcome the raised issues, such as applying validation rules to date fields, assigning each metadata creator with a unique ID, using a predefined list of values to avoid inconsistencies, etc. In [19], Király has suggested a method to assess several quality aspects of metadata, such as scalability, reusability, and adaptability. The proposed approach was designed to be implemented in Europeana, a digital platform for cultural heritage qualified as Big Data. Other metadata quality frameworks were suggested in [20–23].

All these researches have made an outstanding contribution to metadata quality assessment and improvement. However, most of the suggested frameworks about metadata quality are not considering the whole metadata management process. They have been addressed only for a specific metadata processing phase or for a certain quality of metadata aspect.

To the best of our knowledge, no contributions have been conducted to address the existing correlation between metadata quality and the metadata management process. Thus, to highlight how metadata quality could be impacted in the metadata management process, we should first talk about the main characteristics of metadata quality, also known as "Metadata Quality Dimensions". In the next section, we present the most common metadata quality dimensions suggested in the literature.

3 Metadata Quality Dimensions

As discussed in the prior section, metadata quality was initially defined in seven metadata quality dimensions [2]. Later, more metadata quality dimensions were suggested in the literature to cover other metadata quality aspects, such as shareability, extendibility, and versionability [20, 24]. In the following, we present the most common metadata quality dimensions that we group into four aspects: Usability, Reliability, Availability, and Regularity (Fig. 2).

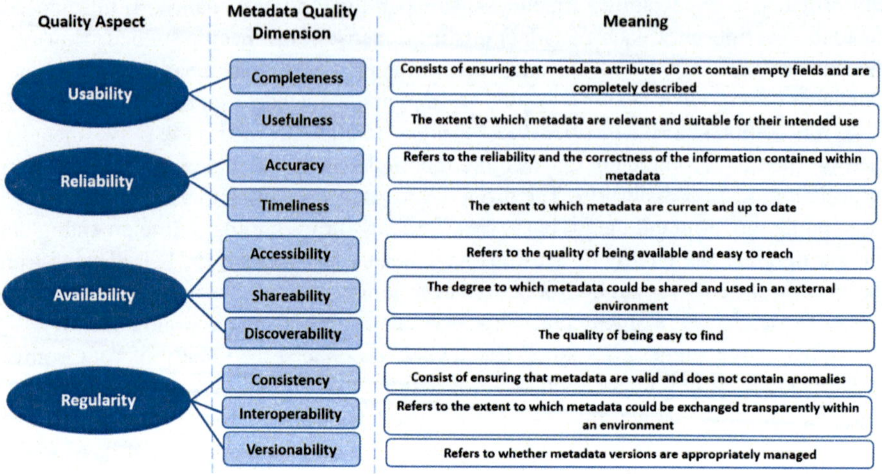

Fig. 2. Metadata quality dimensions

These metadata quality dimensions are highly related. Indeed, improving a specific metadata quality dimension could have an impact on other metadata quality aspects. Thus, data managers should know the existing dependencies between the metadata quality dimensions before improving metadata quality. Also, metadata quality dimensions are contextual and could be interpreted differently depending on the context of use and the metadata processing phase. It is worth noting that metadata quality assessment should not be limited to a particular phase of the metadata management process. Indeed, these metadata quality dimensions should be considered throughout the different metadata processing phases.

The next section presents the different phases of the metadata management process called "Metadata Value Chain". Then, we highlight how these metadata quality dimensions could be projected throughout the whole metadata value chain.

4 Metadata Value Chain in Big Data Context

The "Metadata Value Chain" (MDVC) is a process that describes the different phases that Metadata goes through within data-driven businesses and allows extracting Metadata's value. The MDVC presented below includes four stages: Metadata Collect, Metadata Storage, Metadata Provisioning, and Metadata Maintenance (Fig. 3).

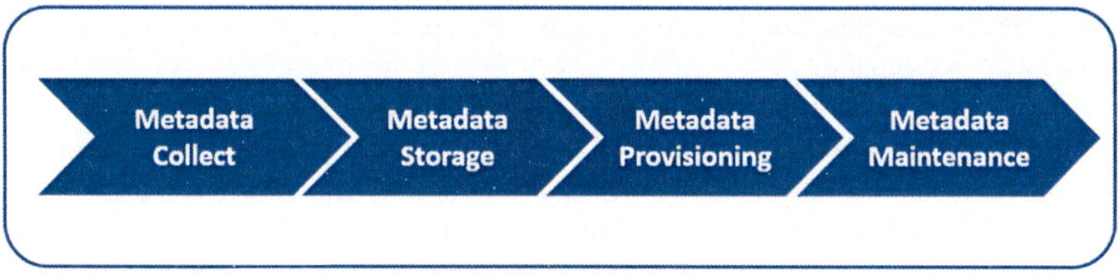

Fig. 3. Metadata value chain

4.1 Metadata Collect

This first phase consists of retrieving metadata information from different metadata repositories. In big data systems, the extracted Metadata are usually various (structured, unstructured) and of different types (text, data, etc.). Once retrieved, metadata are extended with additional information related to the collection process and the corresponding semantic layer. Also, other metadata are generated to support metadata processing. Even if most big data ingestion tools could be used for metadata collection, some technologies are more specific to Metadata, such as Apache Tika, Open Calais, or Apache Atlas, a widely used metadata framework [25].

After metadata extraction and generation, the obtained Metadata are of different types. Indeed, as metadata is used for several purposes, metadata types have evolved to support the whole business process. A brief explanation of the most common metadata' types follows:

- **Descriptive Metadata:** refers to Metadata used to describe their related objects. It gathers all information that identifies data, such as name, data type, version management information, etc.
- **Structural Metadata:** refers to Metadata used to describe how data is structured and formatted, such as Datawarehouse Metadata, HDFS metadata, etc.
- **Administrative Metadata:** refers to Metadata used to manage data such as acquisition information, accessibility rights, etc.

4.2 Metadata Storage

Once metadata are collected and extracted, they are stored in a separate database called "Metadata Repository". Metadata is stored in a data lake, a central repository that allows integrating data from multiple sources. Actually, for structured data, there are three typical structures according to which Metadata are stored and managed [9]:

- **Centralized Structure:** All metadata are stored in a central metadata repository. Hence, a uniform method for metadata access and management should be used across all metadata. Such a structure does not need a metadata exchange mechanism as all metadata are stored in a central database. However, it requires complex maintenance and slows the access speed.
- **Distributed Structure:** Metadata is stored in different local databases. This structure is usually adopted since it enables quick access to Metadata and uses other metadata management models and tools. However, this storage mode requires a

metadata exchange mechanism to ensure metadata interoperability and produce metadata redundancy.

- **Federal Structure:** This method was suggested to combine the advantages of the first two structures. In addition to the local databases, it uses a shared meta-database that provides a unified representation of metadata and manages metadata redundancy. However, exporting metadata from such a structure turns out to be more complex.

Hadoop Distributed File System (HDFS) is widely used for unstructured data to store large amounts of data and metadata. HDFS storage is a distributed storage system that provides a high-access performance. In HDFS, metadata are stored separately in a Namenode and are linked to Datanodes where data are stored.

4.3 Metadata Provisioning

This process makes the stored Metadata available for internal use, such as data analytics or data visualization processes. On the other hand, metadata could also be published for external users or consuming applications using specialized APIs such as open data portals. This process requires, however, formatting metadata using metadata publishing formats such as CWM (Common Warehouse Metamodel), XMI (XML Metadata Interchange), or OWL (Web Ontology Language).

4.4 Metadata Maintenance

It is the final stage among the essential MDVC phases. This stage aims to support all processes and changes related to Metadata while maintaining high-quality metadata. Metadata is used to store information related to any data processing. The stored Metadata is not static and is continuously updated and enriched with new entries. Thus, this phase is extended throughout the big data value chain and should be maintained as long as data are processed. Advanced metadata management tools are widely available for large data environments, such as Cloudera Navigator and Apache Atlas. These tools cannot be practical unless used as part of a metadata management strategy that involves the required resources and meets the business requirements.

It is worth noting that the Metadata Value Chain presented above is generic and comprises the metadata processing phases that are common to any metadata management process. Thus, other processing phases could be integrated for contextual reasons, such as business requirements or data types. It is also important to mention that metadata quality dimensions should be considered in each phase of the MDVC to ensure reliable and usable metadata quality. The following section highlights how the most common metadata quality dimensions could be projected throughout the different metadata value chain phases.

5 The Projection of the Metadata Quality Dimensions Throughout the Metadata Value Chain

This section highlights how metadata quality dimensions could be employed to support the different phases of the MDVC. For this, as shown in Table 1, we present the metadata quality dimensions that should be considered in each step of the MDVC. Moreover, some solutions to overcome the raised quality issues are suggested.

Table. 1. The projection of the metadata quality dimensions throughout the metadata value chain

MDVC phase	Metadata quality dimensions	Explanation
Collect	Accuracy	Metadata accuracy is the first quality aspect that should be inspected when collecting Metadata. Indeed, non-accurate entries do not add any value to the analysis process and may even bias the data analytics results. In a big data context, metadata should be accurate in two senses: **Accurate Provenance:** The collected Metadata are usually coming from multiple sources that are not always reliable. Thus, before collecting data, it is necessary to ensure the credibility of the metadata provenance. Therefore, information like which organization the Metadata comes from, how well they are experts on metadata standards and classifications, and what transformations were applied to Metadata should be considered when assessing metadata accuracy **Accurate Content:** On the other hand, metadata should accurately describe their related objects. This could be ensured using some data inspection methods such as sampling and profiling techniques
	Usefulness	In big data environments, metadata does not contain only the basic description of data. Metadata is usually used to store information related to data storage, data processing phases, and even some data analytics results. Also, metadata may contain some information related to the data's original context that is not relevant to the current environment. Hence, the collected Metadata should be cleansed and filtered to keep only the relevant information. Several methods could be used to clean metadata, such as parsing, outlier detection techniques, and statistical clustering. Also, the collected Metadata may contain some redundancies as they are gathered from different sources. This could be overcome using metadata mapping techniques that allow aligning different metadata models
	Completeness	Once metadata are collected, they should be completed with the appropriate information related to the metadata source and the collection process. Some missing data was not specified in the original context because they are considered assumed knowledge or not applicable [26]. Thus, metadata should be completed in terms of metadata attributes on the one hand (ensuring that the Metadata contains all the required features) and metadata values on the other hand (ensuring that metadata values provide all the needed information). In [17], some guiding questions and good practices that would help data managers to improve metadata completeness for big data are suggested. Among these good practices, we can cite determining the critical metadata elements that should be complete, attributing a completeness score to these metadata elements, evaluating the impact of low metadata completeness on metadata usability, etc.
	Timeliness	The collection of Metadata is not necessarily synchronized with the collection of data. Some metadata takes time to generate, describe, and catalog, especially when human intervention is involved [2]. This asynchronism may significantly impact the metadata management process, especially when Metadata holds critical information or when time-sensitive analysis should be performed. In such a case, assessing the risks associated with metadata unavailability should be performed before collecting Metadata

(*continued*)

Table. 1. (*continued*)

MDVC phase	Metadata quality dimensions	Explanation
Storage	Consistency	One of the most common metadata quality issues faced in the storage phase is metadata inconsistency. As the metadata sources use different metadata models, the gathered metadata are usually of other formats that can create anomalies, especially when all Metadata is stored in the same repository. Besides, inconsistency may be related to metadata values that do not use the same formats or range of values (e.g., inconsistencies with the design of date and time attributes). Thus, uniform metadata models and standards should be determined and harmonized for all metadata. Also, a conflict analysis mechanism should be implemented to estimate the impact of such anomalies on the potential use of data
	Interoperability	Metadata that is stored in a distributed way generally uses different metadata models and schemas. Thus, to ensure interoperability between metadata repositories, a metadata exchange mechanism should be put in place. CWM is commonly used to enable metadata exchange in distributed heterogeneous repositories. However, using distributed metadata storage raises new challenges that metadata models cannot overcome, such as metadata inconsistencies and redundancy. To deal with these issues, some authors have suggested new architectures that provide unified frameworks [27, [28] to manage distributed Metadata and enable effective metadata interoperability
Provisioning	Shareability	Even if Metadata is of high quality within a local repository, metadata may lose its quality once distributed in another environment. Thus, «Shareability» has been introduced as a new metadata quality dimension characterizing Metadata that maintains a high quality and could be effectively used out of its local repository. To ensure metadata shareability, metadata should be associated with knowledge entities that support its elucidation. This could be achieved using RDF annotations, especially ontologies representing Metadata in a readable and machine-processable format
	Discoverability	One of the main challenges faced by data managers concerns finding the needed information. The reason why data could not be easily located is generally related to the fact that Metadata is not structured and specified to make data discoverable. To enhance metadata discoverability, data managers should ensure that Metadata are appropriately linked to their related objects and are adequately documented. Also, it is highly recommended to use ontologies, which makes Metadata, and thus data, more likely to be discovered, read and used. Moreover, metadata should use controlled vocabulary and be linked to semantic annotations to improve semantic search visibility
	Accessibility	Ensuring metadata accessibility is a high priority since inaccessible and unreadable Metadata could not be of great value. As we have mentioned earlier, Metadata is not only used for internal purposes. Indeed, metadata could be published for external users or consuming applications such as open data portals. Thus, depending on the intended use of Metadata (internal or external access), data providers should manage access permissions to provide the appropriate level of access to Metadata without impacting metadata security and usability

(*continued*)

Table. 1. (*continued*)

MDVC phase	Metadata quality dimensions	Explanation
Maintenance	Timeliness	Its high variability usually characterizes big data. Thus, metadata should be continuously updated to maintain data's timeliness. Most of the available metadata management tools for big data allow propagating data updates on the related Metadata. However, not all metadata updates could be automated since some metadata are more connected to the business process and need human monitoring to be managed
	Versionnability	Metadata versioning refers to creating new metadata models when significant changes are applied to Metadata. It is essential to mention that, even if a new metadata version is adopted, the previous versions must be archived and made available within the data system for many reasons, such as recovering metadata from older versions in case of anomalies, tracking metadata upgrades, enhancing a previous version, comparing metadata versions, etc. Thus, a version control system must be implemented to provide a shared workspace to handle metadata versionnability and manage metadata update conflicts

6 Conclusion

Ensuring metadata quality is of great importance, especially for big data, since most big data tools and processes highly use metadata. This work's purpose is to highlight how metadata quality dimensions could be applied to the metadata management process in a big data environment. First, we have presented the metadata management process's different phases: metadata collection, metadata storage, metadata provisioning, and metadata maintenance. Then, we have highlighted the metadata quality dimensions that should be considered in each phase of the metadata management process. Finally, we have shown the importance of metadata governance for metadata management, and we have presented some good practices to achieve successful metadata governance. As future work, we aim to suggest a metadata governance framework for big data environments, which will provide optimized big data management and lead to better decision-making.

References

1. Immonen, A., Paakkonen, P., Ovaska, E.: Evaluating the quality of social media data in big data architecture. IEEE Access **3**, 2028–2043 (2015)
2. Bruce, T., Hillmann, D.: The continuum of metadata quality: defining, expressing, exploiting. ALA Ed., January 2004
3. Holom, R.-M., Rafetseder, K., Kritzinger, S., Sehrschön, H.: Metadata management in a big data infrastructure. Proc. Manuf. **42**, 375–382 (2020)
4. Kapil, G., Agrawal, A., Khan, R.: A study of big data characteristics. In: International Conference on Communication and Electronics Systems, p. 4, October 2016

5. Alaoui, I.E., Gahi, Y., Messoussi, R.: Full consideration of big data characteristics in sentiment analysis context. In: 2019 IEEE 4th International Conference on Cloud Computing and Big Data Analysis (ICCCBDA), Chengdu, China, April 2019
6. Alaoui, I.E., Gahi, Y.: The impact of big data quality on sentiment analysis approaches. Proc. Comput. Sci. **160**, 803–810 (2019)
7. El Alaoui, I., Gahi, Y., Messoussi, R.: Big data quality metrics for sentiment analysis approaches. In: Proceedings of the 2019 International Conference on Big Data Engineering (BDE 2019) - BDE 2019, Hong Kong, Hong Kong, pp. 36–43 (2019)
8. Dong, R., Su, F., Yang, S., Xu, L., Cheng, X., Chen, W.: Design and application on metadata management for information supply chain. In: 2016 16th International Symposium on Communications and Information Technologies, Qingdao, China, pp. 393–396, September 2016
9. Xiao, B., Zhang, C., Mao, Y., Qian, G.: Review and exploration of metadata management in data warehouse. In: 2015 IEEE 10th Conference on Industrial Electronics and Applications (ICIEA), Auckland, New Zealand, pp. 928–933, June 2015
10. Golosova, M., Aulov, V., Kaida, A.: Metadata handling for Big data projects. J. Phys. Conf. Ser. **1117**, 012007 (2018)
11. Megdiche, I., Ravat, F., Zhao, Y.: Metadata management on data processing in data lakes. In: Bureš, T., et al. (eds.) SOFSEM 2021. LNCS, vol. 12607, pp. 553–562. Springer, Cham (2021). https://doi.org/10.1007/978-3-030-67731-2_40
12. Smith, K., et al.: 'Big metadata': the need for principled metadata management in big data ecosystems. Presented at the Proceedings of Workshop on Data analytics, June 2014
13. Shankar: Big data - news, views and reviews: framework for metadata management for big data analytics. Infoivy, April 2013. http://www.infoivy.com/2013/08/framework-for-metadata-management-for.html
14. Reiche, K.J., Hofig, E.: Implementation of metadata quality metrics and application on public government data. In: 2013 IEEE 37th Annual Computer Software and Applications Conference Workshops, Japan, pp. 236–241, July 2013
15. Ochoa, X., Duval, E.: Automatic evaluation of metadata quality in digital libraries. Int. J. Digit. Libr. **10**, 67–91 (2009)
16. Kiraly, P., Buchler, M.: Measuring completeness as metadata quality metric in Europeana. In: 2018 IEEE International Conference on Big Data (Big Data), Seattle, WA, USA, pp. 2711–2720, December 2018
17. The UNECE Big Data Quality Task Team: A Suggested Framework for the Quality of Big Data, December 2014
18. Rousidis, D., Garoufallou, E., Balatsoukas, P., Sicilia, M.-A.: Metadata for big data: a preliminary investigation of metadata quality issues in research data repositories. Inf. Serv. Use **34**(3–4), 279–286 (2014)
19. Király, P.: Measuring metadata quality. Ph.D. thesis, Georg-August, Gottingen (2019)
20. McMahon, C., Denaxas, S.: A novel framework for assessing metadata quality in epidemiological and public health research settings, vol. 2016, pp. 199–208, July 2016
21. Comparison of metadata quality in open data portals using the Analytic Hierarchy ProcessScienceDirect. https://www.sciencedirect.com/science/article/abs/pii/S0740624X16301319. Accessed 05 Mar 2021
22. Király, P.: Towards an extensible measurement of metadata quality. Presented at the Proceedings of the 2nd International Conference on Digital Access to Textual Cultural Heritage, June 2017
23. Sice, P., Elvin, G., Riachy, C., Shang, Y., Ogwu, S., Zink, C.: Online screening of X-System music playlists using an integrative wellbeing model informed by the theory of autopoiesis. IEEE Access **8**, 182307–182319 (2020)

24. Is 'Quality' Metadata 'Shareable' Metadata? The Implications of Local Metadata Practices for Federated Collections
25. Sawadogo, P., Darmont, J.: On data lake architectures and metadata management. J. Intell. Inf. Syst. **56**(1), 97–120 (2020). https://doi.org/10.1007/s10844-020-00608-7
26. Tani, A., Candela, L., Castelli, D.: Dealing with metadata quality: the legacy of digital library efforts. Inf. Process. Manag. **49**(6), 1194–1205 (2013)
27. Sui, Z., Tian, Y., Wang, X., Wu, L.: A framework for distributed metadata management of mineral information resources with access control. In: 2013 21st International Conference on Geoinformatics, Kaifeng, China, pp. 1–4, June 2013
28. Xue, W., Zhu, M.: Efficient dynamic management of distributed metadata. In: Zhu, R., Ma, Y. (eds.) Information Engineering and Applications. LNEE, vol. 154, pp. 354–362. Springer, London (2012). https://doi.org/10.1007/978-1-4471-2386-6_46

Weather Forecast Using Sliding Window Algorithm Based on Hadoop and MapReduce

Hamza Jdi(✉) and Noureddine Falih

Polydisciplinary Faculty Sultan, Moulay Slimane University, Beni Mellal, Morocco

Abstract. Weather has a huge impact on every day human life like planning outdoor events, agriculture, aviation, etc. It is very important to apprehend it in every way possible and be able to predict it as accurately as possible. In recent years, we have accumulated huge amounts of data, which make it presumptive to extract as much information as possible. In this paper, we propose an approach based on Hadoop and MapReduce technologies using Sliding Window Algorithm to forecast the weather in the region of Beni Mellal-Khenifra.

Keywords: Big data · Data analytics · Hadoop · MapReduce · Weather forecasting · Sliding window algorithm

1 Introduction

The increase of data produced by the various types of observational sensors and devices made storing, exploiting and extracting information from data an extremely challenging task [1, 2], which makes it imperative to explore new technologies other than the traditional databases. The prediction of weather conditions is key in different applications and sectors like agriculture. Hadoop [3] offers new ways of storing data by using Hadoop Distributed File System (HDFS) [4, 5] that allows data to be stored and accessed in parallel [6]. The data used is from the National Climatic Data Center concerning the city of Beni Mellal located in Morocco. The dataset contain various meteorological elements, which are perfect to analyze with Hadoop and MapReduce [7, 8]. In this work, we are particularly interested in temperature and wind speed which are important elements when it comes to current weather and predicting future forecasts.

The weather conditions of a given day have a correlation with the weather conditions present in the previous week and previous year. That is the reason behind using the Sliding Window algorithm to predict weather conditions and exploiting past week and 14 days is the same period of the previous year.

The Aim and Objective of the Project is to predict a day's temperature and wind speed using the previous year and previous week data with the help of Hadoop and MapReduce.

Y. Maleh et al. (Eds.): ICI2C 2021, LNNS 357, pp. 122–132, 2022.
https://doi.org/10.1007/978-3-030-91738-8_12

This endeavor signals the start of a new era where technologies are combined and used to make such predictions. Its benefits are numerous as they stretch all the way from planning to making real time adjustments to adapt choices in any given field and minimize if not eliminate possible consequences.

2 Related Work

[9] The authors propose a module for weather temperature prediction composed of three parts, initial training of data and information by data recuperation using data mining concepts and techniques such as Bayesian network, K means, Anomaly or Outlier detection, Clustering analysis. After obtaining the data and information form the first part, training of hybrid approach, a combination of supervised learning and neural networks (unsupervised learning) takes place. Finally, the prediction of weather is achieved.

The result shows an improved accuracy of predicted weather with the hybrid model based on both supervised and unsupervised learning compared to supervised and unsupervised learning used separately.

[10] The authors published a model for weather forecasting using recurrent neural network with the help of Long Short–Term Memory (LSTM) algorithm.

The following is the presented model:

Collection of weather parameters such as temperature, dew point, wind speed and pressure followed by data pre-processing which is the part where unwanted data is removed. Subsequently, LSTM is applied and the training of the model using weather parameters takes place. Before the model gives the predicted weather output, testing occurs.

[11] The authors used Fuzzy K-nn Algorithm to predict one-day visibility of the weather using data from World Meteorological Organization relying on Case-Based Reasoning (CBR). The proposed algorithm compute the degree of similarity to find similar cases from a case base. The authors chose to make one-day visibility prediction that makes it hard to evaluate the overall performance of the algorithm.

[12] The authors present an interesting system that forecasts weather inspired by biology which is the artificial immune system (AIS). This system has three stages. The first stage is antigenic presentation. The second stage is where the implementation and design of the immune learning algorithm take place. The third and final stage is where we get the output of the actual weather forecast results. To provide comparisons, the authors build a traditional artificial neural network (ANN). The results are very promising and they are in favour of the forecasting system based on AIS with an accuracy rate that goes as high as 95.1%.

[13] In this paper, we take a look at regional-based temperature prediction concerning the following districts Kerala-Malappuram, Calicut, and Thrissur, based on various machine learning models namely Multiple Linear Regression (MLR), Artificial Neural Network ANN, and Support Vector Machine SVM. To compare these models the authors used data collected from central Kerala, the result shows that MLR and regression-based SVM prove to be more accurate.

[14] We have in hand a numerical weather prediction (NWP) using Datamining algorithm C5.0 with K-means clustering applied on meteorological datasets gathered from the MERRA database from 1980 to 2020. To predict the following weather parameters temperature, relative humidity, pressure, wind speed, and rainfall the obtained results are compared taking into consideration accuracy, sensitivity, and specificity with the following machine learning methods: Naïve Bayes (NB), C4.5, GA, ANN, SVM, LSTM, and RBF techniques.

The results show that C 5.0 with K-means is more accurate, sensitive, and specific than the other compared techniques.

3 Proposed Approach

3.1 Methodology

In this project, we gathered data from National Climatic Data Center (NCDC) concerning the city of Béni Mellal station ID 60191099999 from 2000 to 2020. It has day-to-day data that consists of fields like dates, station id and parameters like temperature, wind speed, sea-level pressure, total precipitation and more.

A one-day weather condition forecast is correlated with the previous week and the previous year in a period that has to be determined. To predict a chosen day's temperature, we will mathematically model [15] the dependencies that consist of deploying one week before the chosen day and 14 days in the same period of the previous year.

If the temperature of July 20, 2019 is to be predicted, the max temperature, min temperature, and wind speed of the previous week as well as 14 days of the same period from the previous year shall be used. That is to say, the period between July 13, 2019 and July 19, 2019 and 14 days of the previous year (from July 6, 2018 to July 19, 2018), and that's when Sliding Window algorithm comes in handy as it will help us choose the best matching week in the previous year.

3.2 Sliding Window Concept

The goal is to find the temperature and wind speed of a specific day using a historical dataset that contains weather parameters. We particularly are interested by maximum temperature, minimum temperature and the wind speed. After choosing the desired day to predict, we extract the previous week from our dataset and store it in an array that has the size of 3×7 that we call current year data (CD), and we extract the previous 2 weeks of the previous year and we store them in an array that has the size of 3×14 called previous year's data "PD". We divide (PD) to 8 windows (W1, W2, W3, W4, W5, W6, W7, W8) as shown in the Fig. 1.

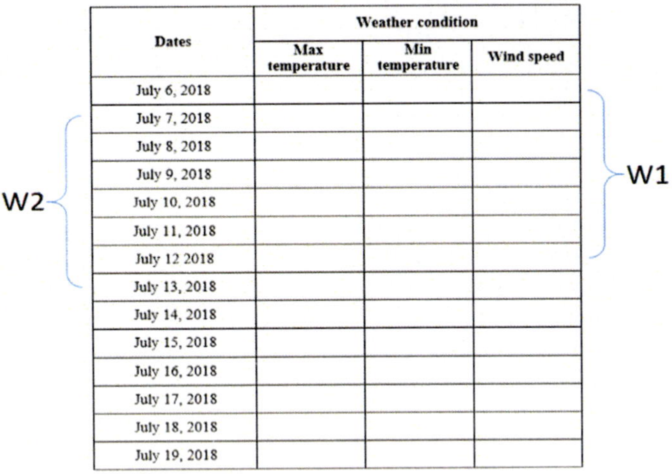

Fig. 1. Sliding window concept

NB: *W*1 represents Window number 1 and *W*2 represent window number 2.

We compute the Euclidean distance between each and every window and compare it to our CD array. We choose the best matching window, meaning the window with the minimal Euclidean distance so we can use it for weather prediction. Afterwards, we proceed to calculate PD variation vector and store it in an array called VP and do the same for variation vector of the current year called VC. VP and VC both has the size of 3×1. The next step would be to calculate the mean of both VC and VP then calculate the predicted variation using the following formula:

$$V = (Mean1 + Mean2)/2$$

V: predicted variation

Mean1: the mean of VC

Mean2: the mean of VP

Finally, we add the predicted variation to the weather conditions of the previous day in order to get the forecast.

The following steps illustrate the algorithm [16].

- Step 1. Take matrix "CD" of last seven days for current year's data of size 7×4.
- Step 2. Take matrix "PD" of fourteen days for previous year's data of size 14×4.
- Step 3. Make 8 sliding windows of size 7×4 each from the matrix "PD" as *W*1,2, *W*3,…, *W*8.
- Step 4. Compute the Euclidean distance of each sliding window with the matrix "CD" as ED1, ED2, ED3, …, ED8.
- Step 5. Select matrix *Wi* as *Wi* = Corresponding Matrix (Min.(ED*i*)) $\forall i \in [1, 7]$.

- Step 6. For k = 1 to n.
    - (i) For WC*k* compute the variation vector for the matrix "CD" of size 6 × 1 as "VC".
    - (ii) For WC*k* compute the variation vector for the matrix "PD" of size 6 × 1 as "VP".
    - (iii) Mean1 = Mean (VC)
    - (iv) Mean2 = Mean (VP) Predicted Variation
    - (v) "*V*" = (Mean1 + Mean2)/2
    - (vi) Add "*V*" to the previous day's weather condition in consideration to get the predicted condition.
- Step 7. End.

4 Experimental Setup

In this project, we used the latest version of Hadoop 3.2.2 on a single-node in a pseudo-distributed mode where each Hadoop daemon runs in a separate Java process installed under Ubuntu operating system version 20.04.2.

4.1 Hadoop Deamons

Hadoop Daemons [17, 18] are a set of processes that run on Hadoop. Hadoop is a framework written in Java, so all these processes are Java Processes.

Hadoop consists of the following Daemons:

- NameNode works on the Master System. The primary purpose of Namenode is to manage all the MetaData. Metadata is the list of files stored in our HDFS (Hadoop Distributed File System).
- DataNode runs on the Slave system. The NameNode directs DataNode for storing the data. DataNode is a program run on the slave system that attends the client read/write
- Secondary NameNode is designed to take the hourly backup of the data so in case of damage the secondary Namenode will secure and save the hourly backup and transfer it to fsimage. the MetaData then is assigned to a new system to create a new Master to allow the cluster to start running again
- Resource Manager [19] is also known as the Global Master Daemon that works on the Master System.
- The Node Manager run on the Slaves System that supervises the memory resource inside the Node and Memory Disk.

4.2 MapReduce

MapReduce is a framework for processing large amounts of data in parallel using a large number of computers (nodes). The main idea of MapReduce is to break down data processing problems into small portions to allow the available nodes to work on them in parallel. Processing can occur on data stored either in a file system (unstructured) or in a database (structured). To minimize communication overhead MapReduce processes data near the place it is stored. The following are the three operations needed in order to execute a MapReduce framework: Map: each worker node applies the map function to the local data, and writes the output to a temporary storage. A master node ensures that only one copy of the redundant input data is processed. Shuffle: worker nodes redistribute data based on the output keys (produced by the map function), so that all data belonging to one key is located on the same worker node. Reduce: worker nodes now process each group of output data, per key, in parallel.

MapReduce is designed to work with of HDFS. Apache Hadoop automatically optimizes the execution of MapReduce programs so that a given Map or Reduce step is run on the HDFS node that contains locally the blocks of Beni Mellal weather data of 2019 and also where we save the output file at the end of the process.

After going through each component separately now will combine and explain how it works together first we compile MapReduce code and export the jar then execute the MapReduce code in the Ubuntu command line and specify the input and output directory.

5 Results and Discussion

Sliding Window algorithm has been used to compute weather conditions using Beni Mellal weather data collected from National Climatic Data Center (NCDC). The algorithm has been run with the help of Hadoop/MapReduce, to predict one year weather conditions, that is, maximal temperature, minimal temperature and wind speed in the year of 2019. The Table 1 shows the month wise average accuracy of predicted weather conditions.

Table 1. Average accuracy of predicted weather condition for the year of 2019

Months	Average accuracy of predicted weather condition		
	Max temperature	Min temperature	Wind speed
January	97.86	97.51	93.15
February	89.79	89.01	85.65
March	93.05	92.91	88.84
April	97.16	97.24	92.91
May	90.95	91.14	85.43
June	98.95	98.63	95.94
July	91.65	91.26	87.84
August	80.06	79.95	75.76
September	79.63	79.93	74.72
October	95.11	95.03	90.49
November	99.08	99.01	97.08
December	96.95	96.80	91.42

Please note that temperatures are in (Fahrenheit), wind speed in (.1 knots), and the accuracy of prediction is in percentage.

After studying and analyzing the three weather conditions taken into consideration (maximum temperature, minimum temperature and wind speed). With the help of Hadoop/MapReduce for the whole year of 2019, the results are compared with the existing weather data to calculate the accuracy of the data predicted as shown in Table1.

We remark the following:

- The average accuracy of predicted weather condition are the highest and lowest in the months of November and September respectively.
- The average accuracy for temperatures is higher than to wind speed accuracy.

We choose to provide detailed results for the two months that are characterized by the highest and lowest average accuracy, the months of November and September respectively, by showing one random week of each month -From the weeks of which the accuracy was verified- for the three predicted weather conditions to verify the consistency of the results shown in Table 1 concerning the two findings.

The chosen two weeks are the following:

- The first week is in November in the period spanning from November 10, 2019 to November 16, 2019.
- The second week is in September in the period spanning from Septembre7, 2019 to September 13, 2019.

5.1 Maximal Temperature Prediction

Table 2 and Table 3 show the day wise maximal temperatures predicted, actual temperatures and the accuracy of the prediction for each of the two selected weeks of the year 2019.

Table 2. Predicted, actual maximal temperature and accuracy of a week of November

Dates	Max temperature		
	Predicted	Actual	Accuracy of the prediction
November 10, 2019	63,756	64,40	99,0
November 11, 2019	63,724	64,40	99,0
November 12, 2019	69,207	69,80	99,2
November 13, 2019	71,027	71,60	99,2
November 14, 2019	66,742	68,00	98,2
November 15, 2019	56,628	57,20	99,0
November 16, 2019	54,946	55,40	99,2

Table 3. Predicted, actual maximal temperature and accuracy of a week of September

Dates	Max temperature		
	Predicted	Actual	Accuracy of the prediction
September 7, 2019	69,915	87,80	79,6
September 8, 2019	68,484	87,80	78,0
September 9, 2019	70,694	89,60	78,9
September 10, 2019	66,349	86,00	77,2
September 11, 2019	65,550	84,20	77,9
September 12, 2019	64,766	82,40	78,6
September 13, 2019	60,160	75,20	80,0

5.2 Minimal Temperature Prediction

Table 4 and Table 5 shows the day wise minimal temperatures predicted, actual temperatures and the accuracy of the prediction for each of the two selected weeks of the year 2019.

Table 4. Predicted, actual minimal temperature and accuracy of a week of November

Dates	Min temperature		
	Predicted	Actual	Accuracy of the prediction
November 10, 2019	42,137	42,80	98,5
November 11, 2019	45,913	46,40	99,0
November 12, 2019	53,037	53,60	99,0
November 13, 2019	42,509	42,80	99,3
November 14, 2019	43,775	44,60	98,2
November 15, 2019	47,412	48,20	98,4
November 16, 2019	43,837	44,60	98,3

Table 5. Predicted, actual minimal temperature and accuracy of a week of September

Dates	Min temperature		
	Predicted	Actual	Accuracy of the prediction
September 7, 2019	51,735	66,20	78,2
September 8, 2019	49,448	62,60	79,0
September 9, 2019	53,727	68,00	79,0
September 10, 2019	52,000	66,20	78,6
September 11, 2019	49,485	62,60	79,1
September 12, 2019	47,515	60,80	78,2
September 13, 2019	46,545	59,00	78,9

5.3 Wind Speed Prediction

Table 6 and Table 7 shows the day wise wind speed predicted, actual wind speed and the accuracy of the prediction for each of the two selected weeks of the year 2019.

Table 6. Predicted, actual wind speed and accuracy of a week of November

Dates	Wind speed		
	Predicted	Actual	Accuracy of the prediction
November 10, 2019	5,926	6,10	97,2
November 11, 2019	3,694	3,80	97,2
November 12, 2019	5,018	5,20	96,5
November 13, 2019	3,394	3,50	97,0
November 14, 2019	6,221	6,40	97,2
November 15, 2019	8,860	9,20	96,3
November 16, 2019	5,386	5,60	96,2

Table 7. Predicted, actual wind speed and accuracy of a week of September

Dates	Wind speed		
	Predicted	Actual	Accuracy of the prediction
September 7, 2019	3,577	4,90	73,0
September 8, 2019	2,863	3,90	73,4
September 9, 2019	4,441	6,00	74,0
September 10, 2019	4,894	6,60	74,2
September 11, 2019	2,156	2,90	74,3
September 12, 2019	3,006	4,00	75,2
September 13, 2019	3,745	5,00	74,9

5.4 Discussion

Table 1 shows the average accuracy of predicted weather conditions for the full year of 2019. Moreover, Tables 2, 3, 4, 5, 6 and 7 give detailed predictions about maximum temperature, minimum temperature and wind speed, as well as accuracy of predictions on a daily basis regarding the chosen two weeks. The readings further confirm the remarks concluded from Table 1.

In the month of November, the average accuracy for both temperature and wind speed is very high relative to the month of September, and that is understandable because of seasonal change where conditions are very unpredictable. If we compare the average accuracy predicted between the temperature and wind speed, we remark that temperature prediction accuracy is very high and that is true in general because temperature prediction is more accurate than any weather parameter.

In regard to the average accuracy of predicted maximum temperature the highest (99.08%) and lowest (89.79) are in the months of November and September respectively. We get relatively the same result concerning the average accuracy of predicted minimum temperature with the highest (99.01%) and lowest (79.93%) in the months of November and September respectively with respect to wind speed yet again the average accuracy of predicted wind speed is the highest (97.08%) and lowest (74.72%) in the months of November and September.

The results are not at all surprising as the sliding window algorithm is more accurate and consistent if the month in hand is in line with the previous month in the past year, meaning we can easily find a window that matches the current window which translates to better accuracy. Furthermore, to improve the accuracy of any month we can add more days to our window which in turn proves the chances of finding a match in the same month of the previous year.

6 Conclusion

The Sliding Window Algorithm proved to be very accurate in predicting weather conditions especially for temperatures even though the prediction in the month of seasonal change, where conditions are very unpredictable, can be less accurate. In future work, we will work with concepts of Machine Learning and Distributed Learning using Apache Spark to solve the low accuracy issue of the months with low accuracy such as September and August.

References

1. Kaur, A.: Big data: a review of challenges, tools and techniques. IJSRSET **2**(2), 1090–1093 (2016)
2. Michael, K., Miller, K.: Big data: new opportunities and new challenges [guest editors' introduction]. Computer **46**, 22–24 (2013). https://doi.org/10.1109/MC.2013.196
3. Holmes, A.: Hadoop in Practice, pp. 4–9 (2014)
4. Shvachko, K., Kuang, H., Radia, S., Chansler, R.: The hadoop distributed file system. In: 2010 IEEE 26th Symposium on Mass Storage Systems and Technologies (MSST), pp. 1–10. IEEE (2010)
5. Borthakur, D.: The hadoop distributed file system: architecture and design. Hadoop Proj. Website **11**, 21 (2007)
6. Lee, K.-H., Lee, Y., Choi, H., Chung, Y., Moon, B.: Parallel data processing with MapReduce: a Survey. SIGMOD Rec. **40**, 11–20 (2011). https://doi.org/10.1145/2094114.2094118
7. Dean, J., Ghemawat, S.: MapReduce: simplified data processing on large clusters. Commun. ACM **51**, 137–150 (2004). https://doi.org/10.1145/1327452.1327492
8. Dean, J., Ghemawat, S.: MapReduce: a flexible data processing tool. Commun. ACM. **53**, 72–77 (2010). https://doi.org/10.1145/1629175.1629198
9. Arunachalam, N., Giles, G., Raghunath, R., Kaviyarasan, V.: A hybrid approach model for weather forecasting using multi-task agent. In: 2015 2nd International Conference on Electronics and Communication Systems (ICECS), pp. 1675–1678 (2015). https://doi.org/10.1109/ECS.2015.7124871

10. Fente, D., Singh, D.: Weather forecasting using artificial neural network, pp. 1757–1761 (2018). https://doi.org/10.1109/ICICCT.2018.8473167
11. Li, K., Liu, Y.-S.: Fuzzy case-based reasoning: weather prediction. In: Proceedings of the International Conference on Machine Learning and Cybernetics, vol. 1, pp. 107–110 (2002). https://doi.org/10.1109/ICMLC.2002.1176719
12. Xu, C., Li, T., Huang, X., Jiang, Y.: A weather forecast system based on artificial immune system. In: Wang, L., Chen, K., Ong, Y.S. (eds.) ICNC 2005. LNCS, vol. 3611, pp. 800–803. Springer, Heidelberg (2005). https://doi.org/10.1007/11539117_112
13. Anjali, T., Chandini, K., Anoop, K., Lajish, V.L.: Temperature prediction using machine learning approaches. In: 2019 2nd International Conference on Intelligent Computing, Instrumentation and Control Technologies (ICICICT), pp. 1264–1268 (2019). https://doi.org/10.1109/ICICICT46008.2019.8993316
14. Murugan Bhagavathi, S., et al.: Weather forecasting and prediction using hybrid C5.0 machine learning algorithm. Int. J. Commun. Syst. **34**(10) (2021). https://doi.org/10.1002/dac.4805
15. Lai, L.L., Braun, H., Zhang, Q.P., et al.: Intelligent weather forecast. In: Proceedings of the International Conference on Machine Learning and Cybernetics, Shanghai, China, pp. 4216–4221, August 2004
16. Tsihrintzis, G.A., Hwang, W.-L., Kapoor, P., Bedi, S.S.: Weather forecasting using sliding window algorithm, p. 3 (2013)
17. White, T.: Hadoop the Definitive Guide, pp. 253–259. O'Reilly Media (2009)
18. Ghazi, M., Gangodkar, D.: Hadoop, MapReduce and HDFS: a developers perspective. Proc. Comput. Sci. **48**, 45–50 (2015). https://doi.org/10.1016/j.procs.2015.04.108
19. Ankam, V.: Big Data Analytics, pp. 24–26. Packt Publishing Ltd. (2016)

Semantic Web Technologies for Internet of Things Semantic Interoperability

Fatima Zahra Amara[1(✉)], Mounir Hemam[1], Meriem Djezzar[1], and Moufida Maimour[2]

[1] Abbes Laghrour University of Khenchela, Khenchela, Algeria
{f.amara,hemam.mounir,meriem.djezzar}@univ-khenchela.dz
[2] Université de Lorraine, CNRS, CRAN, 54000 Nancy, France
moufida.maimour@univ-lorraine.fr

Abstract. The rapid development of the Internet of Things has transformed our daily life by reaching all areas that we can imagine and does not stop evolving. It exchanges any information at anytime from anywhere. The Internet of Things is a set of physical objects carrying embedded technologies to gather data from their environment. The huge amount of raw data produced by different IoT devices is heterogeneous with different types and formats. As a result, sharing and reusing raw IoT data is a challenging task for IoT applications. The lack of interoperability in IoT systems creates a difficult issue that prevents them from functioning properly. To overcome this problem, data modelling and knowledge representation with semantic web technologies can be an adequate solution. In this paper we present a survey on the state of the art of current solutions to the problem of semantic interoperability in IoT systems and how semantic web technologies can be used to achieve the conceptual Internet of Things.

Keywords: Internet of Things · Semantic web technologies · Ontology · Linked data · Semantic web of things

1 Introduction

The Internet of Things (IoT) devices are growing rapidly due to the high speed of Internet connectivity and the rapid improvement in wireless technologies. Current statistics predict that there will be around 41 billion IoT devices by 2027 [10], with irregular growth because of the dynamic and pervasive nature of the IoT. This massive number of interconnected and heterogeneous IoT devices gather, generate, and share a massive volume of IoT data streams while also communicating autonomously without human intervention. As a consequence, data is proliferating, so the presence of heterogeneity between them is inevitable.

The concept of the "Internet of Things" implies that all things are harmoniously interconnected so that they can communicate and are also easily accessible over the Internet to provide services to end users [20]. Therefore, the interest and enormous IoT raw data that is on the web and the large number of interconnected things, lead to scalability, heterogeneity and various interoperability

Y. Maleh et al. (Eds.): ICI2C 2021, LNNS 357, pp. 133–143, 2022.
https://doi.org/10.1007/978-3-030-91738-8_13

problems [31]. This creates a range of problems that can be considered as a major obstacle to the visionary IoT and its complete implementation and deployment in our everyday lives. This means that providing interoperability between the "things" of the IoT is one of the most fundamental requirements to support the addressing, tracking and discovery of objects as well as the representation, storage and exchange of information [3].

Several researchers propose the use of semantic web technologies as an appropriate solution for providing semantic interoperability in Internet of Things systems. The World Wide Web Consortium (W3C) has proposed a set of standards for the semantic web. Semantic web technologies allow an explicit description of things and their data collected from various sources in different formats. As a result, semantics provide a common understanding of the meaning of data and facilitate its integration by converting heterogeneous data into a reliable and usable knowledge base using the same vocabulary. This allows for unambiguous exchange and easy integration of data that come from different IoT sources. Also, IoT data can be published over the Internet which permits its enrichment and reuse. As a result, new innovative IoT applications and systems can be developed. In this paper, we look at how integrating semantic web technologies into Internet of Things systems can help to deal with the lack of semantic interoperability and achieve the true vision of IoT. This paper is organised as follows. After the introduction we present the background concepts along with the basic use of semantics in the IoT. Afterward, we present a set of ontologies and frameworks that are being developed in this context. We conclude the paper with the identification of open issues and future directions for researchers to work on in this field.

2 Background

2.1 Semantic Web Technologies

The Semantic Web was introduced by the authors of [8] in May 2001. It was introduced as an expansion of the traditional web in which data has a well-defined meaning to allow people and machines to collaborate. It provides a common structure and tools for manipulating, sharing, reusing, and integrating information from disparate data sources. Several components, such as languages and knowledge representation formalisms, have been identified through various research works to achieve the Semantic Web approach. They were developed by the W3C as part of the Semantic Web:

Resource Description Framework (RDF) [4] is the first W3C standard for publishing and linking data to enrich resources on the Web. It provides a mechanism to describe things with an RDF triple that makes a relationship between its identifier and other web resources.

RDF Schema (RDFS) [9] offers a higher level of abstraction by providing a data modeling vocabulary for the RDF data. It aims to define metadata schemes. It defines the meaning, characteristics, and relationships of a set of properties.

Ontology Web Language (OWL) [13] is designed to represent complex knowledge and relationships between things. It describes and represents a specific domain of knowledge by defining classes of resources or objects and their relationships. It defines ontologies by expanding the vocabulary and properties of RDF.

SPARQL Protocol And RDF Query Language (SPARQL) [22] is a standardized query language published by the W3C in 2013, based on the use of RDF triples that process data on the Web. It is a protocol for searching, adding, modifying, and deleting IoT data stored in RDF triples using a set of queries.

Semantic Web Rule Language (SWRL) [25] is a standard language based on OWL-DL and the Rule Markup Language (RuleML). It expresses rules for reasoning and inferring new knowledge from an existing IoT data. Then it enriches the RDF triples by adding supplementary predicates.

2.2 Linked Data

Tim Bernes-Lee coined the term in 2006, "The Semantic Web is not just developed to put the data in the Internet, but to create links that allow individuals or machines to explore data. When you have linked data, you can find other, interconnected, data" [7].

Describe resources and publish them according to semantic web formats, facilitate their access and their interpretation by software agents [14]. There are several projects that index ontologies and vocabularies on the Web.

Linked Open Vocabularies (LOV) [32] is a getaway created by The Open Knowledge Foundation to reuse semantic vocabularies on the web [48]. The Linked Open Vocabularies for the IoT (LOV4IoT) [33] present a collection of 740 specific vocabularies for IoT domains at the time of writing. Schema.org [43] is a collaborative community activity with the mission of creating, maintaining and promoting structured data schema on the Internet, on web pages, in electronic messages, and beyond. However, it does not include an IoT specific concept.

2.3 From IoT to SWoT

The Internet of Things, coined for the first time in 1999 by Kevin Ashton, was born out of the convergence of wireless technology and the Internet. It is defined as a network composed of millions of private, public, academic, com-mercial and government networks, ranging from local to global levels, connected through a wide range of electronic, wireless and optical network technologies [36].

The evolution of the IoT gave birth to two branches: The Web of Things that provides an Application Layer that simplifies the development of IoT applications composed of multiple devices across different platforms and application domains [40]; and the Semantic Web of Things [17] that incorporates the Semantic Web technology into the IoT applications to solve the issues of the lack of interoperability among IoT systems related to the heterogeneity of the huge amount of IoT data. While the semantic web's purpose is to have and create coherent content that machines and humans can process, the IoT can take advantage of this vision to reflect and explain things and to annotate the generated data.

3 Basics Uses of Semantics in the IoT

Semantic annotations can provide machine-interpretable descriptions of what the data represents, where it comes from, how it relates to its environment, who provides it, and what the quality, technical, and non-technical attributes [3]. Interoperating and integration of data and information is the more important and demanding task [47]. So, the IoT can benefit from interoperability, which is the first purpose of creating information by annotating data. Semantic annotation is a process that aims to attach complementary data (meta-data) to other data. One of the main principles of the semantic web is that it is necessary to associate web resources with information in order to be used by software agents and to promote the exploitation of these resources [21,34]. Semantic interoperability consists of several layers that must be implemented to build an interoperable environment [35]. It describes the content's meaning and provides a data abstraction.

The semantic web framework proposes formalisms for knowledge representation that can deal with heterogeneity. Also, it allows the integration of heterogeneous data within the same structure and transparent access to applications that can rely on data collected for its specific purpose. Moreover, it offers interpretation by delivering a global context of information or for the results of deduction rules on already existing knowledge. So, it permits for different IoT platforms from multiples clients to work in collaboration [35].

4 Ontologies for the IoT

The Ontology is the most important component of the Semantic Web, providing semantic interoperability and serving as a repository for information and knowledge about things and their types. This corresponds to answering the 4W1H questions (what, where, when, who, and how) [50]. Ontologies assist in providing simple semantic annotation of raw IoT data. Also, it ensures interoperability by providing an abstraction layer [41], and produce knowledge artifacts. An OWL ontology contains descriptions of classes, properties and individuals [24]. Several ontological models have been proposed in the framework of the Semantic Web of Things. In the following, we present the most eminent of them.

Semantic Sensor Network Ontology (SSN) [11], developed by the W3C Semantic Sensor Network Incubator Group, is actually the most popular ontology. It describes sensors and their observations. The procedures involved, the features of interest, the observed properties and actuators. A wide variety of technologies and use cases can be supported by SSN, including the Web of Things. Several ontologies extend and build the SSN ontology.

Wireless Semantic Sensor Network (WSSN), proposed as an extension of the SSN ontology in [5]. They added new Wireless Sensor Network (WSN) concepts to provide communication and transformation of sensing data like communication, data streams and state.

Sensor Core Ontology (SCO) [44] extends the SSN ontology with modules, including the component module, service module and context module. Three essential classes are introduced within the context module: space, time and theme.

M3 Ontology [18] extends the concepts of the SSN ontology. It describes sensor metadata such as observation, units of measurement and domains of application. In order to facilitate the description of sensor data, The M3 ontology's main feature is the ability to link domain ontologies and rules in order to derive meaningful knowledge from sensor data.

OntoSensor ontology [42] extends the concepts from SensorML and SUMO (Suggested Upper Merged Ontology). Its purpose is to identify sensors, their behaviors, relationships, additional characteristics and performance by adding new concepts. It supports data discovery and processing. But, it is unable to describe sensor observations properly.

Commonwealth Scientific and Industrial Research Organization (CSIRO) [12] is a generic ontology that represents the capabilities and properties of the sensor aspects. It describes the functional, physical, and metrological aspects of the sensors. It does not describe the values or their units, the observations and results of the objects can not be described either.

OneM2M Ontology [39] its main purpose is to provide the interoperability among different IoT technologies. It describes the meaning of sensors and actuators in order to achieve semantic interoperability.

IoT ontology [30] describes the knowledge of devices. By describing concepts such as intelligent entities, physical entities, control entities, electronic devices and networks. It is based on SSN Ontology, DOLCE Ultralite Upper ontology (DUL), quantities, units, Dimensions and Data Types Ontologies (QUDT).

IoT-O ontology [27] is supposed to be extended to address specific application needs and actual-life devices and services. It is intended to model knowledge about IoT systems and applications. It is based on various ontologies such as SSN and SAN [46].

Smart Onto Sensor [2] is an ontology developed for smartphone-based sensors. Based on SensorML and SSN. The intended purpose of the SOS is to develop an ontological model consisting of formal conceptualization of smartphone resources and sensors, including their categories, taxonomy, relationships and metadata regarding sensor characteristics, performance, and reliability [2].

[49]*'s Ontology,* is a comprehensive ontology for knowledge representation in the IoT. It describes IoT services, observations and their measurements. The description ontology is developed using a knowledge-driven approach and aims to capture most of the important concepts and their relationships in the IoT domain [6].

IoT-Lite ontology [6], is an instantiation of the SSN ontology that is designed for IoT knowledge. IoT-Lite focuses on the concepts of discovery and their relationships, which can provide answers to a wide range of application queries. However, it does not offer any reasoning.

Sensor, Observation, Sample, and Actuator Ontology (SOSA) [28], is a lightweight event-centric ontology that is built on top of SSN. The SOSA ontology provides a formal but lightweight general purpose specification for modelling the interaction between the entities involved in the acts of observation, actuation, and sampling.

5 IoT Frameworks Enhancing Semantic Interoperability

The Semantic IoT framework is a set of layers responsible for aggregation, persistence, analytics and serving of data [38]. In this section, we systematically present some existing IoT relevant frameworks supporting semantic interoperability in IoT systems.

BiG-IoT [23,29] relies on semantic addressing and IoT interoperability problems by developing the BiG-IoT API to realize the real Internet of Things (IoT) ecosystems. It is a web platform which unifies several platforms and numerous middleware systems. The project uses Schema.org as a vocabulary of concepts. It is easy to create applications and services for heterogeneous platforms with an increasing level of semantic interoperability with an API that has a well-defined architecture.

FIESTA IoT. Federated Interoperable Semantic IoT Testbeds and Applications (FIESTA IoT) is a research project as well as an "Innovation Action" funded by the European Union's Horizon 2020 program. It treats the topic of "Future Internet Research and Experimentation". The project consists of large-scale experiments in the IoT domain that will utilize data and resources from heterogeneous IoT platforms [1]. The FIESTA project allows researchers and experimenters to share and reuse data from diverse IoT testbeds using semantic technologies. These experiments provide a variety of tools and best practices for improving the interoperability of heterogeneous IoT platforms.

VICINITY. This project aims at interfacing cloud-based platforms from various application domains by providing "interoperability as a service" for the IoT [16]. The major goal of the VICINITY is to increase semantic interoperability. Through the use of the standard W3C Web Language Ontology. The VICINITY ontology network is composed of cross-domain ontologies addressing the modelling of general concepts like time, space, and the Web of Things. It represents the information for exchanging IoT description data between peers. Domain-oriented ontologies aim to cover a wide range of domains such as health, transportation, and so on.

INTER-IoT. The main goal of this project is to design, implement and test a framework, that will allow interoperability among different IoT platforms [26].

The ultimate goal of the INTER-IoT is to provide an interoperable system architecture for the seamless integration of numerous IoT architectures in different application areas. At different levels, interoperability will be provided: system, network, middleware, services and data levels. It allows the developer to create effective and efficient smart IoT applications on different heterogeneous IoT platforms.

Open-IoT. The ICFOSS promotion is a community-enabled knowledge sharing platform for sharing, discussing and collaborating on open-source hardware and software for the IoT [37]. The Open-IoT project focuses on increasing semantic interoperability [45]. A middleware platform was created which allows semantic integration of applications in the cloud. It uses the sensor network (SSN) ontology to model the information and for the semantic integration of various IoT systems. Also, it uses Linked Data to enrich the data and infer it.

SymbIoTe (Symbiosis of Smart Objects Across IoT Environments). The architecture of symbIoTe is designed around a hierarchical IoT stack and covers multiple IoT platforms. It implements a flexible and secure interoperability middleware across IoT platforms. The main goal of the project is to create IoT applications on IoT platforms as well as dynamic and adaptive smart spaces that they can collaborate [15]. It provides a semantic IoT search engine for connected smart objects registered by platform providers. In addition, an abstraction layer unifies the usage of those resources across different platforms.

M3 (Machine-to-Machine Measurement) Framework. The M3 Framework project is directed to deal with the lack of semantic interoperability in the IoT. The project's framework employs M2M to semantically annotate and reason about M2M data gathered from disparate IoT devices, systems, and domains. To increase the level of interoperability at the syntactic, but especially at the semantic level, the M3 is framed by the following layers [19,20]:

- Perception layer, consists of physical IoT devices, and smart objects.
- Data acquisition layer, focuses on the collection of raw IoT data from devices and converting them in a unified way.
- Persistence layer, responsible to store M3 in a database for the storage of semantic sensor data, which is known as the triple store.
- Knowledge management layer, responsible of finding, indexing, designing, reusing and combining domain-specific knowledge, such as ontologies and datasets to update the M3 domain ontologies, datasets and rules.
- Reasoning layer, infers new knowledge using reasoning engines and M3 rules.
- Knowledge query layer, executes SPARQL queries on inferred sensors data.
- Application layer, uses an application to analyze results and display them to end users.

6 Conclusion

The Internet of Things (IoT) gathers a tremendous amount of raw data from its environment via amalgamated smart objects. The collected raw data is heterogeneous. Hence, computers can not handle them. Since the major goal of IoT research is to allow physical world information and knowledge to be obtained and used omnipresent by IoT applications that require a good understanding of the environment to make smart decisions. Therefore, semantic interoperability has become a cornerstone of the Internet of Things development, while the IoT applications demand a good understanding of the data, because, with good data, we make a good decision, especially in critical domains such as Smart Health, Smart Homes, Smart Energy Management... etc. In this paper, we presented the ultimate existing solutions to provide semantic interoperability among IoT systems. This can serve for other researchers as a guideline to know more. After data has been enriched with semantic annotations, the combination of semantic web technologies and the internet of things allows for better use of the data. The enriched IoT data descriptions create a consistent context for IoT applications to make appropriate and precise decisions. Today, we are borrowing the same concerns. Reaching complete standardization and generalization is one of the main problems to be tackled in the future to achieve full interoperability at the semantic data level. Also, the current IoT applications require the proposal of relevant solutions to deal with the lack of semantic interoperability in real-time for streaming raw data.

References

1. Agarwal, R., et al.: Unified IoT ontology to enable interoperability and federation of testbeds. In: 2016 IEEE 3rd World Forum on Internet of Things (WF-IoT), pp. 70–75. IEEE (2016)
2. Ali, S., Khusro, S., Ullah, I., Khan, A., Khan, I.: Smartontosensor: ontology for semantic interpretation of smartphone sensors data for context-aware applications. J. Sens. **2017**, 1–26 (2017)
3. Barnaghi, P., Wang, W., Henson, C., Taylor, K.: Semantics for the internet of things: early progress and back to the future. Int. J. Semant. Web Inf. Syst. (IJSWIS) **8**(1), 1–21 (2012)
4. Beckett, D., Berners-Lee, T., Prud'hommeaux, E., Carothers, G.: RDF 1.1 turtle. World Wide Web Consortium, pp. 18–31 (2014)
5. Bendadouche, R., Roussey, C., De Sousa, G., Chanet, J.P., Hou, K.M.: Extension of the semantic sensor network ontology for wireless sensor networks: the stimulus-WSNnode-communication pattern. In: 5th International Workshop on Semantic Sensor Networks in Conjunction with the 11th International Semantic Web Conference (ISWC), pp. 16-p (2012)
6. Bermudez-Edo, M., Elsaleh, T., Barnaghi, P., Taylor, K.: IoT-lite: a lightweight semantic model for the internet of things. In: 2016 INTL IEEE Conferences on Ubiquitous Intelligence & Computing, Advanced and Trusted Computing, Scalable Computing and Communications, Cloud and Big Data Computing, Internet of People, and Smart World Congress (UIC/ATC/ScalCom/CBDCom/IoP/SmartWorld), pp. 90–97. IEEE (2016)
7. Berners-Lee, T.: Linked data-design issues (2006). http://www.w3.org/DesignIssues/LinkedData.html
8. Berners-Lee, T., Hendler, J., Lassila, O.: The semantic web. Sci. Am. **284**(5), 34–43 (2001)

9. Brickley, D., Guha, R.: RDF schema 1.1, W3C recommendation (2014). http://www.w3.org/TR/rdf-schema (2017)
10. http://www.businessinsider.com/internet-of-things-report. Accessed 14 Dec 2020
11. Compton, M., et al.: The SSN ontology of the W3C semantic sensor network incubator group. J. Web Semant. **17**, 25–32 (2012)
12. Compton, M., Neuhaus, H., Taylor, K., Tran, K.N.: Reasoning about sensors and compositions. SSN **522**, 33–48 (2009)
13. World Wide Web Consortium, et al.: Owl 2 web ontology language document overview, December 2012
14. Djezzar, M., Hemam, M., Maimour, M., Amara, F.Z., Falek, K., Seghir, Z.A.: An approach for semantic enrichment of sensor data. In: 2018 3rd International Conference on Pattern Analysis and Intelligent Systems (PAIS), pp. 1–7. IEEE (2018)
15. Gojmerac, I., Reichl, P., Žarko, I.P., Soursos, S.: Bridging IoT islands: the symbiote project. e & i Elektrotechnik und Informationstechnik **133**(7), 315–318 (2016)
16. Guan, Y., et al.: An open virtual neighbourhood network to connect IoT infrastructures and smart objects-vicinity: IoT enables interoperability as a service. In: 2017 Global Internet of Things Summit (GIoTS). pp. 1–6. IEEE (2017)
17. Guinard, D., Trifa, V.: Towards the web of things: web mashups for embedded devices. In: Workshop on Mashups, Enterprise Mashups and Lightweight Composition on the Web (MEM 2009) and Proceedings of WWW (International World Wide Web Conferences), Madrid, Spain. vol. 15, p. 8 (2009)
18. Gyrard, A., Bonnet, C., Boudaoud, K.: Enrich machine-to-machine data with semantic web technologies for cross-domain applications. In: 2014 IEEE World Forum on Internet of Things (WF-IoT), pp. 559–564. IEEE (2014)
19. Gyrard, A., Datta, S.K., Bonnet, C., Boudaoud, K.: Standardizing generic cross-domain applications in internet of things. In: 2014 IEEE GlobeCom Workshops (GC Workshops), pp. 589–594. IEEE (2014)
20. Gyrard, A., Serrano, M.: A unified semantic engine for internet of things and smart cities: from sensor data to end-users applications. In: 2015 IEEE International Conference on Data Science and Data Intensive Systems, pp. 718–725. IEEE (2015)
21. Handschuh, S., Staab, S.: Annotation for the Semantic Web, vol. 96. IOS Press, Amsterdam (2003)
22. Harris, S., Seaborne, A.: SPARQL 1.1 query language. SPARQL recommendation. World Wide Web Consortium, March 2013. http://www.w3.org/TR/sparql11-query
23. Hatzivasilis, G., et al.: The interoperability of things: interoperable solutions as an enabler for IoT and Web 3.0. In: 2018 IEEE 23rd International Workshop on Computer Aided Modeling and Design of Communication Links and Networks (CAMAD), pp. 1–7. IEEE (2018)
24. Hemam, M.: An extension of the ontology web language with multi-viewpoints and probabilistic reasoning. Int. J. Adv. Intell. Paradigms **10**(3), 247–265 (2018)
25. Horrocks, I., et al.: SWRL: a semantic web rule language combining OWL and RuleML. W3C Member Submiss. **21**(79), 1–31 (2004)
26. Inter-IoT. http://www.inter-iot.eu. Accessed 14 Dec 2020
27. IoT-O. http://www.irit.fr/recherches/MELODI/ontologies/IoT-O.html. Accessed 24 Dec 2020
28. Janowicz, K., Haller, A., Cox, S.J., Le Phuoc, D., Lefrançois, M.: SOSA: a lightweight ontology for sensors, observations, samples, and actuators. J. Web Semant. **56**, 1–10 (2019)

29. Jell, T., Baumgartner, C., Bröring, A., Mitic, J.: BIG IoT: interconnecting IoT platforms from different domains—first success story. In: Latifi, S. (ed.) Information Technology - New Generations. AISC, vol. 738, pp. 721–724. Springer, Cham (2018). https://doi.org/10.1007/978-3-319-77028-4_91
30. Kotis, K., Katasonov, A.: An IoT-ontology for the representation of interconnected, clustered and aligned smart entities. Technical report, VTT Technical Research Center, Finland VTT Technical Research Center, Finland (2012)
31. Lee, I., Lee, K.: The internet of things (IoT): applications, investments, and challenges for enterprises. Bus. Horiz. **58**(4), 431–440 (2015)
32. Linked open vocabularies (LOV). http://lov.okfn.org/dataset/lov. Accessed 17 May 2020
33. LOV4IoT. http://www.lov4iot.appspot.com/?p=ontologies. Accessed 14 Dec 2020
34. Nilsson, M., Palmér, M.: Semantic web meta-data for e-learning: some architectural guidelines. In: 11th World Wide Web Conference (WWW 2002) (2002)
35. Noura, M., Atiquzzaman, M., Gaedke, M.: Interoperability in internet of things: taxonomies and open challenges. Mob. Netw. Appl. **24**(3), 796–809 (2019)
36. Nunberg, G.: The advent of the internet. 12th April, Courses (2012)
37. OpenIoT. http://www.openiot.in. Accessed 14 Dec 2020
38. Palavalli, A., Karri, D., Pasupuleti, S.: Semantic internet of things. In: 2016 IEEE Tenth International Conference on Semantic Computing (ICSC), pp. 91–95. IEEE (2016)
39. Park, H., Kim, H., Joo, H., Song, J.: Recent advancements in the internet-of-things related standards: a oneM2M perspective. ICT Express **2**(3), 126–129 (2016)
40. Pliatsios, A., Goumopoulos, C., Kotis, K.: A review on IoT frameworks supporting multi-level interoperability-the semantic social network of things framework. Int. J. Adv. Internet Technol. **13**, 46–64 (2020)
41. Poggi, A., Lembo, D., Calvanese, D., De Giacomo, G., Lenzerini, M., Rosati, R.: Linking data to ontologies. In: Spaccapietra, S. (ed.) Journal on Data Semantics X. LNCS, vol. 4900, pp. 133–173. Springer, Heidelberg (2008). https://doi.org/10.1007/978-3-540-77688-8_5
42. Russomanno, D.J., Kothari, C., Thomas, O.: Sensor ontologies: from shallow to deep models. In: Proceedings of the Thirty-Seventh Southeastern Symposium on System Theory, SSST 2005, pp. 107–112. IEEE (2005)
43. Home - schema.org. http://schema.org. Accessed 15 May 2020
44. Shi, Y., Li, G., Zhou, X., Zhang, X.: Sensor ontology building in semantic sensor web. In: Wang, Y., Zhang, X. (eds.) IOT 2012. CCIS, vol. 312, pp. 277–284. Springer, Heidelberg (2012). https://doi.org/10.1007/978-3-642-32427-7_38
45. Soldatos, J., et al.: OpenIoT: open source internet-of-things in the cloud. In: Podnar Žarko, I., Pripužić, K., Serrano, M. (eds.) Interoperability and Open-Source Solutions for the Internet of Things. LNCS, vol. 9001, pp. 13–25. Springer, Cham (2015). https://doi.org/10.1007/978-3-319-16546-2_3
46. Spalazzi, L., Taccari, G., Bernardini, A.: An internet of things ontology for earthquake emergency evaluation and response. In: 2014 International Conference on Collaboration Technologies and Systems (CTS), pp. 528–534. IEEE (2014)
47. Sriram, R.D., Sheth, A.: Internet of things perspectives. IT Prof. **17**(3), 60–63 (2015)
48. Vandenbussche, P.Y., Atemezing, G.A., Poveda-Villalón, M., Vatant, B.: Linked open vocabularies (LOV): a gateway to reusable semantic vocabularies on the web. Semant. Web **8**(3), 437–452 (2017)

49. Wang, W., De, S., Toenjes, R., Reetz, E., Moessner, K.: A comprehensive ontology for knowledge representation in the internet of things. In: 2012 IEEE 11th International Conference on Trust, Security and Privacy in Computing and Communications, pp. 1793–1798. IEEE (2012)
50. Zhang, D., Wang, L., Xiong, H., Guo, B.: 4W1H in mobile crowd sensing. IEEE Commun. Mag. **52**(8), 42–48 (2014)

Composition of Large Modular Ontologies Based on Structure

Sajida Mhammedi(✉), Hakim El Massari, and Noreddine Gherabi

LaSTI Laboratory, National School of Applied Sciences, Sultan Moulay Slimane University, Khouribga, Morocco
sajida.mhammedi@usms.ac.ma,
{h.elmassari,n.gherabi}@usms.ma

Abstract. Ontologies are commonly used in data integration because of their expression of knowledge as a formal description of a subject of interest. Typically, the construction of large ontologies necessitates the reuse of existing ones. Modularity is a fundamental need for collaborative ontology engineering and web reuse of distributed ontology. In this scenario, combining ontology modules to get a global ontology is crucial since it needs a solution to the mismatches between the comparative concepts. This paper proposes an approach to automatically compose ontology modules. The technique suggested is based on the proposed approach on similarity measures computed against concept names, attributes, and relationships, in order to allow the merging of ontology modules into a global one.

Keywords: Modular ontologies · Concepts merging · Similarity measures composition · Concept structure

1 Introduction

It's no secret that the volume of data available is growing exponentially, resulting in the definition of big data as a large volume of both unstructured and structured data. This massive influx of data from many sources outstrips the capabilities of traditional software and database approaches. Furthermore, sharing and reusing knowledge in big data applications, especially when data is unstructured, remains a challenging problem to solve. Semantic integration in big data may be the answer to this problem. Data must be represented as knowledge to be properly managed [1, 2].

Ontologies are the foundation of semantic Web technology; they describe knowledge as a set of concepts within a domain and the relationships that allow for global interoperability. Ontologies are designed to collect information about a specific domain and offer a standard language as well as an acceptable conceptual representation for reuse and sharing. However, as this ontology grows in size, it is getting increasingly hard to create, utilize, comprehend, manage, transform and classify [3].

Towards these ends, researchers had paid considerable attention to modular ontologies given that many tasks concerning ontologies integration tasks, such as mapping ontology from Database [4] and merging of separately generated ontologies

Y. Maleh et al. (Eds.): ICI2C 2021, LNNS 357, pp. 144–154, 2022.
https://doi.org/10.1007/978-3-030-91738-8_14

into a single, require modularity. Using modular representations makes it much easier for understanding, reasoning, extending, and reusing modular representations is much easier.

This paper is a continuation of our previous works [5] that have proposed an automatic ontology learning approach, which requires OWL ontology based on data from Couchbase. The ontology that results serves as a conceptual view of the data stored in the Couchbase database. In this work, we present a method for composing ontology modules into a global ontology automatically.

The remainder of this paper is as follows: We start in Sect. 2 by explaining similar work, drawing on previous research in related fields, and synthesize constraints and shortcomings that we plan to address. in Sect. 3, we expose our research context. Our approach to construct ontology modules will be stated in Sect. 4, Finally, Sect. 5 presents future work and draws the conclusion.

2 Related Work

The composition of modular ontologies is not a new topic of study. As a result, we will only cover recent solutions in this area.

Approaches	Ontologies of Input	Strategy	User involvement
(2015) [6]	2 or +	Simple merge	Manual
(2016) [7]	2 or +	Full merge	Semi-automatic
(2017) [8]	2 or +	Full merge	Automatic
(2017) [9]	2	Full merge	Automatic
(2018) [10]	2 or +	Asymmetric merge	Automatic
(2019) [11]	2	Asymmetric merge	Automatic
(2020) [12, 13]	2 or +	Full merge	Automatic

It is worth mentioning that nowadays, current ontology integration techniques are becoming more general, allowing for the integration of more than two ontologies. Furthermore, they are working to fully automate and scale these techniques, unlike earlier works that required a lot of human interaction to either confirm the proposed correspondences in the matching stage or the actions to be done in the merging step.

We do also observe that the authors of the investigated methods do not take into account the variation in granularity levels of the assembled modules, which might lead to issues with information sharing owing to the varying degrees of the definition of an object in reality.

To overcome these constraints, it is necessary to take into account complicated correspondences while looking for overlaps between modules to compose, and calculate practical similarity measures to identify the closeness between concepts.

3 Research Context

This work comes with the scope of a Big Data integration approach [1, 5]. The goal is to construct an ontology for the integration of large data where large data are viewed as data from multiple sources. Each source includes extremely large amounts of data, grows and develops independently of the other. This approach is based on three main steps. The initial step is the homogenization of data, which consists in wrapping each source of data in a Couchbase NoSQL database. After that, by using the transformation rules from Couchbase [5] to the OWL3 language, we provide an ontology matching each data source. Then, the generated ontologies must be made into a global one. These steps are highlighted in Fig. 1. In the scope of this work, each ontology module represents a viewpoint covered by a data source that concerns the model domain.

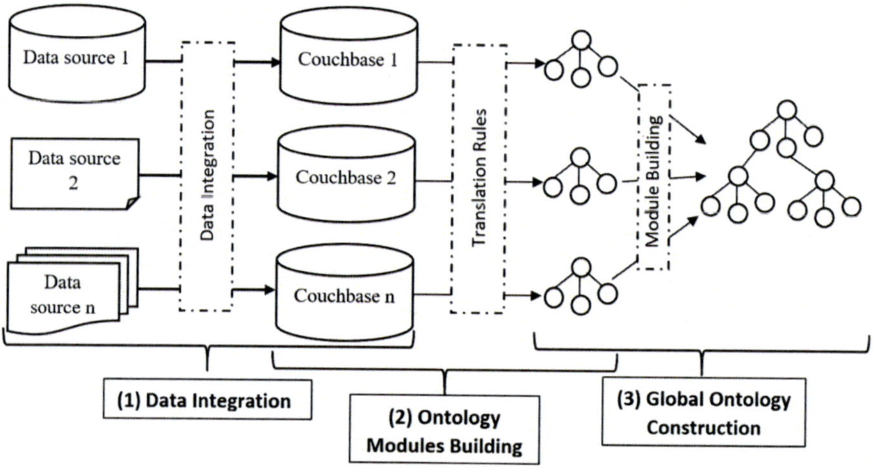

Fig. 1. Ontology-based big data integration.

4 Global Ontology Construction

This section outlines our method to merge the modules that have been obtained in the last work [5] to achieve global ontology. Measuring the similarity between sources of ontology concepts and preprocesses merging operations. The composition of ontology modules requires structural similarities between concepts. We suggest a new solution in this context to integrate ontology modules based on syntactic and semantic similarity steps, which combine concept names comparison, attributes, and interrelationships.

Two ontology modules Ont1 and Ont2 are to be combined in a global version (Ont3). The global ontology Ont3 obtains the reference ontology and is then combined with concepts, attributes, and relationships of the ontology of input. In terms of concepts and relationships, the ontology reference is known as the richer module. The integrated ontology module (Ont2) is compared to the reference ontology (Ont1), and the similarity value between 0 and 1 is calculated for any conflict identified between the concept of reference ontology (COnt1) and the concept of input ontology (COnt2). We may find concepts that are relatively identical, but vary according to a particular attribute or relationship. We use mathematical Equations to calculate the similarity value to obtain an accurate degree of similarity between concepts.

We combine syntactically with semantic matching to calculate the similarities of two given concepts belonging to various ontology modules. This syntactic match matches strings that characterize concept names as well as concept attributes, while the semantic match is based on the similarity of relationships.

We use a distance function over a pair of strings to calculate the syntactic match. The authors in [14] affirm that the edit distance in [15] is useful to discover a match of the object name in data integration settings. In a scaled version, this distance is also the best among different matching algorithms, based on string comparisons for names and records based on field comparisons. We use the Levenshtein distance to return the number of changes in character used to transform one string into another (LEV). The smaller this difference is, the less modifications are required, the closer the strings are.

Our algorithm relies on an ontology reference module (Ont1) for comparison and extends this with new concepts, attributes and relations from the input ontology module (Ont2). The global ontology (Ont3) is initiated by the ontology module of reference (Ont1) and improved by entities from the ontology module (Ont2).

4.1 Step (1): Detection of Overlaps

Step 1.1: The Reference ontology module and the input ontology module In terms of concepts and relationship numbers, the two ontology modules to be combined are compared. The ontology module of reference (Ont1) obtains the richer module as regards numbers of concepts and relationships and is then copied into the ontology resulting (Ont3) to keep these two modules unchanged. The other ontology (Ont2) module is assumed to be the ontology module of input to be combined with Ont1.

Step 1.2: Ont2 concepts loading in order to start comparisons, the structure of the input ontology module Ont2 is loaded.

Step 1.3: In the Ont1 ontology module, look up the concept's name Each Ont2 concept's name is sought for in Ont1. When the concept's name is discovered, Con tinue to Step 1.4. Otherwise, proceed to step 1.5.

Step1.4: Attributes of concept in the Ont1 ontology module Once the term nominating the Ont2 concept is identified in the Ont1 ontology module, the Ont2 concept attributes are compared to the Ont1 ontology concept one by one. The algorithm checks for a single-tone match that exists between the attributes of each input concept, and the concept attributes of the reference ontology. If the same attributes are achieved, go to stage 2.9. If at least one of the attributes has a disparity, go to step 1.6.

Step1.5: Define the concept name in the Ont1 ontology attributes. Once COnt2 is not discovered in concepts of the Ont1 ontology module, we look into it amongst the attributes of the Ont1. If there is a difference between the COnt2 concept name among the Ont1 attributes and COnt1 attributes from COnt2, proceed in parallel to 2.7, 2.8, and 2.9 steps. If the name of the COnt2 concept is not identified in the attributes of Ont1 and COnt1 attributes match the COnt2 attributes, move to step 3.13.

Step 1.6: In the names of the Ont1 concept, define concept attributes. If the concept attribute of O2 cannot be located in the attributes of Ont1 ontology module, we shall find it amongst the names of the concept Ont1 ontology. When you find the name of the characteristic, move to step 3.14. Else go to step 2.8.

4.2 Step (2): Similarity Measures Calculation

Step 2.7: Calculate the similarity of names The input concept's name is compared to each Ont1 concept's name to see how close they are. We apply Eq. (1) [16] to measure the similarity between the concept name of the first (COnt1) in the reference ontology module (Ont1), and the second (COnt2) in the input ontology module (Ont2).

$$\boldsymbol{SimName(COnt1, COnt2) = 1 - \frac{LEV(COnt1_{name}, COnt2_{name})}{MAX(LENGTH(COnt1_{name}), LENGHT(COnt2_{name}))}} \quad (1)$$

Where COnt1 and COnt2 are, respectively, the Ont1 and Ont2 ontology module concept, COnt1name and COnt2name are, respectively, the terms used to designate the Ont1 and the Ont2 ontology module concepts.

Move to step 2.10 at the end of this step.

Step 2.8: Compute attributes similarity This stage calculates a structural resemblance between the input idea, and each Ont1 ontology module concept in terms of its characteristics. The first Formula calculates the similarity between attribute names, and the second calculates the similarity between global attributes between the two concepts.

In order to calculate the similarity of attribute names, we compare an attribute from (COnt2At) the input ontology module to all attributes of a concept from the reference ontology module (COnt1At j) using the Eq. (2) [8]. For the similarity between attribute names, we compare an attribute of the input ontology module (COnt2At) by using the Eq. (2) [16], to all attributes of a concept from the reference ontology module (COnt1At j),.

Where j is the index of the COnt1 attribute being processed, and k is the number of COnt1 attributes.

$$SimName(CO2_{At}, CO1_{Atts}) = max_{j=1}^{k}\left(1 - \frac{LEV(CO2_{Atname}, CO1_{Atnamej})}{max(LENGHT(CO2_{Atname}), LENGHT(CO1_{Atnamej}))}\right) \quad (2)$$

Equation (2) evaluates the similarity of each concept in the Ontology reference module (COnt1) to the name of a COnt2 attribute, and then provides the highest values corresponding to the COnt2 attribute's closest comparable attribute, that has been processed. In order to determine the similarity of attributes between two concepts, one belongs to the Ont2, while the other belongs to the Ont1, we compare each attribute of the concept of the input ontology module (COnt2Ati) with all of the concept's attributes of a reference ontology Module (COnt1Atts), and then, we calculate their average. For this purpose, we utilize Eq. (3).

$$SimAtts(COnt2_{Atts}, COnt1_{Atts}) = \frac{\sum_{i=1}^{n} SimName(COnt2_{Ati}, COnt1_{Atts})}{n} \quad (3)$$

N is the number of COnt2 attributes.

Equation (3) calculates the average similarity of characteristics between two concepts. One belongs to the ontology input module (COnt2), while the other belongs to the reference ontology module (COnt1). It examines the COnt1 (COnt1Atj) property, which is the most comparable to the COnt2 emission (COnt2Ati). This is done for each input concept attribute (COnt2). In order to get the similarity of the average attributes, the recovered values are then summed and divided by the attributes number of the COnt2 concept.

The following are the next steps:

- If this step is called from step 1.5, move to step 2.10.
- Move to step 3.15; if this step is called from step 1.6, and COnt1 is similar to COnt2, and COnt1 attributes are included in COnt2 attributes, go to step 3.15.
- If that step is called from step 1.6 and the name of the COnt1 differs from the name of COnt2, and the attributes of the COnt1 are in COnt2, and the number of similar attributes is higher than n/2 where n is the number of attributes of COnt2, move to step 3.17.

Step 2.9: *compute* the *similarity relationship* The relationship similarity in the input concept is measured against each of the reference ontologies concept in the object property association as well as with "IS-A" taxonomic relationships. The comparison of relationships calculates semantic matching.

Equation (4) calculates the hierarchical similitude, and Eq. (7) calculates the objectProperty relationship similarity.

– Hierarchical Similarity Computation: We based on [17] to compute the hierarchical similarity (Algorithm 1), this method of similarity measuring uses the Dijkstra algorithm, this similarity calculates the hierarchical similarity of two concepts that belong to the same hierarchy, in which the similarity of the two nodes relies on its depth and the weight of the arcs which represent the relations between the nodes of the studied ontology (Fig. 2).

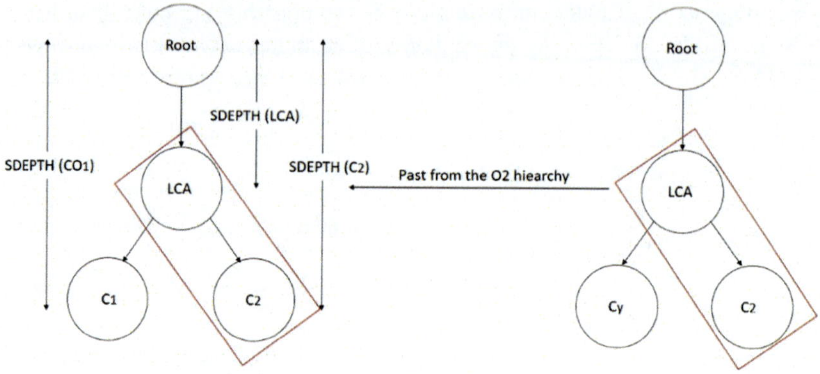

Fig. 2. The hierarchical similarity measure applied against two ontology modules

In order to compute the similarity in terms of the location in the hierarchy where each concept is situated, Knowing the fact that both concepts belong to various ontologies, we must assume that COnt2 and COnt1 concepts are part of the same hierarchy in order to be able to apply this technique. To do so, we must ascend through the Ont1 and Ont2 hierarchies to reach the idea of intersection. In both Ont1 and Ont2, this concept must have the same name and characteristics. We label LCA concept to represent the least common ancestor of COnt1 and COnt2, and we paste the Ont2 ontology module's sub-hierarchy LCA-COnt2 extract into the Ont1 ontology module.

The hierarchical similarity is then achieved by means of Eq. (4).

$$SimHier(COnt1, COnt2) = \frac{1}{DEG * SDIS(COnt1, COnt2) + 1} \tag{4}$$

where

$$\begin{aligned} SDIS(COnt1, COnt2) = W[SDEPTH(COnt1)] + W[SDEPTH(COnt2)] \\ - 2 * W[SDEPTH(LCA)] \end{aligned} \tag{5}$$

The calculation process of the shortest path *SDEPTH* defined in [17] Where, SPath1 represents the shortest path between two nodes.

COnt1 is the Ont1 concept, and COnt2 is the Ont2 concept inserted into the Ont1 hierarchy and LCA is selected where both Ont1 and Ont2 have the same name and attributes.

And W [$SDEPTH_i$] is defined as follows:

$$W[SDEPTH_i] = \sum\nolimits_{j=1}^{k} W_j[m, n] \tag{6}$$

m and n represent two nodes directly connected in $SDEPTH_i$ and k represents the set of arcs in $SDEPTH_i$.

Algorithm 1 : Hierarchical similary
Brown the Ont1 ontology module
FOR EACH concept of Ont2 **DO**
FOR EACH concept of Ont1 **DO**
If COnt1== COnt2 then
SDIS=0
Else if COnt1 is directly connected to COnt2 then
$SDIS = \mathrm{W[COnt1, COnt2]}$
Else
Define $SDEPTH_1$ and $SDEPTH_2$
Calculate $SDIS(COnt1, COnt2)$
End if
Calculate $SimHier(COnt1, COnt2)$

The Similarity of ObjectProperties

The links of the objectProperty are used to compute the similarity associations among two concepts.

The following measure is applied [Form (7)].

$$SimRel(COnt2, COnt1) = \frac{Common(Rel(COnt1), Rel(COnt2))}{Rel(COnt2)} \tag{7}$$

Where objectProperty links are Rel(COnt1) and Rel(COnt2) of COnt1 and COnt2, respectively, Rel(COnt1) offers objectProperty connections existing in Ont1 as well as in Ont2 ontologies.

Step 2.10: Computing the global similarity

The global similarity probability is achieved by summing the similarity of attributes, names similarity, the hierarchical similarity and the objectProperties similarity of the input concept, then computing their average. For each Ont2 ontology concept, this computation is conducted against each Ont1 ontology module's concept. Equation (8) provides the final similarity value (Sim).

$$Sim(COnt2, COnt1) = \frac{SimName + SimAtts + SimHier + SimRel}{4} \tag{8}$$

The higher the value of Sim, the more similar the concepts are. Go to step 2.11 at the end of this step.

Step 2.11: Verify the limit values in this stage. The maximum similarity value of the current input concept (SV) is checked (COnt2).

The following are three alternatives:

- Go to step 17 if SV is below the minimal level.
- If SV exceeds the highest limit values, or varying between the extreme limits value such that COnt2 is similar to only one concept in Ont1 and furthermore COnt2 has a parent concept, go in parallel to steps 3.14, 3.16 and 3.17
- If SV is varying between the minimum and maximum limit value and COnt2 is the same as more than one concept in Ont1, go to step 3.19.

4.3 Step (3): Updating Reference Ontology Module

Step 3.12: Add an objectProperty to the Ont1 ontology module, Once the Ont2 concept name is revealed in the attributes of the Ont1. The objectProperty relationship in Ont1, is added, so that: the domaine gets COnt1, and the range gets COnt2, and the objectPropert name gets the word "has" plus COnt2 name.

Step 3.13: COnt1 and COnt2 are merged into COnt1. If COnt2 concept name is missed in the Ont1 attributes and attributes of the two concepts are equal. Go to step 3.20

Step 3.14: Insert an objectProperty into the Ont1 ontology module. Once the attribute of the COnt2 concept is missed in the names of the Ont1 ontology module concepts, in Ont1, an objectProperty relationship is added so that: the domaine get COnt2, the range get COnt1 and the objectProperty name get the word "has" plus COnt1 name.

Step 3.15: in this step COnt1 attributes must be updated if: (COnt1 name = COnt2 name and COnt1 attributes are found in COnt2 attributes) OR (SV is higher than the maximum sill) OR (SV is between the minimal and maximum sill). Then COnt2 attributes that do not already exist in COnt1 are added to COnt1. Go to step 3.20.

Step 3.16: in this step COnt1 relationships must be updated if: (COnt1 name = COnt2 name and COnt1 attributes = COnt2 attributes) OR (SV is higher than the maximum sill) OR (SV is between the minimal and maximum sill). Then COnt2 relationships are added to COnt1. Go to step 3.20.

Step 3.17: Update COnt1 hierarchy if: (COnt1 name # COnt2 name and COnt1 attributes are found in COnt2 attributes and the number of identical attributes shared by COnt1 and COnt2 exceeds n/2, where n is the number of COnt2 attributes) OR (SV is higher than the maximum sill) OR (SV is between the minimal and maximum sill). Then include the IS-A relationship in the Ont1 ontology. Go to step 3.20.

Step 3.18: Add to Ont1 ontology a new concept if SV is less than the minimum threshold. Insert COnt2 in the Ont1 ontology module. Go to step 3.20.

Step 3.19: Choose the Ont2 concept that is most close to the Ont1 ontology If SV ranges from minimum to maximum and COnt2 is the same as more than one concept in Ont1, in terms of attributes of similarity and relationships, pick the most similar COnt2 concept. Go to step 3.20.

Step 3.20: go to the end if the current concept is the last one in the Ont2 ontology module. If it is not, go to step 1.3.

5 Conclusion and Future Work

Ontology plays a significant role in representing data-derived knowledge. In addition, it also aids in the retrieval of hidden meanings that are not readily accessible from data sources. As the use of ontologies is spreading increasingly and ontologies are expanding considerably, the requirement to build ontologies with a reusable viewpoint is becoming more and more important. The aim of this work is to deal with modular ontologies composition by merging ontology modules into a global one. Our technique is based on similarity measurements based on concept names, attributes, and relationships. The aim is to establish a new technique for the combination of modular ontology. Similar methods have been discussed and the technique for performing the merging phases. In future work, we intend to provide a method for updating the resulting ontology while taking module changes into consideration.

References

1. Khan, Z.M.A., Saeidlou, S., Saadat, M.: Ontology-based decision tree model for prediction in a manufacturing network. Prod. Manuf. Res. **7**, 335–349 (2019). https://doi.org/10.1080/21693277.2019.1621228
2. Al-Aswadi, F.N., Chan, H.Y., Gan, K.H.: Automatic ontology construction from text: a review from shallow to deep learning trend. Artif. Intell. Rev. **53**(6), 3901–3928 (2019). https://doi.org/10.1007/s10462-019-09782-9
3. Seidenberg, J., Rector, A.: Web ontology segmentation: analysis, classification and use. In: Proceedings of the 15th International Conference on World Wide Web - WWW 2006, p. 13. ACM Press, Edinburgh (2006)
4. Gherabi, N., Addakiri, K., Bahaj, M.: Discovering new technique for mapping relational database based on semantic web technology. In: International Conference on Communications and Information Technology - Proceedings, pp. 27–32, 6285808 (2012)
5. Mhammedi, S., El Massari, H., Gherabi, N.: Cb2Onto: OWL ontology learning approach from couchbase. In: Gherabi, N., Kacprzyk, J. (eds.) Intelligent Systems in Big Data, Semantic Web and Machine Learning. AISC, vol. 1344, pp. 95–110. Springer, Cham (2021). https://doi.org/10.1007/978-3-030-72588-4_7
6. Ziemba, P., Jankowski, J., Wątróbski, J., Wolski, W., Becker, J.: Integration of domain ontologies in the repository of website evaluation methods. Présenté à 2015 Federated Conference on Computer Science and Information Systems, 11 October 2015 (2015)
7. Caldarola, E.G., Rinaldi, A.M.: An approach to ontology integration for ontology reuse. In: 2016 IEEE 17th International Conference on Information Reuse and Integration (IRI), p. 384–393 (2016)
8. Chatterjee, N., Kaushik, N., Gupta, D., Bhatia, R.: Ontology merging: a practical perspective. In: Satapathy, S., Joshi, A. (eds.) Information and Communication Technology for Intelligent Systems (ICTIS 2017) - Volume 2. ICTIS 2017. Smart Innovation, Systems and Technologies, vol. 84, pp. 136–145. Springer, Cham (2018). https://doi.org/10.1007/978-3-319-63645-0_15
9. OIM-SM: a method for ontology integration based on semantic mapping - IOS Press. https://content.iospress.com/articles/journal-of-intelligent-and-fuzzy-systems/ifs161553

10. Stoilos, G., Geleta, D., Shamdasani, J., Khodadadi, M.: A novel approach and practical algorithms for ontology integration. In: Vrandečić, D., et al. (eds.) The Semantic Web – ISWC 2018 ISWC 2018. Lecture Notes in Computer Science, vol. 11136, pp. 458–476. Springer, Cham (2018). https://doi.org/10.1007/978-3-030-00671-6_27
11. Lou, W., Pi, R., Wang, H., Ju, Y.: Low-cost similarity calculation on ontology fusion in knowledge bases (2020). https://journals.sagepub.com/doi/10.1177/0165551519870456
12. Babalou, S., König-Ries, B.: Towards building knowledge by merging multiple ontologies with CoMerger: a partitioning-based approach. arXiv:2005.02659 [cs] (2020)
13. Babalou, S., Konig-Ries, B.: Towards multiple ontology merging with CoMerger (2020)
14. Cohen, W.W., Richman, J.: Learning to match and cluster large high-dimensional data sets for data integration (2002)
15. Levenshtein, V.I.: Binary codes capable of correcting deletions Insertions and Reversals. Sov. Phys. Dokl. **10**, 707 (1966)
16. Extraction of semantic relationships starting from similarity measurements: In: Proceedings of the Ninth International Conference on Enterprise Information Systems, pp. 603–606. SciTePress - Science and Technology Publications, Funchal (2007)
17. Daoui, A., Gherabi, N., Marzouk, A.: An enhanced method to compute the similarity between concepts of ontology. In: Noreddine, G., Kacprzyk, J. (eds.) ITCS 2017. AISC, vol. 640, pp. 95–107. Springer, Cham (2018). https://doi.org/10.1007/978-3-319-64719-7_9

Detection and Prediction Using Similar Trajectory Measurements

Benali Mouad(✉) and Amnai Mohamed

Laboratoire en Recherche Informatique (laRI), University Ibn Tofaïl, Kenitra, Morocco
{mouad.benali,mohamed.amnai}@uit.ac.ma

Abstract. To extract knowledge about the movement of different types of mobile agents (e.g., humans, animals, vehicles) and dynamic phenomena (e.g., hurricanes) requires new exploratory data analysis methods for datasets massive movement. Different types of moving objects share similarities and express differences in dynamic behavior and the nature of their movement. Extracting such similarities can contribute significantly to the dynamic phenomena of prediction, modeling and simulation. Therefore, with the development of a quantitative methodology, this research intends to study and explore the similarities in the dynamics of moving objects using measures of the similarities in knowledge discovery.

Keywords: Detection · Prediction · Similarity · Trajectory · Euclidean distance · Dynamic time warping · LCSS

1 Introduction

A trajectory is a name given to the path a given moving object takes through space as a function of time. In discrete mathematics, a trajectory is a sequence $(f^k(x))_{K \in N}$ of values calculated by iteratively applying the mapping from f to an element x. In general, a trajectory refers to the ordered set of intermediate states assumed by a dynamical system due to a time evolution. It is interesting to note that the trajectory of a moving object depends on the referential that is being used, since there is no absolute space.

Due to the large size and high dimensionality of many time series. Data reduction leads not only to much smaller storage space, but also to much faster data processing. Among the strategies for data reduction, one can find:

- Selection of a subset of attributes: removes redundant or irrelevant attributes, or dimensions;
- Dimensional reduction: usually employs signal processing techniques to obtain reduced versions of the original data;
- Numerical reduction: data are replaced, or estimated by smaller alternative representations, such as histograms, clustering, and sampling.

Since a time series can be viewed as a high dimensionality data sequence, each time point can be considered a dimension. A dimensional reduction can be the primary

Y. Maleh et al. (Eds.): ICI2C 2021, LNNS 357, pp. 155–163, 2022.
https://doi.org/10.1007/978-3-030-91738-8_15

focus. For example, to compute the relationship between two-time series curves, reducing the time series from length (or dimension) n to k can lead to a reduction from $O(n)$ to $O(k)$ in terms of computational complexity.

In recent years, research on moving object databases has attracted a lot of attention in many applications such as location-based services, traffic management, hurricane research, sales trend, etc. Some moving objects in the real world share the same movement patterns, which is an important characteristic and can help people solve real problems. We will use this technology to solve an example of a problem.

We aim to predict and identify the best product sold using the data collected to make good product sales for a company. In this paper, we will see the definition of trajectory and the measures to study the similarity of the trajectories to detect and predict the best product to sell.

This paper is structured as follows: Section 2 contains related work on trajectory similarity. Section 3 presents the theoretical part concerning the similarity of the trajectories through the distance of some tests. Then, in Sect. 4, we describe the experimental process and discuss the results and reliability of this work. Finally, Sect. 5 is reserved for the conclusion of this paper.

2 Related Work

There are two methods to measure trajectory similarity. The first one is to work with the whole trajectory [2–5], and another is to work with a part of the trajectory [6–8]. Ryatidamavandi et al. [6] propose two different strategies, such as Locality-Sensitive and Distance-Based Hash function, to group trajectories. Fan and Yao [3], propose a function that takes as input two discrete trajectories of the same length generated by a road-based-location sign. Wang et al. [8] analyze and compare dix widely used measures of trajectory similarity: Euclidean Distance-DTW, piece-wise DTW, edit distance with a real penalty, edit distance on real sequence, and LCSS. Tooshey and Duckman [7] implement and compare four standard trajectory similarity measures: LCSS, Frechet distance, DTW, edit distance. Later, Mao and et al. [5] evidence the lack of similarity measures robust to noisy data. Similarly, Vlachos, Gunopulos, and Kollios [2] suggest a distance function for data with a high number of outliers. Furtado et al. [4] propose a new distance function robust to movement uncertainty generated by the heterogeneity in the sampling.

3 Problematic

Making detection and prediction from the similarity of trajectories is not an easy thing to do. First of all, we need to know what a trajectory is? How can we calculate the similarity of trajectories? And how will we use this similarity to do detection and prediction? In this section, we present the concepts of (i) trajectory. Then, (ii) measures of similarity to study the similarity of the trajectories that will help us detect and predict accompanied by a test in Sect. 4.

3.1 Trajectory and Data Trajectory

A trajectory is a moving object's path, while a data trajectory is data collected from a moving object. In general, this data trajectory is a set of sequence points (2d spatial + time) or (3d spatial + time) (Fig. 1).

Fig. 1. Trajectory 2d spatial + time.

Using suitable measurements, we can plot a trajectory that represents the moving object with a constant speed between two consecutive points (Fig. 2).

Fig. 2. Trajectory 2d spatial + time.

3.2 Euclidean Distance

Euclidean distance is one of the most used means of measuring the similarity of objects. Euclidean distance between two points a and b where each point had a defined spatial location {latitude, longitude} is defined by Eq. 1.

$$d_E(a,b) = \sqrt{(a.lat - b.lat)^2 + (a.long - b.long)^2} \tag{1}$$

To measure the Euclidean distance between two trajectories A and B, we average the Euclidean distance between the points within each trajectory, as proposed by [1] Eq. 2.

$$d_{EM}(A,B) = \frac{1}{L}\sum\nolimits_{k=1}^{L} dE(ai,k,bj,k) \quad (2)$$

Where $dE(ai,k,bj,k)$ is the Euclidean distance between two points at position k. L is a length of trajectories A and B. As long as the Euclidean distance is small as long as the similarity becomes greater.

The Euclidean distance contains two disadvantages (i) the two trajectories require the same size Fig. 3, which represents two trajectories in 2d with a different size, we cannot calculate the Euclidean distance because the point at position t' we have just one point of the second trajectory, (ii) the two trajectories must have the same speed Fig. 4 which represents two similar trajectories in 2d, we can present them in a single point 10d, and by canceling the Euclidean distance we will have a significant distance which is a false result, and that is because between t_0 and t_1 the first object has accelerated and between t_1 and t_2 second object has accelerated.

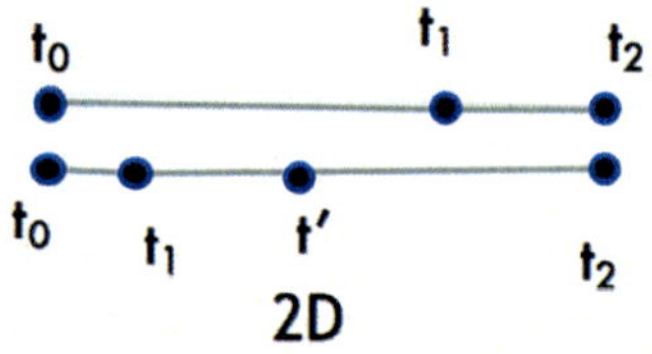

Fig. 3. Two trajectories with different size

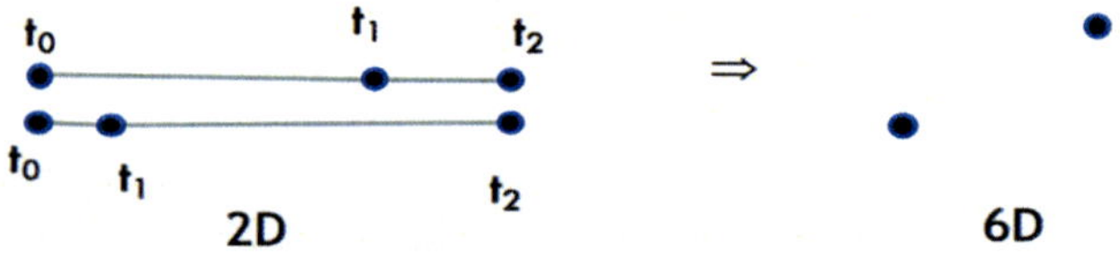

Fig. 4. Two trajectories with different speeds.

3.3 Principal Component Analysis (PCA)

This technique helps us find patterns in high-dimensional data by transforming a set of data possibly related into a set of data linearly independent. These are known as principal components. For the trajectory case, we can reduce the dimensionality of the original spatial trajectories keeping the main component with the highest eigenvalue. Using PCA will help us to interpret our data. It simplifies the complexity of big data while maintaining trends and patterns.

3.4 Alignment Trajectory

Alignment Trajectory is a technique that will help us calculate Euclidean distance, taking as input two trajectories with different sizes and time intervals. Figure 5 explains the trajectory alignment technique. It takes like input two trajectories, interval time, and time window. In our example, the two trajectories are A and B, begin time, and end time are 1:00 and 2:20 for a time window of 10min. Then, we keep only time slots that have points from both trajectories, which in our example are 2, 3, 7, and 8. Therefore, with this technique, we solve the problem of "two trajectories with different sizes".

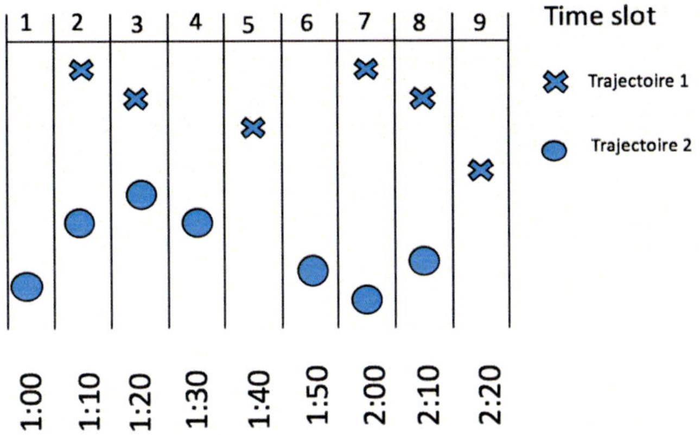

Fig. 5. Alignment trajectory.

3.5 Dynamic Time Warping (DTW)

This algorithm was developed to manage the trajectories of different quantities of measured points. It works on pairs of points and uses a point of a trajectory several times in the pair distance calculations if the other moves "too fast". DTW solves the problem of two trajectories with a different speed based on the Euclidean distance Fig. 6.

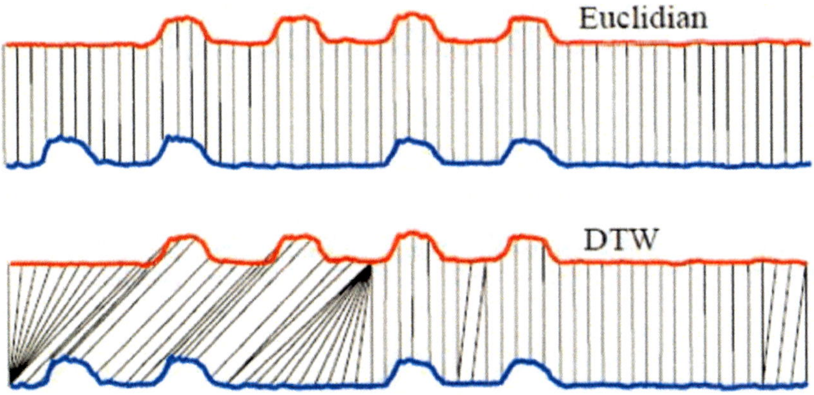

Fig. 6. Two trajectories with different speeds using Euclidean and DTW.

3.6 Longest Common Sub-Sequence (LCSS)

The longest common subsequence (LCSS) measures similarity where trajectories can be stretched, while certain points can remain unmatched to provide an accurate similarity analysis [2]. We, therefore, use LCSS to group trajectories according to general trends. Figure 7 represents an example of LCSS, which shows the number of data points paired between the two trajectories.

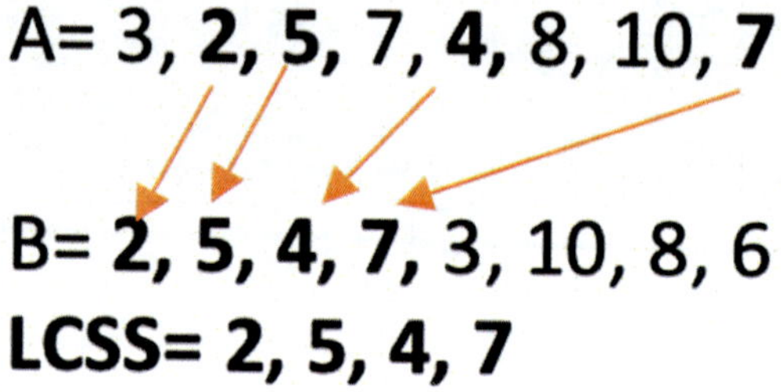

Fig. 7. Example of LCSS

LCSS distance between two trajectories A and B is defined by Eq. 3. Where LCSS (A, B) is the longest common Sub-Sequence between the two trajectories A and B, min (A, B) represents the length of the smallest path, which is in our example Fig. 7 equal 8.

$$D_{LCSS}(\mathrm{A},\mathrm{B}) = 1 - \frac{LCSS(\mathrm{A},\mathrm{B})}{\min(A,B)} \tag{3}$$

4 Experiment and Discussion

In this section, we detail the process to predict and identify the best product to sell based on the similarity of the trajectories. The process has three steps (i) data input, (ii) Calculate distance and (iii) output. For the first step, the process receives an optimal weekly product sales trajectory and a data set containing all weekly product sales trajectories to compare with. The data used for this work are weekly sales transaction data set found in the UCI Dataset Repository to perform our sales-based time series analysis. Dataset Contains weekly purchased quantities of 800 products (P1 to P800) over 52 weeks (1 year). The second step, is to calculate distance. This step takes optimal weekly product sales trajectory and a set of trajectories to compare to. Then, we calculate our distance using the three metrics mentioned in Sect. 3 Euclidean distance, DTW and LCSS. Finally, the last step is the output of result data to identify the best product to sell. The process selects the shortest metric distances of the three measurements Euclidean, DTW and LCSS. To facilitate the process exploitation we have implemented a graphical user interface (Fig. 8), which Dataset field defines the dataset containing the product set, the optimal product field define the trajectory of the optimal weekly product sales, N product field defines a top number of trajectories with the lowest distance.

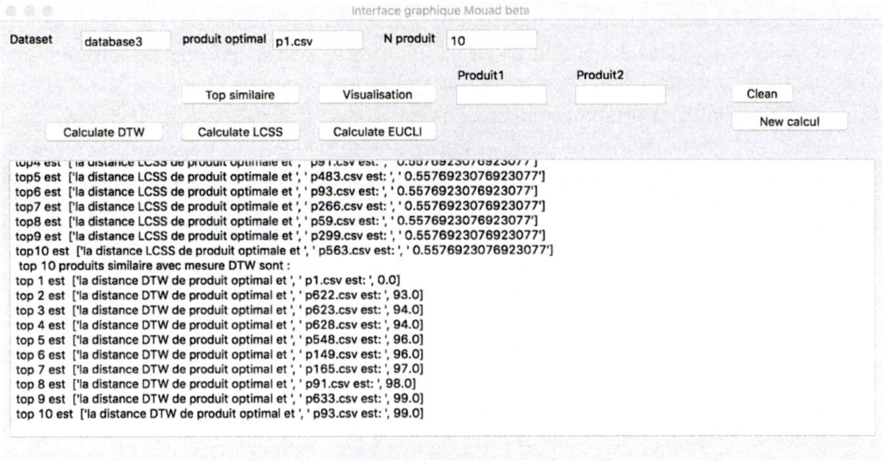

Fig. 8. Interface user.

Now, we will discuss the validation of the results. As shown in Fig. 8, the process shows us the 10 lowest distances for each measure. Top 1, which is the product “p1” it’s distance equal 0 for the three measures, because we took it as the optimal weekly product sales trajectory. For that, in Fig. 9 we can see just one trajectory while we have two trajectories (because the two trajectories are the same with the same coordinate).

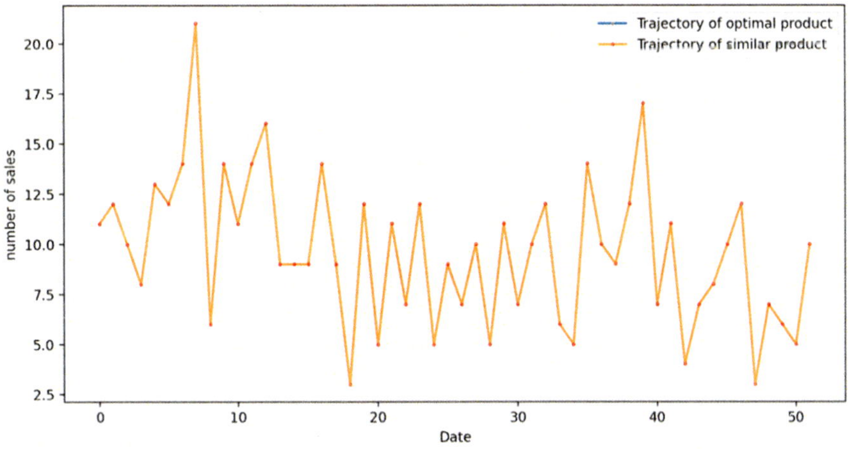

Fig. 9. Visualization of the trajectory of the product Optimal and the product P1.

From top 2 to top 10 we have a different result for the three measures. So, the question arises, what measure should be used to achieve a suitable result. Thus, the Euclidean distance must be eliminated because we have trajectories with time series in the same way LCSS must also be eliminated. After all, the trajectories depend on time series. Therefore, the appropriate measure in this type of analysis is DTW because it is a measure that calculates the similarity of trajectories whose speed may be different and which depend on time series. According to the result obtained in Fig. 8, the distance DTW between optimal product and P622 is the smallest distance Fig. 10 shows us a visualization of the trajectory of product optimal with the second most similar product, product number 622. However, the result obtained for similar top trajectories DTW means the best product can be sold with a priority in the order of top1 to top10.

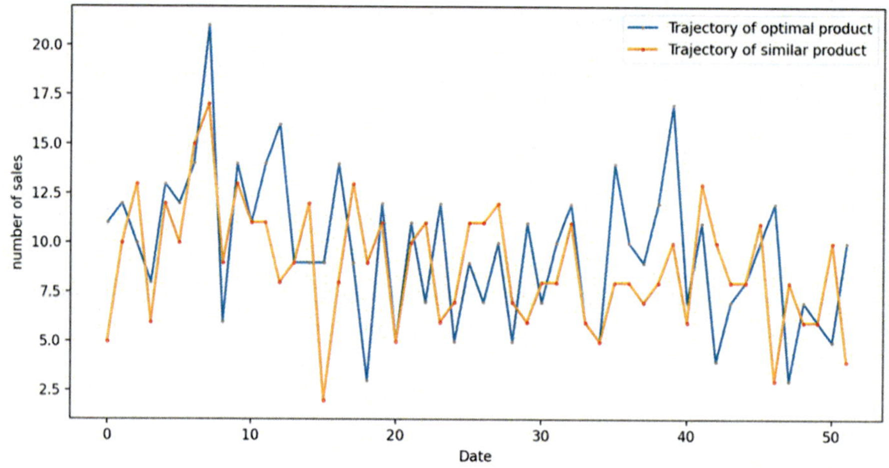

Fig. 10. Visualization of the trajectory of the product Optimal and the product P622.

5 Conclusion

This work proposed a process that determines the trajectories most similar to optimal product sales through similarity measures. The methodology used is based on using the three measures of the similarity calculation (Euclidean, DTW, LCSS) to obtain these similar trajectories. Finally, we can find the best product that looks like the optimal product for sale that represents our best product with these measurements.

References

1. Hu, W., Xie, D., Fu, Z., Zeng, W., Maybank, S.: Semantic-based surveillance video retrieval. IEEE Trans. Image Process. **16**(4), 1168–1181 (2007)
2. Buzan, D., Sclaroff, S., Kollios, G.: Extraction and clustering of motion trajectories in video. In: Proceedings of the 17th International Conference on Pattern Recognition, ICPR 2004, vol. 2, pp. 521–524, August 2004

3. Fan, H., Yao, W.: A trajectory prediction method with sparsity data. In: 2017 IEEE International Symposium on Parallel and Distributed Processing with Applications and 2017 IEEE International Conference on Ubiquitous Computing and Communications (ISPA/IUCC), pp. 1261–1265. IEEE (2017)
4. Furtado, A., Alvares, L., Pelekis, N., Theodoridis, Y., Bogorny, V.: Unveiling movement uncertainty for robust trajectory similarity analysis. Int. J. Geogr. Inf. Sci. **32**, 1–29 (2017)
5. Mao, Y., Zhong, H., Xiao, X., Li, X.: A segment-based trajectory similarity measure in the urban transportation systems. Sensors **17**(3), 524 (2017)
6. Rayatidamavandi, M., Zhuang, Y., Rahnamay-Naeini, M.: A comparison of hash-based methods for trajectory clustering. In: 2017 IEEE 15th International Conference on Dependable, Autonomic and Secure Computing, 15th International Conference on Pervasive Intelligence & Computing, 3rd International Conference on Big Data Intelligence and Computing and Cyber Science and Technology Congress, pp. 107–112. IEEE (2017)
7. Toohey, K., Duckham, M.: Trajectory similarity measures. SIGSPATIAL Spec. **7**(1), 43–50 (2015)
8. Wang, Y., Wang, L., Li, Y., He, D., Liu, T.Y., Chen, W.: A theoretical analysis of NDCG type ranking measures. CoRR abs/1304.6480 (2013)

Check for updates

Classification of Arrhythmias from ECG Using Fractal Dimensions and Wavelet Theory

Ben Ali Sabrine(✉) and Aguili Taoufik

University Tunis El Manar, Tunis, Tunisia

Abstract. The statistics indicate a dramatic increase in mortality due to cardiovascular failures worldwide. Since an electrocardiogram (ECG) is one of the essential tools for diagnosing cardiovascular diseases, it is a test that records the electrical activity of the heart muscle and how it contracts. This study has been presented to analyze this highly inexpensive and available signal. Noise reduction in ECG signals is one of the main problems that appear during the analysis of the heart's electrical activity. The most troublesome noise sources contain frequency components within the ECG spectrum. Such noises are difficult to remove using typical filtering procedures. In this work, we show that the wavelet transforms very useful for denoising such types of signals. ECG signal is a self-similar object. So, fractal analysis can be implemented for the proper utilization of the gathered information. The fractal dimension is the best representative of the ECG signal, which can account for its hidden complexity. This paper introduces a new technique for the honest classification of arrhythmias from ECG signals using the fractal dimension.

Keywords: ECG Signal · Fractal dimension · Wavelet theory · Classification of cardiac diseases

1 Introduction

Cardiovascular disease is the most common cause of death worldwide. Therefore, the diagnosis of these dangerous diseases seems a vital task. The statistics indicate that cardiovascular failures cause a growing number of deaths. According to the World Health Organization, about 17.3 million people died of this complication in 2008 [1]. This figure is expected to reach 23.3 million people by 2030. About 80% of death cases have occurred in countries with low or moderate per capita income, and most of them have been caused by misdiagnosis or late diagnosis of the disease. Since ECG is one of the most cost-effective and accurate tools for diagnosing cardiovascular diseases, this tool can analyze heart behavior [2]. The electrocardiogram (ECG) shows the electrical activity of the heart. It brings together a set of waves that are directly correlated with this activity. The P wave identifies the QRS complex and the T wave, where the P wave reflects atrial depolarization; the QRS complex reflects ventricular depolarization and the T wave reflects ventricular repolarization.

These waves occupy frequency spectra and temporal locations that are generally different for the same signal from the same healthy subject or cardiac pathology. Analysis of ECG recordings can diagnose and locate many pathologies but is not

Y. Maleh et al. (Eds.): ICI2C 2021, LNNS 357, pp. 164–172, 2022.
https://doi.org/10.1007/978-3-030-91738-8_16

sufficient to detect specific abnormalities. For this we need a more reliable technique that allows us to analyze an ECG signal. This study aimed to investigate the application of wavelet denoising in the noise reduction of ECG signals. In particular, the influence of the wavelet function selection and the decomposition level choice on the denoising process. The ECG signals present fractal properties. As a result, the fractal dimension, which represents the non-linear dynamism, is an appropriate method for analyzing and extracting the characteristics from these signals [3, 4].

2 Materials and Methods

The block diagram of the proposed method has been demonstrated in Figure 1. As shown in Figure, the entire methodology is divided into three basic steps: preprocessing, calculating the fractal dimension and classification. In the preprocessing step, the sampling frequencies have been normalized to a base frequency that improves the algorithm's speed and the accuracy of the classification step. After, the de-noising and Baseline wander removal of ECG signals have been done based on the wavelet transform. The second step of the proposed model serves to calculate the fractal dimension. The last step of this proposed method is dedicated to the classification of arrhythmias. Details of these steps are explained in the following sub-sections.

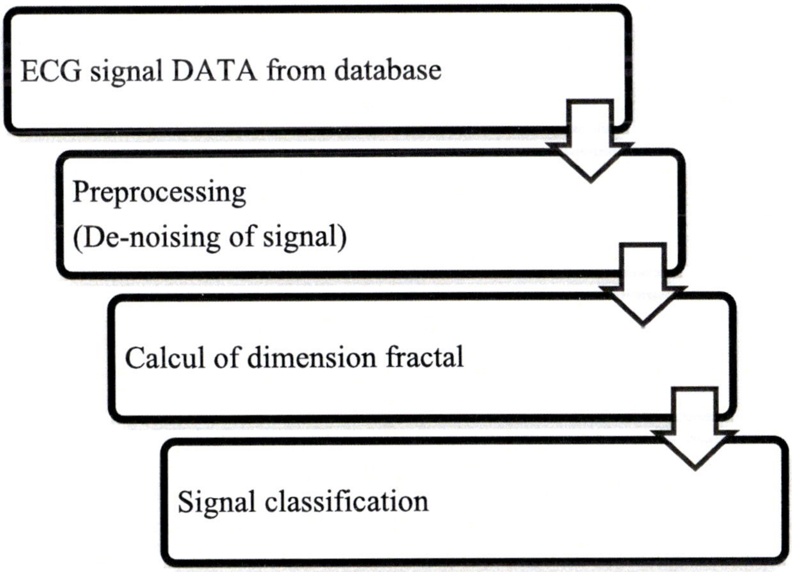

Fig. 1. Block diagram of the proposed method of ECG analysis

2.1 Database

The MIT/BIH physiological database comprises different databases, the most widely used of which is the basis of MIT/BIH arrhythmia [5]. This database contains 48 ECG signal recordings of 47 different cases. Each recording lasts about 30 min, the sampling frequency is 360 Hz and the resolution is 200 samples per mV. All the records include more than 109,000 beats classified into 15 different types of heartbeats. The different pathologies are divided according to the MIT-BIH basis into arrhythmia class. The most abundant class is normal rhythms (NOR), with 70% of the beats.

2.2 Preprocessing

The preprocessing of the ECG signal consists of sampling the QRS complex and determining the reference point relative to the extremum of this complex (R). We used the list of points of reference of the signals provided by the base MIT_BIH. Each reference point (R), made it possible to determine the start of the QRS (Q = R – 50 ms) and its end (S = R + 100 ms) (fig. 2). Therefore the QS interval is 150 ms, or 54 samples for a sampling frequency of 360 Hz [6]. This pretreatment is performed for each complex labeled as heart disease of type LBBB, P, NOR, RBBB, AP, FPN, FVN and PVC from the MIT_BIH base [6].

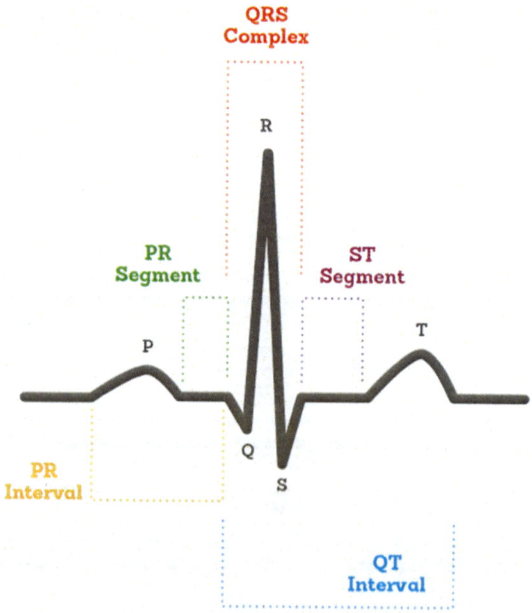

Fig. 2. Determination of the QS interval

In this step, it is necessary to remove noise from input signals. Noise removal in the preprocessing of the ECG signal includes different strategies for each noise source [7]. We will de-noise the ECG signal using the wavelet transform. The first step of the wavelet denoising procedure is selecting mother wavelet ψm,n (t), which forms a set of functions (family of wavelets). In further analysis following mother wavelet was used:

$$\psi_{m,n}(t) = 2^{\frac{m}{2}}\psi(2^{-m}\,t - n) \tag{1}$$

Where n coefficient of time translation
m coefficient of scale (compression).

The next step is the selection of a number of decomposition levels of signal xi(t). As a result of this transformation, so-called approximations (cAj) and details (cDj) are obtained [8] (Fig. 3).

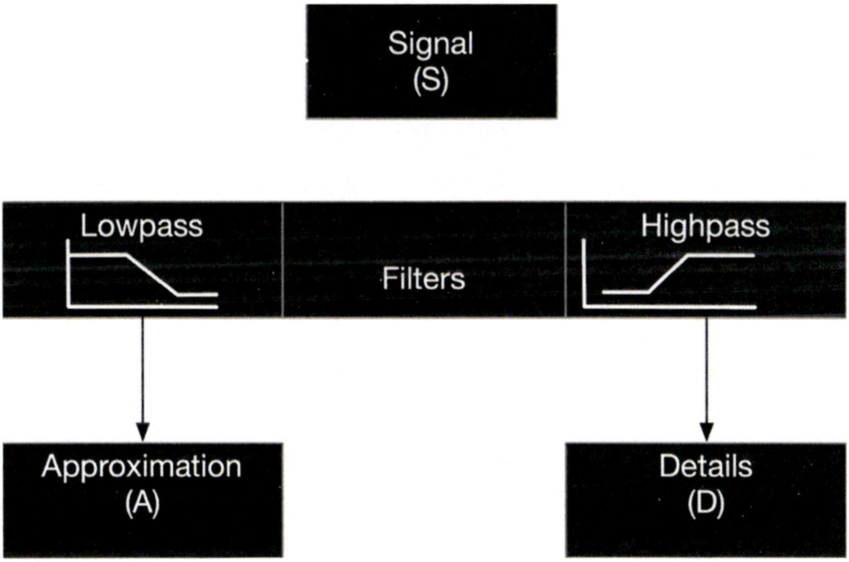

Fig. 3. Decomposition of signal into detail and approximation

Approximations (cAj) correspond to low-frequency components present in signal, while details (cDj) contain high-frequency components. Next levels of decomposition of signal x(t) – detail coefficients cDj+1 and approximation cAj+1 are obtained analogously, when in place of original signal approximation coefficients of j-th decomposition level are analyzed (Fig. 4).

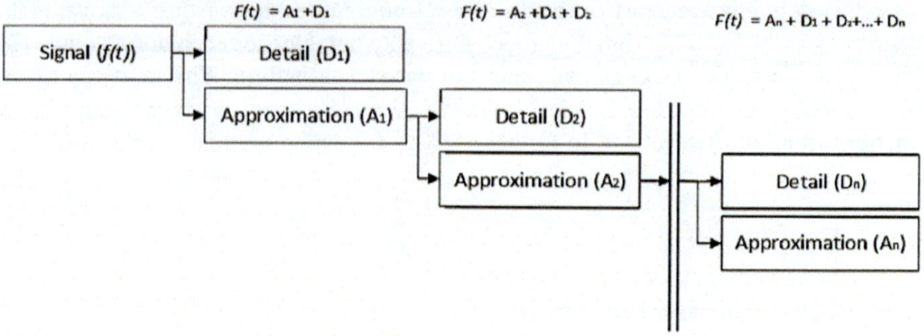

Fig. 4. Diagram of multiresolution analysis of signal

The final step of the analysis is the reconstruction of signal xi(t) based on approximation coefficients chosen i-th level of decomposition (cAi) and modified detail coefficients from i-th (cDi) as well as higher levels of decomposition:

$$X_i(\mathrm{t}) = \sum\nolimits_n cA_{m,n}\, \varphi\, \mathrm{k}\ (2^{-m}\ \mathrm{t} -\ \mathrm{n}) + \sum\nolimits_{m=m0}^{mk} \sum\nolimits_n c\,D_{m,n} \psi_{m,n}(2^{-m}\ \mathrm{t} -\ \mathrm{n}) \quad (2)$$

Where $\varphi(t)$ is scaling function from k-th level of decomposition, ψm,n (t) are wavelet functions for m = m0…mk levels of decomposition. An example of ECG signal denoising by this theory is shown in the following figure (Fig. 5):

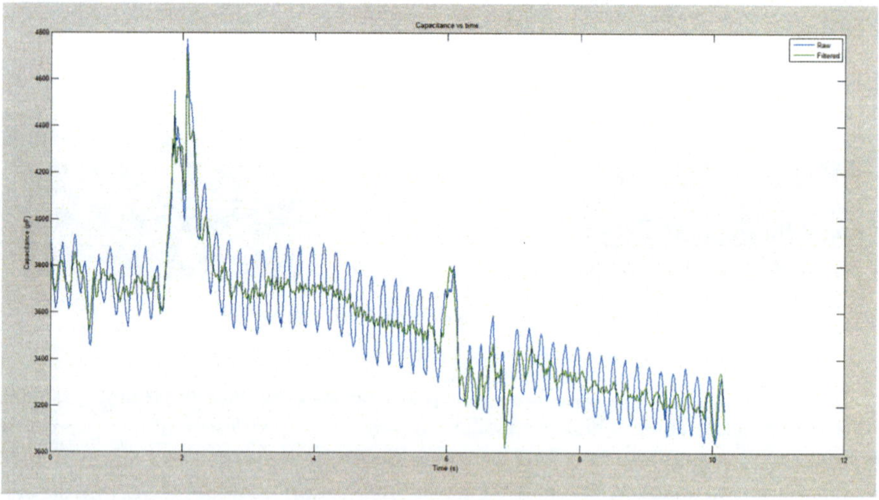

Fig. 5. Enlarged fragment of original (raw) and denoised (filtered) ECG signal from single lead

2.3 Fractal Dimension

The fractal dimension is one of the important characteristics of fractal objects. It is considered the universal index of the degree of their irregularity [9]. "Fractal is strictly self-similar if it can be expressed as a union of sets, each of which is an exactly reduced copy of the full set (Sierpinski triangle, Koch flack). The most fractal looking in nature does not display this precise form. Natural objects are not a union of exact reduced copies of whole. A magnified view of one part will not precisely reproduce the whole object, but it will have the same qualitative appearance. This property is called statistical self-similarity or semi-self-similarity, as the ECG signal of a human heart is a self-similar object. So, it must have a fractal dimension" [10].

2.3.1 Methods of Calculating the Fractal Dimension

There are different algorithms for calculating the Fractal Dimension, such as Katz [12], box-counting [13], Higuchi [14], Regularization [15], etc. Each of them has its advantages and disadvantages. Using the Katz'algorithme for example, the fractal dimension is obtained directly from the time series and can be defined as follows:

$$DFK = (Log(L))/(Log(d)) \tag{3}$$

Where L is the total length of the ECG time series and d is the (Euclidean) distance between the first point in the series and the point that provides the furthest distance concerning the first point. For waveforms, which are ordered sets of pairs of points (x, y), the total length, L, is simply the sum of the distances between successive points, that is:

$$L = sum(dist(i, i+l)) \tag{4}$$

Where $dist(a, b)$ means the distance between two pairs of points $a = (x, y)$ and $b = (u, v)$.

Monotonically advancing waveforms have natural starting points. So, the plane extent or the diameter of a waveform can be considered as the furthest distance between the starting point (point 1) and any other point (point i) of the waveform, namely

$$d = max(dist(l, i)) \tag{5}$$

There is also the Hausdorff' algorithm covering the curve with balls B_i whose diameter (diam B_i) is less than or equal to ε. The sum of the diameters of the balls provides an approximation of the measure of the whole. As the intersections of these balls are not empty, we therefore consider the minimum of these measurements. To obtain the best approximation of E by these overlaps, we tend ε towards 0. The Hausdorff dimension is given by:

$$DF_H = \lim_{\infty \to 0} \frac{\ln N\varepsilon}{\ln\left(\frac{1}{\varepsilon}\right)} \tag{6}$$

Where $N_\in$ is the minimum number of balls of diameter ε, necessary for paving the curve.

These methods and others have been studied and reviewed by Kourosh Kiani and Farzane Maghsoudi [16]. It has been included that Higuchi is an accurate method and a good representative of disease.

2.3.2 Higuchi Algorithm

Fractal Dimension is a descriptive measure that has been proven useful in quantifying the complexity or self-similarity of biomedical signals. The ECG signal of a human heart is a self-similar object, so it must have a fractal dimension that can be extracted using mathematical methods to identify and distinguish specific states of pathological heart conditions [11]. As we said in the previous paragraph, there are different algorithms for calculating the Fractal Dimension. It has been included that Higuchi is an accurate method and a good representative of the disease.

Assume time series of $x = \{x(1), x(2), \ldots, x(N)\}$, the Fractal Dimension is calculated as follows:

a) Construct K new time series $x_{\mathrm{m}}^{\mathrm{k}}$ are defined as:

$$x_m^k = \{x(m), \ldots, x(m + \lfloor (N - m)/k \rfloor k)\} \tag{7}$$

Where $m = 1, 2, \ldots, k$ shows the initial time value, K denotes the discrete-time interval between points.

b) Compute the length of each new time series as previously defined:

$$L_m(k) = 1K/\{(N-1)/\lfloor (N-m)/K \rfloor K\, \Sigma |x(m+ik) - x(m+(i-1)k)|\} \tag{8}$$

Where $(N - 1)/\lfloor (N - m)/K \rfloor K$ is a normalization factor

c) Compute the length of the curve for the time interval K:

$$L(k) = 1/K\, \Sigma_{m=1} k\, L_m(k) \tag{9}$$

d) Finally, according to the following equation, D represents the Fractal Dimension curve

$$log(L(k)) = D\, log(1 / K) + \bar{b} \tag{10}$$

3 Results and Discussion

The ECG signal of a human heart is a self-similar object. In this context, we have used fractal analysis. After filtering the ECG signal using the wavelet theory, we calculate the fractal dimension (FD) based on the Higuchi algorithm by Matlab. In this paper, fractal dimension is used to distinguish between the Electrocardiogram (ECG) signals of a healthy person and persons with Premature Ventricular Complex (PVC), Premature Atrial Contracture (PAC) and Paroxysmal Supraventricular Tachycardia (PSVT) from the raw ECG data. According to the results and concerning the fractal dimension, the ECG signals can be classified as follows:

Normal: FD > 1.56
PAC: 1.37 < FD $\leq$ 1.56
PVC: 1.3 < FD $\leq$ 1.37
PSVT: 1 < FD $\leq$ 1.3

In fact, according to the fractal dimension value, we can distinguish between a disease and a normal case. In addition, we can estimate if it is a PVA, PAC, or PVST. So, the main objective of this study is to classify a variety of arrhythmias with high accuracy based on a fractal dimension.

4 Conclusion

In the present study, wavelet filtration was applied to ECG signals denoising. The wavelet method's advantage is the possibility of receiving good quality signal for a beat to beat. This paper aims to calculate fractal dimension (FD) to classify different cardiac diseases. The use of wavelets and fractal analysis tools is appropriated to the irregular signal analysis because the local regularity is important in the description of these signals.

References

1. Mendis, S., Puska, P., Norrving, B.: World Health Organization. Global Atlas on Cardiovascular Disease Prevention and Control. World Health Organization, Geneva (2011)
2. Kirk, K.J., O'Shea, J., Ruhf, L.K.: ECG interpretation made incredibly easy! (2011). Chris Burghardt.5
3. Sedielmaci, I., Reguig, F.B.: Detection of some heart diseases using fractal dimension and chaos theory. In: 2013 8th International Workshop on Systems Signal Processing and their Applications (WoSSPA), no. 1, pp. 89–94 (2013)
4. Oweis, R., Hijazi, L.: A computer-aided ECG diagnostic tool. Comput. Methods Programs Biomed. **3**, 279–284 (2006)
5. http://www.physionet.org/physiobank
6. de Chazal, P., O'Dwyer, M., Reilly, R.B.: Automatic classification of heartbeats using ECG morphology and heartbeat interval features. IEEE Trans. Biomed. Eng. **51**(7), 1196–1206 (2004)
7. Mary, H.M., Singh, D.: Fractal dimension of electrocardiogram: distinguishing healthy and heart-failure patients. J. Electrocardiol. **1**, e21–e37 (2013)
8. Quaye-Ballard, J.A., et al.: Assessment of vegetation dynamics in upper east region of Ghana based on wavelet multi-resolution analysis. Model. Earth Syst. Environ. **6**(3), 1783–1793 (2020). https://doi.org/10.1007/s40808-020-00789-8
9. Kak, S.: Fractals with optimal information dimension. Circuits Syst. Signal Process. 1–11 (2021)
10. Islam, N., et al.: Detection of some major heart diseases using fractal analysis. Int. J. Biom. Bioinform. (IJBB) **4**(2), 63 (2010)
11. http://www.wahl.org/fe/HTML_version/link/FE4W/c4.htm
12. Kania, M., Fereniec, M., Maniewski, R.: Wavelet denoising for multi-lead high resolution ECG signals. Meas. Sci. Rev. **4**, 30–33 (2007)
13. Raj, V.N., Venkateswarlu, T.: ECG signal denoising using undecimated wavelet transform. In: 2011 3rd International Conference on Electronics Computer Technology, no. 1, pp. 94–98 (2011)

14. Maghsoudi, F., Kiani, K.: A powerful novel method for ECG signal de-noising using different thresholding and dual tree complex wavelet transform. In: Proceedings of the 2th International Knowledge-Based Engineering and Innovation (2015)
15. Donoho, D.L., Johnstone, I.M.: Adapting to unknown smoothness via wavelet shrinkage. J. Am. Stat. Assoc. **1**, 1200–1224 (1995)
16. Kiani, K., Maghsoudi, F.: Classification of 7 arrhythmias from ECG using fractal dimensions. J. Bioinform. Syst. Biol. **2**(3), 53–65 (2019)

Benchmarking Classification Algorithms for Measuring the Performance on Maintainable Applications

Zineb Sabouri(✉), Yassine Maleh, and Noreddine Gherabi

Lasti Laboratory Khouribga, Sultan Moulay Slimane University, ENSA Khouribga, Beni-Mellal, Morocco
zineb.sabouri@usms.ac.ma, n.gherabi@usms.ma

Abstract. Classification is one of the most important techniques, is applicable to handle a large amount of data. It classifies each element of a data set into a set of predetermined classes. Various classification algorithms are being used to organize various data sets. This paper provides a study of various classification methods, including their advantages and disadvantages. Three different types of classification methods were considered: Decision Tree, K-Nearest Neighbor, and Naive Bayes. The study compares the classification methods in terms of accuracy, speed, performance, efficiency, scalability, and other criteria.

Keywords: Decision Trees · Classification · Naive Bayes · Classification Techniques · K-Nearest Neighbor

1 Introduction

Diabetes is one of the most significant global health issues of the twenty-first century. It has a direct signal of high blood sugar, and has affected hundreds of millions of people. Early detection of diabetes is crucial, because of the critical complications that it can cause. There have been various studies on the detection of diabetes, many of which have recently been developed to predict diabetes patients using machine learning algorithms. Research has concentrated on diabetes identification through Logistic regression, DNN, SVM, Decision Tree, and Naive Bayes [1]. They compare the accuracy of each classifier by obtaining some inspiring results.

Supervised learning is one of two principal aspects of machine learning that enables the model to classify outcomes accurately. This means it analyses the training data and uses labeled data to predict the correct outcome by producing the mapping function between input and output variables. Supervised learning is separated into two sub-classes: classification and regression. In regression, the output variable is a continuous range of values, while in classification, outputs are categorical.

Classification is a procedure for classifying a given dataset. It is used to study and examine the existing sample data and predict future behavior to that sample data. The classification task is divided into two phases. The first is the training stage, in which the classification model is created. In The second phase, the data is tested and the accuracy of classification patterns is archived. There are numerous types of classification tasks:

Y. Maleh et al. (Eds.): ICI2C 2021, LNNS 357, pp. 173–179, 2022.
https://doi.org/10.1007/978-3-030-91738-8_17

- Binary classification is the process of classifying the given datasets into two class labels.
- Multi-class classification is the process in which the given datasets are categorized into different class labels.

In this paper, we suggest a survey of the most regularly used classification techniques. This study compares three well-known classification algorithms (Decision Tree, K-Nearest Neighbor, and Naive Bayes) to demonstrate the accuracy and benefits of each algorithm for classification in terms of scalability, performance efficiency, speed, and other issues. In addition, this study compares the advantages and disadvantages of one algorithm against the other.

The remaining content of this work is organized as follows: In Sect. 2, the literature review of related work is presented. Section 3 contains a description of the selected algorithms for this research. Section 4 covers the methodology used and the comparative analysis of these classification algorithms, while in Sect. 5, the paper is concluded.

2 Literature Review

Recently, extensive research has been conducted to make comparative studies for different classification techniques used for various purposes.

For example, Sossi Alaoui S, Farhaoui Y, and Aksasse B [1] proposed a comparison of four algorithms based on different criteria as accuracy and execution time. Also, they use the Weka tool to present the results after implementing the classification algorithms that put the data into predefined groups using the same dataset, which is related to diabetes on women.

Moreover, Shafi'i Muhammad Abdulhamid, Maryam Shuaib, and Oluwafemi Osho [2] focused on evaluating the performance of some classification algorithms, including Naïve Bayes, Bayesian Logistic Regression, Voted Perceptron, Logit Boost, Lazy Bayesian Rule, NNge, Rotation Forest, Logistic Model Tree, REP Tree, Radial Basis Function (RBF) Network, Multilayer Perceptron, Hidden Naïve Bayes, Random Tree and J48, that are used to classify emails as spam or not spam. They used a weka data mining tool to measure the performance of these methods in terms of Precision, Accuracy, Receiver Operator Characteristics Area, F-Measure, and other characteristics.

Also, Shruthi U, Nagaveni V, and Raghavendra B K [3] presented a study that compares five different types of machine learning classification methods for plant disease identification. In this survey, they noticed that Convolutional Neural Networks have a high level of accuracy and can detect a greater number of diseases of multiple crops.

Furthermore, Bendi Venkata, Ramana, and Raja Sarath Kumar Boddu [4] applied different classification algorithms: Naive Bayes, J48, Neural network, Rule-based and KNN on health datasets to evaluate their classification accuracy, sensitivity, precision, specificity, and ROC area.

In addition, Omar E. Elejla, Bahari Belaton, and All [5] compared five classification algorithms to detect the ICMPv6-based DDoS attacks. Several classification algorithms, including Naïve Bayes, Decision Tree, K-Nearest Neighbors, Support Vector Machine, and Neural Networks, have been practically evaluated and applied to identify IPv6 attacks. These classification algorithms are compared in classification accuracy, true positive rate, false-positive rate, precision, f-measure, and training time.

3 Classification Techniques

3.1 Decision Trees

The decision trees Algorithm is a well-known classification technique that belongs to the supervised machine learning family. This algorithm aims to learn simple decision rules based on the conditions present in the data sets to develop a model that predicts the class or value of a target variable. It's a classifier with a tree structure, with each internal node representing a feature, each branch representing a decision rule, and each leaf node representing the class label [6].

According to the attribute's value tested at each iteration, the tree is built by a recursive partition of each node, using the features from the most useful data for the specific mission. This is made for categorical decision trees, by evaluating metrics, such as the Entropy and the Gini index. The process stops when the elements of a node have the same value for the target variable [7, 9].

3.2 Naïve Bayes

Naïve Bayes is a simple probabilistic supervised learning algorithm depending on the Bayes Theorem [7]. It is renowned for outperforming even very advanced classification methods [8]. It provides a group of probabilities by calculating the frequency of values and their combinations in a given dataset [11, 12]. However, the Naïve Bayes classifier is based on a strong assumption that the attribute values are independent given the target value. So, it is characterized as Naive because this independent assumption is never true in real-world applications [8, 9].

The Bayes' theorem is a mathematical formula for calculating conditional probabilities of events:

$$p(A|B) = \frac{p(B|A) \cdot p(A)}{p(B)}$$

$p(A|B)$: Probability of A given that B is true.
$p(B|A)$: Probability of B given that A is true.
$p(A)$: Independent probability of A.
$p(B)$: Independent probability of B.

3.3 K-Nearest Neighbor

K-Nearest Neighbor algorithm is among the most simple, nonparametric lazy algorithms and is commonly used in classification techniques with various applications

[13]. The kNN classifier measures the similarity between the new point and each training data by searching for the group of K most identical neighbors to estimate the value or class of the new point [14, 15]. This means that a value is attributed to the new point based on proximity to the points in the training set [8, 9].

The calculating distance is defined by different methods, which the most commonly known are Euclidian and Hamming distance wherein Euclidian term, the distance between two elements, X = (x1,x2,……,xn) and Y = (y1,y2,….,yn) is denoted by d(X, Y) [7]:

$$d(x,y) = \sqrt{\sum_{i=1}^{n}(y_i - x_i)^2}$$

4 Comparison Among K-NN, Naive Bayes and Decision Trees Techniques

The comparative study of the algorithms mentioned above has been done relative to some important parameters, as shown in Table 1 [10].

Table 1. Differences among frequently used classification methods. (+++ plus signs represent the best and + plus sign the worst performance)

Comparison parameters	Decision Trees	Naïve Bayes	K-Nearest Neighbor
Proposed by	Quinlan	Duda and Hurt	Cover and Hart
Deterministic	Yes	No	No
Speed of Learning	+	++++	++++
Speed of classification	++++	++++	+
Performance with the presence of missing value	+	++++	+
Tolerance to irrelevant attributes	+++	++	++
Noise tolerance	++	+	+
Effectiveness on	Large data	Huge data	Small data
Dealing with discrete binary/continuous attributes	All	Not continuous	All
Tolerance with parity problems	++	+	+
Attempts for incremental learning	++	++++	++++
Clarity on Classification prediction	++++	++++	+
Support Multi classification	++++	+++	++++
Accuracy	++	+	+++

5 Advantages and Disadvantages

Each technique has some advantages and disadvantages. We give some benefits and limitations of these methods.

Table 2. Classification techniques advantages and disadvantages.

Algorithms	Advantages	Disadvantages
Decision Trees	- Easy interpretation - Normalization of data is not required - Assignation of specific values to each problem - Familiar with programmers	- Unstable nature - Consumes more memory - High complexity
Naïve Bayes	- Good performance - Short computational time - Resolution of regression and classification issues	- Assumption is the independence of features - Big quantity of records is needed - storage of the training samples
k-Nearest Neighbor	- Easy implementation - Rapid training - Suitable for multi-modal classes	- Determination of values of the parameter - Computation complexity - Memory limitation - Poor run-time performance

6 Discussion

In this work, the discussion was held on three popular machine learning classification algorithms such as decision trees, naïve bayes and k-nearest neighbour by describing their working mechanism and showing their strengths and weaknesses as shown in Table 1 and Table 2.

In terms of naive bayes, it needs a small amount of memory during the classification and the training phases. It can examine all features without taking dependencies into account, so it is better to use it with more than two class labels. Like: spam filtering, recommender systems, text classification.

The next option is decision tree that it can be understood by non-programming users if the tree has a small number of leaves, and it can improve efficiency and clarity with fewer classes. Such as, the problems that can be represented as a set of (if –then) rules. Decision tree faces a lot of challenges as example: overfitting and instability.

The last option is the K-nearest neighbor that it uses a lot of storage space during execution and training stages. It is highly used in recommended system, and search applications. Knn has the highest accuracy, after that, decision tree has good accuracy and naive bayes has the lowest accuracy among them. The comparative study has shown that none of the discussed classification techniques can satisfy all the criteria because each learning algorithm is suitable for a particular data set and its area of implementation.

7 Conclusion

The discussed classification techniques can be used on various data sets, including health, financial, and other types. It's difficult to predict which technique is better because each algorithm has its characteristics and implementation challenges.

This study aimed not to figure out which technique is superior to others but to know that the selection of the correct technique depends on the required application conditions and domain. Future work will focus on machine learning techniques to measure semantic similarity and improve old Works [16, 17]. We will exploit other methods such as AHP [18], AWOT [19], FUZZY AHP [20, 21] for algorithm selection.

References

1. Sossi Alaoui, S., Farhaoui, Y., Aksasse, B.: A comparative study of the four well-known classification algorithms in data mining. In: Ezziyyani, M., Bahaj, M., Khoukhi, F. (eds.) Advanced Information Technology, Services and Systems. AIT2S 2017. LNNS, vol. 25. Springer, Cham (2018). https://doi.org/10.1007/978-3-319-69137-4_32
2. Shuaib, M., Osho, O., Ismaila, I., Alhasan, J.K.: Comparative analysis of classification algorithms for email spam detection. Int. J. Comput. Netw. Inf. Secur. **10**(1), 60 (2018)
3. Shruthi, U., Nagaveni, V., Raghavendra, B.K.: A review on machine learning classification techniques for plant disease detection. In: IEEE 5th International Conference on Advanced Computing & Communication Systems (ICACCS). Coimbtore, India (2019)
4. Ramana B.V., Kumar Boddu R.S.: Performance comparison of classification algorithms on medical datasets. In: 2019 IEEE 9th Annual Computing and Communication Workshop and Conference (CCWC), pp. 0140–0145 (2019)
5. Elejla, O.E., Belaton, B., Anbar, M., Alabsi, B., Al-Ani, A.K.: Comparison of classification algorithms on ICMPv6-based DDoS attacks detection. In: Alfred, R., Lim, Y., Ibrahim, A., Anthony, P. (eds.) Computational Science and Technology. Lecture Notes in Electrical Engineering, vol. 481. Springer, Singapore (2019). https://doi.org/10.1007/978-981-13-2622-6_34
6. Bhargava, N., Sharma, G., Bhargava, R., Mathuria, M.: Decision tree analysis on j48 algorithm for data mining. Proc. Int. J. Adv. Res. Comput. Sci. Softw. Eng. **3**(6) (2013)
7. Gorade, M.S.M., Deo, A., Purohit, P.: A study of some data mining classification techniques. Int. Res. J. Eng. Technol. (IRJET) **4**, 3112–3115 (2017)
8. Nejjahi, R., Gherabi, N., Marzouk, A.: Towards classification of web ontologies using the horizontal and vertical segmentation. In: Noreddine, G., Kacprzyk, J. (eds.) International Conference on Information Technology and Communication Systems. ITCS 2017. Advances in Intelligent Systems and Computing, vol. 640, pp. 70–81. Springer, Cham (2018). https://doi.org/10.1007/978-3-319-64719-7_7
9. Soofi, A.A., Awan, A.: Classification techniques in machine learning: applications and issues. J. Basic Appl. Sci. **13**, 459–465 (2017)
10. Sen, P.C., Hajra, M., Ghosh, M.: Supervised classification algorithms in machine learning: a survey and review. In: Mandal, J.K., Bhattacharya, D. (eds.) Emerging Technology in Modelling and Graphics. AISC, vol. 937, pp. 99–111. Springer, Singapore (2020). https://doi.org/10.1007/978-981-13-7403-6_11

11. Saritas, M., Yasar, A.: Performance analysis of ANN and Naive Bayes classification algorithm for data classification. Int. J. Intell. Syst. Appl. Eng. **7**(2), 88–91 (2019)
12. Vembandasamy, K., Sasipriya, R., Deepa, E.: Heart diseases detection using Naive Bayes algorithm. IJISET-Int. J. Innov. Sci. Eng. Technol. **2**, 441–444 (2015)
13. Sharma, S., Agrawal, J., Agarwal, S., Sharma, S.: Machine learning techniques for data mining: a survey. In: 2013 IEEE International Conference on Computational Intelligence and Computing Research, pp. 1–6 (2013)
14. Kuhkan, M.: A method to improve the accuracy of k-nearest neighbor algorithm. Int. J. Comput. Eng. Inf. Technol. **8**(6), 90–95 (2016)
15. Okfalisa, I., Gazalba, M., Reza, N.G.I.: Comparative analysis of k-nearest neighbor and modified k-nearest neighbor algorithm for data classification. In: 2017 2nd International Conferences on Information Technology Information Systems and Electrical Engineering (ICITISEE), pp. 294–298 (2017)
16. Daoui, A., Gherabi, N., Marzouk, A.: An enhanced method to compute the similarity between concepts of ontology. In: Noreddine, G., Kacprzyk, J. (eds.) International Conference on Information Technology and Communication Systems. ITCS 2017. Advances in Intelligent Systems and Computing, vol. 640, pp. 95–107. Springer, Cham (2018). https://doi.org/10.1007/978-3-319-64719-7_9
17. Daoui, A., Gherabi, N., Marzouk, A.: A new approach for measuring semantic similarity of ontology concepts using dynamic programming. J. Theor. Appl. Inf. Technol. **95**(17), 4132–4139 (2017)
18. Bourhim, E.M., Cherkaoui, A.: Selection of optimal game engine by using AHP approach for virtual reality fire safety training. In: Abraham, A., Cherukuri, A., Melin, P., Gandhi, N. (eds.) Intelligent Systems Design and Applications. ISDA 2018 2018. Advances in Intelligent Systems and Computing, vol. 940. Springer, Cham (2020). https://doi.org/10.1007/978-3-030-16657-1_89
19. Bourhim, E.M., Cherkaoui, A.: Exploring the potential of virtual reality in fire training research using A'WOT hybrid method. In: Bindhu, V., Chen, J., Tavares, J.M.R.S. (eds.) International Conference on Communication, Computing and Electronics Systems. LNEE, vol. 637, pp. 117–127. Springer, Singapore (2020). https://doi.org/10.1007/978-981-15-2612-1_11
20. Bourhim, E.M., Cherkaoui, A.: Usability evaluation of virtual reality-based fire training simulator using a combined AHP and fuzzy comprehensive evaluation approach. In: Jacob, J.I., Shanmugam, S.K., Piramuthu, S., Falkowski-Gilski, P. (eds.) Data Intelligence and Cognitive Informatics. Algorithms for Intelligent Systems. Springer, Singapore (2021). https://doi.org/10.1007/978-981-15-8530-2_73
21. Bourhim, E.M., Cherkaoui, A.: Efficacy of virtual reality for studying people's pre-evacuation behaviour under fire. Int. J. Hum. Comput. Stud. **142**, 102484 (2020). ISSN 10715819

Application of Machine Learning Techniques for Credit Risk Management: A Survey

Adil Oualid[1(✉)], Abderrahim Hansali[2], Youssef Balouki[1], and Lahcen Moumoun[1]

[1] MISI Laboratory, FSTS, Hassan 1st University, Settat, Morocco
{a.oualid,youssef.balouki,lahcen.moumoun}@uhp.ac.ma
[2] LAREGO Laboratory, ENCG, Cadi Ayyad University, Marrkech, Morocco
a.hansali@uca.ma

Abstract. Artificial Intelligence (AI) techniques will significantly impact the financial services industry, with various implications, ranging from the redefinition of processes, products, and services to transforming the way we interact with customers. Despite the still tentative benefits of AI, given the impact already being seen in the financial markets and the potential disruptive power of the entire banking industry, financial institutions worldwide are making large-scale investments in artificial intelligence. The banks that will reap the most benefits from MI innovations are most prepared and willing to make changes and adapt their approaches to people, processes, and data. In the context of innovations in finance, this paper has two main dimensions. The first dimension aims to present an overview of financial innovations and applications. The second dimension of this work seeks to illustrate the use of MI techniques, more particularly machine learning algorithms in risk management.

Keywords: Fintech · Artificial intelligence · Machine learning · Applications in banking · Credit risk management

1 Introduction

Machine Learning (ML) techniques will have a huge impact on the financial services industry, with several implications such as (i) redefining processes, products, services and (ii) transforming the way we interact with customers. A study by the Bank of International Settlements (BIS) indicates that artificial intelligence brings substantial revolutions in the financial segment, redesigning models and mechanisms of communication and interaction between market participants [1]. However, banks are currently deriving limited benefits from AI, given the difficulties in effectively adopting and incorporating this new technology into business, mainly due to the need for change in management models [2]. According to a study by KPMG [2], banks that overcome this management difficulty will have significant competitive advantages. Despite the still timid benefits of AI, considering the impact already being observed in financial markets and the potential disruptive transformation power of the entire banking industry, financial institutions worldwide are making large-scale investments in artificial intelligence [3]. In this context, Plaschke et al. [4] suggest that artificial

Y. Maleh et al. (Eds.): ICI2C 2021, LNNS 357, pp. 180–191, 2022.
https://doi.org/10.1007/978-3-030-91738-8_18

intelligence, and in particular machine learning algorithms, will transform the finance function, requiring strategic positioning to compete in a market that can create substantial value in the coming years. More particularly, as McKinsey 4 Global Institute estimates, artificial intelligence tools can generate an additional value of more than USD 250 billion in banking [5]. According to Jubraj et al. [6], corroborating the KPMG (2020) analysis results, the banks that will gain the most benefits from AI innovations are most prepared and willing to adapt their approaches concerning people, processes and data. In the context of AI adoption, according to a study prepared by McCauley et al. [7], financial services companies are among the pioneers, just behind information technology and telecom companies and ahead of consumer goods and retail companies. Notably, the value gains from financial innovation may also be accompanied by increased potential risks to the market. For example, as Muniesa and Lenglet [8] suggest, financial innovation played a prominent role in the credit crisis of the late last decade, fueling the debate about the responsibilities of the banking services industry.

Machine learning-based solutions make a decisive contribution to risk management on several levels: improving data quality, perfecting and accelerating the risk modeling process and risk anticipation; and increasing the reliability and speed of reporting.

Since the financial crisis, regulators have been paying particular attention to risk management, expecting financial institutions to have transparent and auditable risk measurement tools. At the same time, start-ups and fintech, encouraged by digital growth and the increasing amount of available data, are increasingly using artificial intelligence solutions. To keep up with these new players, banks are being driven to develop more reliable models to reduce decision time and improve profitability.

Quantitative modeling techniques are being used to get more out of the data, reduce costs and increase overall profitability. Every model has inherent flaws: it is essential to ensure that these errors are reduced. At the same time, machine learning methods are developing at a rapid pace. The processing capabilities of computers now allow the use of methods discovered years ago, such as Machine Learning, decision tree-based algorithms (such as Decision Trees, Random Forest and Gradient-Boosting Machines).

In the context of innovations in finance, this paper has two main dimensions. The first dimension aims to present an overview of financial innovations, focusing on credit risk management. Thus, financial innovations arise in a context in which assumptions, such as the absence of transaction costs or the existence of complete markets, are violated. The second dimension of this work seeks to illustrate ML techniques for banking risk management applications. The paper is structured as follows. In the next section, a research methodology is conducted. Subsequently, the ML methods that will be explored in this study are briefly described, aiming to illustrate the use of machine learning techniques for credit risk management in the banking industry. The results of the study are discussed and finally, some comments are presented, including future directions.

2 Research Methodology

In this section, we present the methodology used in this survey. A systematic mapping study (SMS - Systematic Mapping Study) [9] was conducted on the current state of ML techniques in credit risk management to assist in developing the research proposal. An SMS is a standard methodology to help conduct research, such as in the medical area. Since 2007, its applicability in the Computing area has grown substantially, especially in Software Engineering [10].

According to Petersen [9], SMS provides a categorization structure, offering a visual summary, mapping, and results. This requires less effort while providing an overview with a better selection of information.

Initially, SMS was geared towards software engineering and was usually recommended for research areas with a range of relevant and high-quality early-stage studies. As ML for fintech is at an early stage and high-quality studies are emerging, SMS can categorize these studies.

The SMS was first organized by defining the research questions. For this, it was necessary to think about the opportunities that ML for the fintech industry produces. The primary categories related to the topic were defined: machine learning, fintech, credit risk. To highlight the necessity of creating an up-to-date and detailed survey, we propose the main research question that we will address in this paper: **What are the main algorithms and methods powered by machine learning for credit risk management?**

To select the papers relating to this survey, we rely on a three-stage selection procedure: (1) identifying search terms, (2) selection of sources, and (3) identifying and applying inclusion/exclusion criteria for selected papers:

- **Search terms:** This stage aims to select the most appropriate keywords in terms of providing our survey. Also, synonyms, alternative spellings of the main elements of the field of threats, and mitigations in the fintech context are included to find relevant papers. Therefore, keywords such as *"Machine Learning for Fintech", "Fintech ML", "ML for fraud detection", "ML for financial risk management", "Supervised learning for fraud detection", "unsupervised learning for Fraud detection",* were used for finding relevant papers.
- To define the search string, the Boolean operations "OR" was used to select optional words and synonyms and "AND" to choose the terms related to population, intervention, and effect, resulting in the search string (("fintech" OR "credit risk" OR "financial risk) AND ("Machine Learning" OR "ML" OR "supervised ML" OR "Unsupervised ML" OR "applications")).
- **Selection of sources:** Table 1 shows the popular digital academic databases selected for searching at this stage. The articles from journals as well as conference proceedings are considered for our survey. Some references for downloaded articles were looked upon, and publications related to interest were also included. Other databases were not used due to either the complexity (or impossibility) of instantiating the search string, presenting a high index of repeated articles with the IEEE and ScienceDirect databases, or performing only one indexing (like Google Scholar for example).

Table 1. Digital databases used in the SRL

Online databases #	Online database	URL
1	IEEE	http://ieeexplore.ieee.org/
2	ScienceDirect	http://www.sciencedirect.com/
3	ACM	http://www.acm.org/
4	SpringerLink	http://link.springer.com/

1- **Inclusion/exclusion criteria:** Table 2 shows the summary of the inclusion-exclusion criteria for the SMS.

Table 2. Summary of the inclusion-exclusion criteria for paper selection

Criterion	Rational
Inclusion 1. The early-stage study should propose an application of ML in fintech	Fintech is our reference field. Thus, we need papers that directly propose ML classification for credit risk management
Inclusion 2. The early phase study must describe ML applications for fintech	Both academic and industrial studies are essential to this study
Inclusion 3: Publication period	Papers published between 2015 and 2020
Exclusion 1. A study that does not focus on ML for fintech	The emphasis of this paper is only on papers that discuss ML for credit risk management
Exclusion 2. Non-English papers	The papers that did not publish in English are excluded
Exclusion 3. The study has not been published between 2015 and 2020	The papers will be excluded after a full-text reading. If the focus is not published between 2015 and 2020

3 The Proposed ML Techniques for Financial Risk Management Applications

Classification's problems are concerned with identifying the class to which a given observation is most likely to belong. For example, in credit scoring applications, one can use machine learning techniques to identify the group of defaulters to which a given borrower is most likely to belong. In fraud analysis applications, one seeks to determine whether a given financial transaction is more likely to be legitimate or fraudulent.

More specifically, this study analyzes machine learning applications that address credit risk management.

3.1 Supervised Machine Learning Techniques

In recent years, especially after the financial crisis, financial institutions have started to pay attention to a more accurate assessment of the credit risk and their customer's ability to pay more. While the assessment and measurement of corporate credit risk are still partly done manually, the assessment of personal and household customers is largely automated. This is due to the high demand for household and personal products faced by financial institutions, making manual processes, imprecise and inefficient methods unsuitable for assessing credit risk. It should also be noted that, given the extremely high volume of credit granted, even small improvements in credit risk prediction models and processes can generate substantial additional revenues for financial institutions [11]. The weaknesses of traditional models used to estimate credit risk are related to their assumptions and limitations regarding the data under study. Traditional models do not perform optimally in analyses based on huge datasets nowadays, where the number of potential explanatory variables of events can be large. For these reasons, financial institutions have sought to find new rapidly adaptive models for credit risk analysis. These data analysis techniques, often based on machine learning or artificial intelligence, are worth exploring for the potential benefits they offer. The modern analytical methods discussed in this paper differ from the traditional models, in particular, because of their non-parametric approach. The models are divided into two different categories, which are discussed in the following subsections. This is done by illustrating the main differences and strengths of the models in a simplified way. This is due to the complexity of the models and the nature of this survey, which examines the utility of the models in managing credit risk rather than their mathematical properties.

Supervised machine learning algorithms refer to non-parametric models used in data analytics that aim to summarize the effect of different factors on the outcome of a given event and to divide different observation units into two different categories based on this outcome. In addition to classifying customers, these models seek to identify the individual parameters of the characteristics of customers seeking credit that significantly impact the customer's ability to pay. This process, where the model "learns" to weigh different features based on their importance, is called model training. Training is a basic prerequisite for the optimal functioning of the model. For example, suppose a machine learning model is to classify credit applicants as accurately as possible into so-called good and bad customers based on the probability of risk. In that case, the model must first understand the factors that influence the probability of risk. There are several guided algorithms for assessing the credit risk. As with traditional statistical methods, these models attempt to divide the population under analysis into two categories as precisely as possible. However, there are differences in how the models work, making specific models more popular for assessing the credit risk. The most popular models are Support Vector Machine (SVM) and Artificial Neural Networks (ANN) [12]. The support vector model, first described by Cortes and Vapnik [13] in their study, is based on the partitioning of observation units by vectors, while the Artificial Neural Network model partitions observation units into different categories, mimicking the functioning of the human brain to find complex relationships between factors and outcomes. In addition to these machine learning models, various decision tree classifiers are popular for tasks requiring binary classification [14]. In decision tree models, observation units are divided into steps using a tree-like graph consisting of nodes and branches. In decision tree models, each node represents a feature of the sample and a branch represents the value of that feature.

Paying back given loans is one of the bank's significant issues since the beginning of financial services. Therefore, banks have always used several methods to assess credit risks.

In this scope Bayraktar et al. [15] proposed credit risk analysis with classification Restricted Boltzmann Machine which relies on machine learning methods such as Classification Restricted Boltzmann Machine and Multilayer Artificial Neural Networks. Depending on the results of the application scores the client and loan lenders decide to approve or deny credits. The proposed methods were applied banking datasets that are blacked. 9 of ŞCD (Şekerbank Credit Dataset) consists of 35 features, of which categorical and 26 are numerical. ACD (Australian Credit Dataset) 8 categorical and 6 numerical It consists of 14 attributes. GCD (German Credit Dataset) consists of 20 features, 17 of which are categorical and 3 are numerical.

In the same perimeter Yu Li [16] proposed a credit risk prediction based on machine learning methods. Used an XGBoost, which is an implementation of gradient boosting algorithm for regression and classification. XGboost produces a prediction model to distinguish bad customs from good ones.

Yu li used data as an observation period chosen as the latest credit report within 2 months before the application date. The performance period is 15 months after the application date. The response variable is defined by two categories: (1) bad samples: 90 days overdue within 15 months +; (2) good samples: no more than 89 days overdue within 15 months. To ensure the predictive modeling effect and the speed of algorithm operation, stratified sampling was carried out to get the model samples, including 45,000 good samples, 5,000 bad samples and 9:1 ratio of good samples to bad samples.

The study concluded that the model discrimination and model stability of the XGBoost model is significantly higher than that of the logistic regression model, which can effectively improve the identification ability of personal fast credit risk.

Shivanna et al. [17] attempted to predict payment defaulter better using Deep Support Vector Machine (DSVM), Boosted Decision Tree (BDT), Averaged Perceptron (AP) and Bayes Point Machine (BPM) on a dataset, comprising of 25 attributes and 30k instances, which was obtained from the repository of machine learning, University of California, Irvine (UCI).

Khatri et al. [18] used several supervised machine learning algorithms for credit card fraud detection. kNN, decision tree, random forest, Naive Bayes and Logistic Regression were applied to a publicly available imbalanced dataset, This particular dataset contains the record of transactions made by European cardholders. It includes the records of 284,807 operations made over two days. The study showed that the sensitivity of the kNN model is greater than that of the Decision tree, but as the time taken by kNN for testing the data is very large, the Decision Tree was chosen over kNN. In the case of fraud detection, a minimum time is taken for prediction. Therefore, the Decision Tree is the preferred model.

3.2 Unsupervised Machine Learning

Unsupervised machine learning models, which in most cases refer to a kind of clustering of observation units, differ from supervised machine learning and artificial intelligence models because of their purpose and principle of operation. While guided models focus on generating predictions about an observation unit, unsupervised models are mainly used to mine data and identify different clusters within a population [19]. Such models include the k-means clustering algorithm and a self-organizing map developed by Kohonen (1998). As both algorithms are very complex, their actual working principles of mathematical properties will not be discussed in this paper. Instead, it is important to understand that unsupervised machine learning and artificial intelligence algorithms are mainly used in credit risk analysis in the context of supervised models. For example, clustering can be used to divide a population into groups before the actual classification based on the risk probability. This can improve the accuracy of the classification process itself.

Bao et al. [19] compare different supervised and unsupervised machine learning and artificial intelligence models in credit risk prediction performance using three different credit databases. The study evaluates the most common and most studied models, whose performance is measured using different measures such as accuracy (ACC) and Matthews correlation coefficient (MCC). Particular emphasis is given to the correlation coefficient of Matthews. The study also assesses the ability of unsupervised machine learning and artificial intelligence models to improve the predictions of supervised models.

4 Comparison and Discussion

It soon becomes apparent that machine learning provides several models, advantages and applications. There is no unanimity among the models. The unanimity is in saying that each problem has its peculiarity that fits a particular model. It is up to the banking institution to choose its champion model through its profile, strategies and tests. Regardless of the choice of model, what should be explained is that machine learning:

- Defines a probability of the credit risk;
- Finds unsupervised behavioral patterns;
- Provides a deeper evaluation of credit variables that helps in the inclusion of important variables and excludes variables that do not prove relevant, with the advantages of providing more, concise, accurate, flexible, and optimizable credit management models—minimizing computational effort, time savings in execution and better accuracy in decisions.
- It provides efficient credit risk analysis, minimizes the risk of failure.

Machine learning simplifies credit operations at the operational level, increases information security, and ensures regulatory compliance of credit contracts by preventing criminal behavior by employees and customers (corruption and money laundering). It improves business decisions as they can nimbly analyze past default rates, geographic data, information from various customer sources to predict possible future outcomes. In addition, machine learning offers credit risk, an alternative approach to digitizing core processes, where algorithms themselves determine risk drivers. Thus, minimizing manual interventions by modeling, simplifying, standardization and automation and enabling analysis of a larger volume of customer information at lower operating costs improves overall data quality, aggregation capabilities, and opportunity reporting. They can innovate in providing reports, replacing paper with an interactive solution that would offer a large amount of credit risk information in real-time. Such reports would enable users to do their analysis, and banks to make decisions more quickly. In addition, real-time credit risk reports based on transparent and consistent data would improve the quality of fact-based decisions, help banks identify potential risks even earlier, and enable a more timely response. Machine learning can align credit management with the company's financial strategy, optimizing the balance sheet. Indicating the best time to increase or decrease a loan portfolio [4].

Table 3 provides a list of the papers reviewed, classifying them by the risk type, including the risk management method addressed, the ML algorithms, and the dataset used in the referred applications.

Table 3. Literature review

Reference	Risk type	Risk management method	ML technique	Dataset
Bao et al. [19]	Credit risk	Credit scoring	Kohonen's self-organizing maps (SOM)	Chinese P2P credit dataset, German and Australian credit datasets
Barboza et al. [20]	Credit risk	Exposure (PD, LGD, EAD)	Neural Network, SVM, Boosting, Bagging, Random Forest	Financial data on American and Canadian companies covering 1985 to 2013
Hamori et al. [21]	Credit risk	Scoring Models	Bagging, Random Forest, Boosting	The German credit dataset
Lessmann et al. [22]	Credit risk	Scoring Models	Multiple algos assessed	Australian credit (AC) and German credit (GC)
Yao et al. [23]	Credit risk	Exposure (PD, LGD, EAD)	SVM	N/A
Yao et al. [24]	Credit risk	Exposure (PD, LGD, EAD)	east squares support vector regression SVR	UK credit card lender
Ala'Raj and Abbod [25]	Credit risk	Scoring Models	Multiclassifer system (MCS)—Ensemble—neural networks (NN), support vector machines (SVM), random forests (RF), decision trees (DT) and naïve Bayes (NB)	German, Australian and Japanese credit cart scoring datasets
P. Caroline CynthiaS and Thomas George [26]	Fraudulent transactions		Local outlier factor, isolation forest, support vector machine, and logistic regression	European cardholders

(continued)

Table 3. (*continued*)

Reference	Risk type	Risk management method	ML technique	Dataset
Bayraktar et al. [15]	Credit risk	Scoring Models	KNN, SVM, C4.5, LR	Şekerbank Credit Dataset, Australian Credit Dataset, German Credit Dataset
Tavana et al. [27]	Liquidity Risk	Scoring Models	ANN, Bayesian Networks	Real bank dataset
Min Tae et al. [28]	Credit risk	Scoring Models	Random Forests, AdaBoost, LR; KNN, NB, Decision Tree, Neural networks	Kaggle Credit Card Fraud Detection
de Roux et al. [29]	Tax fraud	Scoring Models	Kernel density estimation	1,367 tax declarations of building projects in the city of Bogotá, Colombia
Xia et al. [30]	Credit risk	Credit scoring	SVM, Bagging, DT, LR	Germany, Australia
Monamo et al. [31]	Bitcoin Fraud	Credit scoring	k-means clustering, Mahalanobis distance and unsupervised Support Vector Machines (SVM)	Bitcoin dataset housed by the Laboratory for Computational Biology at the University of Illinois
Awoyemi et al. [32]	Credit risk	Credit scoring	Naïve Bayes (NB), k-nearest neighbor (KNN) and logistic regression. Hybrid technique	European cardholder dataset
Sanford et al. [33]	Operational Risk	Operational Risk Losses	Bayesian networks	N/A
Shivanna et al. [17]	Credit risk	Credit scoring	Deep Support Vector Machine (DSVM), Boosted Decision Tree (BDT), Averaged Perceptron (AP) and Bayes Point Machine (BPM)	Downloaded from UCI website
Keswani et al. [34]	Credit risk	Credit scoring	Logistic Regression, K-nearest neighbor, SVM, and decision tree	The dataset used is captured from university libre de Bruxelles Web
Khatri et al. [18]	Credit risk	Credit scoring	Decision Tree, kNN, Random Forest, Naive Bayes, Logistic Regression	Imbalanced dataset by European cardholders

5 Conclusion

This paper examines the various ML methods and models used to assess credit risk through their risk type and method. At the same time, it has sought to answer the research question: what are the benefits of using machine learning methods and models to manage credit risk? The paper has first introduced the components and concepts that make up credit risk, followed by a specific focus on assessing credit risk and the methodology used in this process. In contrast to previous research literature, the purpose of this paper has not been to explore the mathematical properties of the advanced models and the underlying theories of the models in-depth or to describe them in detail. Instead, in addition to answering the research question, the paper aimed to develop an understanding of the application of machine learning techniques for credit risk.

There are several potential further research questions, as the paper discusses different ML models through a literature review. A natural extension of the topic would be to empirically investigate the performance of the models using differently weighted data sets, which would allow, for example, to investigate the effects of different distributions of variable classes on accuracy. It would also be interesting to explore other centralized solutions for assessing credit risk, which financial institutions could potentially use in their role as customers.

References

1. BIS: Big tech in finance: opportunities and risks. BIS Annu. Econ. Report. Bank Int. Settlements, vol. Access on (2019)
2. KPMG: Living in an AI World. Financial Services Report, Relatório Técnico. KPMG, 2020. Disponível em (2020). https://advisory.kpmg.us/content/dam/advisory/en/pdfs/2020/financial-Serv. Accessed 12 June 2021
3. McWaters, R., Galaski, J.R.: The new physics of financial services: understanding how artificial intelligence is transforming the financial ecosystem. World Econ. Forum. (2018)
4. Plaschke, B.F., Seth, I., Whiteman, R.: Bots, algorithms, and the future of the finance function. McKinsey, pp. 1–9 (2018)
5. Franke, M.: Machine learning and artificial intelligence. In: Managing Airline Networks, no. February 2019, pp. 153–159 (2020). https://doi.org/10.4324/9780429318795-15
6. Jubraj, E., Graham, R., Ryan, T.: Redefine banking with artificial intelligence. Intell. Bank 1–20 (2020)
7. Ito, H., McCauley, R.N.: Currency composition of foreign exchange reserves. J. Int. Money Financ. **102**, 102104 (2020). https://doi.org/10.1016/j.jimonfin.2019.102104
8. Muniesa, M., Lenglet, F.: Responsible InnovResponsible innovation in finance: directions and implicationsation in practice. Responsible Innov. Manag. Responsible Emerg. Sci. Innov. Soc. 185–198 (2013). https://doi.org/10.3390/wsf2-00836
9. Petersen, M., Feldt, K., Mujtaba, R., Mattsson, S.: Systematic mapping studies in software engineering. In: 12th International Conference on Evaluation and Assessment in Software Engineering (EASE), vol. 12, pp. 1–10 (2008)
10. Bailey, J., Budgen, D., Turner, M., Kitchenham, B., Brereton, P., Linkman, S.: Evidence relating to object-oriented software design: a survey. In: First International Symposium on Empirical Software Engineering and Measurement (ESEM 2007), pp. 482–484 (2007). https://doi.org/10.1109/ESEM.2007.58

11. Abellán, J., Castellano, J.G.: A comparative study on base classifiers in ensemble methods for credit scoring. Expert Syst. Appl. **73**, 1–10 (2017). https://doi.org/10.1016/j.eswa.2016.12.020
12. Wang, M., Ku, H.: Utilizing historical data for corporate credit rating assessment. Expert Syst. Appl. **165**, 113925 (2021). https://doi.org/10.1016/j.eswa.2020.113925
13. Vapnik, V.: The Nature of Statistical Learning Theory. Springer Science & Business Media, Heidelberg (2013)
14. Lee, T.-S., Chiu, C.-C., Chou, Y.-C., Lu, C.-J.: Mining the customer credit using classification and regression tree and multivariate adaptive regression splines. Comput. Stat. Data Anal. **50**(4), 1113–1130 (2006). https://doi.org/10.1016/j.csda.2004.11.006
15. Bayraktar, M., Aktaş, M. S., Kalıpsız, O., Susuz, O., Bayracı, S.: Credit risk analysis with classification restricted Boltzmann machine. In: 2018 26th Signal Processing and Communications Applications Conference (SIU), pp. 1–4 (2018). https://doi.org/10.1109/SIU.2018.8404397
16. Li, Y.: Credit risk prediction based on machine learning methods. In: 2019 14th International Conference on Computer Science & Education (ICCSE), pp. 1011–1013 (2019). https://doi.org/10.1109/ICCSE.2019.8845444
17. Shivanna, A., Agrawal, D.P.: Prediction of defaulters using machine learning on Azure ML. In: 2020 11th IEEE Annual Information Technology, Electronics and Mobile Communication Conference (IEMCON), pp. 320–325 (2020). https://doi.org/10.1109/IEMCON51383.2020.9284884
18. Khatri, S., Arora, A., Agrawal, A.P.: Supervised machine learning algorithms for credit card fraud detection: a comparison. In: 2020 10th International Conference on Cloud Computing, Data Science & Engineering (Confluence), pp. 680–683 (2020). https://doi.org/10.1109/Confluence47617.2020.9057851
19. Bao, W., Lianju, N., Yue, K.: Integration of unsupervised and supervised machine learning algorithms for credit risk assessment. Expert Syst. Appl. **128**, 301–315 (2019). https://doi.org/10.1016/j.eswa.2019.02.033
20. Barboza, F., Kimura, H., Altman, E.: Machine learning models and bankruptcy prediction. Expert Syst. Appl. **83**, 405–417 (2017). https://doi.org/10.1016/j.eswa.2017.04.006
21. Hamori, S., Kawai, M., Kume, T., Murakami, Y., Watanabe, C.: Ensemble learning or deep learning? Application to default risk analysis. J. Risk Financ. Manag. **11**(1) (2018). https://doi.org/10.3390/jrfm11010012
22. Lessmann, S., Baesens, B., Seow, H.-V., Thomas, L.C.: Benchmarking state-of-the-art classification algorithms for credit scoring: an update of research. Eur. J. Oper. Res. **247**(1), 124–136 (2015). https://doi.org/10.1016/j.ejor.2015.05.030
23. Yao, X., Crook, J., Andreeva, G.: Support vector regression for loss given default modelling. Eur. J. Oper. Res. **240**(2), 528–538 (2015). https://doi.org/10.1016/j.ejor.2014.06.043
24. Yao, X., Crook, J., Andreeva, G.: Enhancing two-stage modelling methodology for loss given default with support vector machines. Eur. J. Oper. Res. **263**(2), 679–689 (2017). https://doi.org/10.1016/j.ejor.2017.05.017
25. Ala'raj, M., Abbod, M.F.: Classifiers consensus system approach for credit scoring. Knowl.-Based Syst. **104**, 89–105 (2016). https://doi.org/10.1016/j.knosys.2016.04.013
26. Caroline Cynthia, P., Thomas George, S.: An outlier detection approach on credit card fraud detection using machine learning: a comparative analysis on supervised and unsupervised learning. In: Peter, J.D., Fernandes, S.L., Alavi, A.H. (eds.) Intelligence in Big Data Technologies—Beyond the Hype. AISC, vol. 1167, pp. 125–135. Springer, Singapore (2021). https://doi.org/10.1007/978-981-15-5285-4_12

27. Tavana, M., Abtahi, A.-R., Di Caprio, D., Poortarigh, M.: An artificial neural network and Bayesian network model for liquidity risk assessment in banking. Neurocomputing **275**, 2525–2554 (2018). https://doi.org/10.1016/j.neucom.2017.11.034
28. Tae, C.M., Hung, P.D.: Comparing ML algorithms on financial fraud detection. In: Proceedings of the 2019 2nd International Conference on Data Science and Information Technology, pp. 25–29 (2019). https://doi.org/10.1145/3352411.3352416
29. De Roux, D., Perez, B., Moreno, A., Villamil, M.D.P., Figueroa, C.: Tax fraud detection for under-reporting declarations using an unsupervised machine learning approach. In: Proceedings of the 24th ACM SIGKDD International Conference on Knowledge Discovery & Data Mining, pp. 215–222 (2018). https://doi.org/10.1145/3219819.3219878
30. Xia, Y., Liu, C., Da, B., Xie, F.: A novel heterogeneous ensemble credit scoring model based on bstacking approach. Expert Syst. Appl. **93**, 182–199 (2018). https://doi.org/10.1016/j.eswa.2017.10.022
31. Monamo, P., Marivate, V., Twala, B.: Unsupervised learning for robust Bitcoin fraud detection. In: 2016 Information Security for South Africa (ISSA), pp. 129–134 (2016). https://doi.org/10.1109/ISSA.2016.7802939
32. Awoyemi, J.O., Adetunmbi, A.O., Oluwadare, S.A.: Credit card fraud detection using machine learning techniques: a comparative analysis. In: 2017 International Conference on Computing Networking and Informatics (ICCNI), pp. 1–9 (2017). https://doi.org/10.1109/ICCNI.2017.8123782
33. Sanford, A., Moosa, I.: Operational risk modelling and organizational learning in structured finance operations: a Bayesian network approach. J. Oper. Res. Soc. **66**(1), 86–115 (2015). https://doi.org/10.1057/jors.2013.49
34. Keswani, B., et al.: Adapting machine learning techniques for credit card fraud detection. In: Khanna, A., Gupta, D., Bhattacharyya, S., Snasel, V., Platos, J., Hassanien, A. (eds.) International Conference on Innovative Computing and Communications. AISC, vol. 1087, pp. 443–455. Springer, Singapore (2020). https://doi.org/10.1007/978-981-15-1286-5_38

Advances in Smart Systems and Networks

Applying Advanced IoT Network Topologies to Enhance Intelligent City Transportation Cost Based on a Constrained and Secured Applicative IoT CoAP Protocol

Mohammed Yachou[1], Taoufiq El Harrouti[2(✉)], Salma Ait Oussous[1], Siham Beloualid[1], Sanaa El Aidi[1], Abdelhadi El Allali[1], Abderrahim Bajit[3], and Ahmed Tamtaoui[4]

[1] Laboratory of Advanced Systems Engineering (ISA), National School of Applied Sciences, Ibn Tofail University, Kenitra, Morocco
{Mohammed.yachou, Salma.aitoussous, sanaa.elaidi}@uit.ac.ma, sihambeloualid@gmail.com, aelallali@gmail.com

[2] Laboratory of Engineering Science (SI), National School of Applied Sciences, Ibn Tofail University, Kenitra, Morocco
taoufiq.elharrouti@uit.ac.ma

[3] Laboratory of Engineering Science, National School of Applied Sciences, Ibn Tofail University, Kenitra, Morocco
abderrahim.bajit@gmail.com

[4] Laboratory of Advanced Systems, National Institute of Posts and Telecommunications (INPT-Rabat), SC Department Mohammed V University, Rabat, Morocco
tamtaoui@inpt.ac.ma

Abstract. The analysis, planning, and modeling of the transportation system today cover several areas. In addition, satisfying demand with minimal losses is a major challenge, which must focus on several objectives, including reducing greenhouse gas emissions and promoting the deployment of renewable energy. These objectives aim to achieve significant energy savings in the future. In this work, based on the electric mobility concept consisting of road and streetcar networks using an all-electric multimodal urban transportation network, we implemented a synchronized and secure IoT smart transportation platform, which is used to improve the transportation cost of the smart city through an electric minibus. The idea is to inform the bus about the optimal path it will take according to criteria. It will take the maximum number of people with the minimum number of stops, with reduced prices, and avoid wasting the bus resources. To answer this problem, we have applied the different network topologies star, tree, mesh, and cluster to see which one meets our need by making a comparative study between them in terms of execution time and memory consumption, using CoAP protocol secured with AES.

Keywords: IoT · Smart transportation platform · Optimal path · Electric mobility · STAR/TREE/MESH/CLUSTER topologies · CoAP IoT Protocol · AES-SHA256

Y. Maleh et al. (Eds.): ICI2C 2021, LNNS 357, pp. 195–205, 2022.
https://doi.org/10.1007/978-3-030-91738-8_19

1 Introduction

Transportation and transit systems are the lifeblood of any nation's economic development. Every modern country has a well-established transportation system with effective traffic management. The movement of goods, industrial products, manpower, and machinery significantly impacts industrial growth. Mismanagement and traffic delays also result in extended wait times, wasted fuel, and money [1]. During these last years, the Rabat-Sale agglomeration has experienced an urban expansion causing an increased demand for mobility in parallel with an increase in the number of vehicles, more than 760,000 cars, motorcycles, trucks, and buses, which has caused an environmental nuisance due to the existing transport systems such as congestion, pollution and travel time. To regularize this situation, many efforts have been made and electric mobility was one of the options via the insertion of the tramway in the interurban transport mode since May 2011. Still, this mode could not be generalized on the whole network of the agglomeration for infrastructure and lack of financial means for the investment.

Our problem in this work is to reduce CO2 emissions, energy consumption, time, and costs of citizens' transportation. And to solve it, we have created a solution or a model of public transport named MBE, which means "the mini electric bus that has a capacity of 50 seats", with a green energy source powering the charging stations, via an IoT platform allowing the optimization of travel conditions by strengthening the competition of electromobility in public transport with the car. The purpose of this model is to analyze and manage data in real-time, monitor its evolution, and make adjustments by helping to establish routes and make travel more efficient [2]. In addition, save energy by planning the optimal path, determining the stop stations, and getting information about the route path if there are changes such as delays, station closures, or reroutes. And also, improve the customer experience as IoT technologies can help provide customers with more accurate, up-to-date, real-time data to plan their trips better and improve communication.

To realize this work, we used the CoAP protocol for communication between nodes. It is a web transfer protocol for Internet of Things communication [3]. The CoAP messages's are encoded in a simple binary format. Moreover, the packets are simple to make and can be processed on the spot without consuming energy in low-power systems [4, 5]. Therefore, the CoAP protocol uses other protocols for security, such as DTLS and IPSec [6]. These protocols are responsible for key management, data privacy, and data authentication. However, DTLS is not suitable for restricted IoT devices [7]. A second choice is to use IPsec with CoAP protocol when used without DTLS. With proper configuration, restricted devices can use hardware encryption embedded in the link-layer [8, 9]. But in our case, we added another security layer based on AES-SHA256. This type of security is a symmetric block cipher algorithm used for data security. It also uses a shortest symmetric key size. In addition, it uses a single common key shared by the sender and receiver to encrypt and decrypt a message [10]. It can also handle data up to 128 bits in length and keys up to 128, 192, or 256 bits in length. In addition, many loops are available depending on the length of the key [11] and AES, like many other block ciphers, uses cipher rounds to perform the encryption transformations. Each round consists of several different blocks that work together to form a function repeated several times [12, 13]. One way to detect faces from the photos we get from the stations is to have a set of templates to compare each photo using image processing to give our program the ability to recognize faces. Still, it is recommended to use the ready-made library to process the images in a real application. In our case, OpenCV or Open-source computer vision, a cross-platform library intended mainly for real-time computer vision [14, 15]. Finally, we have studied the impact of the different topologies star, tree, mesh and cluster on the performance of our platform in terms of execution time and memory consumption to choose the best one.

2 General Architecture

Our platform is based on IoT to facilitate the collection of people in optimal conditions. To reduce the queue time, our strategy is to make the collection stations dynamic, i.e., to bring the means of transport closer to the population to encourage this type of transport. To do this, we have implemented an information system capable of facilitating communication between people and devices in real-time via an IoT tool. In addition, our work focuses on the general problem of face detection in a video surveillance context to detect the number of people's faces in each station, 50 people maximum in our case to choose the optimal path. Thus, we have designed and implemented this intelligent and secure transportation IoT platform using dynamic network topologies based on the following architectures: STAR, TREE, MESH, and CLUSTER-3, depending on the coverage area and the type and timing of support.

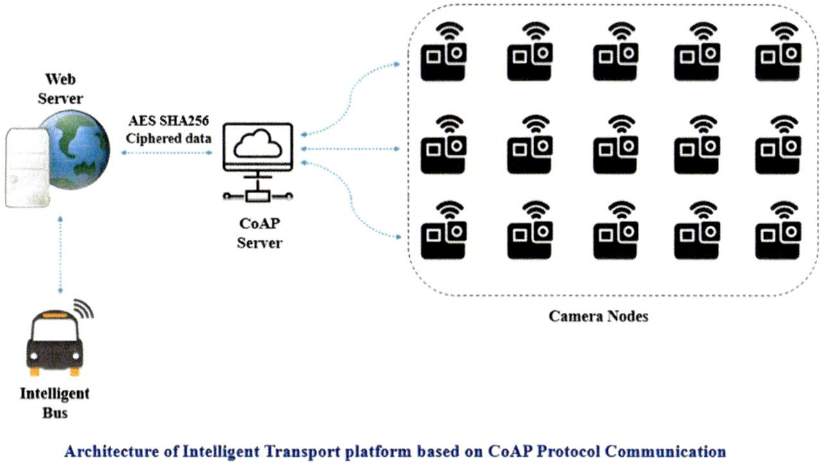

Fig. 1. Intelligent and secure transportation IoT platform.

For a better understanding, we consider Fig. 1, which illustrates the architecture of the platform. So, each station contains an IoT camera node, which takes photos. Then, the image will be sent to the CoAP server, which is the intermediary that passes messages between the IoT camera nodes and the webserver. Next, the obtained results will be sent to the webserver, who will calculate the number of faces for each station, and detect the optimal path that the intelligent bus MBE can use, a high density of faces must characterize the path used. The route of the MBE is carried out between points starting (pickup stations) and arrival points (tramway stations). Moreover, to make these IoT objects communicate, it is necessary to use a communication protocol, so we used the CoAP protocol, and for data security, we added a security layer based on AES-SHA256.

3 Methodology

Given the problems of congestion and pollution caused by the car fleet of the agglomeration, our work had the ambition to try to regulate this situation by using the MBE model, with a green energy source feeding the charging stations. So our approach aims in the first step to model the public transport system; and in the second step, to optimize the transport cost and the energy factor in environmental terms.

Concerning the data collection, we used IoT camera nodes for face detection using the OpenCV framework. And to make the nodes communicate with each other, we used the CoAP communication protocol since it consumes less memory and takes less time to execute compared to MQTT from previous work [16]. In addition, to secure our data, we added a security layer, AES-SHA256. The choice of security method is a choice of use that determines the use of a method or another and not a security question. For our case, it was AES, because it is faster on and supports a large amount of data to encrypt, unlike RSA, which is used to encrypt smaller amounts of data.

4 The Proposed Approach

As already mentioned before, our platform guides the MBE smart bus to get the optimal path by taking a maximum number of people avoiding the maximum stops. And to achieve this work, we have worked on two approaches. Figure 2 illustrates how these approaches work by algorithms.

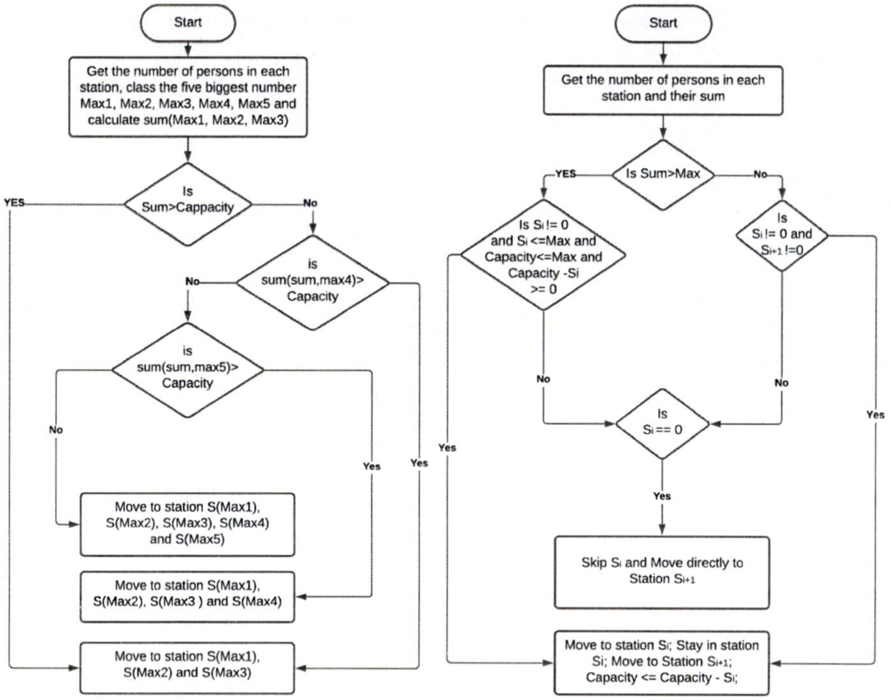

Fig. 2. Approach 1's and approach 2's algorithms

For the first approach, after detecting the number of people in each station, the bus will move to the stations one by one and stops in each one except the station with 0 person till to fill the capacity, which is 50. It will move to the nearer tram station, so we can get the optimal path. And Fig. 3 shows two examples of optimal path's results for this approach. In the left result, as shown the bus moves from one station to another. Station 9 becomes full and moves to the nearest station tram, which is s.tram 2. The same treatment has been done for the second example on the right, so the bus gets its maximum in station 13 and moves to s.tram 3, the nearest station.

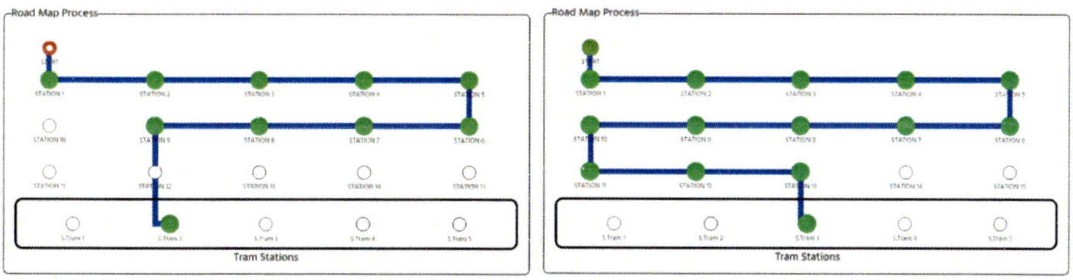

Fig. 3. Approach 1's optimal path examples

In the second approach, after the webserver detects the number of persons in each station, we aim to move to three stations at minimum or to 5 stations at maximum, as shown in right algorithm Fig. 2. The web server will class the five biggest numbers then. It will calculate the sum of the three biggest numbers. If the sum is smaller than the capacity it will calculate the sum of the four biggest. In this case, If the sum is bigger than the capacity it will move to the station's position of the three biggest number of person, else it will calculate the sum of the five biggest; if the sum is bigger than the capacity, it will move to the station's position of the four biggest number of person, else If the sum is smaller or equal the capacity it will move to the station's position of the five biggest number of person, If the sum equals the capacity it will move to the station's position of the three biggest number of person. We take the example shown in Fig. 4. After collecting information about the number of persons in each station, we found that the optimal path is from the start point to stations 2, 3, and 7.

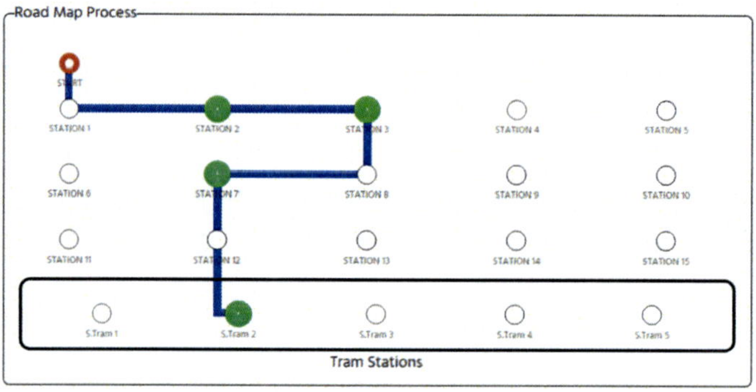

Fig. 4. Number of person bigger than 50 for 3 stations maximum example.

Moreover, the second example is a test for 4 stations at maximum in this case. So, as shown in Fig. 5, the optimal path is from start point to Station 2, 4, 9 and 7.

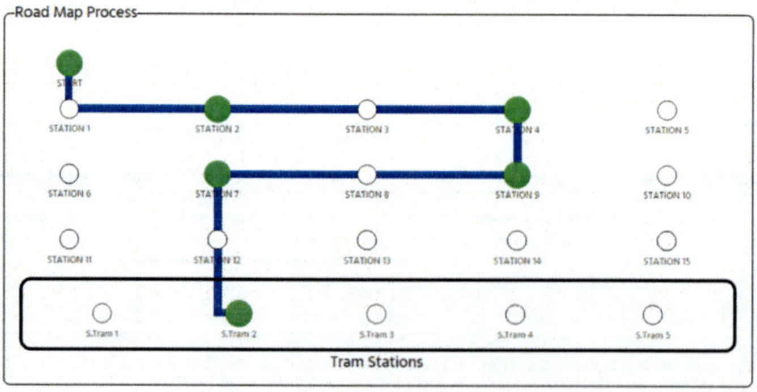

Fig. 5. Number of person bigger than 50 for 4 stations maximum example.

For the last example, as shown in Fig. 6, the optimal path is from the start point to Station 1, 4, 9, 10 and 11.

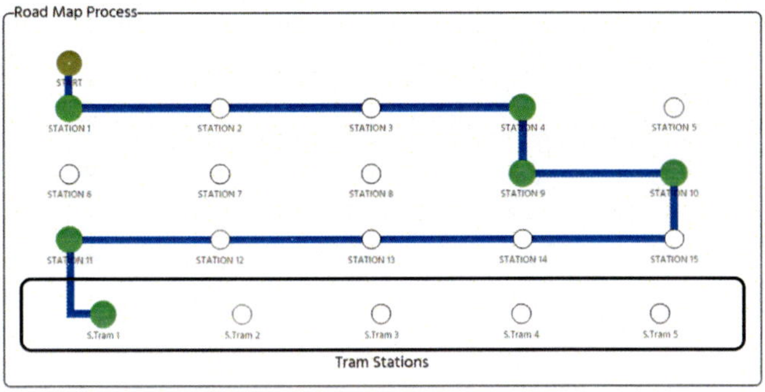

Fig. 6. Number of person bigger than 50 for 5 stations maximum example.

5 Discussion and Results

This work aims to choose the optimal path for the bus intelligent MBE applying two approaches, with different types of topologies and using the CoAP protocol with and without a security layer AES-SHA256.

Firstly, according to the previous works about CoAP and MQTT protocols, we noticed that CoAP is better than MQTT since it consumes less memory and time than the other in secure and non-secure cases.

Moreover, encoding or encryption is the process of transforming data to be unreadable. Cryptography uses a variety of encryption algorithms, including AES and RSA. Due to their efficient, secure, and high-performance use in these resource-limited situations, AES has become widely employed in many devices, particularly in resource-constrained environments. Additionally, according to the results obtained in those articles [17–19], we can suggest that AES is better than RSA due it uses the smallest key and has a positive impact on the computing resources, for example using it, can conserve more energy from all the nodes IoT and extend the life of the network such that it consumes less memory and time. So, based on these studies, we have chosen to work with the CoAP protocol, which is suitable for our platform and uses the AES-SHA256 security. Secondly, we have analyzed the impact of different topologies Star, Tree, Mesh, and Cluster in terms of execution time and memory consumption with and without security layer.

According to the results obtained in Fig. 7 and 8, Mesh topology takes more time to execute and consumes higher consumption memory than the rest of topologies. Moreover, as shown in Fig. 9, the average execution time for Mesh topology is higher than star, tree and cluster. However, for the average memory consumption, we noticed not much difference in consumption between the four topologies.

Additionally, to these comparative results, we can consider mesh topology the best, even if it consumes higher memory and time. It has more advantages than the others, like if one of the routers falls, the nodes can continue the transmission of the payload to the server because the nodes are connected to multiple routers at once.

Thirdly, in this platform, we applied two approaches and showed the results of each one in the last part. From the results, we can remark that to fill the capacity of the MBE bus by the optimal path, applying the second approach is shorter than the one in the first approach, which means that the time and the energy consumed in the second approach are less than the first one. Finally, That leads us to presume that approach 2 is better than approach 1 because, through it, we can realize the goals that we want to be obtained from this work, such as reduce CO2 emissions and gases, the price by looking for the optimal path, the time of the route; and the respect of the environment.

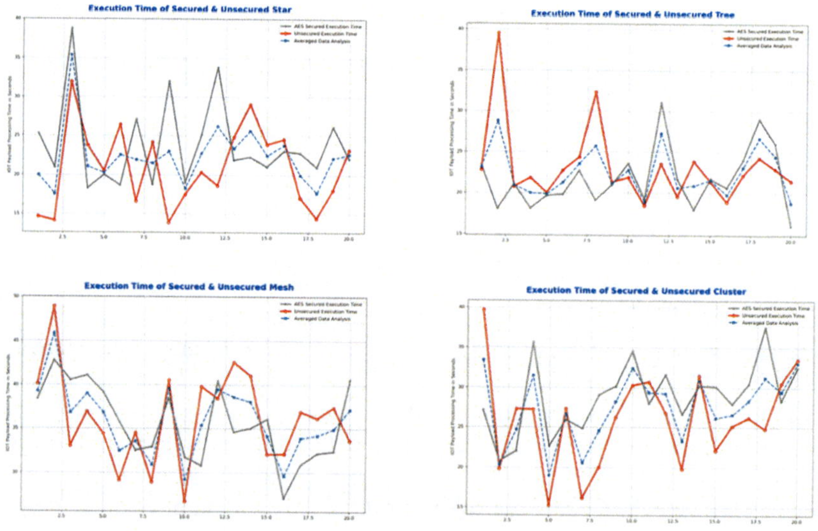

Fig. 7. Execution time by topology and security

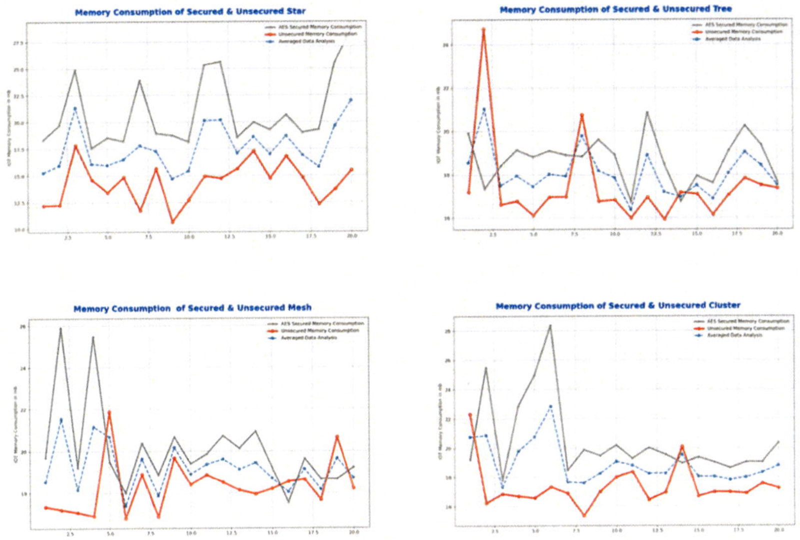

Fig. 8. Memory consumption by topology and security

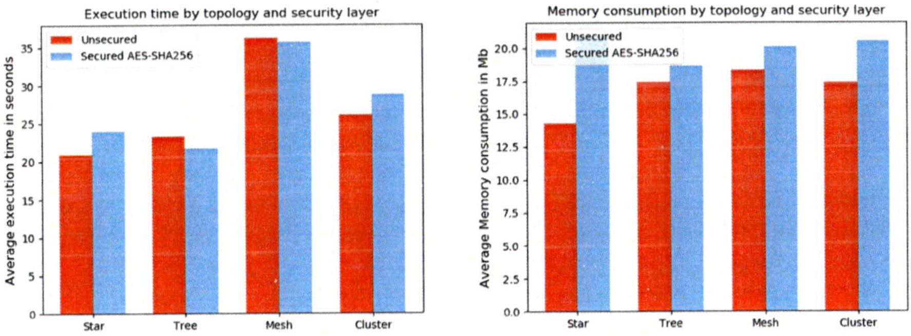

Fig. 9. Average execution time and memory consumption by topology and security

6 Conclusion and Perspectives

In this work, we have created an intelligent and secure transportation IoT Platform for guiding the MBE bus to take the optimal path, using two approaches that had been tested in a real-time environment to find the suitable approach. And to make this work more efficient and secure, we have used the AES algorithm for encryption and decryption transmission payloads. In the future work, we will implement ECC encryption on our platform to compare AES with ECC and choose the best security layer for our platform.

References

1. Chavan, S.S., Deshpande, R.S., Rana, J.G.: Design of intelligent traffic light controller using embedded system. In: 2009 Second International Conference on Emerging Trends in Engineering & Technology, pp. 1086–1091 (2009). https://doi.org/10.1109/ICETET.2009.76
2. Barodi, A., Bajit, A., El Aidi, S., Benbrahim, M., Tamtaoui, A.: Applying real-time object shapes detection to automotive traffic roads signs. In: 2020 International Symposium on Advanced Electrical and Communication Technologies (ISAECT), pp. 1–6 (2020). https://doi.org/10.1109/ISAECT50560.2020.9523673
3. Bellavista, P., Zanni, A.: Towards better scalability for IoT-cloud interactions via combined exploitation of MQTT and CoAP. In: 2016 IEEE 2nd International Forum on Research and Technologies for Society and Industry Leveraging a better tomorrow (RTSI), pp. 1–6 (2016). https://doi.org/10.1109/RTSI.2016.7740614
4. Kayal, P., Perros, H.: A comparison of IoT application layer protocols through a smart parking implementation. In: 2017 20th Conference on Innovations in Clouds, Internet and Networks (ICIN), pp. 331–336. Paris (2017). https://doi.org/10.1109/ICIN.2017.7899436
5. Rahman, A., Dijk, E. (ed.): Group Communication for the Constrained Application Protocol (CoAP) (2014). https://www.hjp.at/doc/rfc/rfc7390.html
6. Peretti, G.: CoAP over DTLS TinyOS Implementation and Performance Analysis, MS Thesis, University of Padova, Italy (2013)
7. Naik, N.: Choice of effective messaging protocols for IoT systems: MQTT, CoAP, AMQP and HTTP. In: 2017 IEEE International Systems Engineering Symposium (ISSE), pp. 1–7 (2017). https://doi.org/10.1109/SysEng.2017.8088251
8. Majumder, S., Ray, S., Sadhukhan, D., Khan, M.K., Dasgupta, M.: ECC-CoAP: elliptic curve cryptography based constraint application protocol for internet of things. Wirel. Pers. Commun. **116**(3), 1867–1896 (2020). https://doi.org/10.1007/s11277-020-07769-2
9. Kayal, P., Perros, H.: A comparison of IoT application layer protocols through a smart parking implementation. In: 2017 20th Conference on Innovations in Clouds, Internet and Networks (ICIN), Paris, pp. 331–336 (2017). https://doi.org/10.1109/ICIN.2017.7899436
10. Lu, C.C., Tseng, S.Y.: Integrated design of AES (Advanced Encryption Standard) encrypter and decrypter. In: Proceedings IEEE International Conference on Application- Specific Systems, Architectures, and Processors, pp. 277–285. San Jose, CA, USA (2002). https://doi.org/10.1109/ASAP.2002.1030726
11. Tsai, K., Huang, Y., Leu, F., You, I., Huang, Y., Tsai, C.: AES-128 based secure low power communication for LoRaWAN IoT environments. IEEE Access **6**, 45325–45334 (2018). https://doi.org/10.1109/ACCESS.2018.2852563
12. Panda, M.: Data security in wireless sensor networks via AES algorithm. In: 2015 IEEE 9th International Conference on Intelligent Systems and Control (ISCO), pp. 1–5 (2015). https://doi.org/10.1109/ISCO.2015.7282377
13. Agrawal, M., Mishra, P.: A comparative survey on symmetric key encryption techniques. Int. J. Comput. Sci. Eng. (IJCSE) **4**(05), 877 (2012). ISSN: 0975-3397
14. Guennouni, S., Ahaitouf, A., Mansouri, A.: Multiple object detection using OpenCV on an embedded platform. In: 2014 Third IEEE International Colloquium in Information Science and Technology (CIST), pp. 374–377 (2014). https://doi.org/10.1109/CIST.2014.7016649
15. Saxena, M.R., Pathak, A., Singh, A.P., Shukla, I.: Real time object detection using machine leariningand opencv. Int. J. Inf. Sci. Appl. (IJISA) **11**(1), 16–19 (2019). ISSN 0974-2255, (Special Issue) © International Research Publication House, Department of Computer Science and Engineering, IMS Engineering College, Ghaziabad, Uttar Pradesh, India

16. Sanaa, E.L., Bajit, A., Barodi, A., Chaoui, H., Tamtaoui, A.: An optimized security vehicular internet of things-IoT-application layer protocols MQTT andCOAP based on cryptographic elliptic-curve. In: 2020 IEEE 2nd International Conference on Electronics, Control, Optimization and Computer Science, ICECOCS 2020, p. 9314579 (2020)
17. El Aidi, S., Bajit, A., Barodi, A., Chaoui, H., Tamtaoui, A.: An advanced encryption cryptographically-based securing applicative protocols MQTT and CoAP to optimize medical-IOT supervising platforms. In: Saeed, F., Mohammed, F., Al-Nahari, A. (eds.) Innovative Systems for Intelligent Health Informatics. IRICT 2020. Lecture Notes on Data Engineering and Communications Technologies, vol. 72, pp. 111–121. Springer, Cham (2021). https://doi.org/10.1007/978-3-030-70713-2_12
18. Fdil, E.M., El Haidi, M., Bajit, A., El Aidi, S., Barodi, A., Tamtaoui, A.: An new constrained protocol S-CoAP applied to optimize COVID-19 medical IOT intelligent and security-based DATA supervising platform. In: Proceedings of the International Symposium on Advanced Electrical and Communication Technologies (ISAECT), Kenitra, Morocco (2020)
19. Bajit, A., Barodi, A., Chaoui, H., Tamataoui, A.: An elliptic-curve based cryptographically optimized vehicular protocols applied to secured applicative protocols MQTT and CoAP. In: Proceedings of the 2020 International Conference on Electronics, Control, Optimization and Computer Science (ICECOCS), pp. 1–6. Kenitra (2020)

Applying Lightweight Elliptic Curve Cryptography ECC to Smart Energy IoT Platforms Based on the CoAP Protocol

Ayoub Ech Chkaf[1], Abdelhadi El Allali[1], Siham Beloualid[1], Taoufiq El Harrouti[2(✉)], Sanaa El Aidi[1], Abderrahim Bajit[1], Habiba Chaoui[1], and Ahmed Tamtaoui[3]

[1] Laboratory of Advanced Systems Engineering (ISA), National School of Applied Sciences, Ibn Tofail University, Kenitra, Morocco
{ayoub.ech-chkaf, siham.beloualid, sanaa.elaidi, habiba.chaoui}@uit.ac.ma, aelallali@gmail.com, abderrahim.bajit@gmail.com

[2] Department of Computer Science, Logistics and Mathematics (ILM) Engineering Science Laboratory, National School of Applied Sciences, Ibn Tofail University, Kenitra, Morocco
taoufiq.elharrouti@uit.ac.ma

[3] National Institute of Posts and Telecommunications (INPT-Rabat), SC Department, Mohammed V University, Rabat, Morocco
tamtaoui@inpt.ac.ma

Abstract. Nowadays, Morocco has adhered to a policy aimed at modernizing its cities in terms of smart city and green city to improve the quality of life, protect the environment, and compete with the world's major cities in this promising field. In this perspective, our current work revolves around the innovation of smart cities via new technologies of interconnection between objects and man in maximum respect that possible of the environment and especially the air quality.

The rate of innovation in creating smart cities will make it feasible to bring IoT technology into cities development. In this article, we develop a smart and secure IoT platform to monitor and control the energy consumption of any area in a smart city. From this, we used the CoAP and MQTT application protocols for data transfer from the Smart meters to the webserver. So, to optimize the performance of our platform, we're going to explore several topologies, including: STAR, TREE, MESH and CLUSTER. Ultimately, to overcome existing limitations related to key management and data transformation security by CoAP and MQTT protocols in each topology, we proposed an effective and secure communication protocol to establish a secure session key between IoT devices and the remote server, using the elliptical curve cryptography (ECC/SHA256) that we compared with another RSA/SHA256 (Rivest–Shamir–Adleman) communication protocol in terms of execution time and memory consumption to see the suitability and performance of the platform and infer the best secure topology in our smart city.

Keywords: Smart city · Smart grid · CoAP · MQTT · Internet of Things (IoT) · ECC/SHA256 · RSA/SHA256 · Security

Y. Maleh et al. (Eds.): ICI2C 2021, LNNS 357, pp. 206–218, 2022.
https://doi.org/10.1007/978-3-030-91738-8_20

1 Introduction

To establish a management model of an energy mix capable of supporting the optimization of the production and use of renewable energies in smart cities with a green touch to bring comfort of life for the inhabitants, we try in this work to appeal to the new interconnection technologies that we consider capable of facilitating the management and use of energy in vital multi-domains. And so, we need to proceed to a reliable. Real time data collection knowing that we are dealing with massive data given the consistency of large cities in people and multiple networks approach, aims at using smart platforms via Internet of Things (IOT) protocols to control networks and energy consumption in a smart city while thinking about the sustainability factor as a goal to ensure a "green life" lifestyle.

Smart Grids is one of the components of Smart City, covering a wide range of domains and applications: transport, distribution, counting, connection of ENRs (renewable energy), telecommunications, electric vehicles, demand management, etc. The interoperability of smart grids must therefore be ensured for simplified deployment in cities [1].

This type of Smart Grid uses IT techniques to optimize the production, distribution, consumption and possibly storage of energy to better coordinate the entire grid, from the producer to the final consumer. It improves the overall energy efficiency by minimizing online losses and, optimizing the efficiency of the means of production used, in relation to instant consumption. So, according to one of the authors, we need IoT devices that generate a large amount of data that needs to be collected, transmitted and analyzed to deliver actionable results. With the help of the Internet of objects, which is used to collect process and analyze the vast amount of data required by artificial intelligence algorithms, cities can be improved at multiple levels, such as helping manage energy production and consumption in a changing zone [2].

This large quantity of information to be analyzed must ensure that it is transmitted safely and fluidly. We use highly effective and secure encryption and decryption protocols for this. This work aims to create an IoT platform and make it more efficient and secure using CoAP, MQTT and two RSA/SHA256 and Elliptic Curve Cryptography (ECC/SHA256).

To study the impact of topologies (STAR/TREE/MESH/CLUSTER) on the operation of our platform according to 3 parameters: execution time, memory usage, and power consumption for each topology in three cases, unsecured, secured by RSA/SHA256, and Secured by ECC/SHA256. These topologies contain virtual End Devices that act as smart meters, routers that act as intermediaries, a coordinator that manages the topology components, and a Web server that gets the data to be analyzed in a synchronous, intelligent, and secure IoT platform to infer topologies and protocol more efficient in each zone in our smart city. In addition to knowing which zone is consuming a lot of energy, it helps us know why, to think about improvement.

In this paper, we focus on the security of the application layer protocol used in our system. The first section of this document will contain the CoAP context and two RSA/SHA256 and ECC/SHA256 cryptosystems, in the second section we will represent the general architecture, in the third section we will represent the different topologies proposed, in the fourth section we will describe these proposed secure CoAP topologies using RSA/SHA256 and ECC/SHA256, the rest of the document will present a comparative study and finally a conclusion (Fig. 1).

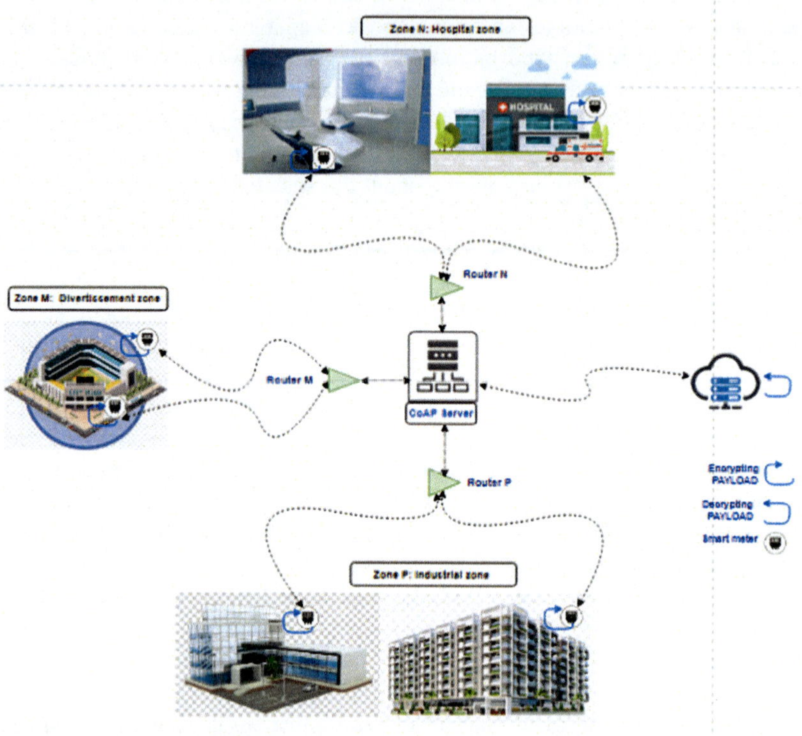

Fig. 1. Supervising smart grid IoT architecture.

2 Supervising Smart Grid IoT Platform

The main objective of our study is to implement an IoT platform for a smart grid in a smart city based on real smart, synchronous and secure time, which is used to collect, analyze and control energy consumption in each area (urban, tourism, industrial, sports, commercial, neighborhood, garden, leisure center…) in our smart city.

We used IoT smart meters nodes for each Zone for this platform. These IoT smart meters measure the amount of energy consumed over a specified period, and then send data to the coordinator (CoAP server). The CoAP server takes the data and sends it to the cloud or web server, to model and analyze the data. This analysis provides an overview of our city's energy consumption and discovers which areas are consuming the most energy and why. This allows us to improve consumption in these areas of the city. We implemented this platform using four different topologies, the first is a star topology where the IoT smart counter nodes are connected directly to the coordinator.

The second is a tree topology. The coordinator is connected to routers that act as intermediaries. In our case, we worked on IoT End devices, where each End device is connected to the router and the router communicates with the coordinator. Then with the website, the third topology is a Mesh topology, which is a change in the topology to a tree. Routers can communicate directly with each other without having to follow the tree structure and the last clustered topology is a combination of the three-star, tree, and mesh topologies.

For payload transmission, we used two RSA/SHA256 and ECC/SHA256 asymmetric encryption algorithms to provide optimal security, ensuring confidentiality, integrity, and authentication. In addition, the purpose of this implementation is to compare the different topologies and the two for payload transmission. We used two RSA/SHA256 and ECC/SHA256 asymmetric encryption algorithms to provide optimal security, ensuring confidentiality, integrity, and authentication. In addition, the purpose of this implementation is to compare the different topologies and the two security protocols used, in terms of execution time and memory consumption.

3 Methodology

To model and optimize the management of energy resources by promoting green resources to power modern life devices in smart cities and to contribute to the improvement of air quality and the reduction of environmental nuisances, we will adopt an IoT platform model capable of meeting the needs of analysis and optimization of resources in collecting reliable and consistent metadata.

In this work, we created an IoT test platform for a smart grid in a real-time, intelligent, synchronous, and secure smart city, using CoAP to have the platform node communicate with each other. In addition, we added two RSA/SHA 256 and ECC/SHA256 cryptosystems for the security layer.

3.1 IoT Communication Protocol

CoAP on the message layer based on message exchange over UDP, which supports 4message types: CON (Confirmable), NO (NonConfirmable), ACK (Acknowledgement), RST (Reset) [5]. The protocol ensures that the message is reliable as a confirmable (CON) message). Message is retransmitted after a default delay and exponential count time. Decreases the count time exponentially until the CoAP client sends an ACK message with the same ID [4–6, 8].

The choice of CoAP protocol in this work based on already done studies, which we have concluded that CoAP is much better than MQTT in terms of run time, memory usage and power consumption, and MQTT, also works when the system is less loaded [3, 7].

3.2 The Cryptographic Algorithms

In cryptography, **RSA** encryption is an asymmetric algorithm, one of the first major advances in two key cryptography (public key and private key), uses the public key for encryption shared with everyone, and the private key that remains private is used for decryption. As RSA is the first algorithm suitable for signing and encryption, it is also widely used in e-commerce protocols [9]. RSA is easier to understand. Signature and decryption are similar, while encryption and verification are similar. This simplifies understanding. Simplicity is more important if you implement the algorithms yourself instead of using a well-known library.

A cryptographic hash is a kind of "signature" for a text or data file. SHA-256 generates a nearly unique 256-bit (32-byte) signature for text. The SHA 256 algorithm is used in digital certificates for data integrity [10]. SHA 256 was developed by N.I.S.T [11].

ECC is the most modern and preferred approach to public-key asymmetric cryptography because of its smaller keys, shorter signatures, and best performance, based on the algebraic structure of the elliptical curves [12].

Elliptic curves equal set of points {x, y} such that:

$$y^2 = x^3 + ax + b \tag{1}$$

In addition, ECC uses EC over the finite field Fp (p is prime, $p > 3$), a set of integer coordinates {x, y}, such that $0 \leq x, y < p$, and staying on the elliptic curve [12]:

$$y^2 = x^3 + ax + b(mod\ p) \tag{2}$$

ECC cryptosystems provide several groups of algorithms, based on the mathematics of elliptic curves over finite fields. For example in this work, we used the Elliptic Curve Integrated Encryption Scheme (ECIES).

ECIES Hybrid Encryption Scheme is a public key encryption scheme, that uses a key-derivation function (KDF) to derive a separate MAC key and symmetric encryption key from the Elliptic Curve Diffie-Hellman Key Exchange (ECDH) shared secret. The Hybrid Encryption Scheme has been standardized in ANSI X9.63 and ISO/IEC 15946-3, and is part of the IEEE P1363a standard.

To provide data encryption by the EC private key and decryption by the corresponding EC public key, Elliptic Curve Integrated Encryption Scheme uses ECC based on asymmetric cryptography and AES symmetric ciphers [13, 24].

ECC is more complicated to implement securely than RSA. This is not necessarily a problem if you are using proven cryptographic libraries from trusted providers. Most cryptographers recommend ECC over RSA in general due to its smaller key length and signature length and high performance, especially with some highly optimized curves [14].

3.3 IoT Network Topologies

In our IoT platform, there are different types of IoT topologies for networking. To decide which network is best, we need to know the pros and cons of each (Fig. 2).

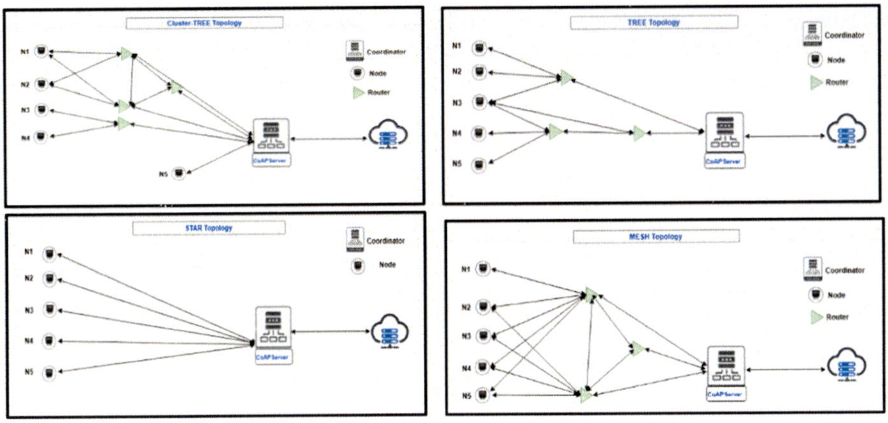

Fig. 2. IoT network topologies

A STAR topology consists of a single network coordinator and one or more terminal devices. All terminals communicate only with the coordinator. This coordinator can be a switch, hub, or router [15]. For example, if a leaf wants to pass data to another leaf, it passes it through the coordinator. This topology is not reliable because if the coordinator fails the whole network becomes non-functional [16]. The Star topology is very secure because individual machines are isolated. The failure of one device node has no effect on the operation of others and among the benefits of this topology is that it establishes central management for network operations and makes it easy to add a new node to the network, and when a new node is added, just update the coordinator.

A TREE topology is a child-parent relationship, or the coordinator is connected to routers. The router can also be connected to either other routers or to leaf nodes in addition to the coordinator. The router thus allows the number of nodes that can join the network to be increased. In addition, extend the network scope, so that nodes that are far from the coordinator can join it. As with the star topology, leaf nodes cannot communicate with each other. Thus, messages can be transmitted either through the coordinator or through the routers. There is only one path between any two connected nodes in the tree topology [17].

MESH topology is a change in the topology to a tree, which has no hierarchical relationship. This is because communication is more flexible, but the leaf nodes still cannot communicate directly. In contrast, routers and the coordinator can communicate directly with each other without having to follow the tree structure. The benefits of this topology are the latency of messages that can be reduced, and Reliability increases with the available alternative paths [18].

CLUSTER Topology is one of the most complicated computer network topologies, because multiple computer network topologies are combined to form a cluster topology. Here, the Cluster network is a combination of Star, Tree and Mesh topology, each topology has its own strength and weakness. Few end devices are connected directly to the coordinator node and other end devices need the help of the parent (router) node to receive the data. So, when different types of topologies are combined, their total effective networks prove their performance. The advantages of Cluster topology are useful for implementing larger networks. And as it is a combination of multiple network topologies. So, it can handle a large volume of traffic [19].

4 The Proposed Approach

Our approach is based on implementing an IoT test platform for a smart grid in an intelligent city, which collects, analyzes, and monitors energy consumption in each area in our smart city. To measure the amount of energy consumed in each zone, we used IoT Smart meters. These IoT smart meters use the CoAP application protocol to transfer their data to the cloud or web server (Fig. 3).

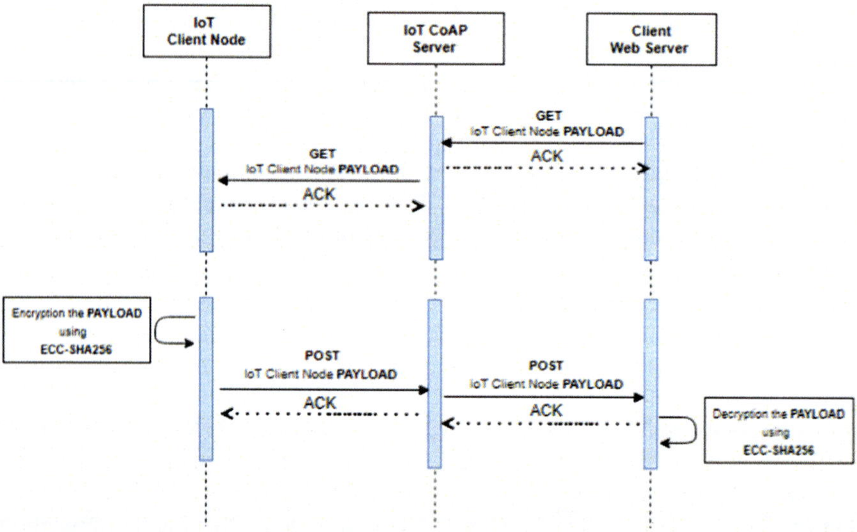

Fig. 3. CoAP secure mechanism.

CoAP is a specialized web transfer protocol with constrained devices (such as microcontrollers) and networks constrained on the Internet of Objects. This protocol is designed for machine-to-machine (M2M) applications such as intelligent energy and building automation, using a request-response paradigm. CoAP is essentially a one-tone protocol that is very similar to the HTTP protocol. And can be considered a highly optimized variant of the Hypertext Transfer Protocol (HTTP) for use in a low-resource integrated domain [20, 21].

MQTT uses a central broker to dispatch messages coming from the publisher to the clients. While CoAP is more suitable for state transfer. In addition, MQTT uses Asynchronous messaging, but CoAP uses Asynchronous and Synchronous messaging. In addition, based on previous work, we have concluded that the use of CoAP for data transfer is preferable to MQTT in terms of execution time, memory occupancy, and energy consumption [22, 23].

Our platform built is based on different areas. Each area uses topologies (Star, Tree, Mesh, and Cluster). The use of these topologies is based on several standards of our city area, such as surface, a number of smart meters, and distance between End devices (Smart meters), routers, and coordinators. You can also refer to the importance of getting the data and make sure you do not lose it, here. It may be suggested to use the Mesh topology, for example. Instead, to enhance the security and availability of data flowing between IoT smart meters, CoAP server, and Web server. We added a security layer using two RSA/SHA256 and ECC/SHA256 asymmetric encryption algorithms to choose the best running time, memory occupancy, and power consumption. Both cryptosystems (RSA and elliptic-curve cryptography) work with private and public keys. Also, provide similar capabilities like key generation, digital signatures, key agreement schemes and encryption schemes.

Generally, it is considered that ECC/SHA256 is the modern and the preferable public-key cryptosystem because of smaller keys, shorter signatures and better performance, but some people disagree. The Elliptic-curve cryptography (ECC) has the following advantages over RSA. The first advantage is smaller keys, ciphertexts and signatures. Typically for 128-bit encryption, a 256-bit EC private key is used. Signatures typically consist of 2 times the private key length or longer (depending on the encoding scheme). By contract, in RSA for 128-bit security, a 3072-bit private key is needed. Ciphertext length is typically the same length as the unencrypted plain text—the second advantage is very fast key generation. In the ECC cryptosystems, key generation consists of random number generation and EC point multiplication and is extremely fast, even for the heaviest curves used in practice. By contrast, in RSA, key generation may be slow, especially for long keys (e.g., 4096-bit and longer keys), due to the prime number generation process, RSA produces large signatures (such as the private key). This is unacceptable for many systems that retain a large number of signatures, such as Public Blockchain.

Finally, Fast ECC encryption and decryption. In fact, the encryption is as fast as the underlying symmetric key encryption (due to the ECIES encryption scheme). In addition, for a synchronous IoT platform with high-security performance, using ECC for encryption and decryption is preferable to RSA because of its smaller key length and signature length and high performance, especially with some highly optimized curves (Fig. 4).

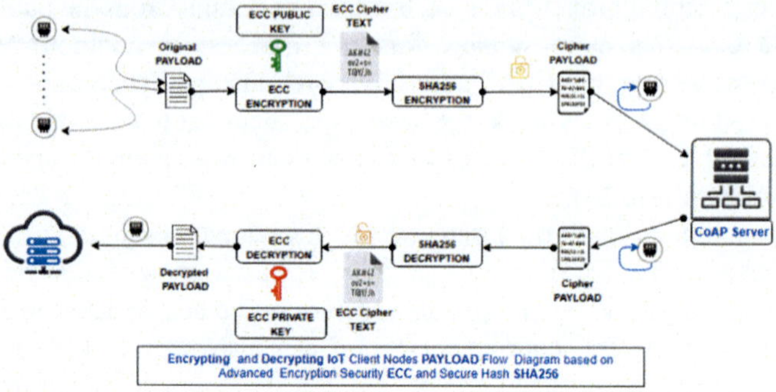

Fig. 4. Encrypting and Decrypting IoT Client Nodes PAYLOAD Flow Diagram based on (ECC) and Secure Hash SHA256

5 Discussion

According to the previous works [22, 23], we found that CoAP is much better than MQTT in terms of execution time, memory occupation and energy consumption, that's why we chose CoAP to implement it in our Smart Grid platform.

We have observed that in both secure and non-secure modes, CoAP consumes less memory and time than MQTT. In addition, CoAP packets are smaller and simple to generate and parse in situ without consuming additional memory in constrained devices. Thus, these results show a significant gain in memory consumption and network response time, which justifies the use of CoAP over MQTT (see Fig. 5).

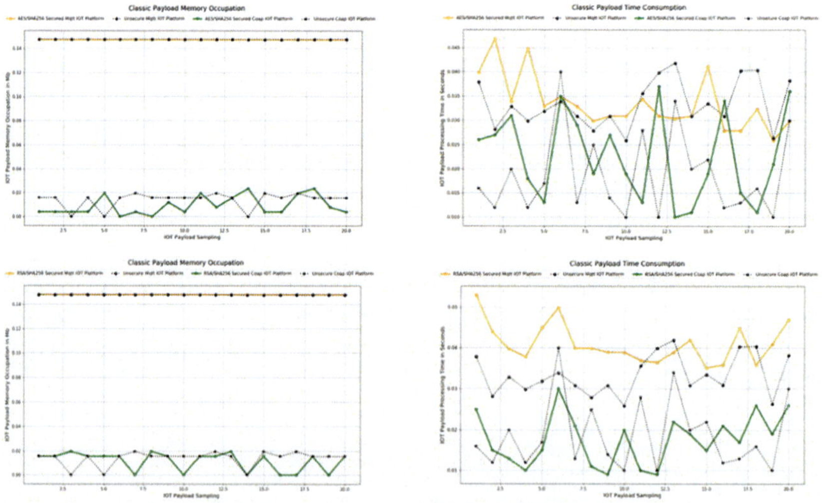

Fig. 5. Comparison of execution time and memory occupation between CoAP and MQTT protocols applied on payloads sent by IoT nodes using symmetric and asymmetric encryption with AES-SHA256 and RSA-SHA256 algorithms

Thus, to ensure data security, we used two ECC and RSA encryption algorithms with SHA256 to choose the best execution time, memory occupancy, and power consumption based on these figures (see Fig. 6). We conclude that ECC is much better for IoT platforms.

From Fig. 6 and Table 1 above, which proves what was said in the preceding paragraphs. Where RSA/SHA256 is found to consume more execution time and memory than ECC/SHA256 in all types of topologies that can be used in each area of our Smart grid, this is due to the features of ECC/SHA256 mentioned above. Then, to determine how much time we saved and how much memory we consumed, we performed a triple operation, which gave us the results listed in Table 2. We found that while it was more challenging to decrypt an encrypted message with ECC/SHA256 than an encrypted message with RSA/SHA256, ECC/SHA256 has saved us from running consuming memory in all topologies. It is also noted that this gain varies from topology to topology, which is normal, since the usage mode and number of components for each topology are different.

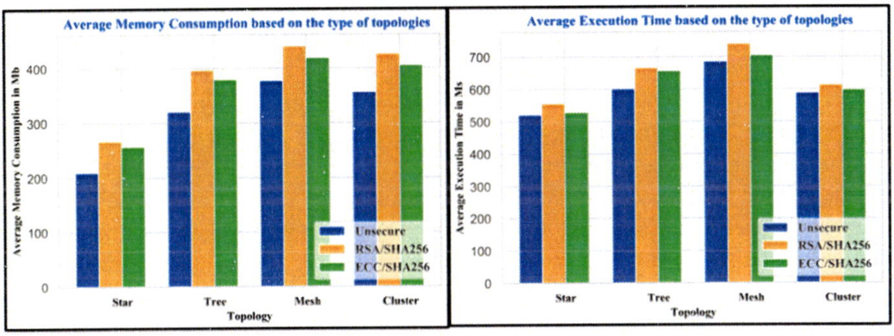

Fig. 6. Comparison of average execution time and average memory occupation between Unsecure and secure in RSA/SHA256 and ECC/SHA256 Topologies.

Table 1. Average of Time consumption and RAM occupation of each security protocol different topologies

		STAR	TREE	MESH	CLUSTER
TIME (Ms)	Unsecure	520.58	600.35	685.68	588.04
	RSA/SHA256	**554.21**	**665.47**	**740.56**	**611.85**
	ECC/SHA256	**527.77**	**655.83**	**704.45**	**597.28**
RAM (Mb)	Unsecure	209.28	320.26	377.08	355.81
	RSA/SHA256	**266.65**	**395.98**	**440.31**	**426.03**
	ECC/SHA256	**256.28**	**378.93**	**418.85**	**404.81**

Table 2. Gain percentage of Time consumption and RAM occupation of ECC compared to Unsecure RSA in different topologies

		STAR	TREE	MESH	CLUSTER
TIME (Ms)	ECC%Unsecure	1.38%	9.24%	2.73%	1.57%
	ECC%RSA	**4.77%**	**1.44%**	**4.87%**	**2.38%**
RAM (Mb)	ECC%Unsecure	22.45%	18.31%	11,07%	13.77%
	ECC%RSA	**3.88%**	**4.30%**	**4.87%**	**4.98%**

For non-secure topologies, we have added them to compare with secure topologies in RSA/SHA256 and ECC/SHA256 in terms of running time and memory occupancy, which shows us that run time and memory consumed in secure topologies, especially those secured in ECC/SHA256, not very large by reporting insecure. For example, in the TREE topology, we see that ECC/SHA256 is almost equal to unsafe in terms of time execution.

So, we implemented our IoT platform with ECC instead of RSA for 2two specific reasons:

The first is that ECC offers us better security of our IoT platform. Decrypting a message when we don't know the private key is more complicated, increasing the number of calculations. Still, the key generation, key agreement, signatures, encryption and decryption (using the ECIES encryption scheme) is fast, so it doesn't generate much processing time compared to RSA.

The second reason for using ECC is memory-free. ECC can provide the same encryption benefits as RSA but with shorter keys. A smaller data size means less communication between the server and the client, resulting in less memory usage. This is highly recommended for IoT platforms where storage space is very limited. Another advantage of ECC over RSA is performance. The ECC allows data to be loaded more quickly, reducing execution time.

6 Conclusion and Perspectives

In this work, we were able to create an intelligent smart grid platform on an intelligent city that is used to monitor and analyze power consumption, we used Smart meters IoT nodes for each area of the city, we chose CoAP to provide communication between the nodes, we used four different topologies to study the impact of each one, and for data security, we chose ECC.

For future work, we will apply this platform to solvethea problem of optimizing the costs of transmission of electrical energy using renewable resources, this platform will be used to propose a typical model to reduce costs and greenhouse gas emissions. And we will implement this platform using ZigBee and compare it with CoAP and deploy another security method in the data storage database, Blockchain storage service, to provide the most efficient, secure, and optimal protocol for this platform.

References

1. Hemanth, G.R., Charles Raja, S., Venkatesh. P.: Introduction to Smart Grid (2021). https://doi.org/10.1002/9781119751908.ch3
2. Şerban, A.C., Lytras, M.D.: Artificial intelligence for smart renewable energy sector in Europe smart energy infrastructures for next generation smart Cities. IEEE Access **8**, 77364–77377 (2020). https://doi.org/10.1109/ACCESS.2020.2990123
3. Larmo, A., Ratilainen, A., Saarinen, J.: Impact of CoAP and MQTT on NB-IoT system performance. Sensors **19**, 7 (2018). https://doi.org/10.3390/s19010007
4. Shelby, Z., Hartke, K., Bormann, C.: Constrained Application Protocol (CoAP), Universitaet Bremen TZI, June 2014
5. Jain, R.: Constrained Application Protocol for Internet of Things Xi Chen, chen857 (at) wustl.edu
6. Silva, J.D.C., Rodrigues, J.J.P.C., Al-Muhtadi, J., Rabelo, R.A.L., Furtado, V.: Management plat-forms and protocols for internet of things: a survey. Sensors **19**, 676 (2019)
7. Alhaj Ali, A.: Constrained Application Protocol (CoAP) for the IoT (2018). https://doi.org/10.13140/RG.2.2.33265.17766
8. Fries, J.: Why are IoT developers confused by MQTT and CoAP? techtarget.com, 04 May 2017. Accessed 28 March 2017
9. Kute, V., Paradhi, P., Bamnote, G.: A software comparison of RSA and ECC. Int. J. Comput. Sci. Appl. **2** (2021)
10. NIST 2002: Secure Hash Standard (SHS), FIPS PUB 180-2
11. Kasgar, A.K., Agrawal, J., Sahu, S.: New modified 256-bi MD5 algorithm with SHA compression function. Int. J. Comput. Appl. **42**(12), 47–51 (2012)
12. Arun, C., Basha, S.H., Sivakumar, D.L., Rizwan, M.M., Kumar, M.P.: Secured image transmission using elliptic curve cryptography (ECC). Engineering, Communication (2020). https://doi.org/10.37896/jxu14.5/400
13. Mahto, D., Yadav, D.K.: RSA and ECC: a comparative analysis. Int. J. Appl. Eng. Res. **12**, 9053–9061 (2017)
14. Gayoso Martínez, V., Hernández Encinas, L., Queiruga Dios, A.:Security and practical considerations when implementing the elliptic curve integrated encryption scheme. Cryptologia **39**, 1–26 (2015). https://doi.org/10.1080/01611194.2014.988363
15. Eric, C., Seth, M., Joshua, F.: GISSP Study Guide (third edition), pp 219–291 (2015)
16. Pramono, S., Putri, A.O., Warsito, E. and Basuki, S.B.: Comparative analysis of star topology and multihop topology outdoor propagation based on Quality of Service (QoS) of wireless sensor network (WSN). In: 2017 IEEE International Conference on Communication, Networks and Satellite (Comnetsat), 2017, pp. 152–157 (2017). https://doi.org/10.1109/COMNETSAT.2017.8263591
17. Celtek, S.A., Durdu, A., Kurnaz, E.: Design and simulation of the hierarchical tree topology based wireless drone networks. In: 2018 International Conference on Artificial Intelligence and Data Processing (IDAP), pp. 1–5 (2018).https://doi.org/10.1109/IDAP.2018.8620755
18. Liu, Y., Tong, K.F., Qiu, X., Liu, Y., Ding, X.: Wireless mesh neworks in IoT networks. In: 2017 International Workshop on Electromagnetics: Applications and Student Inovation Competition, pp. 183–185 (2017). https://doi.org/10.1109/iWEM.2017.7968828
19. Basu, S.: What are the 5 main network Topologies? Explained with Diagram, September 2019. https://www.how2shout.com/what-is/what-are-the-5-main-network-topologies-explained-with-diagram.html

20. Zhang, Z., Li, P., Zhao, S., Lv, Z., Du, F., An, Y.: An adaptive vision navigation algorithm in agricultural IoT system for smart agricultural robots. Comput. Mater. Contin. **66**(1), 1043–1056 (2021)
21. Esposito, C., Ficco, M., Gupta, B.B.: Blockchain-based authentication and authorization for smart city applications. Inf. Process. Manag. **58**(2), 102468 (2021)
22. El Aidi, S., Bajit, A., Barodi, A., Chaoui, H., Tamtaoui, A.: An advanced encryption cryptographically-based securing applicative protocols MQTT and CoAP to optimize medical-IOT supervising platforms. In: Saeed, F., Mohammed, F., Al-Nahari, A. (eds.) Innovative Systems for Intelligent Health Informatics. IRICT 2020. Lecture Notes on Data Engineering and Communications Technologies, vol. 72, pp. 111–121. Springer, Cham (2021). https://doi.org/10.1007/978-3-030-70713-2_12
23. Sanaa, E.L., Bajit, A., Barodi, A., Chaoui, H., Tamtaoui, A.: An optimized security vehicular internet of things -IoT-application layer protocols MQTT and COAP based on cryptographic elliptic-curve. In: 2020 IEEE 2nd International Conference on Electronics, Control, Optimization and Computer Science, ICECOCS 2020, p. 9314579 (2020)
24. Zanella, A., Bui, N., Castellani, A., Vangelista, L., Zorzi, M.: Internet of things for smart cities. IEEE Internet Things J. **1**(1), 22–32 (2014)
25. Maryoosh, A.: Data security for cloud computing based on elliptic curve integrated encryption scheme (ECIES) and modified identity based cryptography (MIBC). Int. J. Appl. Inf. Syst. (IJAIS) **10**, 7–13 (2016)

Service Selection in Cloud Computing Environment by Using Cuckoo Search

Sana Kouchi(✉) and Hassina Nacer

MOVEP Laboratory, Department of Computer Science, University of Science and Technology Houari Boumediene (USTHB), Algiers, Algeria
{skouchi,hnacer}@usthb.dz

Abstract. Cloud computing opens a new dimension and becomes a foundation technology for developing and integrating Web applications offered as cloud services. However, the number of these services is expected to grow in the future, then, it is required for providers to overcome this situation by being able to offer cloud services with several qualities. That is why the cloud services selection is becoming a big challenge. In this paper, we design an optimization approach based on a meta-heuristic by using Cuckoo Search Algorithm (CSA) with a combination with the Lévy flight behavior to solve the optimal service selection optimization problem in the cloud computing environment considering QoS constraints. The proposed algorithm consists of three phases, initialization phase of the initial population then the evaluation of this population and the last phase of the search for relevant services using Lévy flight. A simulation design demonstrates that a strong selection using the CSA can be achieved in the cloud environment.

Keywords: Cloud computing · Service selection · QoS · Cuckoo Search Algorithm

1 Introduction

The evolution of information technology requires a new computing paradigm that supports the delivery of IT services without the need to install them at local sites with minimal cost. Cloud computing offers the same model described above.

According to the National Institute of Standards and Technology (NIST), cloud computing is a model that provides convenient, on-demand network access to a shared pool of configurable computing resources (e.g., networks, storage servers, applications, and services) that can be rapidly provisioned and available without much management effort or interaction with the service provider [14].

The cloud can be effectively deployed in many ways and four main deployment models are [5]: (1) Public cloud that provides cloud services for public use (e.g., Google Cloud Platform), (2) Private cloud that provides cloud services only for the organization itself and the main usage is managing of sensitive data, (3) Hybrid cloud combines the functionalities of both the public cloud and private

Y. Maleh et al. (Eds.): ICI2C 2021, LNNS 357, pp. 219–228, 2022.
https://doi.org/10.1007/978-3-030-91738-8_21

cloud, and (4) Community cloud shares the cloud infrastructure among several organizations or individuals to address the common concerns [5].

Cloud computing offers a multitude of services that are classified into four main models as follows: the Software as a Service (SaaS), Platform as a Service (PaaS), Infrastructure as a Service (IaaS), and Everything as a Service (XaaS) [5].

There are several cloud providers such as IBM, HP, CISCO and Amazon... etc., and each provider develops its own cloud reference architecture, in which we have also proposed a cloud architecture that allows interoperability in multi-cloud environment [15].

In addition, "SaaS" cloud services offer great opportunities for consumers to find the best services that meet the functional and non-functional needs (QoS) required by a consumer. However, this raises new challenges in how to select the best SaaS service that meets the consumer's need from a set of services found through the discovery phase. For each service requested by the consumer, we can discover several services that have the same functionality but differ in their non-functional settings (QoS). This creates a problem in selecting SaaS services.

Furthermore, the challenge of selecting the best services that meet the consumer's needs in terms of QoS constraints has become an NP-hard [5] issue for which selection is not obvious, because it involves selecting reliable, high-quality cloud services from many functionally similar SaaS services alternatives on the user-identified QoS guarantee. For this reason, selecting cloud services based on QoS is a major problem to solve in the cloud computing environment.

Extensive and intensive studies in this area have resulted in many selection techniques, especially metaheuristic methods, e.g., Genetic Algorithm (GA), Particle Swarm Optimization (PSO), Artificial Bee Colony algorithm (ABC). The Cuckoo Search Algorithm (CSA), another metaheuristic that reproduces the breeding strategy of the parasitic cuckoo bird.

The ability to apply CSA to locate the global optimum of optimization problems has been studied in the literature, and the global search capability of CSA is more efficient than that of GA and PSO [15].

The main idea of this article is oriented towards a modified CSA to solve the problem of selecting SaaS services in cloud computing.

In this solution, we integrate the Lévy flights principle to CSA to accelerate the search strategy of SaaS services, instead of using the randomness principle for the search phase in the classical CSA. The rest of this paper is organized as follows: Related work is described in Sect. 2, Sect. 3 describe the definition and mathematical model of the cloud service selection problem. Section 4 presents the cuckoo behavior and Lévy flights, and we give the proposed solution in Sect. 5. The simulation results are provided in Sect. 6. Finally, in the last section, we will conclude the paper and suggest some indications for future researches.

2 Related Work

In recent years, several approaches proposed in the literature based on bio-inspired metaheuristic (swarm intelligence optimization algorithms) to solve a selection of SaaS service problems in the cloud.

In [11] the authors proposed a solution based on a parallel discrete particle swarm optimization algorithm (PDPSO), and clustering method is adopted to distribute numerous service compositions according to the Map/Reduce mechanism. Then optimal solution or optimal solution of each thread is gotten through the parallel calculation, but this solution may fall into the local optimum. [7], an additional enhancement to the Artificial Bee Colony (ABC) approach was proposed for the QoS-aware services selection problem to balance the exploration and exploitation mechanisms of the ABC algorithm. The proposed solution is based on 2 steps: a neighborhood selection and replacement strategy and a permutation process are used to improve the behavior of ABC, the disadvantage of this solution is that the response time is slow. An improved Genetic Algorithm (GA) is also adopted in [6], the authors introduced elitism and elite-based learning mechanism, to accelerate the global convergence greatly, the disadvantage of this solution is that the response time is very slow due to the steps of the GA algorithm. In [9], the authors introduced an improved fruit fly (FOA) simulated with service composition to handle large-scale services available in the cloud. The disadvantage was that the execution time required was high, according to the services available.

We conclude that most researchers solve the selection problem but do not take into account the response time to the client's request.

In this study, we propose a solution based on the Cuckoo Search meta-heuristic combined with Lévy flights considering the global constraints of the customer.

The proposed solution provides real-time response to consumer needs because of the advantages of CS (low execution time, speed of convergence), which means that our approach improves the search and minimizes the response time.

3 Cloud Service Selection Problem Formalization

In the current cloud market, cloud services can be provided by different cloud service providers depending on various parameters. We should select the best services from the list of candidate services (services that meet the functional requirements) with the different quality of service (QoS) properties from different providers that meet the consumer's requirements.

More formally, we model the cloud service selection problem as follows. Where:

- $S = \{S_1, ..., S_n\}$ denotes the n candidate cloud SaaS service that provides the same functionality.
- $QoS(S_i) = \{qos_1, ..., qos_m\}$ indicate QoS criteria of S_i, there are totally m criteria.

- $UR = \{CF; Const\}$ consumer query, with CF a set of functional constraints, Const a set of non-functional constraints.
- $Const(S_i) = \{Const_1, ..., Const_d\}$ represents the set of constraints defined by the consumer, with $0 < d < m$.
- $W = \{W_1, ..., W_m\}$ represents a set of weight provided by a user for each QoS criteria, with $\sum_{j=1}^{n} w_j = 1$.
- U(S) the objective function (fitness value) to maximize, calculated by Eq. 3.

Service attributes are generally divided into positive QoS and negative QoS. The higher the value of positive QoS, the higher the value they provide, such as service reliability and throughput. On the other hand, for negative QoS, the smaller the value, the higher the utility value, such as service response time and cost. To facilitate the calculation of the objective function value of the selection service, we normalize the attribute values of different cloud services to a number between 0 and 1.

For positive QoS, Eq. 1 is used to normalize [1]:

$$QoS_i^+(S) = \begin{cases} \frac{q_i - q_i^{min}}{q_i^{max} - q_i^{min}}, & \text{if } q_i^{max} \neq q_i^{min} \\ 1, & \text{if } q_i^{max} = q_i^{min} \end{cases} \tag{1}$$

For negative QoS, Eq. 2 is used to normalize [1]:

$$QoS_i^-(S) = \begin{cases} \frac{q_i^{max} - q_i}{q_i^{max} - q_i^{min}}, & \text{if } q_i^{max} \neq q_i^{min} \\ 1, & \text{if } q_i^{max} = q_i^{min} \end{cases} \tag{2}$$

Where q_i^{max} and q_i^{min} are the maximum and minimum values of the kt Attribute of service S_i.

Equation 3 is used to evaluate the fitness of the solution.

$$U(S) = \sum_{k=1}^{m} QoS_i^{-/+}(S) * W_i \tag{3}$$

Where $QoS_i^{-/+}$ is the normalized value of each service attribute, and m is the number of service attributes.

4 Cuckoo Behavior and Lévy Flights

4.1 Cuckoo Search Algorithm (CSA)

Cuckoo Search (CS) is an efficient algorithm based on swarm intelligence, has proposed and developed by Xin-She Yang [13]. CS is inspired by the breeding behavior of cuckoo birds. Subsequent studies proved that CS is simple and effective in solving global optimization problems, and that it is much more efficient than particle swarm optimization (PSO) and genetic algorithms (GA). This is because the number of parameters is smaller, which allows a better adaptation to a larger class of optimization problems.

Also, it has been proven that CS satisfy the global convergence requirements and hence usually converges to a globally optimal solution, as it ensures a balance between exploration and exploitation. However, the search process for CS can be time-consuming, due to the associated random walk behavior.

Although various applications of CS exist, there are few theoretical studies of this algorithm [4,9].

We will outline the basic ideas and steps of the CS algorithm.

- Each cuckoo lays one egg at a time and dumps it in a randomly chosen nest.
- The best nests with high-quality eggs (solutions) will be carried out over to the next generations.
- The number of available host nests is fixed, say n, and the host can discover an alien egg by a probability P_a [0, 1].
- The host bird can either throw the egg away or abandon the nest in order to build a completely new nest in a new location.

4.2 Lévy Flight

Lévy's flights are random walks whose directions are random, and their step lengths are derived from the Lévy distribution. These Lévy flights are performed by animals and insects, and it is characterized by a series of straight flights followed by sudden turns. Compared to normal random walks, Lévy flights are more efficient in exploring large-scale search areas. That is mainly due to the fact that the variances of Lévy's flights increase much faster than that of the normal random walk. Lévy flights can reduce the number of optimization algorithm iterations by about 4 orders compared to the normal random walk [10].

5 Cloud Service Selection Based on Cuckoo Search

To optimize the selection, finding the optimal solutions (the best SaaS services) competently and accurately relies entirely on the inherent search process. The search process used for SaaS service selection is based on the mapping of non-functional criteria of the query specified by the consumer and the service qualities published by the providers after the discovery phase.

In this section, we propose a solution based mainly on metaheuristics more precisely on the CS to solve the SaaS service selection problem in cloud computing (see Fig. 1).

In this solution, we use an accelerated search process, to improve the convergence rate of CSA while keeping the attractive features of CSA.

To improve the convergence rate of CSA, we use in our algorithm the Lévy flight for its advantages. Lévy's flights are more efficient for exploring large-scale search areas.

In our case (Cloud SaaS Services Selection), each egg in a nest represents a SaaS service and each cuckoo can lay a single egg (which represents a SaaS), the goal is to use the best service to replace a less good service in a nest. In

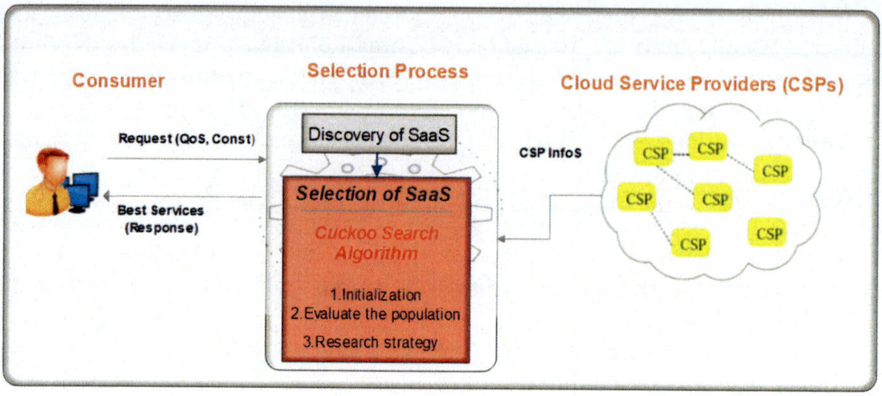

Fig. 1. Architecture of the proposed selection system

this case, there is no more distinction between egg, nest or cuckoo, and each nest corresponds to an egg that also represents a cuckoo that is modelled with a cloud service.

The principle of our proposed algorithm is:

1. **Initialization:**
   The initial population consists of a number n of randomly generated host nests in the search space. In our algorithm the number of host nests, cuckoos and eggs are equal.
2. **Evaluate the population:**
   The efficiency of each nest. (i.e. the performance of the candidate solution) is evaluated by the Eq. 3.
3. **Research strategy**
   - we generate new cuckoos (solutions) from the existing cuckoos (solutions) by performing Lévy Flight. This consists in using the following Eq. 4 evolution law for each cuckoo:

$$x_{new_i} = x_i + \alpha \otimes Lévy(\beta) \tag{4}$$

   Where $\alpha > 0$ is the step size, which should be related to the scales of the problem of interests. In most cases, we can use $\alpha = 1$. The product $\otimes$ means entry wise multiplications. Where $\alpha > 0$ is the step size, depending on the scale of the given problem, $\oplus$ means entry-wise multiplications.
   - Randomly select a host nest $x_j (j \neq i)$ with a cost function F_j, then compares whether the fitness function of the new nest with the randomly chosen one.

$$F_{new_i} < F_j \tag{5}$$

   In this case, the solution x_{new_i} replaces the solution x_j in the population.

- The best nests with high quality eggs will be transferred to the next generations.
- Abandon cuckoo's nests have a probability p_a of being discovered and build new nests.

The flow of the proposed algorithm is given in Algorithm 1.

Algorithm 1: Cuckoo Search Algorithm with Lévy Flights

```
Input: n (number of nests); MaxIt; QoS_i^{-/+}; Const;
Output: Best Services
begin
    Objective Function f(x), x = (x_1, x_2, ..., x_d)
    1/ Generate initial population of n host nests x_i(1, 2, ..., n)
    2/ Evaluate fitness
    while (t ≤ MaxIt) do
        It = It + 1
        3/ Get a Cuckoo randomly by Lévy Flights
        Evaluate its quality/Fitness F_i
        4/ Choose a random nest j;
        if (F_i ≥ F_j) then
            5/ Replace j by the new solution;
        Worst nest is abandoned with probability (p_a) and new nest is built via
        Lévy flights;
        Keep the best solution (or nests with quality solutions);
        Rank the solutions and find the current best service;
    Postprocess result and visualisation;
```

6 Experimental Result

In this section, in order to verify the effectiveness of our proposed CSA algorithm. We compared CSA with Particle Swarm Optimization (PSO) and Genetic Algorithm (GA).

All experiments were performed in the environment of Microsoft Windows 10 (64 bit) system, Intel® core i7 CPU 2.90 GHz and 12 GB RAM, and all algorithms were implemented in JAVA. We considered four QoS attributes of the service: response time, reliability, availability, and cost. Then, we simulated real service data. The dataset contains 2507 candidate cloud services are saved in SQL Server 2014 In the experiments, we have used $\alpha = 1$ and $p_a = 0.25$ we have also tried to vary the number of host nests (or the population size n), in a range of n = [100, 2500]. Finally, we evaluated the algorithm's average execution time.

In the first, we simulated the CSA by varying the number of services to estimate the response time, the result is shown in the Fig. 2, We observe that at n = 600 the execution time is 1.25 s and at n = 1000 the response time is 2.5 s, this increase in time is due to the creation of nests, despite the time consumed for the creation of nests the response to the client's request is done in real-time and as fast, which makes that the response time of this algorithm is acceptable and also reliable.

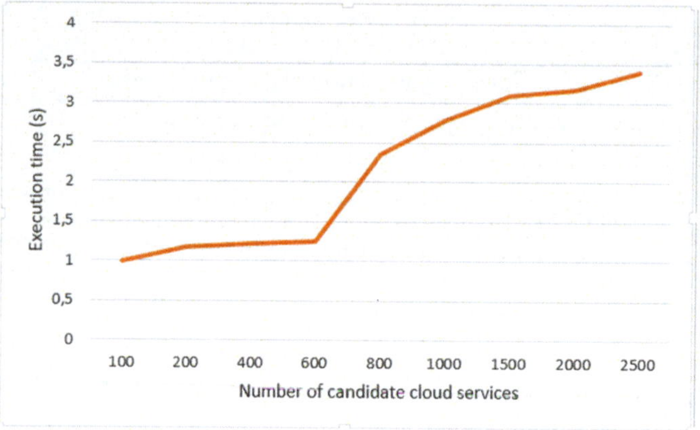

Fig. 2. Performance of CSA over candidate services

In the second simulation, we compare the CSA with PSO and GA for a various number of candidate cloud services in terms of response time. After implementing these algorithms using JAVA, we performed extensive simulations and each algorithm was run at least 10 times to perform meaningful statistical analysis. The algorithms stop when it reaches the indicated number of iterations. To evaluate the algorithms, we varied the number of candidate cloud services in a range from 100 to 2500.

Figure 3 shows the comparison of the execution time of the solution obtained by the three algorithms under different numbers of candidate services.

We observe that when the number of candidate web services increases, the execution time of CSA is less than that of PSO but close to GA. The response time of CSA is considerably low, this is due to its convergence time in addition to its ability to find the best service after a reduced number of iterations and its search range is wide, and it can find better quality solutions in a short execution time. We can see that the CS is much more efficient in finding the best service with a higher success rate, the primary reasons are two folds: A fine balance of randomization and intensification, and a less number of control parameters, so we can conclude that our proposed CSA has better performance.

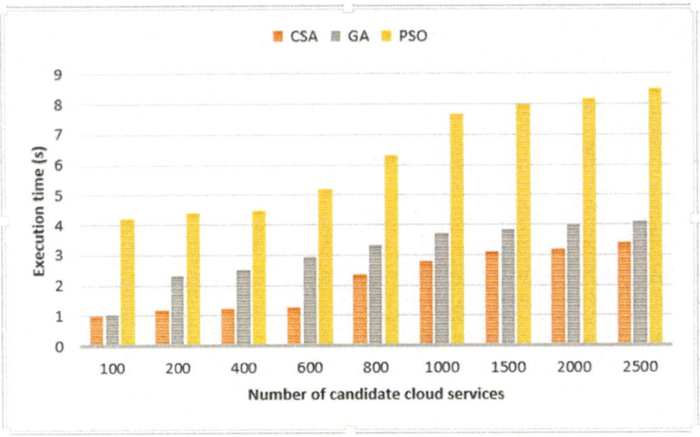

Fig. 3. The performance of the algorithms on varying cloud services.

7 Conclusion

This paper, transformed the cloud service selection optimization problem into a polynomial mathematical model to find the optimal solution. We adapted relatively novel CSA swarm intelligence metaheuristics in combination with Lévy flights for cloud service selection problems. The proposed algorithm has been validated and compared to other algorithms such as genetic algorithms and particle swarm optimization.

Experimental results showed that the proposed algorithm is superior to other algorithms in terms of response time and indicates that the CSA can tackle this kind of problem. This is since CSA has fewer parameters to adjust than in PSO and GA. Therefore, CSA is more generic and robust for the problem of service selection in a cloud computing environment compared to many other metaheuristics algorithms. As a part of future research, could focus on the development of multi-objective evolutionary algorithms to address the cloud service selection problem.

References

1. Dahan, F.: An effective multi-agent ant colony optimization algorithm for QoS-aware cloud service composition. Digit. Object Identifier (2021). https://doi.org/10.1109/ACCESS.2021.3052907
2. Oracle Homepage. https://www.oracle.com. Accessed 20 July 2021
3. Kouchi, S., Nacer, H, Beghded-Bey, K.: Towards a reference architecture for interoperable clouds. In: 8th International Conference on Electrical and Electronics Engineering (ICEEE), pp. 229–233 (2021). https://doi.org/10.1109/ICEEE52452.2021.9415944
4. Li, J., Xiao, D., Lei, H., Zhang, T., Tian, T.: Using cuckoo search algorithm with Q-learning and genetic operation to solve the problem of logistics distribution center location. Mathematics **8**(2), 149 (2020). https://doi.org/10.3390/math8020149

5. Mansouri, N., Ghafari, R., Zade, B.M.H.: Cloud computing simulators: a comprehensive review. Simul. Model. Pract. Theory 102144 (2020). https://doi.org/10.1016/j.simpat.2020.102144
6. Chen, M., Wang, Q., Sun, W., Song, X., Chu, N.: GA for QoS satisfaction degree optimal web service composition selection model. In: 6th International Conference on Behavioral, Economic and Socio-Cultural Computing (BESC) (2019). https://doi.org/10.1109/besc48373.2019.8962994
7. Dahan, F., Mathkour, H., Arafah, M.: Two-step artificial bee colony algorithm enhancement for QoS-aware web service selection problem. IEEE Access **7**, 21787–21794 (2019)
8. Zhou, J., Gao, L., Yao, X., Zhang, C., Chan, F.T., Lin, Y.: Evolutionary algorithms for many-objective cloud service composition: performance assessments and comparisons. Swarm Evol, Comput **51**, 100605 (2019)
9. Abed-alguni, B.H., Paul, D.J.: Hybridizing the cuckoo search algorithm with different mutation operators for numerical optimization problems. J. Intell. Syst. **29**, 1043–1062 (2018). https://doi.org/10.1515/jisys-2018-0331
10. Mareli, M., Twala, B.: An adaptive cuckoo search algorithm for optimization. Appl. Comput. Inform. **14**(2), 107–115 (2018). https://doi.org/10.1016/j.aci.2017.09.001
11. Yimin, Z., Guojun, S., Xiaoguang, Y.: Cloud service selection optimization method based on parallel discrete particle swarm optimization. Chin. Control Decis. Conf. (CCDC) (2018). https://doi.org/10.1109/ccdc.2018.8407473
12. Zhang, Y., Cui, G., Wang, Y., Guo, X., Zha, S.: An optimization algorithm for service composition based on an improved FOA. Tsinghua Sci. Technol. **20**(1), 90–99 (2015)
13. Yang, X.-S., Karamanoglu, M.: Swarm Intelligence and Bio-inspired Computation: An Overview. Elsevier, Amsterdam (2013)
14. The NIST Definition of Cloud Computing, Special Publication 800-145 (2011)
15. Yang, X.-S., Deb, S.: Engineering optimization by cuckoo search. Int. J. Math. Model. Numer. Optim. **1**(4), 330–343 (2010)

Markov Decision Processes with Discounted Rewards: Improved Successive Over-Relaxation Method

Abdellatif Semmouri[1](✉), Mostafa Jourhmane[1], and Bahaa Eddine Elbaghazaoui[2]

[1] Faculty of Sciences and Techniques, Sultan Moulay Slimane University, Campus Mghilla, Beni Mellal, Morocco

[2] Faculty of Sciences, Ibn Tofail University, Kenitra, Morocco

Abstract. In mathematics, discrete-time Markov chains are stochastic control processes. They provide a mathematical framework for modeling decision making in situations where outcomes are partly governed by randomness and partly commanded under the control of a decision maker. In order to reduce the computational efforts of complexity, the action elimination procedures have been applied with considerable success and are appeared in the literature for solving this problem.

We focus our study on the complexity of some classical problem of computing optimal policy computation in Markov decision problems (MDPs) with discounted reward criterion. It is well known that those problems are solvable in polynomial time via the dynamic programming tool. In this context, we give an overview on the main variants of the value iteration scheme which is a part of successive approximation methods. Then, we improve the successive over-relaxation algorithm (SOR) by establishing a new test of non-optimal decision. Finally we give an example for clarification and demonstrating our proposal.

Keywords: Markov decision processes · Discounted criterion · Sub-optimal actions · Stochastic optimization

1 Introduction

The Markov chain framework is essentially one of the directions of stochastic processes, enclosing a rich and fascinating set of results and dealing with models that are characterized by the probabilistic aspect that evolve over time. Although the results are applied to many areas of the real world, they are best understood initially in terms of their mathematical character and interrelationships. Markov chains have successfully modeled a wide range of decision problems related to the uncertainty that is governed by chance arising from the randomness. They give a mathematical tool for solving these situations in many fields in the real life such as systems engineering, industrial engineering, management science, finance, and artificial intelligence. Thus, the main objective is to optimize some performance

Y. Maleh et al. (Eds.): ICI2C 2021, LNNS 357, pp. 229–238, 2022.
https://doi.org/10.1007/978-3-030-91738-8_22

of the system which is described by a utility function in order to calculate an optimal policy for the proposed problem via the dynamic programming method based on the Bellman optimality equation.

Multitude tests of sub-optimal actions were provided in this regard after computing lower and upper bounds on the optimal objective function of discrete finite discounted MDPs over the infinite horizon. In this context, authors contributed by various results for solving these problems as accelerating methods or action elimination techniques. As soon as an action in a certain state is identified as non-optimal control in an MDP, it can be eliminated from the computation planning after this epoch. This idea was firstly used by MacQueen [1,2]. He demonstrated how non-optimal decision can be discarded in discrete MDPs with finite state and action spaces. Furthermore, he showed that non-optimal actions can be tested if terminals of the optimal utility function are determined. Therefore, the bounds found can be exploited for reducing the action space size at each iteration as soon as it is possible. Next, Porteus [3] gave an extension of MacQueen test in the general framework of the MDPs under the discounted criterion. He established new bounds on the optimal discounted gain. His approach is based on to begin with a functional equation structure for a process rather than to show such a structure is induced by appropriate way from an alternative starting point. For reminding, we briefly present the Grinold [4] contribution which was inspired from MacQueen and Porteus results. He confirmed that lower and upper bounds on the extreme objective function can be found via linear programming or policy iteration method. Since the action elimination techniques are the main focus of many interested authors in this background, the efforts of other authors such that Hastings and Mello [5], Puterman and Shin [6], Sadjadi and Bestwick [7], White [8], Semmouri and Jourhmane [9,10], Semmouri Jourhmane and Elbaghazaoui [11] etc. are valued and appreciated by participating to this challenge.

On our side, we will propose an ameliorated algorithm based on the successive over-relaxation (SOR) scheme and a new test for eliminating actions which are not part of optimal decisions at the current stage of a discounted return in Markov decision problems over the infinite horizon in this work.

The structure of this paper, is organized as follows: In Sect. 1, we presents some works in the same context of our manuscript. Next, the Sect. 2 focuses on the model of Markov decision process and the criterion of total expected discounted reward over infinite horizon. Thus, we give an overview of the main schemes of value iteration algorithm and particularly SOR algorithm. Finally, we provide an efficient iterative algorithm based on the SOR scheme and a new action elimination procedure in Sect. 3 and we end our work by conclusion in Sect. 4.

2 Preliminaries

2.1 Markov Decision Process Model

Consider a discrete-time Markov Control Model represented by the tuple:

$$\left(S; A = \bigcup_{i \in S} A(i); Q = \{p_n(j/i, a)\}; r = \{r_n(i, a)\}; \gamma\right)$$

where

- S is a finite state space in which the prosses evolution takes place;
- A the union of all available finite action sets which control the state dynamics;
- $p(./.,.)$ denotes the state transition probability functions;
- r(.,.) gives the reward function defined on state transitions;
- $\gamma \in (0, 1)$ is the discount factor (Fig. 1).

dd

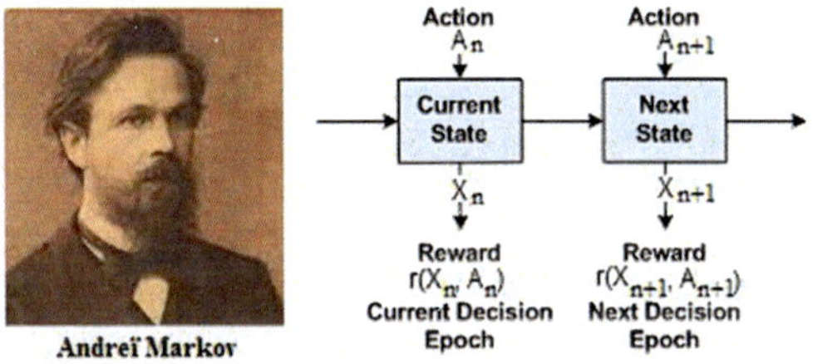

Fig. 1. Markov Decision Process Model

For many details, readers are advised to see [9–12].

2.2 Expected Discounted Reward Criterion

The goal is to determine a policy that at each decision time point, it maximizes the performance of the system over the infinite horizon. This performance is expressed as the expected total discounted reward utility function over the planning horizon. Hence, the decision maker has to find the optimal balance between immediate reward and future reward (see Semmouri, Jourhmane and Belhallaj [12], Howard [13], Bellman [14], Bertsekas [15], White [16] and Puterman [17]).

Definition 1. *If policy $\pi \in \Pi$ is used and state $i \in S$ is the starting state, the expected total γ-discounted reward over infinite horizon, given discount factor $\gamma \in [0, 1)$, is denoted by $J_\gamma(i, \pi)$ and defined by*

$$J_\gamma(i, \pi) := \mathbb{E}_\pi^i[\sum_{n=1}^{\infty} \gamma^{n-1} r(X_n, A_n)] \tag{1}$$

where $r(.,.)$ are the rewards, as functions of current state and action chosen.

Interchanging the expectation and the summation in (1) by using the Lebesgue's dominated Convergence Theorem, the expected total discounted reward over the planning horizon may also be defined as the total discounted expected reward

$$J_\gamma(i,\pi) = \sum_{n=1}^{\infty} \gamma^{n-1}\mathbb{E}_\pi^i[r(X_n,A_n)] = \sum_{n=1}^{\infty}\gamma^{n-1}\sum_{j\in S}\sum_{a\in A(j)} \mathbb{P}_\pi^i(X_n=j,A_n=a)r(j,a)$$

Definition 2. *The optimal value function is given by*

$$J_\gamma^\star(i) := \sup_{\pi\in\Pi} J_\gamma(i,\pi), \quad i\in S \tag{2}$$

and a policy $\pi^\star \in \Pi$ *is said to be optimal if* $J_\gamma^\star(i) = J_\gamma(i,\pi^\star)$, $\forall i\in S$.

Since the state space and action spaces are finite, the existence and the computation of optimal policies are investigated in Markov decision theory.

3 Results and Experiments

3.1 A Brief Overview on the Successive Over-Relaxation (SOR)

In numerical analysis framework, the method of Successive Over-relaxation Method (in short SOR) is a variant of the Gauss-Seidel method for solving a system of linear equations. The convergence of this algorithm is generally faster. This method was discovered simultaneously by David M. Young, Jr. and Stan Frankel in 1950 with the aim of automatically solving linear systems with computers. Next, the over-relaxation methods have been used to solve Markov decision problems as a variant of the standard Value Iteration Algorithm (VIA).

For $\omega \in]0,\omega^\star]$, we define the quantities $\gamma_i(\omega)$, $i\in S$ by the recursion formula:

$$\gamma_i(\omega) = \omega\left[\max_{a\in A(i)}\left\{r(i,a) + \gamma\sum_{j=1}^{i-1} p(j/i,a)\gamma_j(\omega) + \gamma\sum_{j=i}^{|S|} p(j/i,a)\right\}\right] + 1 - \omega$$

and

$$\rho(\omega) = \max_{i\in S}\gamma_i(\omega)$$

Consider the Bellman operator $\Phi : \mathbb{R}^{|S|} \to \mathbb{R}^{|S|}$ given for all vector $V = (V_i)_{i\in S} \in \mathbb{R}^{|S|}$ by

$$(\Phi_\omega V)_i = \omega\left[\max_{a\in A(i)}\left\{r(i,a) + \gamma\sum_{j=1}^{i-1} p(j/i,a)(\Phi_\omega V)_j + \gamma\sum_{j=i}^{|S|} p(j/i,a)V_j\right\}\right] + (1-\omega)V_i \tag{3}$$

It is established by Denardo [18] and Reetz [19] that Φ_ω is a contraction mapping with contraction factor $\rho(\omega^\star) = 1 + (\gamma-1)\omega^\star$ and $\rho(\omega^\star) \le \gamma \prec 1$.

Since the value vector $J_\gamma^\star$ is the unique fixed point of Φ_ω, it follows the SOR algorithm:

Algorithm 1: Successive Over-Relaxation (SOR)

Input: MDP
Output: The optimal value vector; An optimal policy
Initialize $\mathbf{V}^0 = 0 \in \mathbb{R}^{|S|}$, **specify** $\varepsilon \succ 0$
$\mathbf{n} \leftarrow \mathbf{0}$
repeat
  for $i = 1$ *to* $|S|$ **do**
    for $a \in A(i)$ **do**
      $Q^n(i,a) = r(i,a) + \gamma \sum_{j \prec i} p(j/i,a) V_j^{n+1} + \gamma \sum_{j \geq i} p(j/i,a) V_j^n$
    $V_i^{n+1} = \omega^\star \max_{a \in A(i)} Q^n(i,a) + (1 - \omega^\star) V_i;$
  $n \leftarrow n + 1$
until $|| V^{n+1} - V^n ||_\infty \prec \varepsilon$;
for $i \in S$ **do**
  $\pi^\star(i) \in \arg \max_{a \in A(i)} Q^n(i,a);$
return V^n, $\pi^\star$.

Example 1. Consider an MDP with the following data

$$S = \{1,2\};\ A(1) = A(2) = \{a_1, a_2, a_3\}.$$

Transition probabilities:

$p(1/1,a_1) = 0.5$; $p(2/1,a_1) = 0.5$; $p(2/2,a_1) = 0.1$; $p(1/2,a_1) = 0.9$;
$p(1/1,a_2) = 1$; $p(2/2,a_2) = 1$; $p(1/1,a_3) = 1$; $p(2/2,a_3) = 1$.

Rewards:

$r(1,a_1) = 7$; $r(1,a_2) = 5$; $r(1,a_3) = 8$;
$r(2,a_1) = 4$; $r(2,a_2) = 6$; $r(2,a_3) = 3$ (Fig. 2).

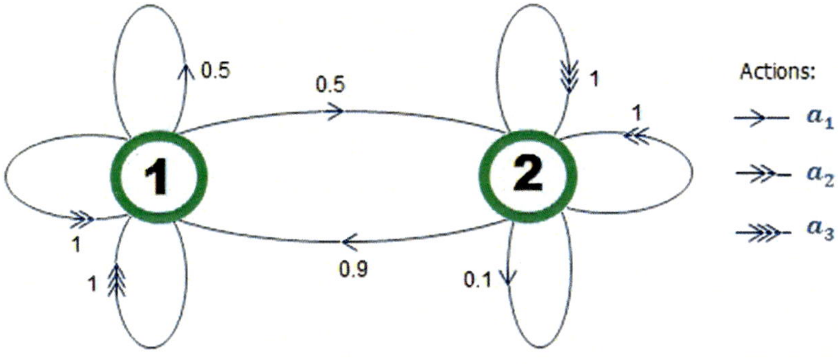

Fig. 2. Graphic representation

Using the standard VIA, the optimal gain value $J_\gamma^\star = (26.6667, 22.3656)$ for $\gamma = 0.7$ is reached at the 61st iteration. On the other hand, this maximum is reached at the 66th iteration for the same value of γ by applying SOR algorithm. This further confirms the efficiency of SOR in terms of speed.

3.2 New Test of Non-optimal Decisions

The suitable choice of $\omega = \omega^\star$, first ensure the convergence of the SOR2 algorithm. Second, this convergence is faster than VIA to the fixed point of the contraction Ψ_ω.

Set

$$R = \max_{i,a} |r(i,a)| \quad and \quad \mu = \rho(\omega^\star)$$

During this subsection, we assume that the reward functions and the transition probabilities are stationary with respect to time. The following lemma gives lower and upper bounds on the optimal value utility J_γ.

Lemma 1. *Let $(V^n)_{n\geq 0}$ be the sequence of vectors given by the iterative method SOR. Then, we get*

$$V_i^n - \frac{\mu^n R}{1-\mu} \leq J_\gamma(i) \leq V_i^n + \frac{\mu^n R}{1-\mu}, \quad for\ all\ i \in S,\ n \geq 0. \tag{4}$$

Proof. The proof of this lemma is demonstrated by the contraction mapping theory. □

The action elimination procedures is a powerful tool that is widely used in solving Markov decision problems. Because it helps to identify non-optimal decisions, it reduces computational complexity. Hence, we provide an efficient test in the following theorem inspired from proposition 6.7.3. in Puterman [17] and (4).

Theorem 1 (New Test of non-optimality). *Let $(V^n)_{n\geq 0}$ be the sequence of vectors generated by Algorithm 1. If*

$$r(i,b) + \gamma \sum_{j\in S} p(j/i,b) V_j^n + \frac{\mu^n(1+\gamma)R}{1-\mu} \prec V_i^n \tag{5}$$

at some stage n, then any markovian policy which selects action b in state i is non-optimal.

Proof. Suppose that the action b is optimal in state i and at epoch n. It is easy to show that the maximum in the Bellman optimality equation is reached by action b in state i. Next, we have

$$J_\gamma^\star(i) = r(i,b) + \gamma \sum_{j\in S} p(j/i,b) J_\gamma^\star(j) \tag{6}$$

From (4) and (6), it follows

$$J^{\star}_{\gamma}(i) \leq r(i,b) + \gamma \sum_{j \in S} p(j/i,b) V^n_j + \frac{\mu^n R \gamma}{1-\mu} \tag{7}$$

Combining (4), (5) and (7), we obtain

$$J^{\star}_{\gamma}(i) \prec V^n_i - \frac{\mu^n R}{1-\mu} \leq J^{\star}_{\gamma}(i)$$

This ends the proof. □

The practical application of our contribution requires the improvement of the SOR algorithm. Therefore, a combination of the test (5) and the SOR algorithm leads to the following algorithm:

Algorithm 2: Improved Successive Over-Relaxation (ISOR)

Input: MDP
Output: The optimal value vector,An optimal policy
Initialize $\mathbf{V^0} = 0 \in \mathbb{R}^{|S|}$, **specify** $\varepsilon \succ 0$
$\mathbf{n} \leftarrow \mathbf{0}$
repeat
  for $i = 1$ *to* $|S|$ **do**
    for $a \in A(i)$ **do**
      $Q^n(i,a) = r(i,a) + \gamma \sum_{j \prec i} p(j/i,a) V^{n+1}_j + \gamma \sum_{j \geq i} p(j/i,a) V^n_j \};$
      $S^n(i,a) = r(i,a) + \gamma \sum_{j \prec i} p(j/i,a) V^n_j + \frac{\mu^n (1+\gamma) R}{1-\mu} - V^n_i;$
      if $S^n(i,a) \prec 0$ **then**
        $A(i) \leftarrow A(i) - \{a\}$
    $V^{n+1}_i = \omega^{\star} \max_{a \in A(i)} Q^n(i,a) + (1-\omega^{\star}) V_i;$
  $n \leftarrow n+1$
until $|| V^{n+1} - V^n ||_{\infty} \prec \varepsilon;$
for $i \in S$ **do**
  $\pi^{\star}(i) \in \arg \max_{a \in A(i)} Q^n(i,a);$
return V^n, $\pi^{\star}$.

Example 2. Consider again Example 1 with discount factor $\gamma = 0.6$. By applying Algorithm 2 (ISOR), we obtain the following numerical results (Table 1):

Table 1. Eliminated actions by the test function S^n

It	$S^n(1,a_1)$	$S^n(1,a_2)$	$S^n(1,a_3)$	$S^n(2,a_1)$	$S^n(2,a_2)$	$S^n(2,a_3)$
1	37.526808	35.399149	38.590638	39.224409	36.462978	33.271489
5	2,076770	1,011377	4,202866	3,808447	3,491562	0,300072
6	0,162438	-0,777077	2,414413	2,187831	1,870946	-1,320543
7	-0,937285	-1,804486	1,387003	1,256839	0,939954	-2,251535
8	-1,569040	-2,394700	0,796789	0,722014	0,405129	-2,786361
9	-1,931964	-2,733759	0,457730	0,414774	0,097888	-3,093601
10	-2,140452	-2,928538	0,262951	0,238274	-0,078611	-3,270100
11	-2,260221	-3,040432	0,151057	0,136881	-0,180004	-3,371494
12	-2,329025	-3,104712	0,086777	0,078634	-0,238252	-3,429741

Each value of $S^n(.,.)$ written in red indicates that the action a selected in state i is identified as a non-optimal action at stage n which must be eliminated. First, actions a_2 and a_1 must be eliminated in state 1 respectively at the 6th iteration and the 7th iteration. Second, actions a_3 and a_2 must be eliminated in state 2 respectively after the 6th iteration and the 10th iteration.

For $\gamma = 0.6$, the optimal value vector $J^\star_\gamma$ is given by

$$J^\star_{0.6} = (20, 15.74468085)$$

In addition, the new action spaces given by Algorithm 2 are:

$$A(1) = \{a_3\} \text{ after } 7^{th} \text{ iteration and } A(2) = \{a_1\} \text{ after } 10^{th} \text{ iteration.}$$

Let n_1 be the smallest number of iterations necessary to eliminate the first action and n_2 be the greatest number of iterations necessary to eliminate all non-optimal actions. By varying the discount factor γ, the behavior of the numbers $n_1(\gamma)$ and $n_2(\gamma)$ is given by the Fig. 3 below:

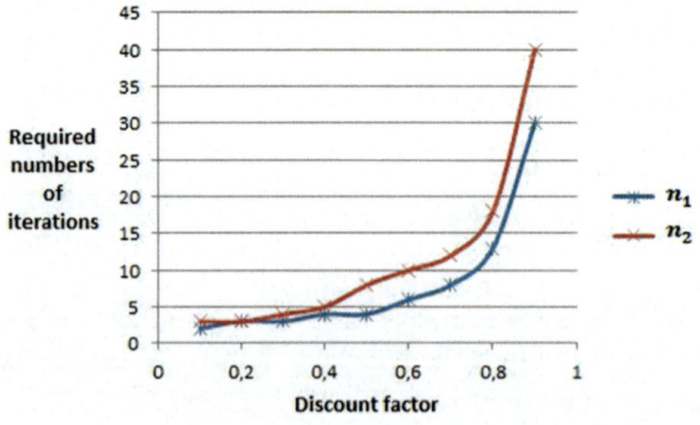

Fig. 3. Variation of $n_1(\gamma)$ and $n_2(\gamma)$

For $0.1 \leq \gamma \leq 0.8$, Fig. 3 clearly shows the efficiency of our contribution. In fact, the maximum number of iterations to eliminate all non-optimal actions is small. Besides, the values of iterations $n_1(\gamma)$ and $n_2(\gamma)$ increase as soon as γ is close to 1.

4 Conclusion and Future Perspective

Dynamic programming remains an effective approach to solve a multitude of stochastic problems which are modeled by Markov decision processes. In this regard, the SOR algorithm as variant of the standard VIA is a meeting point for many research authors. Each improvement aims to accelerate VIA or to reduce the complexity of space calculations in some sense. Hence, the amelioration of VIA is constantly increasing to establish important contributions via the dynamic programming tool.

We hope that our effort is distinguished from preceding works in the same literature of the tests. In Macqueen and Porteus ones, we have to store the vectors V^{n-1}, V^n and V^{n+1} for calculating V^{n+1}. The SOR method is more interesting with regard to memory management. Indeed, we can overwrite the value of V_i^n as we go along and only store a single vector of size $|S|$ during calculations, e.g. the vector:

$$(V_1^{n+1}, \ldots, V_i^{n+1}, V_{i+1}^n, \ldots, V_{|S|}^n)$$

instead of three vectors. When the size of the state space is very large, the power and the efficiency of our result appear very important. In spite of this challenge, most of non-optimality tests that have been proposed until now are inspiring from the MacQueen and Porteus tests as essential references.

In the future perspective, we will apply our theoretical model in Artificial Intelligence by improvement of the Q-learning function. Thus, we can contribute in reducing the computational complexity as best as possible.

Acknowledgements. We are grateful to Professor Dr. C. Daoui Sultan Moulay Slimane University, Beni Mellal, Morocco for his help and encouraging and Mr. Lekbir Tansaoui, ELT teacher, co-author and textbook designer in Mokhtar Essoussi High School, Oued Zem, Morocco for proofreading this paper. We also would like to thank referees for careful reading of the manuscript, useful comments and suggestions for improving this paper.

References

1. MacQueen, J.B.: A modified dynamic programming method for Markovian decision problems. J. Math. Anal. Appl. **14**(81), 38–43 (1965). https://doi.org/10.1016/0022-247X(66)90060-6
2. MacQueen, J.B.: A test for suboptimal actions in Markovian decision problems. Oper. Res. **15**(3), 559–561 (1967). https://doi.org/10.1287/opre.15.3.559

3. Porteus, E.L.: Some bounds for discounted sequential decision processes. Manag. Sci. **18**(1), 7–11 (1971). https://doi.org/10.1287/mnsc.18.1.7
4. Grinold, R.C.: Elimination of suboptimal actions in Markov decision problems. Oper. Res. **21**(3), 848–851 (1973). https://doi.org/10.1287/opre.21.3.848
5. Hastings, N.A.J., Mello, J.M.C.: Tests for suboptimal actions in discounted Markov programming. Manag. Sci. **19**(9), 1019–1022 (1973). https://doi.org/10.1287/mnsc.19.9.1019
6. Puterman, M.L., Shin, M.C.: Modified policy iteration algorithms for discounted Markov decision problems. Manag. Sci. **24**(11), 1127–1137 (1978). https://doi.org/10.1287/mnsc.24.11.1127
7. Sadjadi, D., Bestwick, P.F.: A stagewise action elimination algorithm for the discounted Semi-Markov problem. J. Oper. Res. Soc. **30**(7), 633–637 (1979). https://doi.org/10.1057/jors.1979.156
8. White, D.J.: The determination of approximately optimal policies in Markov decision processes by the use of bounds. J. Oper. Res. Soc. **33**(3), 253–259 (1982). https://doi.org/10.1057/jors.1982.51
9. Semmouri, A., Jourhmane, M.: Markov decision processes with discounted cost: the action elimination procedures. In: ICCSRE 2nd International Conference of Computer Science and Renewable Energies, Agadir, Morocco, pp. 1–6. IEEE Press (2019). https://doi.org/10.1109/ICCSRE.2019.8807578
10. Semmouri, A., Jourhmane, M.: Markov decision processes with discounted costs over a finite horizon: action elimination. In: Masrour, T., Cherrafi, A., El Hassani, I. (eds.) A2IA 2020. AISC, vol. 1193, pp. 199–213. Springer, Cham (2021). https://doi.org/10.1007/978-3-030-51186-9_14
11. Semmouri, A., Jourhmane, M., Elbaghazaoui, B.E.: Markov decision processes with discounted costs: new test of non-optimal actions. J. Adv. Res. Dyn. Control Syst. **12**(05-SPECIAL ISSUE), 608–616 (2020). https://doi.org/10.5373/JARDCS/V12SP5/20201796
12. Semmouri, A., Jourhmane, M., Belhallaj, Z.: Discounted Markov decision processes with fuzzy costs. Ann. Oper. Res. **295**(2), 769–786 (2020). https://doi.org/10.1007/s10479-020-03783-6
13. Howard, R.A.: Dynamic Programming and Markov Processes. Wiley, New York (1960)
14. Bellman, R.E.: Dynamic Programming. Princeton University Press, Princeton (1957)
15. Bertsekas, D.P., Shreve, S.E.: Stochastic Optimal Control. Academic Press, New York (1978)
16. White, D.: Markov Decision Processes. Wiley, Hoboken (1993)
17. Puterman, M.L.: Markov Decision Processes: Discrete Stochastic Dynamic Programming. Wiley, New York (1994)
18. Reetz, D.: Solution of a Markovian decision problem by successive overrelaxation. Z. Oper. Res. **17**(1), 29–32 (1973)
19. Denardo, E.V.: Contraction mappings in the theory underlying dynamic programming. SIAM Rev. **9**(2), 165–177 (1967)

Speech Spectral Subtraction in Modulation Domain

Nesrine Abajaddi[1(✉)], Badia Mounir[2], Laila Elmaazouzi[2],
Ilham Mounir[2], and Abdelmajid Farchi[1]

[1] IMII Laboratory, Faculty of Sciences and Technics, University Hassan First, Settat, Morocco
nesrine.abajaddi@uhp.ac.ma
[2] LAPSSII Laboratory Graduate School of Technology, University Cadi Ayyad, Safi, Morocco

Abstract. This paper is for investigating speech enhancement in modulation. We want to discover how important the modulation domain is for spectral subtraction compared to the acoustic domain. We use a new technique based on spectral subtraction in the modulation domain to remove noise from temporal modulation envelopes. Furthermore, we used coherent demodulation based on the spectral center of Gravity approach, which assures that modulators are free of distortion. The experiment findings of various acoustic noises non-stationary suggest that our proposed approach can enhance speech intelligibility compared to other existing methods in the temporal domain.

Keywords: Coherent demodulation · Modulation domain · Center-of-Gravity Spectral subtraction · Coherent demodulation

1 Introduction

To our knowledge, none of the well-known signal processing techniques has been observed to improve the intelligibility of speech when compared to untreated degraded speech [1], especially with low signal-to-noise ratio levels. Naturally, speech is a modulated signal, which can be described as a collection of small frequency subbands of amplitude-modulated (AM) signals. The modulator "envelope" and carrier "fine structure" are specified for each subband. Many physiological and psychoacoustic investigations, on the other hand, have demonstrated the importance and interest of the modulation domain in speech signal processing and speech enhancement. Many psychoacoustics researchers used simple stimuli like clicks and sinusoids to study the human auditory system. To better understand the auditory system, the researchers adopted more complex stimuli than the two previously mentioned. Among them are the modulated stimuli used for the first time by Zwicker [2]. He proved that the ear could distinguish both amplitude and frequency modulations and perceives amplitude modulation much better. Drullman et al. [3] also used low-pass filters and high-pass filters to investigate the speech signals intelligibility with temporally changed spectral

Y. Maleh et al. (Eds.): ICI2C 2021, LNNS 357, pp. 239–246, 2022.
https://doi.org/10.1007/978-3-030-91738-8_23

envelopes. They discovered that modulation frequencies (4–16 Hz) contribute to speech comprehension, and that the extremely slow and very fast components of the word contain no important linguistic information. Based on this observation, the speech signal could be described as the sum of a series of amplitude modulated (AM) signals in small frequency subbands that cover the signal bandwidth [3]. Furthermore, it seems to be well acknowledged that the algorithms of noise suppression like Minimum Mean Square Error estimation (MMSE), spectral subtraction [4] work significantly more than the short-time Fourier transform (STFT) in the modulation domain [5].

This paper introduces a low-frequency modulation processing-based speech enhancement approach. Speech is considered to be a combination (modulator "envelope"/carrier "fine structure" from the standpoint of spectral modulation analysis). Each modulator can be investigated, charged, and analyzed apart from the carrier, or recombined with its associated carrier. Since non-coherent sub-band demodulation causes bothersome audible artifacts due to the distortion in bandwidth, the current work uses coherent demodulation with spectral center of gravity (COG) to estimate the modulator and carrier [6]. Coherent demodulation provides that the distortion in modulators is free. The modulators' noise dominated in low frequency was identified using the modulation domain SNR, which is like the modulation channel selection scheme [4]. Because the low modulation frequency is critical to maintaining speech intelligibility, the noise in low-frequency of modulators was identified using the modulation domain SNR. Furthermore, Artefacts in processed speech are likewise minimal since speech structure is preserved throughout the modulation processing using a coherent demodulation approach. In addition, we used a simple method (spectral subtraction) on the dominated modulators to improve speech by noise suppression.

The following is the document's structure. The approach and the experiments are explained in the first part. The results will be presented in the second segment. Finally, there is a conclusion.

2 Methods

2.1 Fundamentals of Demodulation

Naturally, speech is a modulated signal defined as a sum of amplitude-modulated (AM) signals in a series of narrow frequency subbands that span the signal bandwidth. The envelope "modulator" and the carrier "fine structure" can be specified for each subband. The modulator can be studied independently of the carrier, or it can be recombined with it. So, we can describe the speech of band speech in discrete-time by:

$$x(n) = \sum_{i=1}^{M} s_i(n) = \sum_{i=1}^{M} m_i(n) \circ c_i(n) \quad (1)$$

Where $x(n)$ is the speech signal, $m_i(n)$ and $c_i(n)$ described the i-th modulator and the subband carrier $s_i(n)$, respectively. M is the subbands number. (o) is the operator that significate the hadamard product. The speech signal modulation decomposition employs two types of approaches (coherent and incoherent). $m_i(n)$ and $c_i(n)$ are approximated from the filter bank (analytic bandpass subband signal (Fig. 1)) $s_i(n)$ using the technique of non-coherent subband demodulation.

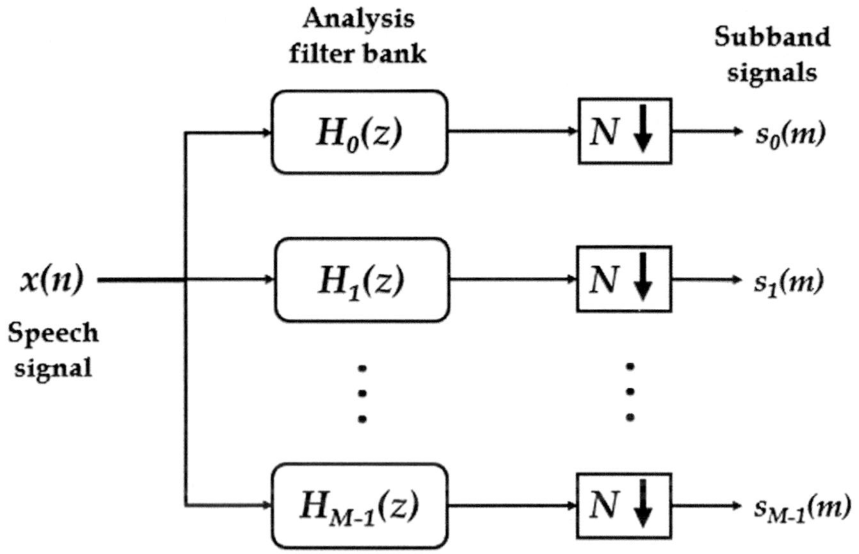

Fig. 1. M-channel Analysis filter bank

Non-coherent demodulation appears to be a simple and uncomplicated method for extracting the modulator from each subband. The modulator (defined as "Hilbert envelope") and carrier are determined in the following format:

$$m_i(n) = |s_i(n)| \tag{2}$$

$$c_i(n) = exp(j\angle s_i(n)) \tag{3}$$

However, because the modulator and carrier are not limited to a specific frequency range, they cannot be modified separately. Because of this, we decided to use coherent subband demodulation. As the instantaneous frequency estimate $f_i(n)$, we employed the

Spectral Center of Gravity approach [7] to follow the evolution of spectral concentration within subband over time. The carrier phase is characterized by the following:

$$\phi_i(n) = \sum\nolimits_{p=0}^{n} f_i(n) \tag{4}$$

Coherent demodulation defines the modulator according to the estimated carrier by:

$$c_i(n) = exp(j\angle s\phi_i(n)) \tag{5}$$

$$m_i(n) = s_i(n) \cdot c_i^*(n) \tag{6}$$

Conceptually, coherent demodulation using spectral COG is presented in Fig. 2.

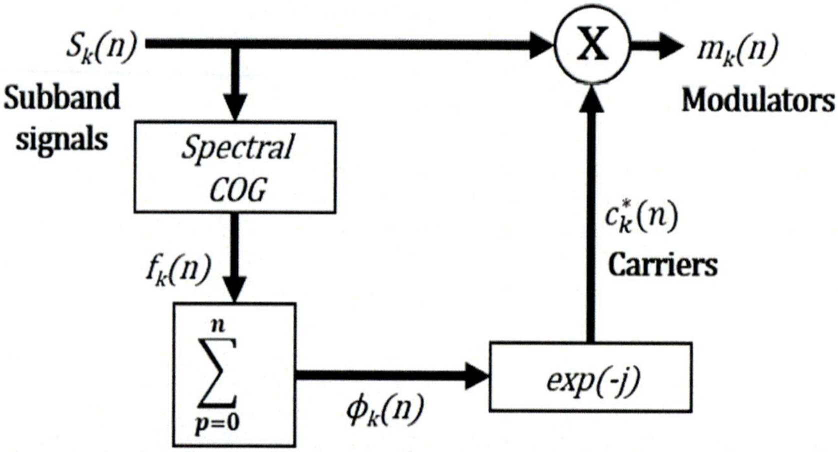

Fig. 2. Coherent demodulation using spectral Center-of-Gravity block diagram.

2.2 Modulators Modification and Synthesis

The five frequency bands considered in this work are as follows (Table 1): Band 1 [100 to 400 Hz]; Band 2 [400 to 800 Hz]; Band 3 [800 to 2000 Hz]; Band 4 [2000 to 3500 Hz] and Band 5 [3500 to 5000 Hz]. The modulators which are extracted using spectral Center of Gravity are modified. We can calculate the modulation spectrum by utilizing STFT on the noisy modulator envelope $m_i(n)$:

$$M_i(p, l) = \sum\nolimits_{n=0}^{L-1} m_i(n + pB)w(n)e^{-j\left(\frac{2\pi ln}{L}\right)} \tag{7}$$

Where p and l indicate the modulation frequency and acoustic frequency index, respectively. L is the FFT length. The acoustic frameshift in samples is denoted by B. The acoustic analysis window function is indicated by $w(n)$. This spectrum of modulation makes it possible to select the noisy target-dominant in the low-frequency modulations, assuming the noise is additive:

$$|M_i(p,l)| = |\hat{M}_i(p,l)| + |D_i(p,l)| \tag{8}$$

We propose to replace $|M_i(p,l)|$ with $|\widehat{M}_i(p,l)|$, where $|\widehat{M}_i(p,l)|$ is a clean modulator estimate, determined using a spectral subtraction rule like Berouti et al. [8] proposed algorithm and described by Eq. 9. Where α is the subtraction factor that determines how much over-subtraction there is.; β is the spectral floor parameter, which is used to set spectral magnitude values below than $(\beta|D_i(p,l)|)$ to that spectral floor; and γ defines the subtraction domain, the noise modulation magnitude spectrum is estimating using a voice activity detector [9] in modulation domain, defined by $|D_i(p,l)|$. After modification of modulators, the signal in the time domain is given by overlap-add synthesis [10].

$$\left|\widehat{M}_i(p,l)\right| = \begin{cases} (|M_i(p,l)|^{\gamma} - \alpha|D_i(p,l)|^{\gamma})^{\frac{1}{\gamma}}, & if(|M_i(p,l)|^{\gamma} - \alpha|D_i(p,l)|^{\gamma})^{\frac{1}{\gamma}} > \beta|D_i(p,l)| \\ \beta|D_i(p,l)| & otherwise \end{cases} \tag{9}$$

Table 1. The five frequency bands.

Bands number	Bands in Hz
Band 1	100–400
Band 2	400–800
Band 3	800–000
Band 4	2000–3500
Band 5	3500–5000

So, our adopted approach is summarized in the following Fig. 3.

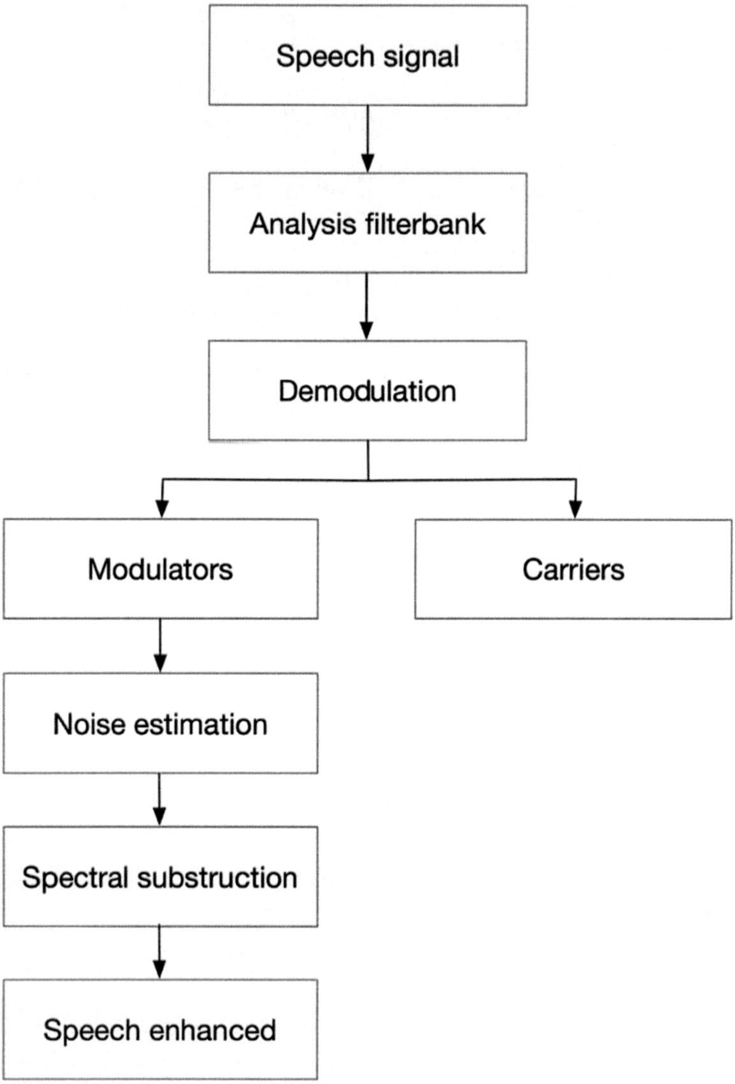

Fig. 3. Block diagram processing of speech in the spectral modulation domain.

3 Experimental Results

3.1 Corpus

We used 40 clean speech sentences from the TIMIT database to test the proposed algorithm's performance. Four male speakers and four female speakers pronounced the sentences uniformly. Noisy samples are generated by combining clean TIMIT sentences with three types of noise: babbling noise, street noise and train noise at the following SNR levels (−5, 0, +5, +10 dB).

3.2 Objective Measure

The suggested technique was compared to the following known techniques: acoustic domain spectral subtraction (Spec- Sub) [8], modulation domain spectral subtraction (Mod- SpecSub) – proposed, and the MMSE method [11]. Extended short term objective intelligibility (ESTOI) and Perceptual evaluation speech quality (PESQ) were utilized as objective measurements.

Table 2 shows the ESTOI and PESQ values obtained using the suggested method and other fundamental methodologies. It's worth noting that the proposed strategy produces the best ESTOI results for all noise levels and types. Compared to noisy untreated speech, the suggested technique improves the ESTOI by 45.3% for SNR 0 dB and 56.6% for SNR > 0 dB. The suggested method also outperforms existing methods in terms of PESQ at all SNR levels and under diverse noise conditions.

Table 2. PESQ and ESTOI improvement averaged over test stimuli (represented as ESTOI (PESQ))

Noise	Input SNR (dB)	Proposed method	Spec- Sub [8]	MMSE method [11]
Babble	−5	0.41 (0.17)	0.11 (0.07)	0.31 (0.15)
	0	0.36 (0.14)	0.21 (0.11)	0.26 (0.03)
	5	0.50 (0.20)	0.19 (0.06)	0.38 (0.03)
	10	0.62 (0.25)	0.08 (0.09)	0.30 (0.12)
Street	5	0.46 (0.19)	0.04 (0.03)	0.40 (0.04)
	0	0.48 (0.13)	0.10 (0.09)	0.44 (0.06)
	5	0.51 (0.16)	0.16 (0.15)	0.27 (0.07)
	10	0.62 (0.21)	0.17 (0.12)	0.38 (0.09)
Train	−5	0.51 (0.12)	0.30 (0.06)	0.45 (0.04)
	0	0.50 (0.13)	0.12 (0.12)	0.33 (0.014)
	5	0.57 (0.16)	0.19 (0.13)	0.20 (0.06)
	10	0.58 (0.23)	0.25 (0.15)	0.27 (0.10)

4 Conclusion

In this paper, to reduce additive noise distortion in noisy speech, we suggested a spectral subtraction approach in the modulation domain based on coherent demodulation (COG). Two objective measures (perceptual evaluation speech quality (PESQ) and extended short-term objective intelligibility (ESTOI)) were used to evaluate the suggested approach. The results show that the proposed method in modulation domain processing can significantly improve speech intelligibility at various signal noise ratio levels (−5, 0, 5 and 10 dB). On average, our proposed method achieves an ESTOI improvement of 45.3% for SNR 0 dB and 56.6% for SNR > 0 dB, and at all SNR levels, it also produces the largest PESQ improvement over other approaches. These results encourage us to study other noises in future work to construct a simple and effective denoising speech system that can be used in various applications like speech recognition systems.

References

1. Hu, Y., Loizou, P.C.: Subjective comparison and evaluation of speech enhancement algorithms. Speech Commun. **49**, 588–601 (2007)
2. Zwicker, E.: Die grenzen der horbarkeit der amplitudenmodulation und der frequenzmodulation eines tones. Acustica **2**, 125–133 (1952)
3. Drullman, R., Festen, J.M., Plomp, R.: Effect of reducing slow temporal modulations on speech reception. J. Acoust. Soc. Am. **95**, 2670–2680 (1994)
4. Wojcicki, K.K., Loizou, P.C.: Channel selection in the modulation domain for improved speech intelligibility in noise. J. Acoust. Soc. Am. **131**, 2904–2913 (2012)
5. Paliwal, K., Schwerin, B., Wojcicki, K.: Single channel speech enhancement using MMSE estimation of short-time modulation magnitude spectrum. In: Interspeech, Florence, Italy, pp. 1209–1212, August 2011
6. Clark, P., Atlas, L.E.: Time-frequency coherent modulation filtering of non-stationary signals. Trans. Signal Process. **57**, 4323–4332 (2009)
7. Loughlin, P.J., Tacer, B.: On the amplitude- and frequency-modulation decomposition of signals. J. Acoust. Soc. Am. **100**(3), 1594–1601 (1996)
8. Berouti. M., Schwartz, R., Makhoul. J.: Enhancement of speech corrupted by acoustic noise. In: Proceedings of the ICASSP, vol. 4, Washington, DC, USA, pp. 208–211, April 1979
9. Loizou, P.: Speech Enhancement: Theory and Practice. Taylor and Francis, Boca Raton (2007)
10. Paliwal, K., Schwerin, B., Wójcicki, K.: Modulation domain spectral subtraction for speech enhancement. In: INTERSPEECH-2009, pp. 1327–1330 (2009)
11. Ephraim, Y., Malah, D.: Speech enhancement using a minimum-mean square error short-time spectral amplitude estimator. IEEE Trans. Acoust. Speech Signal Process. **ASSP-32**(6), 1109–1121 (1984)

COBIT 5 Concepts: Towards the Development of an Ontology Model

Lamia Moudoubah(✉), Khalifa Mansouri, and Mohammed Qbadou

SSDIA Laboratory, ENSET Mohammedia, Hassan II University of Casablanca, Casablanca, Morocco

Abstract. Information systems governance (ISG) defines the objectives assigned to the information system (IS), plans, and implements processes related to the management of information system's life cycle. These activities are based on monitoring and measuring the performance of these processes in relation to the objectives underlying the use of IS. Among the ISG methods, COBIT 5 (Control Objectives for Information and Related Technology) is the most popular and can be applied to an implemented information system in commercial operation. It allows the definition of indicators for controlling and steering the IS. COBIT 5 defines governance best practices for IT professionals and managers. COBIT 5 provides a comprehensive set of metrics, indicators, processes and control objectives for a wide range of IT processes. The objective of this paper is to develop an ontology modeling the concepts of the COBIT 5 framework. The method used is the OWL-API transformation to move our UML meta-model into an OWL ontology meta-model under the "Eclipse" tool. The obtained results are validated and evaluated thanks to the inference engine integrated in the "PROTÉGÉ" tool.

Keywords: ISG · COBIT5 · MDA · UML · Ontology · OWL-API

1 Introduction

A descriptive definition of governance is that it refers to the way in which a system is directed and controlled [1]. Thus defined, governance is the combination of steering, i.e. ensuring that today's decisions adequately prepare for the future, and control, i.e. measuring the deviation from what was planned. Peter Weill [2] guides the definition of ISG by focusing on the concept of decision: it is a monitoring process that aims to control the decisions to be made as well as the underlying risks, and to guide the decisions in order to increase the value and minimize the risks for the organization [1].

In ISG, there are many frameworks, each with its own perspective. Thus, each framework provides an appropriate level of detail in its domain [3]. One of the difficulties of IT governance is to make field approaches, necessarily detailed and specialized, coexist with strategic piloting syntheses [3]. The main reference frameworks are the following:

COSO (Committee of Sponsoring Organizations of the Treadway Commission) published in 1992 a reference framework for internal control to help companies evaluate and improve their internal control system [3]. **Balanced Scorecard (BSC)** is a representation that helps to clarify the vision and strategy of a company and to translate

Y. Maleh et al. (Eds.): ICI2C 2021, LNNS 357, pp. 247–256, 2022.
https://doi.org/10.1007/978-3-030-91738-8_24

it into action plans [3]. It provides feedback on the functioning of internal processes as well as external constraints, allowing for continuous improvement of strategy and performance [3]. **ISO/IEC 27000 standard** presents the vocabulary and definitions of security domain, applicable to each of standards; ISO/IEC 27001 standard describes information systems security management policy within a company, which serves as a reference for certification; ISO/IEC 27002 standard constitutes the good practice guide for IS security; ISO/IEC 27003 standard is intended to be an implementation guide; ISO/IEC 27004 standard will be a new standard for the management of indicators and measures in the field of IS security; ISO/IEC 27005 standard will be a new standard on risk management for IS security; ISO/IEC 27006 standard summarizes the requirements applicable to external auditors in their certification mission on ISO 27001 [3]. **ITIL (Information Technology Infrastructure Library)** is a series of books describing good practices for IT service management [3]. Its approach is more focused on "what to do" than on "how to do it" [3]. **CMMI** framework is intended for companies seeking to master their system or software development factory [3]. In this sense, CMMI is aimed either at very large accounts with a large research department, or at large service companies or integrators [3].

COBIT 5 framework has given rise to a whole series of works and publications [3]. In the first versions, V3 and earlier, the main publication was audit guide. From version 4 onwards, management guide has become the main descriptive work of COBIT 5 [3]. The four **domains of COBIT 5** group together coherent sets of processes. PO domain represents strategic dimension of IT governance. IA domain groups all processes that have an impact on resources, from acquisition to implementation: it includes both projects and operations. DS domain is dedicated to services offered to customers of the Information Systems Department (ISD). Finally, SE domain largely covers control, audit and monitoring dimension of the whole [3].

We have focused our study on COBIT 5 repository, because COBIT 5 has become the integrator of best practices in information technology and the general repository of ISG that helps to understand and manage the risks and benefits associated with them.

COBIT 5 good practices are the result of a consensus of experts. They are very much focused on control in the mastery sense and less on execution. They are intended to help optimize IT investments, ensure service delivery and provide metrics to refer to in order to assess malfunctions. COBIT 5 is continuously updated and harmonized with other standards. In the world of computer science, concept of ontology, which is borrowed from philosophy, has been used in several fields, mainly in artificial intelligence, knowledge representation, natural language processing, semantic web (SW) and software engineering [4]. Technically speaking, an ontology, as used in computer science, can be minimally defined as a conceptual artifact that gathers a vocabulary of terms, specifications of the meaning of these terms and a description of how they are interrelated to each other [4]. Terms in question normally correspond to what are called "classes", "concepts" or "things" [4]. The objective of this paper is to design an ontology based on COBIT 5 framework. The latter is an ISG solution that will clarify the domain's vocabulary and the dependencies between the different elements of COBIT 5. We have adopted OWL-API which consists in transforming UML meta-models into OWL model. Results of our ontology modeling are validated with the Hermit inference engine.

In the remainder of the article, the structure is as follows: A synthesis part of the related work is presented in Sect. 2. Then the description of the used method in Sect. 3. The proposed ontology is introduced in Sect. 4, the results of our solution in Sect. 5, to end with a conclusion in Sect. 6.

2 Previous Work

In [5], authors proposed an ontology-based approach for representation and integration of Archimat meta-model and COBIT 5 concepts and properties, for automatic analysis of the resulting ontology, as a way to systematically answer a set of questions regarding collection of evidence for level 1 assessment of capability of COBIT 5 processes [5]. According to [6], a solution for semantic integration of information from different domain models in the context of automated IT management is proposed [6]. Thus, they formulated an ontology based on the COBIT 5 IT governance framework for integration at a conceptual level and discuss the characteristics of a knowledge-based extensible execution system [6]. Authors of [7] propose a meta-model that provides Modeling and integration of COBIT 5 and other ISG practices requires a uniform representation [7]. A common language ensures that everyone is following the same set of goals, issues, and priorities [7]. In [8], the authors discuss the generation and analysis of the ontological meta-model of the COBIT 5 framework. The works cited earlier have shown that the COBIT 5 framework has attracted a series of works and publications, each work providing meta-models, ontologies or ontological meta-models from a higher level of abstraction. Based on our knowledge, some of the meta-models and ontologies provided in previous works lack details, validation and evaluation. Our work fills this gap and ensures a clear and validated ontology from a higher level of concretization, through a global conceptualization of the components and a rich, logical and semantic structure.

3 Used Method

3.1 MDA (Model Driven Architecture)

Publicly introduced the MDA solution, which is a version of the MDE that brings together all of the OMG's proposed principles [9]. Since it is built on the premise that design models are durable over time while development technology is continuously evolving and can easily become outdated, MDA is an approach that focuses on the use of models for realization of computer systems. Unlike approaches that focus on code or the design of system was only a temporary move, MDA is an approach that focuses on the use of models for the realization of computer systems [9]. MDA approach is based on the following three notions [9]: (Model; Meta-model; Model transformation). MDA is based on a set of principles suggested by OMG, which is modeled after MOF [9]. UML (Unified Modeling Language) is one of these principles that has developed over time: UML [10, 11] is a graphical modeling language commonly used by object-oriented programming [9] systems. It's a programming language dependent on the OMG's MOF specifications [9].

3.2 UML Meta-modeling

UML meta-modeling approach which is used by OMG to describe UML [10]. Meta-modeling approach means that "a meta-model is used to specify the model that comprises UML" [11]. This approach supposedly "offers advantages of being more intuitive and pragmatic for most implementors and practitioners." [11].

3.3 OWL

OWL: W3C Web Ontology Language (OWL) is a Semantic Web language designed to represent rich and complex knowledge about things, groups of things, and relations between things [11]. OWL is a computational logic-based language such that knowledge expressed in OWL can be exploited by computer programs [11]. OWL documents are called ontologies [11]. Ontologies are formalized vocabularies, often covering a specific domain and shared by a community of users [11]. We have used the OWL language to present the proposed ontology under the "PROTÉGÉ" tool.

3.4 OWL-API

OWL API has been around for more than a decade, and is under active development with a team of dedicated software engineers. For people interested in some of the background on design decisions, it is worth it to take a look at the main OWL API publication in the Semantic Web Journal in [12].

3.5 Ontology

Computer science researchers have adopted the term ontology to refer to a computational artifact that formally models the relevant entities and relationships of a certain system of interest [15]. Studer et al. stated, "based on the fusion of the previous definitions, that an ontology, in computer science, is a formal and explicit specification of a shared conceptualization" [14]. A "conceptualization" refers to an abstract vision of the world, in relation to a given domain, conceived as a set of concepts, their definitions and the relations between them [13, 14]. "Explicit" means that the concepts used and the constraints on their relationships must be explicitly defined [13, 14]. Their relationships must be explicitly defined [16]. "Formal" implies that the ontology must be represented in a machine-readable format. And finally, "shared" means that the ontology must result from an agreement, capture consensual knowledge, and be accepted by a group [16].

4 Proposed Ontology

We implemented the following steps for the representation of our ontology: A careful analysis of the domains and extraction of its main concepts and relations based on the COBIT 5 documentation [17–21], followed by its representation in a UML meta-model inspired by previous works [6–8]. Then, generation of JAVA code of our UML meta-model which will be transformed into an OWL ontological meta-model using the OWL-API in the "Eclipse" tool. In Table 1, we have presented the COBIT 5 concepts and their definitions.

Table 1. Needs Identification

Concept	Definition
Entreprise	The entity concerned by the COBIT 5 transformation
Governance_goal	Implementation of COBIT 5 would require an adapted data model, specific to IT department, designed in an IT governance logic, in order to achieve the governance objectives [22]
Company_goal	According to COBIT 5, "information systems governance is responsibility of management and board of directors and consists of the command and operating structures and processes that drive enterprise IT to support and extend the strategies and objectives of the enterprise" [22]
Objectif_IT	The main objective of COBIT 5 processes is to ensure that IT infrastructures (and applications they support) will best meet efficiency and effectiveness requirements of the business [22]
Process	A process is considered to be driven when objectives have been assigned to it and indicators have been defined to achieve the objectives [22]
Objective_Process	For each process, the associated objectives are detailed [22]
Metrics_Processes	For each process, we detail the associated metrics [22]
Practice	Each process area contains objectives to be achieved, as well as a description of practices that will be used (planning processes, providing resources, assigning responsibilities, training people…) [22]
Need_Stakeholders	As a repository for information systems governance, the scope of COBIT 5 goes beyond that of information systems management to encompass all IS stakeholders in enterprise [22]
Activity	A process is defined as a set of correlated activities that transforms incoming elements into outgoing elements, activities themselves being described in procedures [22]
KPI	COBIT 5 provides key performance indicators and key success factors for each of its processes
project_transformation	Implementation lifecycle included in COBIT 5 is an excellent approach to address complexity and challenges typically encountered in any transformation project
Facilitating_goal	COBIT 5 also encourages a holistic approach to understanding and bringing together various service management resources and capabilities
COBIT5_Life_Cycle	The implementation lifecycle included in COBIT 5 is an excellent approach to address the complexity and challenges typically encountered in any transformation project
Facilitator	COBIT 5 defines 7 categories of enablers
Policy	Facilitator among the 7 categories of facilitators
Service	Facilitator among the 7 categories of enablers
Information	Facilitator among the 7 categories of facilitators
Structure	Facilitator among the 7 categories of facilitators
Person	Facilitator among the 7 categories of facilitators
Culture	Facilitator among the 7 categories of facilitators

Conceptualization and implementation: in a second step, we studied the existing terms in order to extract all the questions that this ontology can answer in the context of COBIT 5-based IT governance:

- How to achieve a better alignment of IT with the business of the company due to its business orientation?
- How to have a comprehensible vision by the management of what IT does?
- How to give a clear distribution of ownership and responsibilities, due to the process-based approach?
- How to get a favorable bias from third parties and regulators?
- How to have a shared understanding by all stakeholders through a common language?

Figure 1 shows the COBIT 5 class diagram.

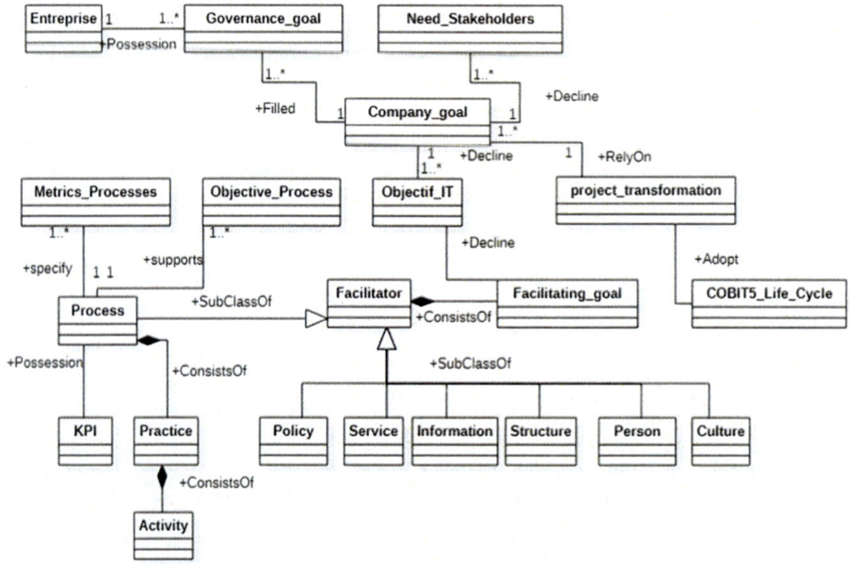

Fig. 1. COBIT 5 meta-model UML Class Diagram.

From UML meta-model of Fig. 1 drawn under StarUML software, we could generate JAVA code. Then we imported it under "Eclipse" tool in order to use OWL-API to generate the ontological meta-model of COBIT 5 UML meta-model. We then proceeded to create OWL Ontology using the OWL-API in Eclipse. To do so, we started by configuring our IDE for OWL-API development, Fig. 2 represents an overview of the configuration code of our IDE.

```java
package msc;

import org.semanticweb.owlapi.apibinding.OWLManager;
import org.semanticweb.owlapi.model.OWLOntologyManager;

public class OWLAPIFIRST {

    @SuppressWarnings("deprecation")
    public static void main(String[] args) {
        OWLOntologyManager man = OWLManager.createOWLOntologyManager();
        System.out.println(man.getOntologies().size());
        }
}
```

Fig. 2. Overview of java code of IDE configuration

The OWLOntologyManager is central class for management of our ontology; it manages the creation, loading and saving of ontologies, application of changes such as annotations or axiom additions. We then developed our code to create our OWL ontology in "Eclipse" tool.

5 Result

Once the UML meta-model was specified, its concepts and relations were transformed into OWL classes. We used the standard OWL language, in order to represent our ontology. Relations between these concepts were transformed into "ObjectProperties", and each class attribute of the diagram was transformed into a "DataType". Figure 3 present the classes and the relations between them.

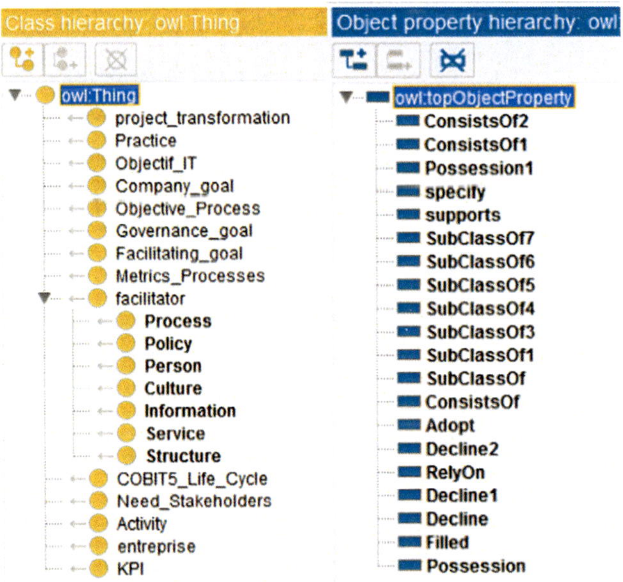

Fig. 3. Overview of proposed Ontology classes and relations

Figure 4 shows the graphical presentation of proposed ontological model 'Onto Graf' to better visualize and understand the different concepts that compose our ontology.

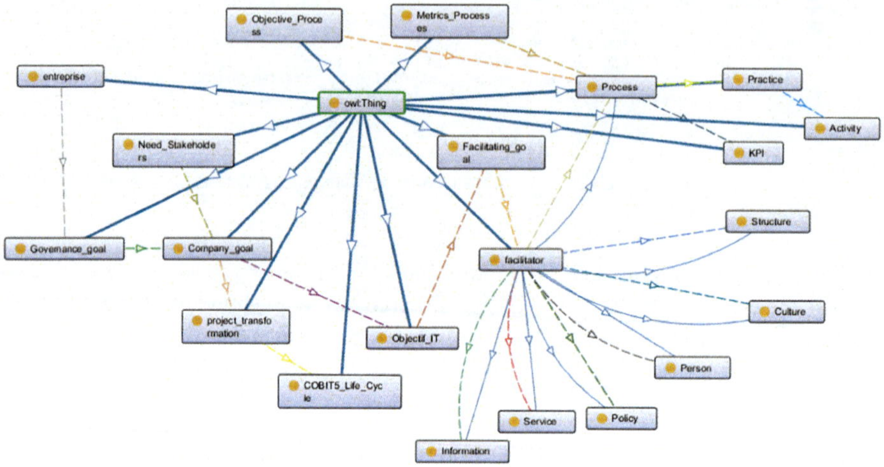

Fig. 4. OntoGraf of proposed Ontology

Validation of ontologies has become a central issue in Knowledge Engineering (KE) [23]. To validate our ontology, we rely on a set of questions that we use the HermiT 1.34.3. 456 [24] inference engine integrated in protégé 5 to answer in order to confirm the structure of our ontology based on the COBIT5 framework. Figure 5 gives an overview of the classes and relations inferred from the HermiT reasoning. This is a demonstration of the success of validation process of our ontology proposed in this paper. And this step creates the distinction between this paper and previous works.

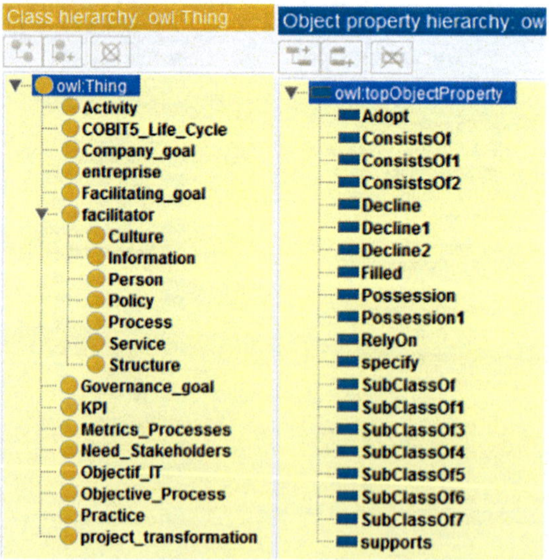

Fig. 5. Classes and Relations inferred from the validation process

6 Conclusion

IT governance includes the assessment, guidance and monitoring of current and future use of IT within the company. It involves taking into account the impact of the company's decisions and activities on society, the environment and economy, and thus its contribution to sustainable development and sustainability. COBIT5 allows and promotes a structured approach to the necessary assessment and decision processes. In this paper, we have contributed by developing an ontology modeling COBIT 5 and which aims to overcome the lack of domain specific concepts and properties of COBIT5. We specified and represented COBIT 5 through a UML meta-model which was transformed into an OWL ontology by OWL-API under 'eclipse' tool and validated thanks to HermiT inference engine installed under 'Protégé' ontology testing tool. The main contribution of this research work is the proposed ontology-based approach capable of answering the questions regarding evidence collection for COBIT 5 process evaluation. The solution provides an integration scheme, capable of addressing the analysis concerns related to evidence collection for COBIT 5 capability assessment. COBIT 5 is a first step towards information systems governance, but there is still a long way to go before a comprehensive approach to the domain is achieved. The ontology proposed in this paper is aimed at different users: management, to provide a means of decision support; direct users, to provide guarantees on security and controls of IT services; auditors and consultants, for whom it provides recognized means of intervention. The ontology proposed in this paper will then be exploited and deployed in a company in our next work.

References

1. Claudepierre, B.: Conceptualisation de la Gouvernance des Systèmes d'Information: Structure et Démarche pour la Construction des Systèmes d'Information de Gouvernance. Dissertation Université Panthéon-Sorbonne-Paris I (2010)
2. Alkhaldi, F.M., Hammami, S.M., Ahmar Uddin, M.: Understanding value characteristics toward a robust IT governance application in private organizations using COBIT framework. Int. J. Eng. Bus. Manag. **9**, 1847979017703779 (2017)
3. Andry, J.F., Setiawan, A.K.: IT governance evaluation using COBIT 5 framework on the national library. Jurnal Sistem Informasi **15**(1), 10–17 (2019)
4. Ouali, S., Mhiri, M., Gargouri, F.: Knowledge engineering for business process modeling. In: ENASE, pp. 81–90 (2017)
5. Margarida, A., Silva, F.: Using ontologies for enterprise architecture and COBIT 5 analysis (2016)
6. Textor, A., Kroeger, R., Geihs, K.: Semantic models for bridging domains in automated IT management: lessons learned. In: 2017 International Conference on Networked Systems (NetSys). IEEE (2017)
7. Almeida, R., Pinto, P.L., Da Silva, M.M.: Using ArchiMate to integrate COBIT 5 and COSO metamodels. In: Proceedings of the 13th European Mediterranean and Middle Eastern Conference on Information Systems (EMCIS), Krakow, Poland (2016)
8. Souza Neto, J.A., Ferreira Neto, A.N.: Metamodel of the it governance framework COBIT 5. JISTEM J. Inf. Syst. Technol. Manag. **10**, 521–540 (2013)

9. Elmounadi, A., Berbiche, N., sefiani, N.: Architecture dirigée par les modèles: Méthodes et outils de transformation de modèles. La cinquième édition des Journées Doctorales en Technologies de l'Information et de la Communication (JDTIC 2013), Kenitra, Maroc (2013)
10. O. M. G. D. Number, S. Associated, and N. M. Files, OMG unified modeling language TM (OMG UML), superstructure, no. August (2011)
11. Vo, M.H.L., Hoang, Q.: Transformation of UML class diagram into OWL Ontology. J. Inf. Telecommun. **4**(1), 1–16 (2020)
12. Lara, A., Henriques, P.R., Gançarski, A.L.: Visualization of ontology evolution using ontodi graph (2017)
13. Machado, L.M.O., Almeida, M.B., Souza, R.R.: What researchers are currently saying about ontologies: a review of recent web of science articles. KO Knowl. Organ. **47**(3), 199–219 (2020)
14. Zhang, N., Wang, Y., Miao, F., et al.: Ontology database construction for medical knowledge base. In: 2019 IEEE/ACIS 18th International Conference on Computer and Information Science (ICIS), pp. 333–337. IEEE (2019)
15. Burov, Y., Mykich, K., Karpov, I.: Building a versatile knowledge-based system based on reasoning services and ontology representation transformations. In: 2020 IEEE 15th International Conference on Computer Sciences and Information Technologies (CSIT), pp. 255–260. IEEE (2020)
16. Forbes, D.E., Wongthongtham, P., Terblanche, C., Pakdeetrakulwong, U.: Ontology engineering. In: Forbes, D.E., Wongthongtham, P., Terblanche, C., Pakdeetrakulwong, U. (eds.) Ontology Engineering Applications in Healthcare and Workforce Management Systems. SSDC, vol. 123, pp. 27–40. Springer, Cham (2018). https://doi.org/10.1007/978-3-319-65012-8_3
17. ISACA, COBIT 5: A Business Framework for The Governance and Management of Enterprise IT. ISACA (2012)
18. COBIT 5: Implementation. ISACA (2012)
19. COBIT 5: Enabling Processes. ISACA (2012)
20. COBIT 5 Process Assessment Model (PAM): Using COBIT 5. ISACA (2013)
21. Cadete, G.R., da Silva, M.M.: Assessing it governance processes using a COBIT5 model. In: Themistocleous, M., Morabito, V. (eds.) Information Systems. LNBIP, vol. 299, pp. 447–460. Springer, Cham (2017). https://doi.org/10.1007/978-3-319-65930-5_36
22. Moisand, D., De Labareyre, F.G.: Cobit: Pour une meilleure gouvernance des systèmes d'information. Editions Eyrolles (2009)
23. Richard, M., Aimé, X., Krebs, M.-O., et al.: LOVMI: vers une méthode interactive pour la validation d'ontologies. In: 26es journées francophones d'Ingénierie des Connaissances (IC) (2015)
24. Yamada, D.B., et al.: Ontology-based inference for supporting clinical decisions in mental health. In: Krzhizhanovskaya, V.V., et al. (eds.) ICCS 2020. LNCS, vol. 12140, pp. 363–375. Springer, Cham (2020). https://doi.org/10.1007/978-3-030-50423-6_27

Advances in Online Learning

A Systematic Study on Tertiary Level Student Tuition Fee Waiver Management During Pandemic Using Machine Learning Approaches

Asif Khan Shakir[1], Shanjoy Sutradhar[1], Ahsanul Hoque Sakib[1], Washim Akram[1], Md. Abu Saleh[1], and Mohammad Zoynul Abedin[2(✉)]

[1] Department of Software Engineering, Daffodil International University, Dhaka, Bangladesh
{asif.swe, shanjoy35-2025, ahsanul35-1838, washim35-1730, abu35-1746}@diu.edu.bd

[2] Department of Finance and Banking, Hajee Mohammad Danesh Science and Technology University, Dinajpur, Bangladesh

Abstract. Machine learning techniques are compressively used in various fields like business, health, education, etc. The growing necessity in the education segment has concreted the approach for plenty of research projects that strongly emphasize student academic achievement and behavior analysis. This research study is about an analysis performed to identify the significant tuition waiver significance of student academic background, accomplishment, and family background. Supervised machine learning approaches gradually, Decision Tree Regression (DTR) and Random Forest Regression (RFR) are utilized in predicting tuition waivers. In addition, Cross-Validation (CV) is used for data overfitting. For better accuracy, DTR and RFR are implemented twice before and after cross-validation. There is enough data to train the models thoroughly and get quite good accuracy and performance. Results appear that the accuracy of DTR and RFR are 74.49% and 77.82%, respectively for before applied CV; and, 71.41%, and 76.90% respectively, for after applied CV.

Keywords: Machine learning · Supervised machine learning · Tuition fee waiver · Cross-Validation (CV) · Decision Tree Regression (DTR) · Random Forest Regression (RFR)

1 Introduction

Every higher educational institute aims to provide tuition waivers to its students. A tuition waiver is granted by each student's chosen individual university and reduces the university expenditure of students. The waiver covers the cost of tuition for a specified number of credits, but may not be used for any other educational expenses. Each university has its policy to get the exemption after matching the tuition qualifications. However, there are many reasons an educational institute can grant a waiver

Y. Maleh et al. (Eds.): ICI2C 2021, LNNS 357, pp. 259–273, 2022.
https://doi.org/10.1007/978-3-030-91738-8_25

[1]. This waiver depends on the academic result, financial problem, freedoms fighter quota, female quota, physically challenged students, etc. Students get interested in concentrating and continuing their education without any interruption.

Educational authorities, when trying to provide a waiver for students, face many problems. Traditionally, it is challenging to make decisions based on student performance, family background, and academic performance to grant tuition waivers traditionally. Furthermore, it requires analyzing thousands of students. This is time-consuming and expensive. Also, the need for experienced people for proper judgment. This research method will help to minimize fatigue and possibly the university's authority on potential errors. The university's administration can effectively determine the ratios of tuition waivers for students using this formula. Also, may be able to exercise caution by examining academic performance and family background.

This research aims to apply machine learning approaches to extract information on academic results such as students' details, family backgrounds, academic performance, etc. This research and implementation aims to identify tuition waivers and analyze the effect of student academic background and family background attributes. Decision Tree Regression and Random Forest Regression are used for the continuous outcome and knowledge extraction. Applications for these algorithms include real estate prices, stock price fluctuations and student test scores, etc. [2]. Cross-validation for data overfitting is also used in this research.

This research paper is arranged systematically as follows: Literature review is embedded in Sect. 2. Section 3 contains a process flowchart that shows the steps of the process workflow. Section 4 presents data collection, preprocessing, and feature engineering formulas of data. The proposed methodology, i.e., decision tree regression and random forest regression explained briefly in Sect. 5. Section 6 contains the results and their analysis. Section 7 presents the conclusion of the paper and the direction of the future work.

2 Literature Survey

Ching-Chieh Kiu [3] proposed a "Data Mining Analysis on Student's Academic Performance through Exploration of Student's Background and Social Activities" using the decision tree, Naïve Bayesian, and neural network. The main aims of this research were to provide performance analysis to identify student's backgrounds, student's social activities, and student's academic performance in predicting student academic performance.

Sujith Jayaprakash et al. [4] this paper referred to make predictions about student performance; Classify features that affect student performance, and clustering students based on their performance. Introducing an internal process for analyzing and predicting student performance early helps the organization and students take immediate action and improve effectiveness. In this research work, they discuss socioeconomic factors, psychological and cognitive factors that affect students' academic achievement and help students to predict at risk.

Ajinaja Micheal Olalekan et al. [5] applied the Bayes theorem and artificial neural networks (ANN) to generate predictive models for the chances of students graduating in a third category institution. Provide an improved forecasting system that can help educational institutions identify and monitor different thresholds for students. Also, it can concentrate on improving students that their threshold is less than graduating at an early stage.

Yue Chen et al. [6] proposed the "Performance of candidates for early admission to university" using Logistic Regression, Bayes Classification, Artificial Neural Network, and Support Vector Machine. Its main aim of predicting students' potential for their academic performance significantly concerns and the statistical process and artificial intelligence technologies were used comparatively to produce prediction models.

Zhuping Wang et al. [7] identified current and potential academic issues among students in the early stages of college and informed them, as well as their parents, about the need to correct students' attitudes. They gather data from students in various departments, including the Office of Academic Affairs, the Library, and others. They introduced an early-warning model of learning after extensive analysis. They used "Principal Components Analysis" to find the original predictors after pre-treatment.

V. Uday Kumar et al. [8] suggested a model for assessing student performance and predicting whether students will pass or fail technical exams administered as part of the recruiting process. K-means and Hierarchical clustering are used in unsupervised learning, while Naive Bayes and Decision Trees are used in supervised learning. First, students' academic performance is examined using K-means and Hierarchical clustering on Inter and BSc data. On technical subjects such as C programming, Operating Systems, Computer Networks, and Database Management Systems, Decision Tree and Naive Bayes were used to determine whether a student can pass or fail.

Dinh Thi Ha et al. [9] investigated using machine learning techniques to predict students' final GPAs based on personal characteristics, university entrance grades, gap years, and first- and second-year academic results. They established a correlation between the factors mentioned above and student academic performance and the most appropriate methodology for predicting student performance through empirical research.

Nongnuch Ketui et al. [10], as part of the data mining strategy, concentrated on preprocessing, modeling, and postprocessing. They used data mining classification techniques to boost student efficiency and focused on classifying the appropriate subject for students in the faculty of science and agricultural technology. A cross-validation test called 10-fold cross-validation is used to divide the dataset in this experiment. Popular assessments such as precision, recall, f-measure, and accuracy scores are used to assess each model's performance.

Abedin et al. [11] discussed the practical application of data mining to predict small business bank customer data. An empirical application of the method shows that it can make accurate predictions and effectively assess their future occupancy rates to managers. They applied in their research paper M5p Algorithm, Jrip Rule Algorithm, M5 Rule Algorithm, and J48 Tree Algorithm.

Fazal Aman et al. [12] In this study, constructed the prediction model, a subjective approach was used to identify academic and socioeconomic characteristics, and then a Logistic Model Trees (LMT), Decision tree-based algorithm was used to learn the intrinsic relationship between the established characteristics and the academic grades of students. The proposed scheme, which controls parents and higher education administrators, achieved a predictive accuracy of 83.48%.

In the same way, supervised and unsupervised machine learning algorithms methods have been used in many other experiments [13–18] to forecast student performance ahead of time to improve student performance.

3 Process Flowchart

Figure 1 represents the flow of research conducted to create the model. The research is first to open the "Tuition Waiver" dataset in Jupyter Notebook. Data preprocessing and feature engineering will apply to this dataset. Then, the appropriate model will be applied Decision Tree Regression, Random Forest Regression basis on this dataset, respectively. Split 90:10 percent for training and testing, also cross-validation will be applying on this. Finally, evaluate the result.

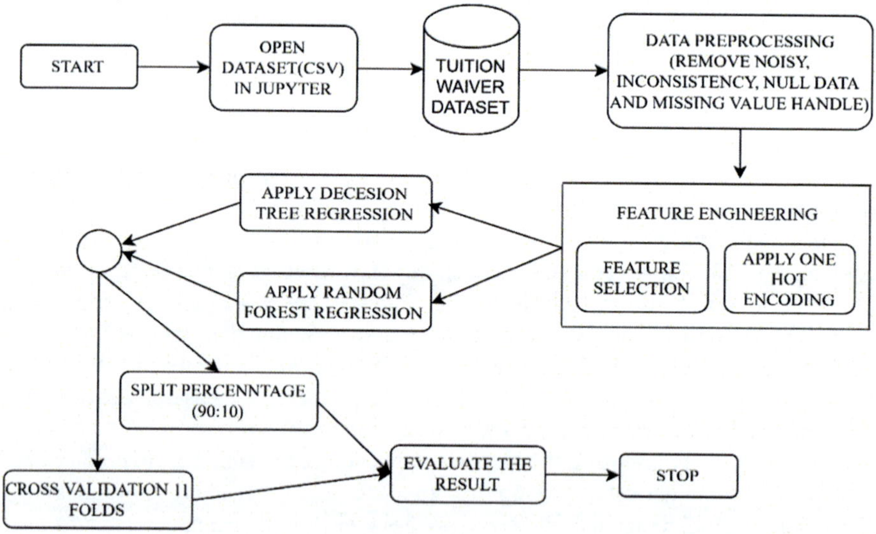

Fig. 1. Flowchart delimitation model creation

4 Data Collection, Preprocessing and Feature Engineering

4.1 Data Collection

In this research, the data are collected from different Daffodil International University's (DIU) departments in Bangladesh. Collecting data, 10818 instances of 12 attributes were considered and consumed 1199 kb of memory. The first eleven attributes are input sets and the twelfth attribute is a result that will be used as a target. Table 1. shows the selected attributes, Description, and type. Table 2, shows the numerical attributes value range, Mean, Standard Deviation (STD) of student's tuition waiver data set. Table 3, shows the values of the nominal attributes. The top ten most unique values listed are nominal features of students' tuition waiver data set.

Table 1. Student's tuition waiver data set and description

Attribute	Description	Type
Department	Department of students	Nominal
Campus	Campus Name of students	
WaiverCategory	In which category are you getting a waiver?	
BAE (Bearing Academic Expense)	Who is bearing your academic expenses?	
OccBAE (Occupation of Bearing Academic Expense)	What is the occupation of the person bearing your academic expenses?	
IncomeSource	What is the source of income of your family?	
AnunualIncome	Amount of Annual Income of your family	Numerical
FamilyMembers	Number of Family Members including you	
CompletedCredits	How many credits you have completed?	
SGPA	What was your result in Fall 2019 (SGPA)	
CGPA	What is your CGPA	
WaiverPercentage	How much waiver you are getting?	

Table 2. Numerical attribute's value range of student's tuition waiver dataset

Type	Attribute	Value range	Mean	STD
Numerical	AnunualIncome	30000–40000000	318445.25	468995.25
	FamilyMembers	1–15	5.04	1.52
	CompletedCredits	0–149	60.25	38.85
	SGPA	0.0–4.0	3.14	0.61
	CGPA	0.0–4.0	3.16	0.53
	WaiverPercentage	0–100	14.06	17.61

Table 3. Nominal attribute's top most ten unique values of student's tuition waiver dataset

Type	Attribute	Value
Nominal	Department	Computer Science and Engineering
		Software Engineering
		Business Administration
		Electrical and Electronic Engineering
		Textile Engineering
		Civil Engineering
		English
		Pharmacy
		Nutrition and Food Engineering
		Law
	Campus	Dhanmondi
		Permanent
		Uttara
	WaiverCategory	No Waiver
		Semester Result
		Permanent Campus Admission
		SSC & HSC Result
		Female Quota
		HSC Result
		Freedom Fighter Quota
		Financial Aid
		Siblings Category
		Diu Scholarship Category
	BAE	Father
		Mother
		Brother
		Parents
		Uncle
		Sister
		Myself
		Husband
		Grandfather
		Father and Uncle
	OccBAE	Banker
		Doctor
		Doing Business
		Doing Government Service
		Doing Job Abroad
		Doing Private Job
		Farmer (Agriculture)
		Retired Person
		Teaching

(*continued*)

Table 3. (*continued*)

Type	Attribute	Value
	IncomeSource	Business
		Private Job Service
		Cultivation and Farming
		Government Service
		Teaching
		Remittance
		Pension
		Shop Business
		House Rent
		Banking Service

4.2 Data Preprocessing and Feature Engineering

In this step, raw data are converted into understandable formats. In this research, the noise, inconsistency, lack of tidying row data, and handling missing values are reduced by forwarding fill and median. In part of Feature Engineering, applying one hot encoding over the nominal data and took topmost ten unique values of nominal attributes through each attribute have ten more unique values. This does not reflect deeply on accuracy. There find out the correlation of the feature. Then, the feature is selected from the correlation for the appropriate model construction.

5 Proposed Methodology

5.1 Decision Tree Regression (DTR)

A decision tree is used to solve classification and regression problems. It's occur overfitting of the process. In this research, Decision Tree Regression (DTR) is used to predict the dependent variable's continuous and numerical target variables [19]. This decision tree predicts response to the data. To predict the response, ask for new data by deciding on a leaf node from the tree's root node. The leaf node contains the response. Regression trees give responses numerical and continuous target outcomes [20, 21].

The DTR-based scheme generally entangles mainly two procedures: - the procedure of the growth and pruning of the tree.

In the tree growth procedure, assume that node n has a collection of K^n training cases. Initially, standard deviation SD (K^n) of the outcome values of the instances in K^n is calculated. Also, assuming, a trail tree exists that allocates K^n into O outcomes, the intention is to search for the most effective trail tree that reduces the standard deviation as much as possible. The standard deviation reduction (ΔSD) is calculated according to Eq. (1),

$$\Delta \text{SD} = \text{SD}(K^n) - \sum\nolimits_i^o \frac{|K_i^n|}{|K^n|} \times \text{SD}(K_i^n) \tag{1}$$

K_i^n indicates the subset of cases that have the ith the outcome of the possible set.

In the pruning procedure, beginning towards the tree's bottom, the algorithm experiments with each non-leaf node in the model if this node could be a re-substitution with the above-mentioned linear model, whether as like as a new leaf of the node or keep uninjured of the subtree. Decisions are formed on which method (i.e., linear model or subtree) the least calculated fault can be found, calculated according to Eq. (2),

$$\text{ERROR} = \frac{N+V}{N-V} \times \frac{\sum_{i=1}^{N} abs\left(V_{act} - V_{pre}\right)}{N} \tag{2}$$

We can see, the calculated fault is the mean absolute difference between the actual target values '*Vact*' of the training instances and the predicted values '*Vpre*'. Where the amount of training instances is 'N' running with the help of the existing node and the number of parameters is 'V' [22, 23].

Figure 2 illustrates the description of Decision Tree Regression (DTR). DTR is performed on students' institutional background, Department, campus, Completed Credit, Waiver Category, SGPA, CGPA, and family background like Bearing Academic Expenses (BAE), Occupation of Bearing Academic Expenses (OccBAE), Income Source, Annual Income, Family Members. The data is partitioned into training and testing data. Training data include 9732 tuples, and testing data include 1086 tuples, simultaneously using 11 folds cross-validation. The following tree has been acquired after cross-validation.

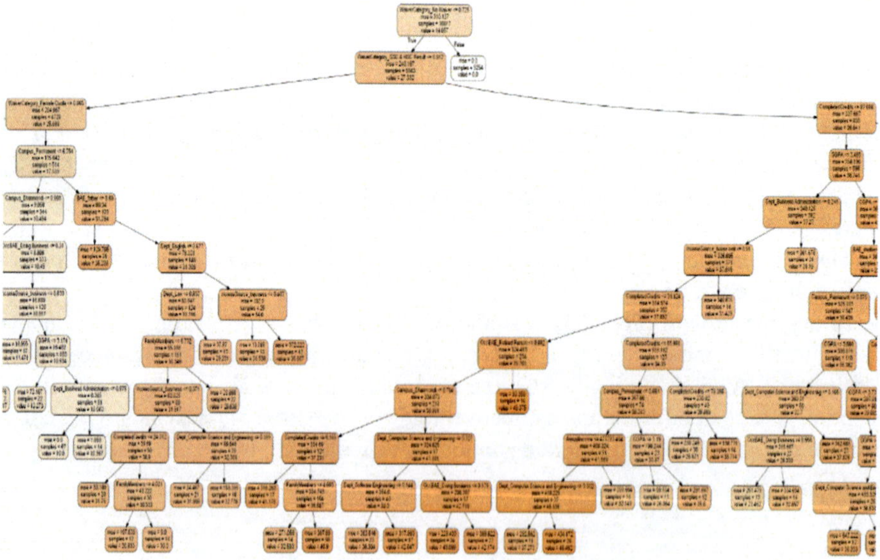

Fig. 2. Description of decision tree regression (DTR)

5.2 Random Forest Regression (RFR)

Regression of Random Forest is a machine learning approach that predicts continuous values, i.e., values across a certain range. Also, RFR follows the Bagging approach from the ensemble technique that assembles a large set of regression trees. "It grows as many trees as possible on a subset of the data and then merges the results of all of the trees. As a result, the problem of overfitting in decision trees is alleviated and the variance. Instead of curve-based algorithms, nonlinear parameters do not affect Random Forest Regression performance.

In contrast, it takes a longer time for training because it creates lots of trees" [24]. A regression tree represents a set of postulates or limitations hierarchically constructed and applied successively from a root to a leaf of the tree [25]. Figure 3 illustrates a simple RFR model. RFR is a four steps procedure [26], let's conducted as follows:

- N data points are select randomly from the training set
- Generate a decision tree associated with these N data points
- Choose the number K of trees to generate and repeat steps 1 and 2
- For a new data point, predict the value of y for the data points of the questions in each of the n-tree trees and set the new data point as the average across all of the predicted y values

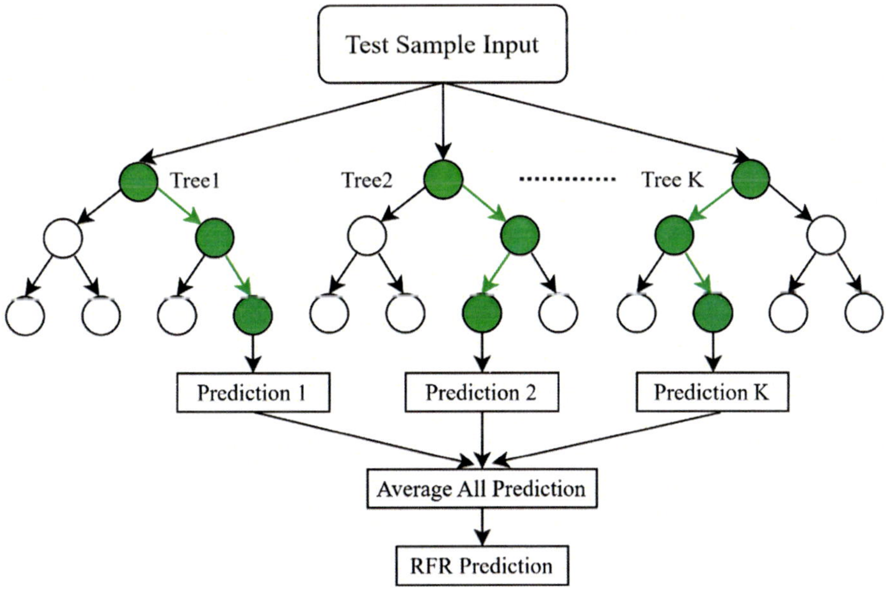

Fig. 3. A simple RFR

Figure 4 illustrates the description of a tree of Random Forest Regression (RFR). The following tree has been acquired after cross-validation. There are showing only a piece of tree.

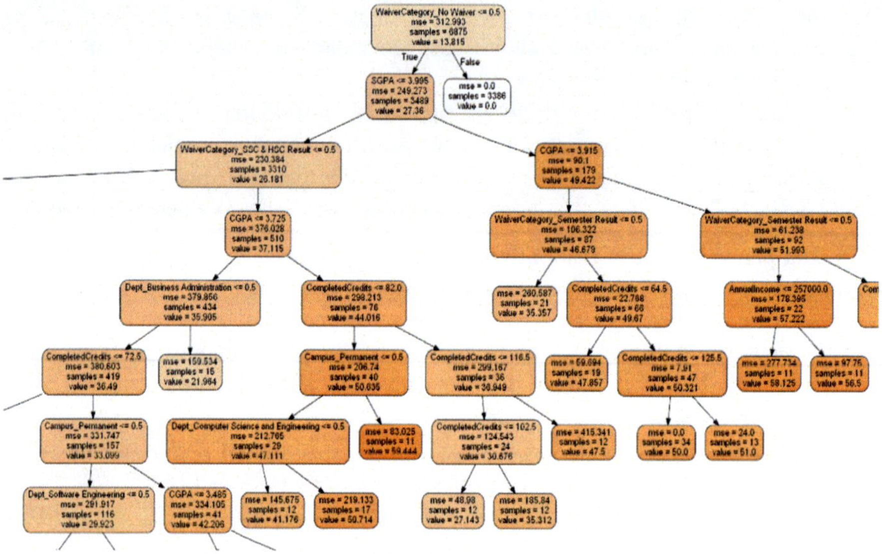

Fig. 4. Description of a tree of random forest regression

6 Result and Discussion

6.1 Result

In this research, the result obtained from these Decision Tree Regression (DTR) and Random Forest Regression (RFR) applied different regression algorithms and each algorithm followed different techniques. Also, the cross-validation (CV) technique is applied for overfitting. The Cross-validation scores are shown in Table 4, Fig. 5 showing accuracy and error score.

Table 4. Cross-Validation score of DTR and RFR

Fold	Decision Tree Regression	Random Forest Regression
1	0.66072572	0.69858279
2	0.69729541	0.70548052
3	0.62938359	0.65396679
4	0.70119085	0.7261528
5	0.7405177	0.79612062
6	0.66477102	0.69095063
7	0.65758588	0.69609722
8	0.67105974	0.70798315
9	0.61272992	0.63670601
10	0.6903235	0.75353595
11	0.67721659	0.70620904

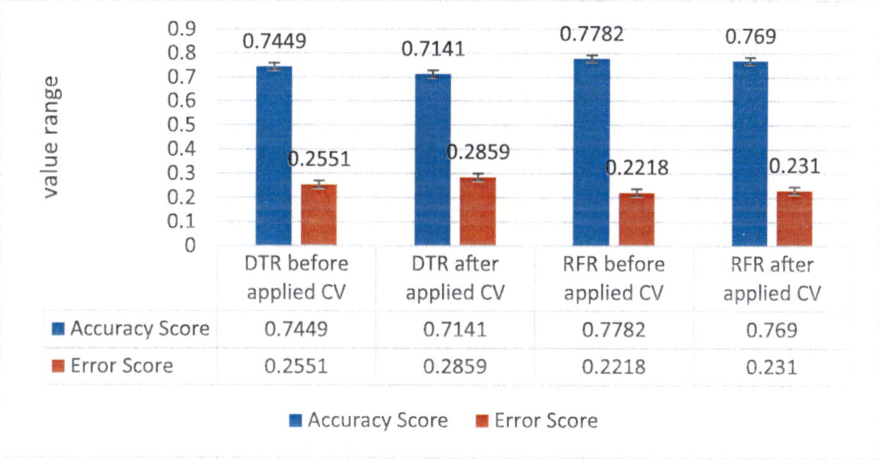

Fig. 5. Accuracy & Error score of DTR & RFR

6.2 Result Analysis

Figure 5 shows the performance and error score of algorithms. The result shows that the RFR performance is better than DTR. RFR accuracy 77.82% and 76.90% are gained before & after applied CV. On the other side, DTR accuracy was 74.49% and 71.41%, before & after applied CV. Also, RFR's Error score is less than the DTR error score. Actual and prediction value differences of DTR & RFR are given in Fig. 6, Fig. 7, Fig. 8, and Fig. 9.

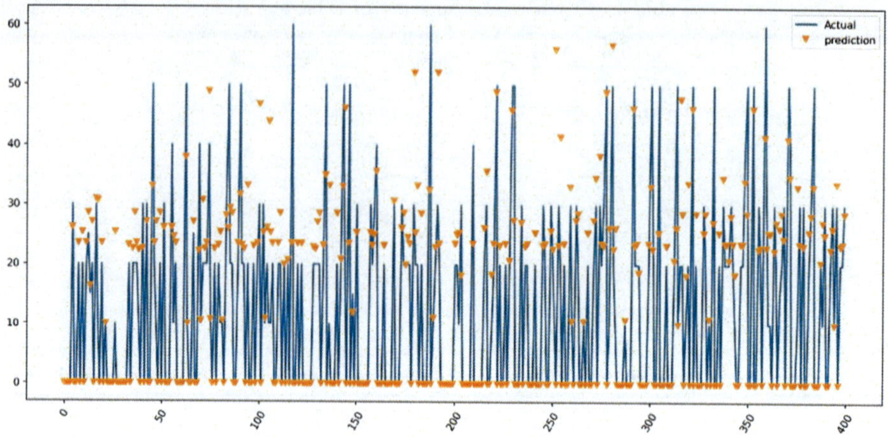

Fig. 6. Actual vs Predicted value before apply CV of DTR

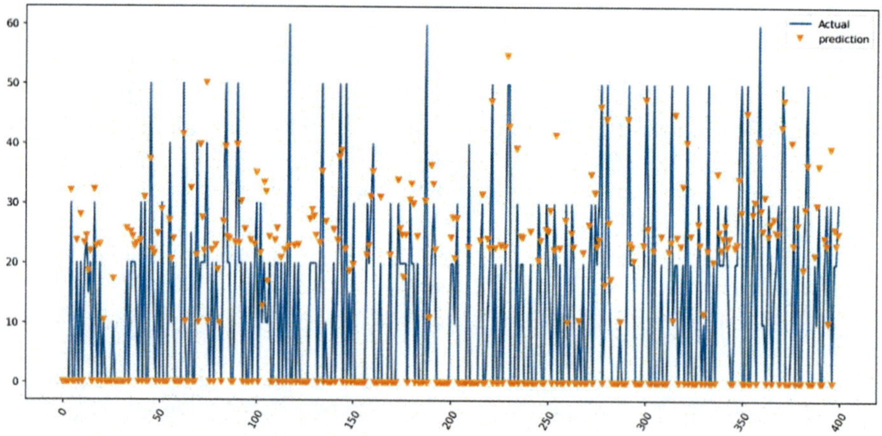

Fig. 7. Actual vs Predicted value before apply CV of RFR

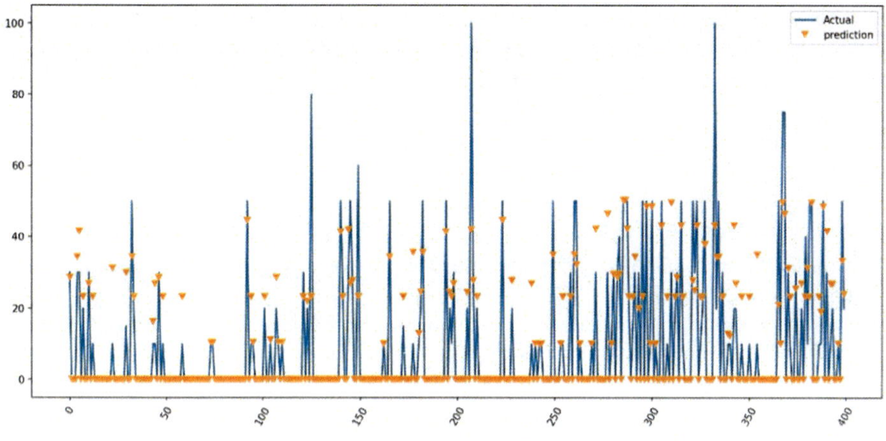

Fig. 8. Actual vs Predicted value after apply CV of DTR

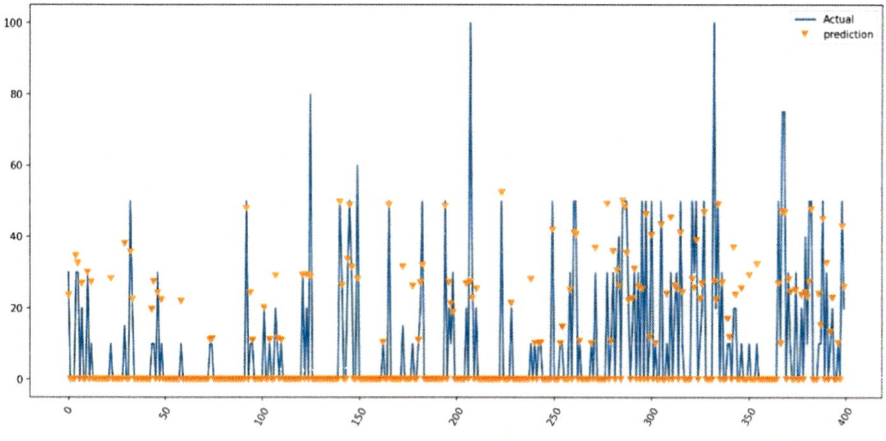

Fig. 9. Actual vs Predicted value after apply CV of RFR

7 Conclusion and Future Work

In this paper, for predicting tuition waiver percentage, data are collected from different university departments. The justification is made from the students' academic performance and family background for early prediction of tuition waivers. Two classic machine learning regression algorithms for information extraction are integrated to train, test, and cross-validate on sample data. The huge amount of data has a significant impact on the train of the models properly. 10818 data properly train the model. As a result, the best accuracy from random forest regression. This research procedure can reduce fatigue and potential errors of university authority. From this model, the university's authority can easily identify student's tuition waiver percentages.

This research work is also suitable for applying unsupervised machine learning techniques. It can explore the correlation and effects using feature analysis and feature selection to provide more accurate prediction models.

References

1. What's the Difference Between a Scholarship and a Tuition Waiver?—startwithfafsa.org. https://www.startwithfafsa.org/2019/05/whats-the-difference-between-a-scholarship-and-a-tuition-waver/. Accessed 18 Mar 2021
2. Modern Machine Learning Algorithms: Strengths and Weaknesses. https://elitedatascience.com/machine-learning-algorithms. Accessed 05 Sept 2021
3. Kiu, C.-C.: Data mining analysis on student's academic performance through exploration of student's background and social activities. In: 2018 Fourth International Conference on Advances in Computing, Communication \& Automation (ICACCA), pp. 1–5 (2018)
4. Jayaprakash, S., Krishnan, S., Jaiganesh, V.: Predicting students academic performance using an improved random forest classifier. In: 2020 International Conference on Emerging Smart Computing and Informatics (ESCI), pp. 238–243 (2020)
5. Olalekan, A.M., Egwuche, O.S., Olatunji, S.O.: Performance evaluation of machine learning techniques for prediction of graduating students in tertiary institution. In: 2020 International Conference in Mathematics, Computer Engineering and Computer Science (ICMCECS), pp. 1–7 (2020)
6. Chen, Y., Pan, C.-C., Yang, G.-K., Bai, J.: Intelligent decision system for accessing academic performance of candidates for early admission to university. In: 2014 10th International Conference on Natural Computation (ICNC), pp. 687–692 (2014)
7. Wang, Z., et al.: Design and implementation of early warning system based on educational big data. In: 2018 5th International Conference on Systems and Informatics (ICSAI), pp. 549–553 (2018)
8. Kumar, V.U., Krishna, A., Neelakanteswara, P., Basha, C.Z.: Advanced prediction of performance of a student in an university using machine learning techniques. In: 2020 International Conference on Electronics and Sustainable Communication Systems (ICESC), pp. 121–126 (2020)
9. Ha, D.T., Loan, P.T.T., Giap, C.N., Huong, N.T.L.: An empirical study for student academic performance prediction using machine learning techniques. Int. J. Comput. Sci. Inf. Secur. **18**(3) (2020)
10. Ketui, N., Wisomka, W., Homjun, K.: Using classification data mining techniques for students performance prediction. In: 2019 Joint International Conference on Digital Arts, Media and Technology with ECTI Northern Section Conference on Electrical, Electronics, Computer and Telecommunications Engineering (ECTI DAMT-NCON), pp. 359–363 (2019)
11. Jovanović, V., et al.: The selection of optimal data mining method for small-sized hotels. In: Synthesis 2015-International Scientific Conference of IT and Business-Related Research, pp. 519–524 (2015)
12. Aman, F., Rauf, A., Ali, R., Iqbal, R., Khattak, A.M.: A predictive model for predicting students academic performance. In 2019 10th International Conference on Information, Intelligence, Systems and Applications (IISA), pp. 1–4 (2019)
13. Hussain, S., Dahan, N.A., Ba-Alwib, F.M., Ribata, N.: Educational data mining and analysis of students' academic performance using WEKA. Indones. J. Electr. Eng. Comput. Sci. **9**(2), 447–459 (2018)

14. Sekeroglu, B., Dimililer, K., Tuncal, K.: Student performance prediction and classification using machine learning algorithms. In: Proceedings of the 2019 8th International Conference on Educational and Information Technology, pp. 7–11 (2019)
15. Francis, B.K., Babu, S.S.: Predicting academic performance of students using a hybrid data mining approach. J. Med. Syst. **43**(6), 1–15 (2019)
16. Nuankaew, W., Thongkam, J.: Improving student academic performance prediction models using feature selection. In: 2020 17th International Conference on Electrical Engineering/Electronics, Computer, Telecommunications and Information Technology (ECTI-CON), pp. 392–395 (2020)
17. Haque, S., et al.: Aspect based sentiment analysis in Bangla dataset based on aspect term extraction. In: Bhuiyan, T., Rahman, Md.M., Ali, Md.A. (eds.) ICONCS 2020. LNICSSITE, vol. 325, pp. 403–413. Springer, Cham (2020). https://doi.org/10.1007/978-3-030-52856-0_32
18. Das, D., et al.: A comparative analysis of four classification algorithms for university students performance detection. In: Kasruddin Nasir, A.N., et al. (eds.) InECCE2019. Lecture Notes in Electrical Engineering, vol. 632, pp. 415–424. Springer, Singapore (2020). https://doi.org/10.1007/978-981-15-2317-5_35
19. Rathore, S.S., Kumar, S.: A decision tree regression based approach for the number of software faults prediction. SIGSOFT Softw. Eng. Notes **41**(1), 1–6 (2016). https://doi.org/10.1145/2853073.2853083
20. Wang, S., Li, H., Li, J., Zhang, Y., Zou, B.: Automatic analysis of lateral cephalograms based on multiresolution decision tree regression voting. J. Healthc. Eng. **2018** (2018)
21. Swetapadma, A., Yadav, A.: A novel decision tree regression-based fault distance estimation scheme for transmission lines. IEEE Trans. Power Deliv. **32**(1), 234–245 (2016)
22. Lin, L., Wang, Q., Sadek, A.W., et al.: Novel Machine Learning Methods for Accident Data Analysis (2018)
23. Sihag, P., Mohsenzadeh Karimi, S., Angelaki, A.: Random forest, M5P and regression analysis to estimate the field unsaturated hydraulic conductivity. Appl. Water Sci. **9**(5), 1–9 (2019). https://doi.org/10.1007/s13201-019-1007-8
24. The Professionals Point: Advantages and Disadvantages of Random Forest Algorithm in Machine Learning. http://theprofessionalspoint.blogspot.com/2019/02/advantages-and-disadvantages-of-random.html
25. Zhou, X., Zhu, X., Dong, Z., Guo, W., et al.: Estimation of biomass in wheat using random forest regression algorithm and remote sensing data. Crop J. **4**(3), 212–219 (2016)
26. Random Forest Regression. Random Forest Regression is a…—by Chaya Bakshi—Level Up Coding. https://levelup.gitconnected.com/random-forest-regression-209c0f354c84. Accessed 15 Mar 2021

Towards a Model of Self-regulated e-learning and Personalization of Resources

Wijdane Kaiss[1,2](✉), Khalifa Mansouri[1], and Franck Poirier[2]

[1] Laboratory Signals, Distributed Systems and Artifical Intelligence, ENSET Mohammedia, University Hassan II of Casablanca, Casablanca, Morocco
[2] Lab-STICC, University Bretagne Sud, Vannes, France
franck.poirier@univ-ubs.fr

Abstract. E-learning systems, also called virtual learning environments, promote education and training activities using modern information and communication technologies. Therefore, e-learning is the application of modern multimedia technologies and the Internet to improve the quality of learning by making resources and services more accessible, as well as exchanges and collaboration at a distance that can help learners in their studies and that can also help teachers to predict the weaknesses, strengths, and level of understanding of learners. Preparing and providing a quality e-learning system and rich learning experience are significant challenges. The lack of interaction, feedback, helping the learner to self-regulate, assessing the degree of knowledge, and adapting teaching methods and resources to the learners' real needs, which imply a lack of motivation and a high dropout rate, diminish the richness of the learning experience. Hence, we propose in this article a robust model of adaptive e-learning environments to optimize learning for each learner, taking into account the heterogeneity of profiles so that students succeed in their learning experiences.

Keywords: E-learning · Personalization · Self-regulated learning · Learner learning experience · Learner success

1 Introduction

Today, online learning has outgrown the traditional classroom environment. Due to the spread of the coronavirus pandemic, LMS (Learning Management System) has experienced unprecedented growth should continue to develop and evolve in the coming years. Most students feel less stress due to online learning and feel more comfortable emotionally [1].

Unfortunately, distance education poses pedagogical issues that reduce the richness of the learning experience. Preparing and providing a quality system and a rich learning experience are significant challenges. The separation of the physical and temporal of the tutor and the learner, and between the learners themselves, can lead to a feeling of isolation and a lack of discussion and interaction and feedback that generally concerns the learners' communication mechanisms with their teachers and peers. These gaps can reduce the richness of the learning experience. At the same time, peer teaching

Y. Maleh et al. (Eds.): ICI2C 2021, LNNS 357, pp. 274–286, 2022.
https://doi.org/10.1007/978-3-030-91738-8_26

contributes to improving student success [2]. For the learner, distance education requires good organization and pursuit of goals so that he/she succeeds in his/her training. Dropout rates from online courses are high due to the inability to self-regulate [3], this may be due in part to the fact that students underestimate the effort involved and fail to plan their work to be successful in online courses properly. Distance education poses another issue: evaluating the student's degree of assimilation of the material taught. This evaluation is essential if we want to accompany the learner and ensure concrete progress. It allows students to determine which parts of the course they need to study. It will enable instructors to know what course content needs to be adjusted.

In addition, e-learning environments do not always present the resources most suited to the real needs of learners such as the language of teaching, what they want to learn, the need to work with peers or individually, the type of questions, and type of learning objects (e.g., video, image, text, etc.). Not all of these needs are automatically detected, which implies a lack of motivation among learners and a high dropout rate that can reduce their success rate in these e-learning systems. Thus, all students learn best when teachers adopt the teaching methods, resources, and learning environment that suit their interests, needs, strengths, and level of knowledge. But even for the best teachers, it is difficult to fully understand each learner's profile and adapt the program accordingly and help them self-regulate. This is why technology could be of great use; therefore, it adds new possibilities for tracking learners' activity and providing them with more immediate feedback on their progress. Hence, the conclusion that one of the main axes is the personalization of learning must be taken into account if we want to see student outcomes improve.

To resolve these issues, we judged that it is essential to include personalization, evaluation, feedback, help the learner self-regulate, etc., while adapting them to the learner's profile (knowledge, preferences, abilities, objectives, etc.) due to their importance. This article proposes a model of e-learning environments, personalizing the pedagogical resources so that students succeed in their learning experiences. Rather than imposing a unique program on all learners independently of their respective abilities and needs, personalized learning allows the development of individualized learning that will motivate them to follow.

The article is structured as follows: The following section reviews the literature on key theories and models, presenting several factors that can contribute to the use, engagement, satisfaction, and success of learners in the e-learning system. Our proposed model is presented in Sect. 3. Subsequently, the proposed approach is described in detail in Sect. 4. Then, we will end our article with a conclusion and a presentation of our future research work.

2 Related Work

A large number of studies over the recent decades have attempted to identify the success factors of e-learning that need to be managed to increase the effectiveness of e-learning systems.

According to [4], the e-learning success model adequately explains and predicts the interdependence of six critical success factors for e-learning systems: course design quality, instructor involvement, student motivation, student-student dialogue, student-instructor dialogue, and self-regulated learning. Several researchers have been interested in aspects of the learner and how they can influence the learning process. Some have looked at personal characteristics such as learning styles [5] or motivation such [6] and claimed that these have an important influence on the success or failure of learners. According to [7] the effectiveness of learner-controlled online learning depends on individual cognitive characteristics and related to motivation such as learning styles, cognitive abilities, attribution of success, and self-regulation abilities.

Motivation is essential to learning and what makes students learn is their conscious engagement in these learning activities because "engagement leads to outcomes such as success" and "motivation underlies engagement", engagement and motivation are not the same [8]. Motivation can be transformed into engagement with an appropriate design of support. According to [9] intrinsic value and self-efficacy are motivation variables, whereas effort and metacognitive regulations are regulation variables, and cognitive and emotional engagements are engagement variables. Supporting students' effort regulation is a unique way to improve the motivation and engagement that influences their success. It is the metacognitive and effort regulation that learners put into their learning that transforms motivation into engagement [9].

The word "self-regulation" refers to a student's ability to regulate his or her thoughts and actions [10]. For example, students can reflect on how they learn and what strategies they will need to succeed in their university studies. To achieve specific learning and performance goals, a self-regulated learner employs motivational, metacognitive, and behavioral processes (for example, help-seeking, goal setting, self-evaluation, and metacognitive monitoring) [11, 12]. Self-regulation processes considered motivating allow learners to initiate and maintain targeted activities focused on objectives while ignoring distractions and setbacks [13].

A self-regulated learner actively engages these processes by using strategies in what [10] have called Self-Regulated Learning Strategies (SRLS) [10, 14]. Examples of SRLS include tutoring, keeping a study journal, and emailing the instructor. Additionally, not all self-regulated learners use the same strategies. Thus, empirical research has shown that implementing various strategies in online courses can promote SRLS in students. Structuring the e-learning environment to encourage self-regulated learning is a central element to fostering students' successful use of SRLS [15, 16]. Researchers have suggested that SRLS are of greater importance in e-learning environments due to their more autonomous nature [17]. Thus, providing self-regulated learning strategies to online instructors to promote the use by students of these strategies in online courses aims to improve their students' academic performance [18].

According to [19] learner success in an e-learning system can be explained by self-regulation and learners' intention to continue using educational platforms. They also show that learners' intention to continue using these educational platforms can be defined by learner satisfaction. Self-regulation can be explained by personal effort and course flexibility. The following factors: perceived usefulness, social interactions, system quality, course and information quality, course flexibility, and diversity of assessments can improve learner satisfaction with educational platforms. According to

[20], parents (family causal factors), teachers (school causal factors), and students (personal causal factors) can influence final performance. However, the combination of these factors affecting academic performance varies from one academic environment to another, from one group of students to another, and even from one cultural setting to another.

Other factors play a role in online learning satisfaction and success related to learner characteristics and course design characteristics of e-learning environments [21]. On the one hand, learner characteristics include self-regulation skills, time management, online learning style, knowledge, better emotional intelligence (i.e., better knowledge of his/her needs, self-regulation of feelings), self-motivation, self-discipline, organization, and planning and self-assessment. On the other hand, professors/instructors need to know the course design characteristics of online learning environments that promote student success and satisfaction, including time management, course alignment, and organization, integration of content with technology to facilitate engagement, instructor facilitation, instructor's comments, prompt feedback which is a primary factor in online satisfaction, and the size of the course which should not be significant. Successful online learning also requires interaction between the learner, the instructor, the learners, and the technology.

The virtual learning environment effectiveness model of [22] posits that two antecedents (design dimension and human dimension) determine the effectiveness of e-learning systems. The human dimension concerns two human entities (the students and the instructor) and their different attributes; and the design dimension includes learning management systems (LMS), self-regulated learning (SRL) and learner control, the quality of course design, and the interaction between human entities.

Learning analytics provides opportunities for students to reflect on learning and develop metacognitive skills. So, to be successful in self-regulation, students must understand their cognitive processes. To this end, linking visualizations to learning goals can help learners and instructors assess whether the goal has been achieved [23].

Interactions with the learning community members that includes learners and instructors, social presence, positive support, engagement, critical thinking, and overall assessment results have a positive impact on the concept of knowledge and skills acquired during the online learner's learning path that represents the backbone of the online learner experience [24]. In addition, forums play an important role in collaborative student learning, also allow students to interact. Students who participate most actively in the discussion forum succeed in the course [25]. Course structure, class size, feedback, prior digital communication knowledge, interface characteristics, content area experience, student roles and pedagogical tasks, differences in student demographics and abilities, etc., are factors that can influence student participation in these online discussion forums [26]. Other research results affirm that all students learn best when the instruction, resources, and learning environment are well-matched with their strengths, interests, needs, and level of knowledge [27, 28]. Hence, the conclusion is that personalization of learning is one of the main axes that must be considered if we want to see student outcomes improve.

Another aspect to take into account is the evaluation of learning. Students' learning depends on how they are evaluated and with what instruments [29]. Thus, the teacher assesses the student to identify the evolution through different levels of learning. On the one hand, depending on the level reached, the instructor proposes actions and material to improve the students' performance [30] evaluations, videos, tests, conferences, among others. Each student, on the other hand, learns at his/her own pace. This requires personalizing the mechanisms used according to each student. Information and communication technologies can help the teacher discover who is learning and who is not motivated. They can suggest actions like when who, and how to evaluate or what activities can be realized.

3 Proposed Model

In this section, we will describe our proposed robust model of e-learning environments that offer the personalization of learning to the specific skills and needs of each learner to improve the learning experience of the students, whose goal is to increase their success rate as well as reducing their dropout rate in these e-learning systems. Figure 1 shows our proposed model. It is then a matter of personalizing the learning objects, adapting teaching methods, helping the learner use self-regulated learning strategies, and the feedback and evaluations provided by the teacher that will be personalized according to the learner's profile. In an e-learning system, they can be applied as components, linking the visualizations to the learning goals to help learners and teachers evaluate if the objective has been achieved. This is a useful practice for teachers and especially for learners since it allows the learner to be accompanied at any time and any place and tests his/her knowledge and skills. Hence the system requires a dynamic combination between the learner's demands (his/her interests, preferences, level of knowledge) and personalization possibilities.

The **Learner**'s need is to learn. Our objective is to help him/her learn, become autonomous, be satisfied, and succeed in his/her learning. In the e-learning system, the learner visualizes the learning objects provided by the teacher, which are personalized according to his/her profile by the e-learning system. He/she has to evaluate his/her knowledge and level of understanding. Also, the learner has to be able to define his/her objectives, to make decisions in his/her program, to apply the right learning strategies, which are cognitive strategies where the learner can take notes, summarize, create mental associations, use images and audio documents, use the mother tongue; metacognitive strategies are there to reinforce memorization and reflect on the learning process by setting goals, reflecting on him/her way of working, having a personal learning journal, self-evaluation; socio-affective strategies bring the learner closer to his/her emotions and feelings by cooperating with peers, encouraging each other, opening up to others.

The role of the **Teacher** is crucial. He/she has to design educational sequences, evaluate the level of the learners, use the educational material provided… E-learning doesn't consist of simply letting learners navigate through a set of resources; it is a matter for the teacher accompanying them in their learning process through the design of educational activities integrated into a coherent scenario. Therefore, the teacher's and the e-learning system's roles are also to guide the learners to identify the best strategies, help them learn and understand, and allow them to self-regulate.

The sub-components of the "**Learner Profile**" component are as follows: history of his/her navigation (his/her learning established based on his/her last interactions with the learning environment); his/her performance (this sub-component updates information about the learner through the results of the suggested evaluations); his/her preferences (based on his/her interaction with the learning environment, we determine the language of teaching, types of learning objects and preferred kinds of questions).

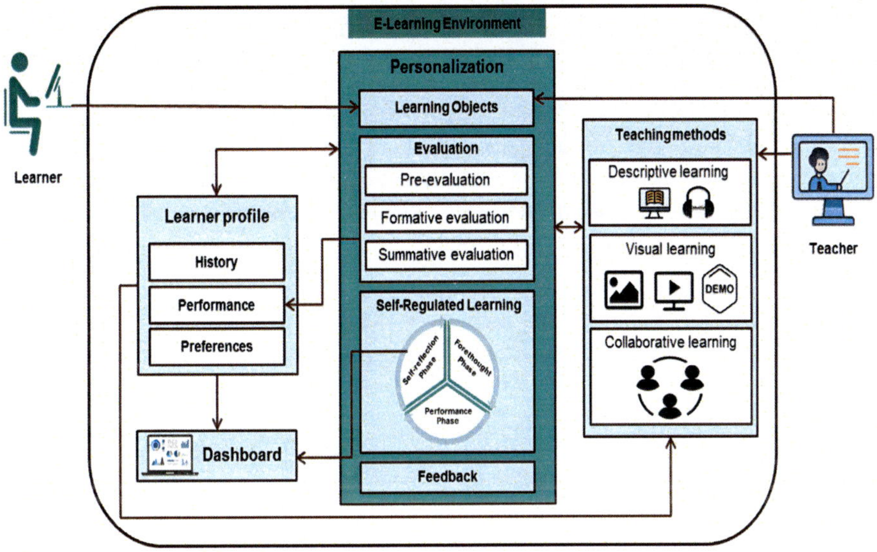

Fig. 1. Our proposed model of e-learning system for successful education.

The "**Evaluation**" component aims to provide the learner with a personalized evaluation based on the information provided in his/her profile (e.g., preferences, type of questions, results obtained in the evaluations). It plays an essential role in how students learn, are motivated to study, and instruct teachers. While teachers, students, and the online learning environment work to achieve the learning outcomes, evaluation plays a critical role in providing helpful information to guide teaching, help students achieve the following steps, and verify progress and accomplishments. Evaluation serves a variety of goals:

- Evaluation in the service of learning: evaluation informs teachers and the online learning environment about what students understand and allows them to know what learning objects to adjust and plan and guide teaching while providing helpful feedback to students.
- Evaluation as learning: evaluation allows learners to become aware of how they learn and adjust and progress their learning by assuming increased responsibility.

- Evaluation of learning: the information gathered from the assessment allows students, teachers, and the online learning environment to be informed about the learning outcomes achieved at a specific time to highlight successes, plan interventions, and keep promoting success.

The evaluation should be planned according to these goals. Assessment in the service of learning, evaluation as learning, and evaluation of learning each have a role in supporting and enhancing students learning. In addition, the effective assessment does not apply the same evaluation strategy to all learners because their knowledge is different. Some students should be evaluated on the entire learning content to assess their overall knowledge; others only need to assess their current knowledge in a few learning steps, which allows them to access the parts of the course that are most appropriate for their level of knowledge. The aim is to orient the learner to the most appropriate resources for his or her needs through a personalized evaluation. Based on the results of the test, a personalized learning itinerary will be provided. The three most popular types of assessment are as follows:

- Pre-evaluation: Before they start learning, it is essential to know what type of students it addresses, evaluating their level of knowledge and preferences. A personalized program can then be created based on the data collected.
- Formative evaluation: used during the first attempt at learning. By keeping track of students' progress and obtaining feedback to identify obstacles and individual difficulties to the diversity of students to provide individualized help and adapt "**Teaching methods**". Three teaching methods enhance the student learning experience: descriptive learning (reading, hearing), visual learning (pictures, videos, and demonstration), and collaborative learning (participatory).
- Summative evaluation: aims to evaluate if the most important knowledge has been acquired when the training is completed to adjust the rest of his/her instruction according to the level achieved.

Also, providing learners with personalized "**Feedback**" on their performance is an effective strategy for promoting learning, with congratulatory messages (e.g., if the learner has successfully achieved his or her goal, a message appears saying "Well done, I'm proud of you") and advisory messages (e.g., if the learner receives a low grade, a message appears saying "Anyone can make mistakes, take extra time next time.").

We then focus on promoting students' use of "**Self-Regulated Learning Strategies**" in an e-learning environment and personalizing them according to their profile. The model of [11] was the most frequently accepted and we will use it. As shown in Fig. 2 of [11], who postulated that SRL occurs in three cyclical phases: forethought, performance, and self-reflection [11, 31]. Each phase of self-regulated learning consists of processes that self-regulated learners engage in (Fig. 3 describes the fishbone diagrams of each process), and each phase influences the processes of the next phase. The third, in turn, influences the first again. The fishbone diagram is a diagram that graphically represents the causes leading to an effect and can also be called a cause and effect diagram [32].

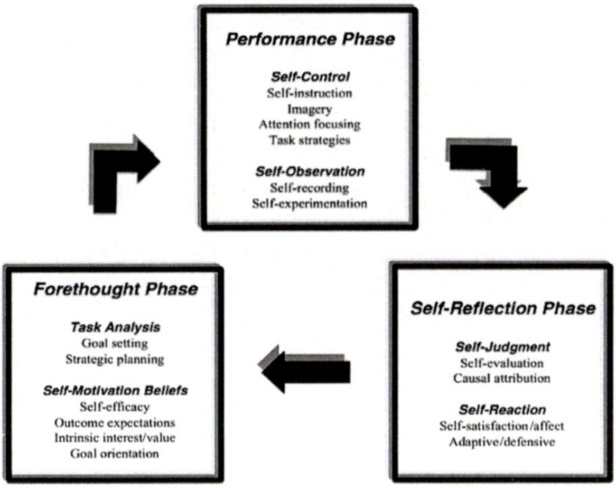

Fig. 2. Phases and subprocesses of self-regulation [33].

The Forethought Phase of self-regulated learning describes the processes by which learners start to launch plans to complete a task or attain a goal (see Fig. 3 (a)). The Performance Phase describes learners' processes to accomplish a task or goal (see Fig. 3 (b)). In the Self-reflection Phase, learners reflect on their progress toward the task or goal, or the results if they have completed the task (see Fig. 3 (c)).

(a)

(b)

(c)

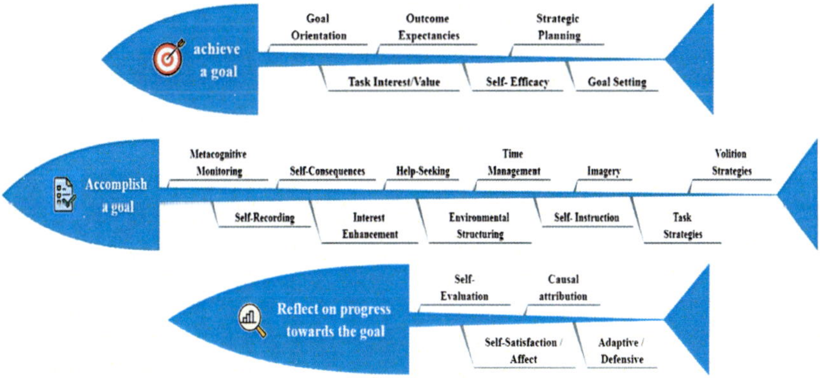

Fig. 3. Fishbone diagrams of the Forethought Phase (a), Performance Phase (b), and Self-Reflection Phase (c) processes [33].

Since students do not necessarily develop SRLS when taking online courses, even though these SRLS are of greater importance in online learning environments due to their more autonomous nature, instructors should consider implementing these strategies to help learners promote positive SRLS [18]. Still, it is difficult for instructors to fully understand each learner's profile to help them use these strategies and personalize these SRLS according to each learner's profile. Therefore, we propose providing e-learning systems with specific strategies to encourage and help students use these effective SRLS, which will be personalized to their profiles. Some students know how to launch plans, but some, for example, do not know how to manage their time. Also, for example, some students want help (Help-Seeking) while others wish to help guide their learning. Therefore, it is our proposition's role to offer personalized self-regulation strategies to e-learning systems to help them self-regulate.

Therefore, to succeed in self-regulation, students must understand their cognitive process [23]. Using goal-oriented visualizations of activity tracking is an interesting experience in analyzing student-centered learning through visualizations [34]. In this case, we add a "Dashboard" component to the e-learning system to allow students to reflect on their activity and compare it with their peers through dashboards. For this aim, the collected information, such as their learning performance, will be displayed in visualization to help them in their future decisions on task or goal completion.

In the next section, a description of the proposed approach is shown in the form of a sequence diagram that shows the scenario behind the background of our model.

4 Scenario of the Proposed Model

In this section, we will describe the context of our research using the personalized learning environment scenario.

Once identified, the teacher adds learning objects, evaluations, and feedback. One identified as well, the learner navigates and consults educational resources. The adapter component uses these traces of the learner's interaction with the learning system to define better the learner's preferences (teaching language, types of learning objects, types of questions, etc.). The adapter's function is to personalize the learning resources according to the learner's profile and consult the traces, results, and profiles of other learners to propose the best configuration for each new learner. The adapter will also help the learner to self-regulate.

The learner must define his/her goal. If he does not do so, the system asks him/her to define it. For example, Yasmine is a student in computer engineering who would like to learn the C language, the system sends Yasmine's objective, the resources (learning objects, evaluations), and the feedback provided by the teacher to the adapter to form a pre-evaluation on the chosen objective. After Yasmine answers the corresponding questions and according to the results obtained (the questions that were not answered correctly), the adapter depends on it to generate personalized feedback to Yasmine, teaching methods adapted to her needs, and a program personalized to her profile (preferences, performance).

She starts learning and to achieve her goal she starts using self-regulation strategies, so the adapter will help her use them to improve her academic performance. For example, she decides to ask for help every week on the parts she did not understand well, and unfortunately, at the end of the semester, Yasmine does not receive a good grade in the course. So she attributes the cause (i.e., causal attribution) to the strategy used (i.e., seeking help did not help her achieve her goal). In the course that follows, Yasmine will decide not to pursue this strategy.

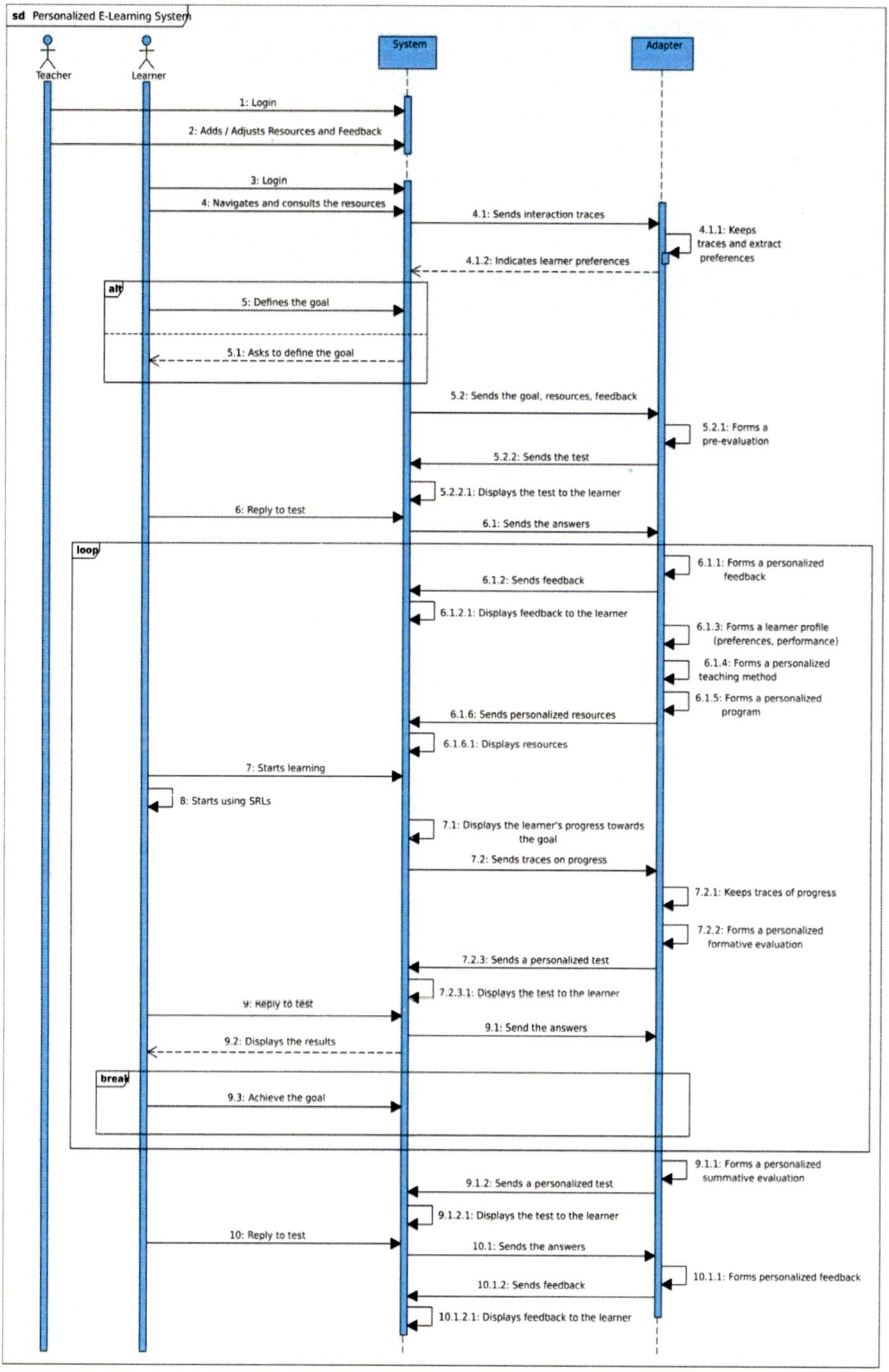

Fig. 4. Personalized E-Learning system sequence diagram.

To successfully self-regulate, the system displays the collected information, e.g. her learning performance, in a goal-oriented visualization so that Yasmine can track her progress towards her goal. Thus, to ensure that Yasmine understands the important concepts of the C language, questions will be researched by the adapter and presented to constitute personalized formative evaluations on this lesson, estimating her level of knowledge at each stage of the learning. The adapter should save Yasmine's progress and tracks in order to use them for future evaluation activities and form personalized feedback according to her profile until she achieves her goal.

Finally, the adapter forms a personalized summative evaluation to Yasmine to affirm that she has succeeded in achieving her goal, understood the important language C concepts, and gave her personalized feedback, e.g., motivational messages, advice, etc. For example, if she succeeded with a good grade, the adapter sends a message saying "Congratulations, it was hard but you did it", this is a motivational message. Figure 4 illustrates a sequence diagram, which shows this scenario behind the background of this personalized e-learning system.

5 Conclusion and Future Work

In this article, we have proposed a new model of online learning environment, helping the learner to self-regulate, and make his/her learning in the e-learning system more accessible, easier, more personalized, more credible, and more desirable which ultimately leads him/her to better success. Thus, we presented the main factors that influence learners' success in their learning experiences that we included them in the model to improve the learning experience and presented a description of the proposed approach in the form of a sequence diagram that shows the scenario behind the background of our model. For future studies, we will experimentally validate this proposed model. Additionally, we suggest further study regarding learner-controlled online learning, how, what, when, and where to learn, which may lead to self-regulation skills and may be a factor in dropout rates of e-learning courses.

References

1. Baudoinm, N., et al.: Le bien-être et la motivation des élèves en période de (dé) confinement Note de synthèse, pp. 1–12 (2020). https://doi.org/10.1111/bjep.12342
2. Bowman-Perrott, L., Davis, H., Vannest, K., Williams, L., Parker, R., Greenwood, C.: Academic benefits of peer tutoring: a meta-analytic review of single-case research. Sch. Psychol. Rev. **42**(1), 39–55 (2013). https://doi.org/10.1080/02796015.2013.12087490
3. Lee, Y., Choi, J.: A review of online course dropout research: implications for practice and future research, pp. 593–618 (2011). https://doi.org/10.1007/s11423-010-9177-y
4. Eom, S.B., Ashill, N.J.: A system's view of e-learning success model, **16**(1), 42–76 (2018)
5. Jahanbakhsh, R.: Learning styles and academic achievement: a case study of Iranian high school girls students. Procedia - Soc. Behav. Sci. **51**(1988), 1030–1034 (2012). https://doi.org/10.1016/j.sbspro.2012.08.282

6. Kyndt, E., Dochy, F., Struyven, K., Cascallar, E.: The direct and indirect effect of motivation for learning on students' approaches to learning through the perceptions of workload and task complexity. High. Educ. Res. Dev. **30**(2), 135–150 (2011). https://doi.org/10.1080/07294360.2010.501329
7. Sorgenfrei, C., Smolnik, S.: The effectiveness of e-learning systems: a review of the empirical literature on learner control, **14**(2) (2016)
8. Martin, A.J.: Motivation and engagement: conceptual, operational and empirical clarity, pp. 0–15 (2012)
9. Kim, C., Park, S.W., Cozart, J., Lee, H.: From motivation to engagement: the role of effort regulation of virtual high school students in mathematics courses, 18, 261–272 (2015)
10. Zimmerman, B.J., Schunk, D.H.: Self-regulated learning and performance: an introduction and an overview (2011)
11. Zimmerman, B.J.: Motivational sources and outcomes of self-regulated learning and performance, 11237 (2011). https://doi.org/10.4324/9780203839010.ch4
12. Zimmerman, B.J.: Investigating self-regulation and motivation: historical background, methodological developments, and future prospects, **45**(1), 166–183 (2008). https://doi.org/10.3102/0002831207312909
13. Schunk, D.H., Meece, J.R., Pintrich, P.R.: Motivation in Education: Theory, Research, and Applications, 4th edn. (2014)
14. Perry Nancy, E., Ahmed, R.: Studying self-regulated learning in classrooms (2011)
15. Barnard-Brak, L., Paton, V., Lan, W.: Self-regulation across time of first-generation online learners. Res. Learn. Technol. **18**(1), 61–70 (2010). https://doi.org/10.3402/rlt.v18i1.10752
16. Ferla, J., Valcke, M., Schuyten, G.: Judgments of self-perceived academic competence and their differential impact on students' achievement motivation, learning approach, and academic performance. Eur. J. Psychol. Educ. **25**(4), 519–536 (2010). https://doi.org/10.1007/s10212-010-0030-9
17. Dabbagh, N., Kitsantas, A.: Supporting self-regulation in student-centered web-based learning environments. Int. J. E-Learning **3**(1), 40–47 (2004)
18. Wandler, J.B., Imbriale, W.J.: Promoting undergraduate student self-regulation in online learning environments (2017). https://doi.org/10.24059/olj.v21i2.881
19. Safsouf, Y., Mansouri, K., Poirier, F.: An analysis to understand the online learners' success in public higher education in Morocco (2020). https://doi.org/10.28945/4518
20. Diaz, A.L.: Personal, family, and academic factors affecting low achievement in secondary school. undefined (2003)
21. Kauffman, H.: A review of predictive factors of student success in and satisfaction with online learning. Res. Learn. Technol. **23**(1063519), 1–13 (2015). https://doi.org/10.3402/rlt.v23.26507
22. Piccoli, G., Ahmad, R., Ives, B.: Web-based virtual learning and a research framework environments: a preliminary assessment of effectiveness in basic it skills training1, **25**(4), 401–426 (2001)
23. Durall, E., Gros, B.: Learning analytics as a metacognitive tool (2014). https://doi.org/10.5220/0004933203800384
24. Hammad, R., Odeh, M., Khan, Z.: eLEM: a novel e-learner experience model, **14**(4), 586–597 (2017)
25. Romero, C., López, M., Luna, J., Ventura, S.: Predicting students' final performance from participation in on-line discussion forums. Comput. Educ. **68**, 458–472 (2013). https://doi.org/10.1016/j.compedu.2013.06.009
26. Yukselturk, E.: An investigation of factors affecting student participation level in an online discussion forum. Turkish Online J. Educ. Technol. **9**(2), 24–32 (2010)

27. Prud, L., Leblanc, M., Paré, M., Fillion, P.: Différencier d'abord auprès de tous les élèves: un exemple en lecture (2015)
28. Przesmycki, H.: La pédagogie différenciée, pp. 2–3 (2004)
29. García Carreño, I.D.: Hacía una evaluación integral con ePortafolio por evidencia y bPortafolio (2012). https://www.researchgate.net/publication/282133572_Hacia_una_evaluacion_integral_con_ePortafolio_por_evidencia_y_bPortafolio. Accessed 06 June 2021
30. Leyva, L., Garrido, Y., Leyva, J.L.L., Varona, R.C., Rodríguez, R.H.R.: Reflexiones sobre la evaluación de la calidad del aprendizaje en la práctica pedagógica en la escuela primaria. undefined (2007)
31. Zimmerman, B.J.: Attaining self-regulation: a social cognitive perspective, pp. 13–39 (2000)
32. Watson, G.: The legacy of Ishikawa. undefined (2004)
33. Zimmerman, B.J., Campillo, M.: Motivating self-regulated problem solvers, pp. 233–262 (2003)
34. Santos, J.L., Govaerts, S., Verbert, K., Duval, E.: Goal-oriented visualizations of activity tracking: a case study with engineering students, pp. 143–152 (2012)

The Preferences and Expectation of Moroccan Teachers from Learning Analytics Dashboards in a Blended Learning Environment: Empirical Study

Amina Ouatiq[1(✉)], Bouchaib Riyami[2], Khalifa Mansouri[1], and Mohammed Qbadou[1]

[1] SSDIA Laboratory, ENSET Mohammedia, Hassan II University of Casablanca, Mohammedia, Morocco
[2] Institut Supérieur du Génie Appliqué IGA Casablanca, Mohammedia, Morocco
bouchaib.riyami@iga.ac.ma

Abstract. Learning analytic dashboards help instructors track, and supervise students in online or hybrid education in order to meet the needs of teachers and better understand their preferences and their expectations of the dashboards, an online questionnaire was conducted for Moroccan teachers in higher education to determine their needs and their uses of a dashboard in a scenario of blended learning, to learn more about the indicators they deem most relevant to their teaching activities.

Keywords: Learning analytics dashboard · Visualization · User's needs · Questionnaire · Indicators

1 Introduction

The majority of Moroccan universities use educational technologies such as Massive Open Online Courses (MOOC), Small Private Online Course (SPOC), Learning Management Systems (LMS) [1] but still struggle with tracking and supervising learners in those platforms, despite the large amount of data that online educational platforms provide [2].

learning analytics which defined by Siemens et al. as "measurement, collection, analysis and reporting of data about learners and their contexts, for purposes of understanding and optimizing learning and the environments in which it occurs" [3]. It came in the use to resolve this problem and helps to track, monitor and supervise the students, using different tools, especially dashboards which defined as "visualization tool built with the purpose of empowering teachers and learners to make informed decisions about the learning process" [4]. Learning analytics dashboards (LAD) give the opportunity to promote awareness, reflection, sensemaking, and impact learning [5].

It's also viewed as a promising way of facilitating self-regulated learning, especially in blended and online settings [6].

Y. Maleh et al. (Eds.): ICI2C 2021, LNNS 357, pp. 287–297, 2022.
https://doi.org/10.1007/978-3-030-91738-8_27

In order to create a dashboard that will help the educators in their teaching activities, allow them to visualize the progress of their students, but also to predict both the struggling and successful students [7], the first step is to ask the four most important questions (Why, For Whom, What, and How) in order to precise the goal of the dashboard, what are the questions that the dashboard should have answers for? Who is the end user (stakeholders) of it, and also what data we need to visualize and how the users will interact with our dashboard? [8].

To answer these questions and know more about the teachers and other participant in the teaching process preferences and what are their needs and how will they use the LAD we made an online questionnaire destined at first place to Moroccan teachers in higher education.

The result of this questionnaire showed us that teachers want to use the dashboards to track their students' progress in the course and in the assessments and evaluation process, as well as the adaptation and improvement of courses and teaching practices, in order to be more relevant to the needs of the students. We also projected the importance of the different indicators coming from the online and face-to-face sessions.

The rest of the article is structured as follows; in the next section we discuss the methodology followed to develop this questionnaire. The third section focuses on the results. In section four we discuss and interpret them. We conclude in the fifth section.

2 Methodology

2.1 Questionnaire Description

In order to help the instructors, support and supervise learners in a hybrid learning environment, and to try to answer the most important question in learning analytics which is what's need to be measured [9]. A questionnaire was created to (1) determine teachers' preferences on a dashboard, (2) learn more about which indicators, they find most relevant to their teaching activities, and (3) to know their needs and uses of a dashboard in a hybrid learning scenario.

We decided to use a questionnaire cause it's the main instrument to collect primary data and give in a coherent and consistent format ready for analysis [10], and because "it can be an effective means of measuring the behavior, attitudes, preferences, opinions and, intentions of relatively large numbers of subjects more cheaply and quickly than other methods." [11].

The questionnaire included 15 questions divided into two categories:

- Teacher profile and their use of new technologies in teaching to know the experience they have in online teaching (how many years and classes they teach remotely or in a hybrid way) and what pedagogical tools and assessment techniques they have been using.
- Dashboard-related questions addressed the use of information provided by the dashboard, the most relevant indicators to both improve teaching activities, and measure the students 'knowledge and predict their success or failure' in a course using indicators from online and face-to-face sessions.

The questions were all in a structed format some were Multiple-Choice Questions, others were Matrix Table Question (Likert scale), with the possibility to add open comments after each question in the second category.

The questionnaire was conducted in French, as it is the official language of higher education in Morocco, between the 20 July and the 20 November 2020.

2.2 Hypotheses

Hypothesis 1 (H1). LADs are used to track students during class and assessments and improve teaching practices achieving an ideal teaching approach.

Hypothesis 2 (H2). (1) Number of assignments submitted, (2) frequency of access to resources, (3) frequency of participation in discussions, (4) previous academic results, (5) student grades, (6) participation in courses, and (7) classroom participation. These indicators which describe the traceability, the exchange, the presence, and the effort of the learner are the most important to know the level and the progress of the learner.

Hypothesis 3 (H3). (1) Average class performance, (2) list of frequently used resources, (3) classroom participation, and (4) time students spend with the system are the cornerstones of a good visualization and monitoring students in an ideal hybrid approach.

2.3 Participants

The questionnaire was designed for teachers, educational and administrative staff involved or willing to get involved in distance or hybrid educational approaches. We reached teachers and other professionals involved in the education system from universities and different higher education institutions (Private and public) such as administrative, technical, and pedagogical staff. The number of responses to our questionnaire was 38. In 38 participants, the largest group of our sample consisted of teachers with 89% (n = 34), followed by technical assistants 7.8% (n = 3), and 5.2% (n = 2) for administration. And a one teaching assistant (2.6%). 5.2% of participants played more than one role in their establishment (Teacher, Technical Assistant).

3 Results

In this section, we will present the results of the questionnaire divided into four subcategories the first linked to the profile of the participant, the second describes their habits of use of ICT in their teaching, the third focuses on the use of the LAD in an online or hybrid class. And the last on information directly concerning the students.

3.1 The Participants' Profile

As we presented in Sect. 2 most of the participants are higher education teachers. Almost half, 47.3% (n = 18) work in establishments with regulated access (establishments whose access is conditioned by admission procedures: preselection, written competition, interview). The remainder is divided between open-access establishment (establishments whose access does not require a competitive examination) 31.5% (n = 12), private higher school 15.7% (n = 6) and 2 (5.2%) which works in university presidencies (administrative or technical staff) as shown in Fig. 1.

42% of teachers are engaged in between two and five online courses, 15.7% between six and ten and 7 (18.4%) have taught more than ten courses. 21 taught only one and only 2.6% of the sample did not teach any courses. It is also found that the vast majority of teachers 92% started online courses for less than a year and only 5.2% spent six years or more teaching online and only one teacher 2.6% who taught between two years. And three years as shown in Fig. 2.

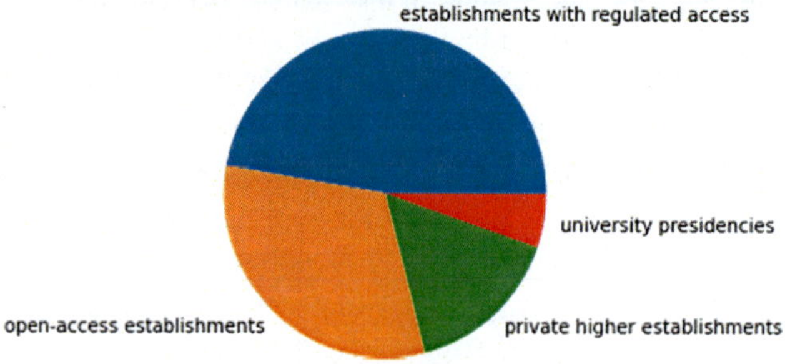

Fig. 1. Type of establishment

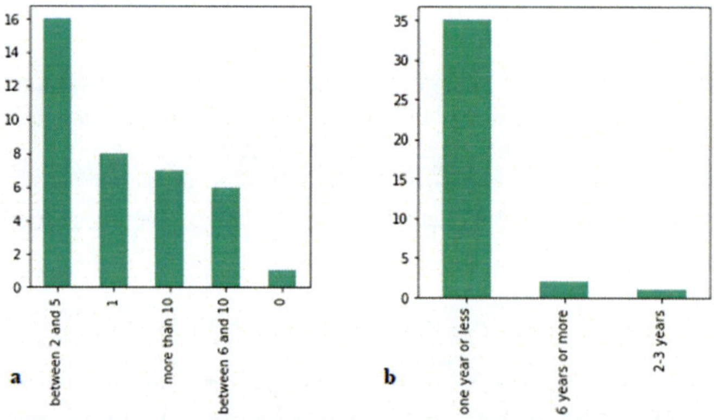

Fig. 2. Number and duration of online courses in which teachers participate

On the other hand, it is found that almost half 44.7% have never received training on new approaches to online teaching, 34.2% take training from time to time and only 10.5% of teachers frequently take training courses. And the same number of teachers rarely receive training in online teaching approaches as shown in Fig. 3.

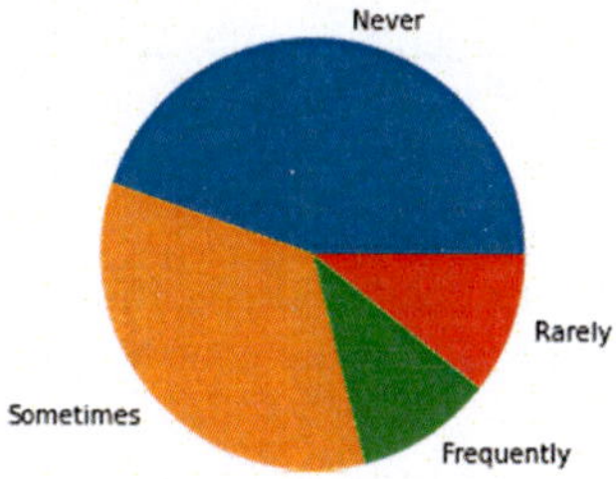

Fig. 3. Frequency of training for teachers

3.2 Evaluation Techniques and Tools Used in Online Education

Teachers use different educational tools to deliver online lessons. We find that 81.5% use videoconferencing, 71%, emails, and 68.4% chat tools, thus they are the three main tools used. Together, they represent 74.2% of all the tools used. Online file storage and sharing services are used by 28.9% of teachers, and 23.6% of them use LMS. Other tools are used, but they only represent 7.8% of the tools.

Online quizzes, assignments, presentations, and group projects feature 69.6% of the evaluation techniques used. Virtual labs, oral assessments, and forum and discussion analysis are the least used forms of evaluations, accounting for only 4.8% of the techniques used by teachers. As shown in Fig. 4.

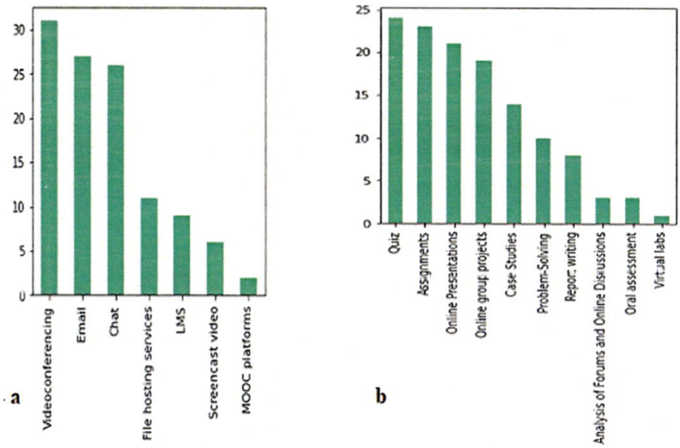

Fig. 4. Evaluation techniques and tools used in online education.

3.3 Learning Analytic Dashboards' Use

Teachers were asked how they will use the information provided by LAD. Monitoring students' performance during assessments and their progress in courses, as well as adapting courses to their needs, are the top three actions taken by teachers with 53.8%.

We also detect that notifying the students in difficulty or late in the course compared to their classmates, and get an idea on the interactions of students with resources in second position with 23.3%. Predicting students' results, comparing their behaviors and their inter-interactions are not actions that interest teachers (Fig. 5). 73.6% of teachers want to use dashboards to track student performance during assessments, and 68.4% chose to track student progress in the course.

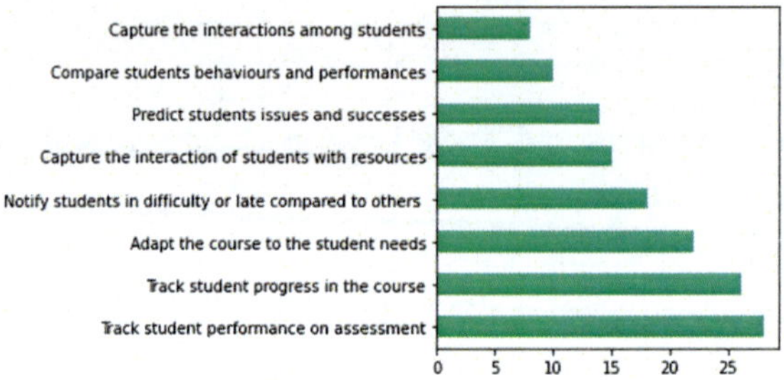

Fig. 5. Learning analytic dashboards' use

However, if we ask for the use of dashboards in teaching activities, we find that 57.8% of teachers will use the dashboard to identify and correct common misunderstandings, 55.2% want to propose specific tasks and assessments for students in difficulty in certain parts of the course, 50% aim to adapt teaching to an individual or a group of students, and 31.5% are for the grouping of students of similar level and create specific training for them (Fig. 6).

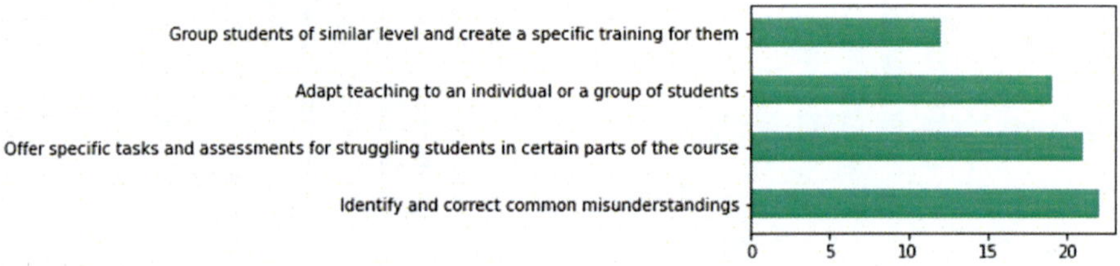

Fig. 6. Dashboards' uses in the improvement of teaching

The correlation between the degree of importance of information in improving courses and teaching practices can be seen in Fig. 7. For example, the list of frequently used resources and the most active discussions on the forum presents the aspects that are of interest to students.

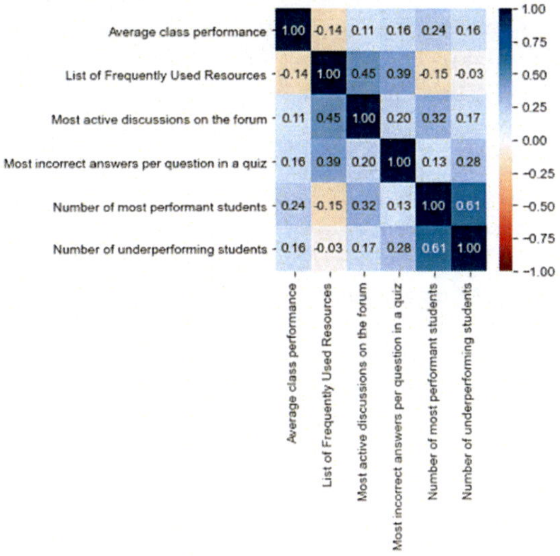

Fig. 7. Correlation between the importance of information in improving teaching practices

3.4 Students' Related Information

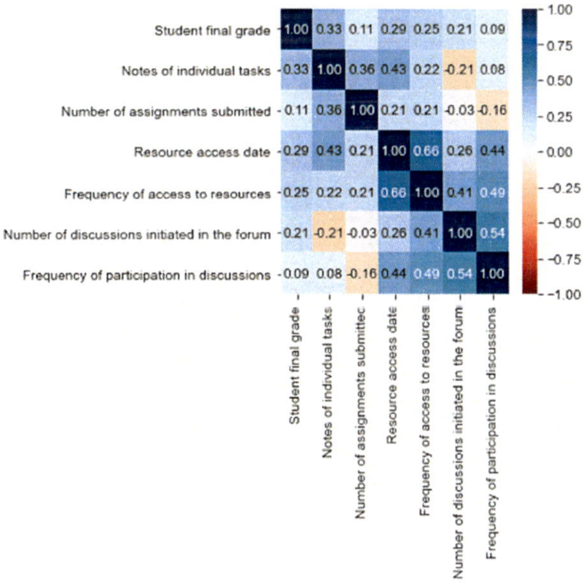

Fig. 8. Indicators to detect the level of knowledge of the student

LADs reflect information about the students and for the good of the students. Therefore, we asked the teachers to specify the degree of importance of some indicators in order to know the level of knowledge of the students, and others to help us to predict the outcome of each student. Figures 8 and 9 represent the correlation between the teachers' choices for the two questions.

The indicators in relation to time such as the date of access to resources, the frequency of access to resources, and the frequency of participation in discussions are interrelated. And also, the homework grades linked with the number of homework submitted and the date of access to the resources.

In order to predict the results of students, preferably group the following indicators: the regularity and time that the student spends on the system, the number of accesses to resources, as well as previous academic performance.

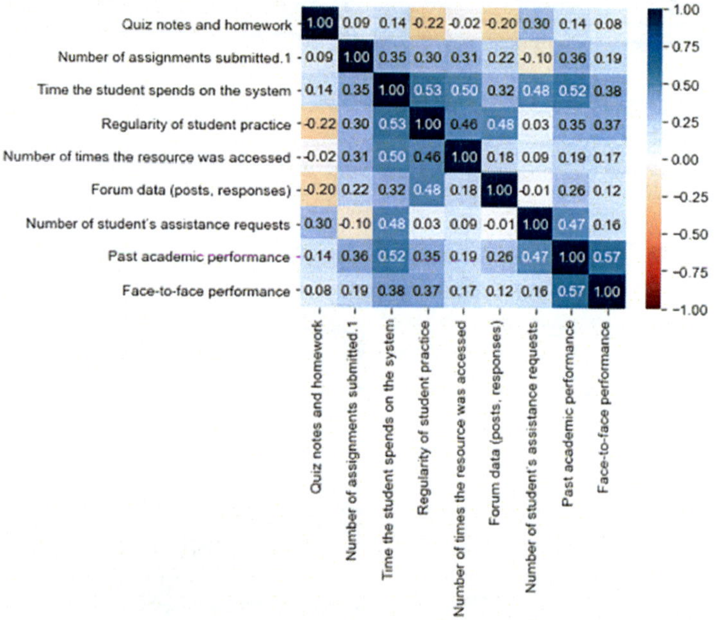

Fig. 9. Indicators to predict the student's result

The categories of students that the teachers are most interested in. They are found to be almost equally interested in all students, as shown in Fig. 10.

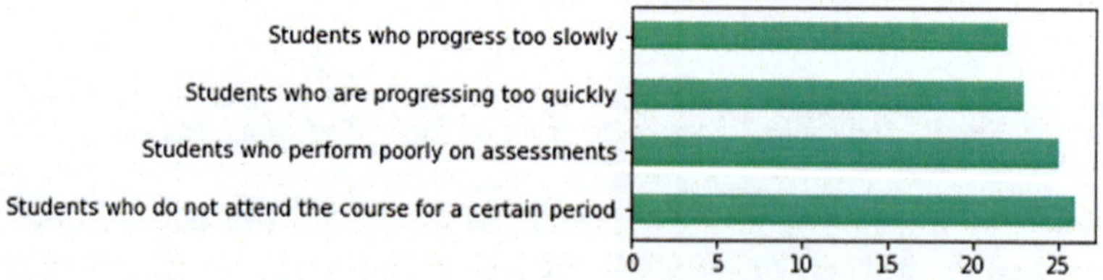

Fig. 10. Students' categories

For the face-to-face part of the hybrid system, teachers are interested in classroom participation and behavioral indicators (Fig. 11).

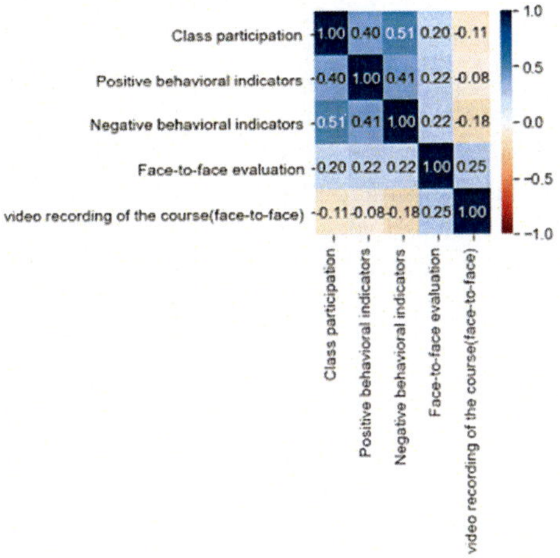

Fig. 11. The correlation between the indices from the face-to-face sessions

4 Discussion

Participants work in different types of institutions, although a good proportion teaches more than one course online, it is found that only 10.5% frequently take training in new approaches of online teaching. More than 90% have only started online courses for a year, we can conclude that the pandemic has a major effect on the digitalization of teaching practices [12], but this movement must of course be supported by continuous training in order to help teachers improve their practices. Video conferencing, synchronous and asynchronous communication tools are the main tools used to provide online courses and communication between instructors and their students. However, teachers should be encouraged to use MOOCs, as they present several opportunities for students to improve their knowledge [13].

Regarding assessment techniques, quizzes, online assignments, online presentations, and group projects are the most used techniques, and this is due to their technical ease. In contrast, virtual labs and oral assessments, and the analysis of forums and discussions that require more time, more effort and technical assistance are used less [14].

Regarding the purpose of the dashboard. The results of the questionnaire confirmed the hypothesis (H1) described in Sect. 2 that monitoring students during the course and assessments will be the main use of the dashboard. But the adaptation of the course to the needs of the students and notifying them in the case of delay or risk of failure was also well appreciated by the teachers which can lead to earlier intervention [15]. To take full advantage of the dashboard and help the teacher improve and adapt their course and teaching practices, the teachers preferred to use the information provided by the dashboard to identify and correct common misunderstandings, suggest student-specific assessments, and tailor courses to their needs.

We also suggested indicators in order to clearly specify their importance and to choose the indicators to visualize and group together to achieve our objective. For example, we can group together the list of the frequently used resources and the most active discussions on the forum to find out which parts the students are most interested in, or a comparison between the best and worst performers, etc. This suggests the objectives of our dashboard; student monitoring, and course adaptation.

Teachers are interested in all the categories of students that have been proposed. One of the participants clarified that if we form specific groups, we risk the marginalization of these students, and he prefers heterogeneous groups.

Teachers find that the final grade, individual task grades, and the date and frequency of access to resources reflect better the level of the student. Yet, in order to predict the results, indicators that describe student behavior and engagement, as well as academic history, are valued to provide insight into student success or failure. This confirms our second hypothesis (H2) that the indicators which describe the traceability, the exchange, the presence, and the effort of the learner are the most important to know the level and the progress of the learner.

Regarding the indications for face-to-face sessions, teachers prefer classroom participation, face-to-face assessment rather than recording and analyzing sessions. That can be justified by the unease of teachers in front of the cameras and the possibility of losing the anonymity and freedom of expression of the students. Our third hypothesis (H3) is in the same context that the performance of students face-to-face or online is the best indicators to build an image on the students and the class in a hybrid environment.

5 Conclusion

This questionnaire presents insights into the information and indicators that teachers working with the hybrid approach prefer, as well as the possible uses of dashboards focusing on current uses of ICTs and new technologies, the use of dashboards, and information directly linked to the student. The preferences and expectations of instructors reveal recommendations to consider when designing a dashboard for Moroccan teachers. Monitoring students, improving and adapting courses are the main expected uses of a dashboard, along with different indicators that represent participation, engagement, and behavioral cues are most valued by teachers.

In addition, recommendations concerning teachers and their needs to frequently follow training courses on new online and hybrid teaching approaches, in order to be able to better serve new techniques; MOOCs, LMSs and use more creative assessment for the benefit of their students. A questionnaire is not sufficient to know all the requirements of teachers, and we aim to conduct interviews to reveal aspects that were not well discussed in this questionnaire, despite it remains an effective way to measure preferences and collect opinions. This part of our work (the questionnaire and future LAD) is designed specifically to educational and administrative staff. In future work we will discuss the preferences of students in dashboards designed for them.

References

1. Riyami, B., Bouaine, A.: MOOC's integration approach: assessment and comparative studies of all Moroccan universities. In: Proceedings of the 5th International Conference on Information and Education Innovations, July 2020, pp. 11–15 (2020)
2. Greller, W., Drachsler, H.: Translating learning into numbers: a generic framework for learning analytics. J. Educ. Technol. Soc. **15**(3), 42–57 (2012)
3. Long, P.D., Siemens, G., Conole, G., Gašević, D. (eds.): Proceedings of the 1st International Conference on Learning Analytics and Knowledge (LAK'11). ACM, New York (2011)
4. Jivet, I., Scheffel, M., Specht, M., Drachsler, H.: License to evaluate. In: Proceedings of the 8th International Conference on Learning Analytics and Knowledge - LAK '18 (2018). https://doi.org/10.1145/3170358.3170421
5. Verbert, K., Duval, E., Klerkx, J., Govaerts, S., Santos, J.L.: Learning analytics dashboard applications. Am. Behav. Sci. **57**(10), 1500–1509 (2013). https://doi.org/10.1177/0002764213479363
6. Rienties, B., Tempelaar, D., Nguyen, Q., Littlejohn, A.: Unpacking the intertemporal impact of self-regulation in a blended mathematics environment. Comput. Hum. Behav. **100**, 345–357 (2019)
7. Schwendimann, B.A., et al.: Perceiving learning at a glance: a systematic literature review of learning dashboard research. IEEE Trans. Learn. Technol. **10**(1), 30–41 (2016)
8. Klerkx, J., Verbert, K., Duval, E.: Learning analytics dashboards. In: The Handbook of Learning Analytics, pp. 143–150 (2017). https://doi.org/10.18608/hla17.012
9. Duval, E.: Attention please! Learning analytics for visualization and recommendation. In: Proceedings of the 1st International Conference on Learning Analytics and Knowledge, pp. 9–17 (2011)
10. Malhotra, N.K.: Questionnaire design and scale development. In: The Handbook of Marketing Research: Uses, Misuses, and Future Advances, pp. 83–94 (2006)
11. McLeod, S.A.: Questionnaire: Questionnaire: Définition, exemples, conception et types. Simplement de la psychologie (2018). https://www.simplypsychology.org/questionnaires.html
12. Draissi, Z., Qi, Z.Y.: COVID-19 outbreak response plan: implementing distance education in Moroccan Universities (2020). https://ssrn.com/abstract=3586783. https://doi.org/10.2139/ssrn.3586783
13. Vázquez-Cano, E., López-Meneses, E., Gómez-Galán, J., Parra-González, M.E.: Prácticas universitarias innovadoras sobre las ventajas educativas y desventajas de los entornos MOOC. Revista de Educación a Distancia (RED), **21**(66) (2021). https://doi.org/10.6018/red.422141
14. Karagöz, Ö., Saka, A.Z.: Development of teacher guidance materials based on 7E learning method in virtual laboratory environment. Procedia. Soc. Behav. Sci. **191**, 810–827 (2015). https://doi.org/10.1016/j.sbspro.2015.04.524
15. Drachsler, H., Greller, W.: The pulse of learning analytics understandings and expectations from the stakeholders. In: Proceedings of the 2nd International Conference on Learning Analytics and Knowledge, pp. 120–129 (2012)

Towards a Smart City Stakeholders Classification: Case of Casablanca Smart City Project

Jihane Tadili(✉) and Hakima Fasly

Strategic Intelligence Research Laboratory, Hassan II University of Casablanca, Casablanca, Morocco

Abstract. Smart cities have gained the support of scientists, organizations, and governments worldwide because they use Information and Communication Technologies (ICT) to suggest innovative solutions for all urban problems.

The Smart City vision evolved as a new urban model during the last decade where all stakeholders promote sustainability, high quality of life, and economic development.

Many researchers argue that government must reconcile the interests of multiple and diverse stakeholders who are sometimes conflicting to achieve the objectives of its Smart City project.

In this context, this article discusses the role of stakeholders in innovating smart cities and suggests a classification of Casablanca Smart City stakeholders project.

Keywords: Smart city · Stakeholders · Casablanca · Classification · Stakeholders engagement

1 Introduction

Eighty percent of the Gross National Product (GNP) worldwide is produced by cities [1]. Indeed; Cities play an important role in the economic and social development of the world and have a significant impact on the environment [2]. The world would arbitrate 43 megacities of more than 10 million inhabitants by 2030, against 31 in 2017 [3]. Moreover, half of the world's population lives in cities. This figure will reach about 5 billion by 2030 [1]. As a result, cities consume about 75% of the global primary energy and generate between 50 and 60% of global greenhouse gas emissions [3].

In this challenging context, smart cities appeared as an innovative solution for all urban development problems using Information and communication technologies (ICT). They gained the support of scientists, urban planners, and governments all over the world.

Many researchers argue that the four pillars of innovation: university, government, industry, and civil society should cooperate and participate in the early innovation of the Smart City process. They introduced civil society as an important stakeholder [4–7].

Y. Maleh et al. (Eds.): ICI2C 2021, LNNS 357, pp. 298–307, 2022.
https://doi.org/10.1007/978-3-030-91738-8_28

However, there is a lack of researches about Smart City stakeholders, especially in the Moroccan and African context. That is why we will suggest classifying Casablanca Smart City stakeholders' projects in this article after explaining their roles.

To this end, we will first present the Smart City concept. Then, we will identify Smart City stakeholders and classify those of the project Casa Smart City. We will finish our work by suggesting new research perspectives.

2 Methodology

We carried out a comprehensive literature review of twenty-seven (27) research articles to clarify the concept of Smart City, the role of stakeholders in each Smart City project's development stages and their classification methods. Then, we classified those of Casablanca city Smart City project based on information that we retrieved from the internal documentation that Casa Numeric (the Smart City department in Casablanca) and the municipality Casablanca (La Commune de Casablanca) shared with us.

3 The Concept of Smart City

There is no consensus among researchers on a definition of the Smart City [7–10]. This is explained mainly by its multidisciplinarity and the interest of organizations of different disciplines and fields of expertise for this concept: Computer Science, Anthropology, Geography, Urbanism, Tourism, Engineering, Sociology, Politics, and Economics [11].

IBM was the first private firm to introduce the term "Smart City" and identify it in ICT in urban management. It also determined three main characteristics: Instrumented, Smart and Interconnected [12].

Other scientists have added other aspects to the definition of Smart City as the faculties and activities of its autonomous, conscious and independent citizens [13], the management of social capital and the preservation of natural resources [14] digital inclusion [8].

To conceptualize the smart city, the authors [13] identified four components of the Smart City: Industry, Education, Citizen Participation, and Technology Infrastructure. Then, another study [15]identified six main components that many scientific research articles about smart cities cited and developed: Smart Economy, Smart Mobility, Smart Environment, Smart People, Smart Living, and Smart Governance.

[9] identified three main dimensions of the Smart City: Technological infrastructure, citizens and institutions. According to the authors, a city is smart when it invests in technological infrastructure and social and human capital to ensure the city's sustainable development and improve the quality of life of its citizens through participatory governance.

4 Stakeholders Engagement in Smart Cities and Their Classification Methods

4.1 Stakeholders Engagement in Smart Cities

Replacing existing forms of governance with the stakeholder model is at the heart of stakeholder theory, [16, 17] cited by [18] discussed the idea of interaction between citizens and organizations as stakeholders with governments.

[19] In e-government, several authors argue that the realignment of development objectives with the needs of the main stakeholders is primordial [19, 20]. It involves key aspects such as identifying stakeholders; understanding their interests and determining how to align their interests with e-government goals [21].

In the urban context, the participation of stakeholders in city management is widely discussed in research on new governance [22].

To become smart, the city should adopt management focused on social capital and relations with stakeholders. To that end, the city must reconcile the interests of multiple and diverse stakeholders who are sometimes conflicting in order to achieve the objectives of its smart city project. The availability of an adequate legal ground should also be addressed to prepare stakeholders' participation in management [22].

According to [23], managing a smart city's stakeholders involves identifying stakeholders, their classification, communication, engagement, accountability, and risk control.

4.2 Smart Cities Stakeholders Classification Models

Several studies have focused on identifying the stakeholders needed for each stage of Smart City development. [24] were able to identify after a literature review several stakeholders that they classified as internal and external stakeholders. For the authors, the internal stakeholders of a Smart City can directly influence or be influenced by the Smart City project. They are rather interested in the financial activities and the efficiency of the project. As for external stakeholders, these can indirectly influence or be influenced by the Smart City project and they are interested in the value and quality of the project.

In more details, internal stakeholders are:

- Energy suppliers: renewable energies are significant in smart cities. Therefore, the involvement of energy suppliers in implementing the smart city is crucial for the project's success.
- Representatives of the ICT sector: Technology is one of the dimensions of a smart city. That is why the ICT sector must contribute to the development of the project.
- Citizens: as a stakeholder, citizens will share their experiences in the urban space and claim any malfunction in the smart city project. Their creativity and initiatives are also important in the initiation of the project.
- The government: The government's responsibility is to plan, initiate, create and manage the smart city project.

- Real estate developers: This stakeholder is interested in innovation and technological progress in the real estate promotion of a smart city.
- Urban planners: sustainable urban development is one of the objectives of the smart city; this is how urban planners play a big role.
- Politicians: their role in the smart city would be to put policies to help set up the project.
- Experts and scientists: they are important for the process of innovation in smart cities.

For external stakeholders, the authors identified:

- Universities: can contribute to smart cities through scientific research and play a significant role in planning and implementing strategies.
- Non-governmental organizations: these stakeholders will be interested in the results of the smart city.
- Political institutions: These can influence the governance of the smart city.
- Media: their role would be to ensure media coverage of the project and its benefits.

[22] opted for another classification according to the role played by the stakeholder in the smart city and were able to identify four categories:

- Facilitator: his role is to create the vision of the smart city, allocate the necessary resources and provide leadership as the representatives of the city's municipality.
- Supplier: to engage scientific researchers and professionals in the smart city innovation process, provide and improve knowledge like universities.
- Creator: his role is to create products and services that meet the objectives set for the smart city as the companies.
- User: he uses the products/services and participates in their improvement by providing feedback on experiences such as citizens, NGOs and visitors.

[25] developed this classification by assigning the types of stakeholder concerned to each role (Table 1):

Table 1. Smart City stakeholder's classification

Role	Stakeholder type
Facilitator	City mayor; the government or the municipality; administrations; businesses
Supplier	Research and Development centers; schools; the universities; incubators and Think/Tanks
Creator	Non-profit organizations; businesses; sellers; industrial
User	Local and national citizens; public and private investors; non-governmental organizations; the users; the residents; the workers; media and journalists

5 Classification of Casablanca Smart City Project Stakeholder

5.1 Presentation of Casablanca Smart City Project

In 2018, Casablanca city in Morocco launched her own Smart City strategy called "the Master Plan for Digital Transformation" containing 70 short and long-term projects approved by the city council in February 2018.

The projects planned for the 2018–2022 period aim to:

- strengthen the links and synergies between the city and its stakeholders, in particular citizens, businesses, public services and associations,
- Establish a favorable and competitive business environment,
- Improve the mode and quality of services for the benefit of Casablanca residents.

This portfolio of projects is divided into four main categories: information systems projects, smart city services; equipment and infrastructure projects, and support projects.

The city has planned to set up a set of smart city services for the benefit of citizens; companies; tourists; administrations and associations.

These services cover several components, including:

- Smart governance: by facilitating citizens' access to city services by their dematerialization. In this sense, an IT solution will be installed to allow citizens to present their requests, monitor them from end to end, and ultimately receive the requested documents in electronic format. For the municipality employees, this new service will allow them to process citizens' requests quickly and under better conditions. The city has also set up the Rokhas.ma platform for the dematerialization and online management of economic and commercial authorizations. This has improved the quality of service provided to citizens and made it easier for civil servants. The platform also ensured transparency by displaying statistics of authorization requests in real-time. The city's tax administration will also be affected by the services of the Smart City. Thus, a Central GIR-CT (Integrated Management of Territorial Collectivities) project was launched in July 2018 with Casablanca, a pioneer city. This project aims to implement an integrated tax information system and sign agreements with all stakeholders: Regional Treasury, Regional Tax Directorate, etc. Citizen participation is one of the main objectives of Smart Governance in the city. Indeed, the information system will make it possible to receive and process citizens' opinions and complaints. Multichannel communication with citizens will be available on the new city portal www.casablancacity.ma as an e-service and the portal's mobile application and the Casa Mdinti application for cleanliness. This project will upgrade administrative processes to establish a "customer" culture.
- Smart people: To develop creativity and participation in public life among citizens, the city has planned to set up a collaborative platform for mobile applications called "Casa Store" to enable citizens and, in particular, IT (individuals or companies) to put their developed applications online

- Smart Living: To improve the quality of life in Casablanca, it is planned to set up services of general interest for the benefit of citizens such as cultural spaces, tourist attractions, and the strengthening of inclusion of digital technology and public safety while setting up new workspaces.
- Smart Mobility: To put an end to the shortcomings of mobility in the city, Casa Smart City project aspires to make the transport system smarter, strengthen inter-modality, organize parking and establish alternative mobility.
- Smart Environment: Casa Smart City project gave great importance to respecting the environment in the city. As a result, an IOT sensor system will be set up for various applications such as waste management, air pollution monitoring, noise mapping, public lighting and water and electricity networks.

5.2 Presentation of Casablanca Smart City Stakeholders' Project

Casa Numeric (the Smart City department in Casablanca) has listed the various contributing stakeholders of the project, namely:

- The municipality: through its council and its municipal administration, the municipality plays a crucial role in the planning and implementing SDTN[1]. Indeed, it participated in the development of the city's overall information system strategy. The municipal council also defined the smart projects to be carried out with approximate budgets and a system of governance while designating the teams who will take care of them. Thus, a Business Unit called Casa Numeric was created as a subsidiary of the local development company "Casa Prestations" to pilot and manage SDTN projects.
- Local development companies: Moroccan law gives local authorities the right to create local development companies called "SDL for Société de Développement Local" in the form of public limited companies to carry out municipal projects and manage other economic activities in the areas of competence of the municipality. The city of Casablanca has six companies that manage various sectors of activity. Some of them dealt only with the implementation of the parts of the project falling within their competence. In contrast, others participated in the planning and implementing the whole project, particularly the company Casa Prestations that carried out studies to determine a business model with the annual amounts of the investment and operating budgets.
- Delegates: Casablanca has several delegated companies that manage the city's services (water, electricity, transport, and cleaning). This is within the framework of the delegated management contracts signed between them and the municipality. Each delegate is responsible for a part of the city's Smart City project by digitizing their services
- City administrations such as the Wilaya, prefectures, regional council, national security and ministries.

[1] The smart city project's strategy of Casablanca city.

- Higher education establishments such as universities, which participate in the development and experimentation of smart services through scientific research and the organization of hackathons.
- Telecom operators: The city council has entrusted the installation of connection networks in the city to one of the telecom operators to set up Smart City services.
- The economical fabric: the private sector also participates in the city's Smart City projects within the framework of public-private partnerships or calls for tenders by project.
- Clusters: Maroc Numeric and eMadina are clusters that take initiatives to digitize the city
- The citizens: All of the city's Smart City projects in Casablanca have been designed for the citizen. The project aims to develop citizen participation and engagement in the city and offer a better quality of life.

5.3 Classification of Casablanca Smart City Stakeholders' Project

To classify these stakeholders and analyze their roles (Fig. 1), we will follow the method proposed by [24] while considering the description of stakeholders' roles in the implementation of Casa Smart City project presented above.

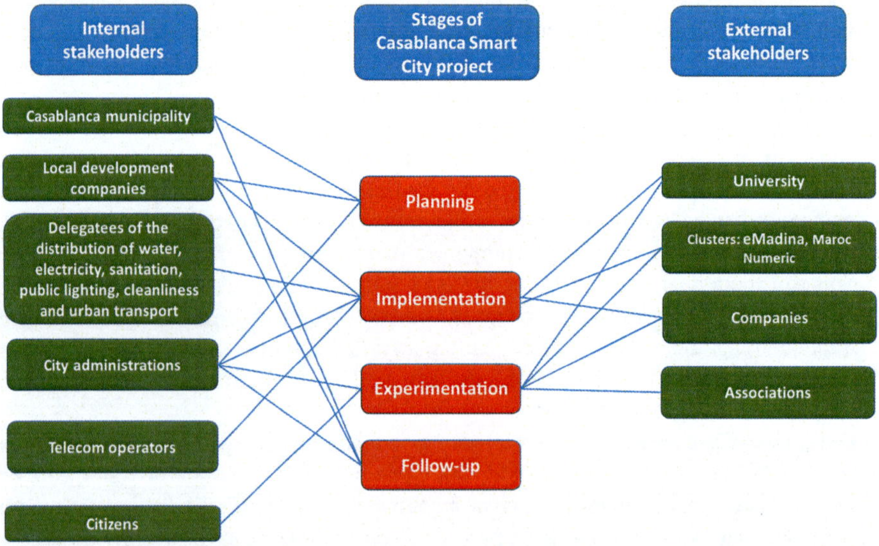

Fig. 1. Casa Smart City project stakeholders' classification model

The city council, the local development societies (SDLs) and the administrations were behind the planning of Casablanca Smart City project. The SDLs are also in charge of implementing the project, notably through Casa Prestations and its business unit Casa Numeric (the Business Unit in charge of the Smart City in Casablanca) and the delegates and telecom operators. The developed Smart City services are experienced mainly by citizens, companies, associations, academia and administrations. Although, the role of the university and the citizens is considered essential in the four stages of development of any Smart City project, these were not solicited during the planning of the project. The clusters, for their part, have only contributed to the implementation and experimentation of the Smart City services, despite their initiatives concerning the project.

Even though the mobilization of the city's stakeholders and the creation of a collective synergy between them is among the objectives of Casa Smart City project, we noted that it was not completed. Moreover, the initiatives taken by external stakeholders remain isolated and the participation of citizens downstream of the project and not upstream.

6 Conclusion

To classify the stakeholders of Casablanca Smart City project, we presented first the Smart City concept and the importance of stakeholders' engagement. Then, we described the Smart City project of Casablanca in Morocco and classified its stakeholders to internal and external ones.

We noticed that stakeholders' engagement in Casa Smart City project had not achieved the maturity stage yet. Since all stakeholders' initiatives are isolated, the city still needs a holistic and integrated collaboration with all stakeholders.

Another important key finding is that citizens and universities were not engaged in the planning of the project.

To address this issue, many researchers supported the concept of Living Lab because it involves university, industry, government, and citizens. A living lab is defined as "a user-centric innovation milieu built on everyday practice and research, with an approach that facilitates user influence in open and distributed innovation processes engaging all relevant partners in real-life contexts, aiming to create sustainable values." [26] Living labs could be an interesting approach because they harmonize the innovation process among the main stakeholders [27].

We acknowledge that our study has some limits because we classified stakeholders based only on information that we retrieved from the internal documentation that Casa Numeric (the Smart City department in Casablanca) and the municipality of Casablanca shared with us. We would have liked to interview all stakeholders for qualitative research to get their feedback about their engagement in Casa Smart City project. However, the lack of time and resources prevented us from doing it. All the same, our research can give researchers interesting insights to develop other classification models for smart cities stakeholders.

Acknowledgments. We would like to express our special thanks to Mr. Najib Ouradi (the director of Casa Numeric) and Ms. Hakima Fasly (the vice-mayor of Casablanca city) for their cooperation.

References

1. Urbanization. https://www.unfpa.org/urbanization. Accessed 30 June 2021
2. Mori, K., Christodoulou, A.: Review of sustainability indices and indicators: towards a new City Sustainability Index (CSI). Environ. Impact Assess. Rev. **32**(1), 94–106 (2012)
3. United Nations. https://www.un.org/development/desa/publications/graphic/world%20urbanization-prospects-2018-more-megacities-in-the-future. Accessed 30 June 2021
4. Cossetta, A., Palumbo, M.: The co-production of social innovation: the case of living lab. In: Dameri, R., Rosenthal-Sabroux, C. (eds.) Smart City. Progress in IS, pp. 221–235. Springer, Cham (2014). https://doi.org/10.1007/978-3-319-06160-3_11
5. Tadili, J., Fasly, H.: Citizen participation in smart cities: a survey. In: Proceedings of the 4th International Conference on Smart City Applications, pp. 1–6 (2019)
6. Lombardi, P., et al.: An advanced triple-helix network model for smart cities performance. In: Regional Development: Concepts, Methodologies, Tools, and Applications, pp. 1548–1562. IGI Global (2012)
7. Tadili, J., Fasly, H.: General smart city experts' perceptions of citizen participation: a questionnaire survey. In: Ben Ahmed, M., Boudhir, A.A., Santos, D., ElAroussi, M., Karas, İR. (eds.) SCA 2019. LNITI, pp. 3–15. Springer, Cham (2020). https://doi.org/10.1007/978-3-030-37629-1_1
8. Hollands, R.G.: Will the real Smart City please stand up? Intelligent, progressive or entrepreneurial? City **12**(3), 303–320 (2008)
9. Nam, T., Pardo, T.A.: Conceptualizing Smart City with dimensions of technology, people, and institutions. In: Proceedings of the 12th Annual International Digital Government Research Conference on Digital Government Innovation in Challenging Times - dg.o 2011, College Park, Maryland, p. 282 (2011). https://doi.org/10.1145/2037556.2037602
10. Vanolo, A.: Smartmentality: the smart city as disciplinary strategy. Urban Stud. **51**(5), 883–898 (2014). https://doi.org/10.1177/0042098013494427
11. Angelidou, M.: Smart cities: a conjuncture of four forces. Cities **47**, 95–106 (2015). https://doi.org/10.1016/j.cities.2015.05.004
12. A strategic view on Smart City technology: the case of IBM Smarter Cities during a recession. ScienceDirect. https://www.sciencedirect.com/science/article/pii/S0040162513002266. Accessed 11 Mar 2019
13. Giffinger, R., Fertner, C., Kramar, H., Meijers, E.: City-ranking of European medium-sized cities. Centre of Regional Science, Vienna University of Technology, pp. 1–12 (2007)
14. Hall, R.E., Bowerman, B., Braverman, J., Taylor, J., Todosow, H., Von Wimmersperg, U.: The vision of a Smart City, Brookhaven National Lab, Upton, NY, USA, BNL-67902; 04042, September 2000. https://www.osti.gov/biblio/773961. Accessed 11 Mar 2019
15. Giffinger, R., Gudrun, H.: Smart cities ranking: an effective instrument for the positioning of the cities? ACE Archit. City Environ. **4**(12), 7–26 (2010)
16. Freeman, R.E.: Strategic Management: A Stakeholder Approach. Cambridge University Press, Cambridge (2010)
17. Carroll, A.B., Näsi, J.: Understanding stakeholder thinking: themes from a Finnish conference. Bus. Ethics Eur. Rev. **6**(1), 46–51 (1997)
18. Wong, K., Fearon, C., Philip, G.: Understanding egovernment and egovernance: stakeholders, partnerships and CSR. Int. J. Qual. Reliab. Manag. **24**, 927–943 (2007)
19. Pardo, T.A., Scholl, H.J., Cook, M.E., Connelly, D.R., Dawes, S.S.: New York state central accounting system stakeholder needs analysis. Center for Technology in Government, Albany (2000)

20. Pardo, T.A., Scholl, H.J.J.: Walking atop the cliffs: avoiding failure and reducing risk in large scale e-government projects. In: Proceedings of the 35th Annual Hawaii International Conference on System Sciences, pp. 1656–1665 (2002)
21. Tan, C.-W., Pan, S.L., Lim, E.T.: Managing stakeholder interests in e-government implementation: lessons learned from a Singapore e-government project. J. Global Inf. Manag. (JGIM) **13**(1), 31–53 (2005)
22. Mayangsari, L., Novani, S.: Multi-stakeholder co-creation analysis in Smart City management: an experience from Bandung, Indonesia. Procedia Manuf. **4**, 315–321 (2015)
23. Rajablu, M., Marth, G., Fadzilah, W., Yusoff, W.: Managing for stakeholders: the role of stakeholder-based management in project success. Asian Soc. Sci. **11**, 111 (2015)
24. Jayasena, N.S., Mallawaarachchi, H., Waidyasekara, K.: Stakeholder analysis for Smart City development project: an extensive literature review. In: MATEC Web of Conferences, vol. 266, p. 06012 (2019)
25. Van der Hoogen, A., Scholtz, B., Calitz, A.: A Smart City stakeholder classification model. In: 2019 Conference on Information Communications Technology and Society (ICTAS), p. 1–6 (2019)
26. Bergvall-Kåreborn, B., Eriksson, C.I., Ståhlbröst, A., Svensson, J.: A milieu for innovation: defining living labs (2009)
27. Ståahlbröst, A., Holst, M.: The Living Lab Methodology Handbook. Luleå University of Technology and CDT, US Food and Drug Administration (2019)

Emerging Learning Environments and Technologies Post Covid-19 Pandemic: What's Next?

Malissa Maria Mahmud[1](✉), Shiau Foong Wong[1], and Othman Ismail[2]

[1] Sunway University, Selangor, Malaysia
{malissam, janicew}@sunway.edu.my
[2] Universiti Teknologi Mara (UiTM), Selangor, Malaysia
othmani@uitm.edu.my

Abstract. The unprecedented COVID-19 pandemic rapidly became a matter of grave concern globally, and positioned numerous industries, sectors and agencies worldwide in a perilous circumstance. Technological advances are racing to stay ahead of the aftermath and adapt to the unparalleled needs, unfolding the emerging and disruptive technologies. In response to the existential threat, educational institutions had to transition and resort to digitized learning to ensure teaching and learning continue. While the connection between digital technologies and learning has been widely recognized in the previous decades, it has yet to be thoroughly investigated with the focus of the post-COVID-19 pandemic. Thus, this study reviews the transformation of digital and technologies disrupted by the COVID-19 crisis in the teaching and learning context, specifically the key areas of educational technologies, such as the widespread availability of open online courses, social media utilization, innovative learning technology, flexible learning, and mobile education. As concluding remarks, the most critical findings and possible future directions are discussed.

Keywords: COVID-19 pandemic emerging technologies · Online teaching and learning

1 Introduction

The inclusion of a modern learning ecosystem over the past few decades has been facilitated by the digital transformation of education systems at all levels. The recent COVID-19 pandemic forced the closing of educational institutions worldwide, causing over 1.4 billion students and 60 million teachers to abruptly change their face-to-face learning habits (Batara 2021). Educational technologies have advanced significantly over the last decade and it is estimated that the global e-learning market will exceed the $300 billion mark in the next few years (Agrawal 2020; Mahmud 2018). Nonetheless, the employment of digitized educational systems remains widely debated among stakeholders, deliberating on the prevalent benefits and drawbacks. According to World Bank statistics, 84% of European Union Member States had Internet access in 2018, compared to 60% in Latin America and 20% in the least developed countries (Amina

Y. Maleh et al. (Eds.): ICI2C 2021, LNNS 357, pp. 308–319, 2022.
https://doi.org/10.1007/978-3-030-91738-8_29

2018). Bates (2020) asserted that the educational framework as an equity, emphasizing for a more universal and low-cost internet access for everyone. In this context, an academic system lacking equitable practices would contribute to disproportioned outcomes socially and economically. Policy analysis for California Education reported that the situation is heightened due to the pandemic, and has affected students' learning, leading to the growing of equity gaps (Hough 2021). Therefore, the importance of emerging learning environments and technologies in the aftermath of COVID-19 is expounded in this study, particularly the numerous studies that have acknowledged issues related to the technological interventions and simultaneously offer possible solutions. Considering the intensifying amount of research articles published retorting to the COVID-19 disruption, we performed a descriptive systematic review of the recent scientific literature to provide an in-depth understanding of quantitative and qualitative studies of various research designs in order to better understand the advancing learning environments, innovations and technologies developed and applied to mitigate the repercussions of COVID-19 pandemic on teaching and learning. The study also consolidates the capacities, shortcomings, opportunities associated with emerging learning environments and technologies.

2 Methodology

Systematic reviews are not the same as literature reviews. They employ a specific method for locating, selecting, interpreting, evaluating, synthesizing, and publishing data that aids in reaching a conclusion (Denyer and Tranfield 2009). Designing a systematic review study necessitates searching of all qualifying papers, scanning of abstracts and full-text, coding appropriate representative of result size heterogeneity, and finally computing and publishing the yielded effect sizes (Pigott and Polanin 2020). Secondary data is examined using periodic evaluations, which retrieve, make, and analyze established information about a topic analytically. This is imperative to avoid the risks of bias (Borrego et al. 2015). Furthermore, systematic reviews are hampered by skepticism about their inherent validity. Therefore, researchers who wish to delve into systematic review should ensure the providence of broad participation and elimination criteria, including unpublished studies alongside published studies, utilizing a structured protocol to safeguard the consistency of coding process. This study adapted the five steps systematic review guidelines and processes adapted from Khan et al. (2003); therefore, the yielded findings are generalizable, and biases are avoided. The first step is to frame the postulated research questions and areas for the purpose of inquiry, within the frame of open online courses, social media utilization, innovative learning technology, flexible learning, and mobile education. Specifically, the primary focus areas of emerging learning environments and technologies were included based on a study conducted by Martin et al. (2020): (1) social media, (2) massive open online courses, (3) special education technology, (4) mobile learning, (5) game-based learning and gamification, (6) adaptive learning, and (7) learning analytics.

The second step is to identify relevant work and literature wherein the related sources within the established scope is discerned. To specify the scope, sources from reputable databases are scrutinized employing the following keywords: *emerging technologies, online teaching and learning, educational technologies, technological intervention, technology, computer, web-based instruction, online, Internet, blended learning, hybrid course, simulation, electronic and multimedia.* Clarivate analytics from reputable databases were used to select academic papers, ensuring that high-quality and high-impact sources were included. These databases are widely recognized as one of the most trustworthy citation index sites for evidence-based, high-quality scientific information. Specifically, the study prioritized high index journal articles from these databases included ResearchGate, a top online resource that allows researchers to obtain access to updated information across professional fields. Google Trends is another academic search engine, which allowed the researchers to identify recent correlations with the current occurrences in emerging learning environments, and technologies post COVID-19. Also, Google Scholar was utilized as it permits the search of scholarly materials related to the research topic. It is one of the most accessible academic databases where numerous materials about a particular topic can be attained. Another database, Science Direct was also utilized, in which it extensively provides a broad scope of scientific information. Finally, the Educational Resource Information Center (ERIC) distributed a broader spectrum of the educational literature for the study. All these databases provided the researchers with a comprehensive approach of garnering enough evidence about emerging learning environments and technologies employed since the breakout of COVID-19.

The third step is to assess the *quality* of the obtained studies. Quality in this context is the established criteria that must be met before the inclusion of the articles as part of the samples. The samples comprise of peer-reviewed published articles, including theses within the operationalized definition, "these technologies are referred to as emerging technologies, a term that indicates that their status and use in educational contexts is still fairly fluid" (Veletsianos 2010, p. 1615). Table 1 below is used to organize and synthesize the samples based on the relevance of apparent criteria as well as findings.

Table 1. Synthesis of findings table

Synthesis Table						
Source	[Finding 1]	[Finding 2]	[Finding 3]	[Finding 4]	[Finding 5]	[Finding 6]
[Author(s) & date of item]						

The scope for the samples' time frame is from March 2020 (post Covid-19) to the present. The fourth step is to summarize included articles to help retort the research question and the fifth step is to make sense of the findings. In this final step, the findings from the review would be further deduce to conclude the transformed educational landscape due to the emerging technologies disrupted by the COVID-19 pandemic.

3 Findings

The use of learning technologies such as extensive open online courses, social media, computer-generated and virtual reality, interactive technologies, accessible educational resources, online synchronous networks, wearable technologies, and mobile technologies has increased over the last few decades study (Bonk 2016). These technologies are denoted as new knowledge due to their roles and applications in educational settings are constantly altering the archetype (Suh and Prophet 2018). The following are descriptive discussions of the new learning environments and developments post COVID-19 attained from the analyses.

3.1 Massive Open Online Courses (MOOCs)

The crippling of educational institutions caused by the COVID-19 crisis has sparked a surge of interest in the online learning. Coursera, edX, and FutureLearn, the top three MOOC providers, each recorded many new users in 2019 (Kaempf and Finn 2021). Following the pandemic, about 30% increase of new users registered on these sites, and Coursera attracted the most students with 40 million new enrollees. Coursera has managed to retain traction through programs such as granting free certificates for more courses and providing free access for students and unemployed people worldwide (Raja and Kallarakal 2020). Udemy, one of the well-known MOOCs, also saw an upright trend attributed to the pandemic. The inclinations towards MOOCs are expected, given that various online courses offered in the preceding years leading up to the pandemic (Zhu et al. 2020). It is claimed that MOOCs would be able to cushion the future educational challenges and it is regarded as one of the feasible solutions recommended (Purkayastha and Sinha 2021; Raja and Kallarakal 2020). The value of MOOCs is enhanced with the COVID-19 crisis, in which it served as a catalyst to offer solution to the disparity in education. Kang (2021) said that MOOCs are able to tackle the social demands which are in-line with the constant development of technologies. A study was conducted by Mahmud et al. (2020) [discovered that there is a significant correlation of the deployed MOOCs and students' performance, and as shown in Table 2, from the eight yielded combined effect sizes, majority formed effective intervention (2020). Table 2 also illustrates the viability of MOOCs in the juxtaposition of medium, large and very large effect sizes yielded from the assessments of various MOOCs platforms (Mahmud et al. 2020). Thus, to stay relevant and competitive, it is essential for educational institutions to explore the idea of offering attractive MOOC content that could be leveraged by students with unique pre-requisites.

Table 2. An enquiry of variables and effect sizes

Combined effect size							
0.187 (Small)	0.556 (Medium)	0.0045 (Small)	0.0096 (Small)	0.8201 (Large)	2.0279 (Very Large)	0.0055 (Small)	0.0709 (Small)

3.2 Social Media

Despite the pushed transition that educators and learners had to face, online learning has developed a panacea for the disruption created by the COVID-19. Nonetheless, switching from the conventional face-to-face to online learning can be a challenging environment for both learners and educators (Mahmud and Wong 2018). In this context, quality education across different online modalities has instigated a transformation of paradigm for teaching and learning (Eutsler et al. 2020). Due to its ubiquity, social media platforms are often used among teachers and students, in particular, Facebook and WhatsApp. WhatsApp group has become the most common means to stay in contact between faculty members and the learning institutions (Al Fadda and Osman 2020). In fact, social media is one of the preferred means for a structured online learning (Mahmud et al. 2018). Since online meeting sites such as Google Classroom and Zoom were initially uncommon to students, most faculty members resorted to Facebook or WhatsApp as their primary method to ensure learning process continues. Facebook (70%) was the most used medium for structured academic communication among students, followed by WhatsApp (60%), YouTube (60%), Wikis (50%), and other social media (20%), such as blogs and Twitter. Affected countries and populations have been impelled to explore fast solutions due to significant and unforeseen closures (Teräs et al. 2020). Social media platforms are used as an avenue for discussions to improve comprehension and engagement (Doucet et al. 2020; Mahmud et al. 2016). Thus, the systematic review confirms that the use of social media amid learning institutions and learners encourages online communication and interactivity, fosters social interaction online and provides the right learning atmosphere (Lai 2020). For numerous organizations and universities, social media has been a critical networking mechanism for disseminating essential knowledge to the public, and now an operable platform for teaching and learning.

3.3 Adaptive Learning

Adaptive learning refers to the delivery of teaching content that changes over time. The changes are contingent to the students' understanding of the subject based on responses and assessments through the graphic display of information (Vanbecelaere et al. 2020). In effect, it offers an educational environment that takes specific differences into account, which can be beneficial during the COVID-19 crisis. As a consequence, it is linked to existing personalized learning patterns. The emergency shift from conventional classes prompted by COVID-19 has resulted in a drop-in instructor-student speaking hour (Ferdig et al. 2020). An absence of communication hours could possibly be reimbursed with adaptive technology. Learners engage with an adaptive technology that can personalize instruction. At this juncture, technology becomes more easily integrated with what teachers are attempting to achieve (Amunga 2021). Socratic pedagogical approaches and better collaborative peer-to-peer learning opportunities will be enabled by adaptive learning technology, particularly in a broader range of situations. For example, Michael McDonald, an Italian educator, uses augmented reality to help people understand or teach English as a foreign language (Ferdig et al. 2020). In a similar vein, adaptive technology provides a forum for teachers to share interactive media, immersive classrooms, and targeted learning tutorials, as well as an educational and staff management system (Teräs et al. 2020). Adaptive technology software uses short videos to teach math and science to primary and secondary institute children to customize learning journeys and teaching. Therefore, technology-based adaptive learning systems or e-learning systems can offer customized and immediate assistance, specific learning resources to cater to the learning outcomes.

3.4 Mobile Technology

As the higher education institutions were shut for an indefinite duration, numerous means were initiated to ensure assigned syllabuses are completed in the time frame (Al Fadda and Osman 2020; Amunga 2021). This has sparked new examples of educational creativity, including technical interventions such as the use of mobile phones. With the slow rate of change of within the public institutions, which tends to use millennia-old lecture-based educational methods, entrenched with outdated classroom pedagogy, the pandemic has created an opportunity. Eutsler et al. (2020) examined 12 years of research using portable devices to help early learners develop their reading skills. Through various electronic learning implementation in the multiple studies, they discovered that tablets were the most widely used platforms for young students. Furthermore, their results revealed a range of efficacy outcomes, especially the reading comprehension. Mobile learning is still an emerging technology with a wide range of applications across cultures, disciplines, and frameworks. Even though the discovery ground is expanding, the infrastructure that supports mobile learning is rapidly changing. In the same way, the field understanding of where and how mobile learning systems can be planned is still being widely researched (Lai 2020). University students mentioned that with the help of mobile learning, they have improved their study skills and interaction (Doucet et al. 2020; Sandars et al. 2020).

3.5 Gamification and Learning

It is not a novel concept to use games for educational purposes. They are widely used to assist students in learning and developing different skills, particularly in the early stages of learning process (Wati 2020). Gameplay has found its way into the worlds of industry, performance, and education, where it has proven to be an essential teaching and motivating tool. Gamification has many educational advantages, including inspiring and engaging students, role-playing, finding answers, teamwork, commitment, and strategic thinking. There are various opportunities for organizing simulation games in groups: the game is easier to manage and, above all, the number of positive interactions between students is more remarkable. From the gamification concept, rivalry has a substantial positive effect on education (Zainuddin et al. 2020). Game-based learning makes teaching and learning more engaging and enjoyable. Consider using a mind map or brainstorming exercise to break up the teaching and force learners to think creatively and communicate with the content. Gamified instruction provides students in academic institutions with an enjoyable and healthy environment to try, struggle, learn and excel while challenging them to experience and apply what they have learned in class.

3.6 Learning Analytics

More than 1 billion learners are currently impacted by the pandemic in 186 countries due to school closures (Matcha et al. 2020). This resulted in a radical transition in education mediums through the phenomenal increase in e-learning, which leverages on interactive channels remotely. The method of gathering, analyzing and documenting data on a technological integration is known as learning analytics. Even though recording learner data is not a new phenomenon, learning analytics has only recently gained widespread adoption among data scientists and educators (Manhanga 2020). Ifenthaler and Yau (2020) discovered that a group of learners' circumstantial awareness and behavioral information from emerging media can be leveraged as a collective assessment data, and facts gleaned from feedback could be to improve future implementations. Matcha et al. (2020) used learning analytics visualizations to personalize learning experiences. Their main aim was to look at how learning analytics visualizations could facilitate self-regulated learning, the kind of data that could be used as a response, and the accuracy of learning projects that reported experiential appraisals. As learning analytics stages turn out to be more nuanced and competitive in the future, learning analytics will significantly impact the future and improvement of K-12 and higher learning, particularly in adapted and online education systems.

3.7 Special Education Technology

To mitigate learning delays due to COVID-19 pandemic, national governments in many parts of the world implemented containment and prevention policies, forcing educational organizations at all levels to transition to emergency remote instruction. The teaching staff had to grow accustomed and move to online instruction methods, digitizing their teaching and appraisal methods. Despite a steep learning curve, educators and trainers around the world responded with grit and ingenuity, devising creative technology-based strategies to help students transition from traditional to online classrooms quickly and efficiently, as well as to promote remote student engagement and learning. Education leaders in Cambodia, for example, devised a plan that integrates SMS, written handouts, and ongoing instructor input, taking advantage of the country's strong cell phone participation. It also guides how to navigate learning services, guarantees that students have access to paper-based learning resources, and includes home visits to track distance learning experiences. Teachers must also provide students with weekly publication materials and meet with them weekly to distribute their identified assignments and issue new ones. In specific educational settings, as in most learning contexts, more common technologies are often recast. This repurposing has become more commonplace as a result of the widespread adoption and societal acceptance of smartphones. Recent research into the use of new technologies for children with special needs looks at how social-collaborative approaches can be aided and the use of robots that facilitate a wide range of play can be developed (Baykal et al. 2020). Given society's recent focus on artificial intelligence (AI), this inquiry into robotics' implementation is unsurprising (Wang and Wang 2021). While artificial intelligence has been used to promote learning for decades, its use in classrooms must be limited. It is less well-known for its use in special education. On the other hand, the types of AI-based approaches used to support students in diverse contexts have been consistent with their ability to aid students suffering disabilities in various ways.

3.8 Cloud Computing

Educational institutions, in general, and higher education institutions, in particular, should seriously consider implementing cloud-computing technologies in order to provide better online learning experience. Cloud computing is a form of computing that uses conventional brick and mortar-based servers; adoption has turned global business information infrastructure into a modern model of online data collection, maintenance, and processing. This new approach allows users to access vast processing resources through the internet easily. In addition, cloud infrastructure can provide IT services in a cost-effective, flexible, readily available, efficient, stable, and seamless way, resulting in increased efficiency and cooperation in the mobile and digital worlds (Liu et al. 2020). Businesses are embracing cloud computing in a variety of industries due to its distinct benefits. For educational institutes, cloud computing can provide a wide range of resources through curriculum delivery and science. By enabling the necessary storage capabilities, applications, and technology, cloud-based services will enhance the quality of learning while also contributing to the long-term stability of education.

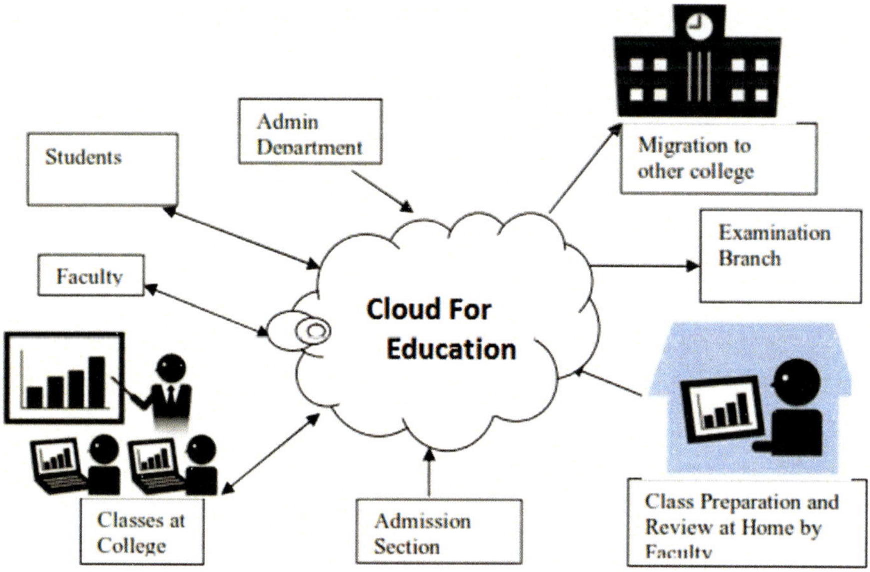

Fig. 1. Cloud computing system for education

Cloud computing adoption in the education sector can help to improve organizational efficiency, stimulate creativity, save money on technology investments, and increase agility. The use of cloud computing is also expected to reduce educational institutes' operating costs. In developing countries, cloud computing offerings can provide a solution to advanced IT resource scarcity and the lack of access to cutting-edge technology. Furthermore, cloud computing offers students and educators 24/7 access to their equipment, mobile devices, and university servers, allowing them to communicate, edit, archive, distribute, and process large volumes of data (Al-Malah et al. 2021). Cloud computing is a crucial feature for e-learning and remote operation. It is also anticipated that the majority of factors of cloud adoption will also apply in various higher education institutions as e-learning is adopted.

4 Discussion and Conclusion

We summarize the study reviews based on the seven critical focus areas of emerging learning environments and technologies which were included based on a study conducted by Martin et al. (2020): (1) social media, (2) massive open online courses, (3) special education technology, (4) mobile learning, (5) game-based learning and gamification, (6) adaptive learning, and (7) learning analytics, and we added another area, which is (8) cloud computing (see Fig. 1) (Fig. 2):

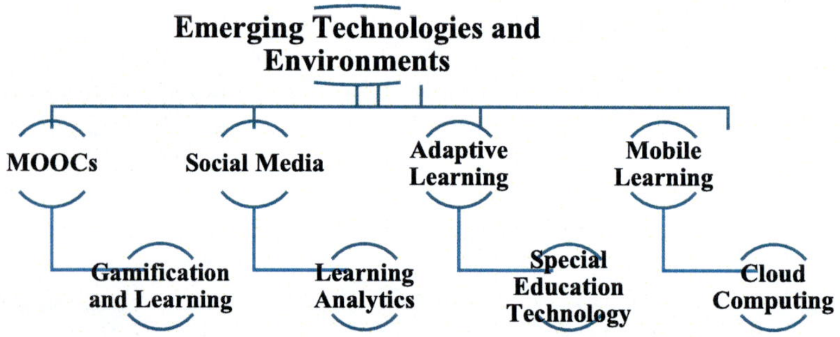

Fig. 2. Emerging technologies and environments post COVID-19

The study gathered systematic reviews on a single subject so that the studies could show the breadth and complexity of research on emerging learning environments and technologies in the aftermath of the COVID-19 crisis. Such research evaluations have become increasingly important as technological innovation waves have occurred over the last few decades. The COVID-19 pandemic, according to the World Economic Forum, has also impacted how many people access and impart education. We need to make some much-needed improvements and innovations to find new solutions to our problems. Educators are accustomed to traditional teaching methods such as face-to-face seminars and are unable to accept any changes. We have no choice but to adapt and welcome the complex situation's adaptation amid this crisis. It will benefit the education sector while also displaying a remarkable amount of innovation. It is impossible to ignore students who do not have access to all online resources.

Thus, emergent learning technology, technically speaking, is not a level playing field; it represents the realities of culture and life, especially the COVID-19 crisis. The included articles offer essential insights into what we know about current research in these fields and policy suggestions and other practical outcomes. This dreadful turn of events has taught us that nothing is assured, and we must be prepared to face challenges. Although the COVID-19 crisis did not teach us how to prepare, we should learn that preparing is the most important thing. We should prepare everything, and Plan B should be ready if Plan A fails. Prioritization of both the critical and complicated conditions that must occur and be scheduled accordingly. The pandemic also demonstrated that students possess specialized skills, such as problem-solving abilities, strategic thinking, and, most importantly, crisis adaptability. Educational institutions must cultivate resilience to ensure and emphasize the incorporation of these skills in their student programs.

References

Al Fadda, H.A., Osman, R.A.: Challenges and strengths of transitioning to online learning during COVID19 university lockdown: case studies from Egypt and Saudi Arabia. CDELT Occas. Pap. Dev. Engl. Educ. **71**(1), 191–212 (2020)

Al-Malah, D.K.A.R., Aljazaery, I.A., Alrikabi, H.T.S., Mutar, H.A.: Cloud computing and its impact on online education. In: IOP Conference Series: Materials Science and Engineering, vol. 1094, no. 1, p. 012024. IOP Publishing, February 2021

Amunga, J.: Leveraging technology to enhance STEM education amidst the COVID-19 pandemic: an overview of pertinent concerns. Technium Soc. Sci. J. **18**, 40–55 (2021)

Baykal, G.E., Van Mechelen, M., Eriksson, E.: Collaborative technologies for children with special needs: a systematic literature review. In: Proceedings of the 2020 CHI Conference on Human Factors in Computing Systems, April 2020. https://doi.org/10.1145/3313831.3376291

Bonk, C.: Keynote: what is the state of e-learning? Reflections on 30 ways learning is changing. J. Open Flex. Dist. Learn. **20**(2), 6–20 (2016)

Borrego, M., Foster, M.J., Froyd, J.E.: What is the state of the art of systematic review in engineering education? J. Eng. Educ. **104**(2), 212–242 (2015)

Doucet, A., Netolicky, D., Timmers, K., Tuscano, F.J.: Thinking about pedagogy in an unfolding pandemic: an independent report on approaches to distance learning during COVID19 school closures. Educ. Int. (2020). https://doi.org/10.1596/33499

Eutsler, L., Mitchell, C., Stamm, B., Kogut, A.: The influence of mobile technologies on preschool and elementary children's literacy achievement: a systematic review spanning 2007–2019. Educ. Tech. Res. Dev. **68**(4), 1739–1768 (2020). https://doi.org/10.1007/s11423-020-09786-1

Ferdig, R.E., Baumgartner, E., Hartshorne, R., Kaplan-Rakowski, R., Mouza, C. (eds.): Teaching, technology, and teacher education during the COVID-19 pandemic: stories from the field. Association for the Advancement of Computing in Education, Waynesville (2020). https://www.learntechlib.org/p/216903/

Hough, H.J.: COVID-19, the educational equity crisis, and the opportunity ahead. Brown Center Chalkboard, 29 April 2021. https://www.brookings.edu/blog/brown-center-chalkboard/2021/04/29/covid-19-the-educational-equity-crisis-and-the-opportunity-ahead/

Ifenthaler, D., Yau, J.-K.: Utilising learning analytics to support study success in higher education: a systematic review. Educ. Tech. Res. Dev. **68**(4), 1961–1990 (2020). https://doi.org/10.1007/s11423-020-09788-z

Kaempf, S., Finn, C.: Teaching international relations through the format of a Massive Open Online Course (MOOC). Int. Stud. Perspect. (2021). https://doi.org/10.1093/isp/ekz025

Kang, B.: How the COVID-19 pandemic is reshaping the education service. Future Serv. Post-COVID-19 Pandemic **1**, 15–35 (2021)

Khan, K.S., Kunz, R., Kleijnen, J., Antes, G.: Five steps to conducting a systematic review. J. r. Soc. Med. **96**(3), 118–121 (2003)

Lai, C.L.: Trends of mobile learning: a review of the top 100 highly cited papers. Br. J. Edu. Technol. **51**(3), 721–742 (2020)

Liu, S., Huang, G., Gui, J., Wang, T., Li, X.: Energy-aware MAC protocol for data differentiated services in sensor-cloud computing. J. Cloud Comput. **9**(1), 1–33 (2020). https://doi.org/10.1186/s13677-020-00196-5

Mahmud, M.M.: Technology and language–what works and what does not: a meta-analysis of blended learning research. J. Asia TEFL **15**(2), 365 (2018)

Mahmud, M.M., Ramachandiran, C.R., Ismail, O.: Social media and classroom engagement: students' perception. J. Media Crit. [JMC] **2**(8), 197–207 (2016)

Mahmud, M.M., Ramachandiran, C.R., Ismail, O.: Social media dependency: the implications of technological communication use among university students. In: Tang, S.F., Cheah, S.E. (eds.) Redesigning Learning for Greater Social Impact, pp. 71–87. Springer, Singapore (2018). https://doi.org/10.1007/978-981-10-4223-2_7

Mahmud, M.M., Wong, S.F., Bakar, M.S.A.: MOOCs: an enquiry of variables and effect sizes. In: Proceedings of the 2020 8th International Conference on Communications and Broadband Networking, pp. 83–88, April 2020

Mahmud, M.M., Wong, S.F.: Facebook and collaborative learning: an empirical study on online assessment. Int. J. Learn. Teach. **4**(2), 107–113 (2018)

Manhanga, L.: Data analytics in information systems (IS) auditing: an examination of the cost-effectiveness of the use of data analytics in information systems auditing. Doctoral dissertation, Capitol Technology University (2020)

Martin, F., Dennen, V.P., Bonk, C.J.: A synthesis of systematic review research on emerging learning environments and technologies. Educ. Tech. Res. Dev. **68**(4), 1613–1633 (2020). https://doi.org/10.1007/s11423-020-09812-2

Matcha, W., Gašević, D., Pardo, A.: A systematic review of empirical studies on learning analytics dashboards: a self-regulated learning perspective. IEEE Trans. Learn. Technol. **13** (2), 226–245 (2020)

Purkayastha, N., Sinha, M.K.: Unstoppable study with MOOCs during Covid 19 pandemic: a study. Libr. Philos. Pract. 1–12 (2021)

Raja, M.A.S., Kallarakal, T.K.: "COVID-19 and students perception about MOOCs" a case of Indian higher educational institutions. Interact. Technol. Smart Educ. (2020). https://doi.org/10.1108/ITSE-07-2020-0106

Sandars, J., et al.: Twelve tips for rapidly migrating to online learning during the COVID-19 pandemic. MedEdPublish 9 (2020). https://doi.org/10.15694/mep.2020.000082.1

Suh, A., Prophet, J.: The state of immersive technology research: a literature analysis. Comput. Hum. Behav. **86**, 77–90 (2018)

Teräs, M., Suoranta, J., Teräs, H., Curcher, M.: Post-Covid-19 education and education technology 'solutionism': a seller's market. Postdigit. Sci. Educ. **2**(3), 863–878 (2020)

Vanbecelaere, S., Van den Berghe, K., Cornillie, F., Sasanguie, D., Reynvoet, B., Depaepe, F.: The effectiveness of adaptive versus non-adaptive learning with digital educational games. J. Comput. Assist. Learn. **36**(4), 502–513 (2020)

Veletsianos, G.: Emerging Technologies in Distance Education. Athabasca University Press (2010). http://www.aupress.ca/index.php/books/120177. Accessed 24 Jan 2010

Wang, X.V., Wang, L.: A literature survey of the robotic technologies during the COVID-19 pandemic. J. Manuf. Syst. (2021). https://doi.org/10.1016/j.jmsy.2021.02.005

Wati, I.F.: Digital game-based learning as a solution to fun learning challenges during the Covid-19 pandemic. In: 1st International Conference on Information Technology and Education (ICITE 2020), pp. 202–210. Atlantis Press (2020). https://doi.org/10.2991/assehr.k.201214.237

Zainuddin, Z., Chu, S.K.W., Shujahat, M., Perera, C.J.: The impact of gamification on learning and instruction: a systematic review of empirical evidence. Educ. Res. Rev. **30**, 100326 (2020)

Zhu, M., Sari, A.R., Lee, M.M.: A comprehensive systematic review of MOOC research: research techniques, topics, and trends from 2009 to 2019. Educ. Tech. Res. Dev. **68**(4), 1685–1710 (2020). https://doi.org/10.1007/s11423-020-09798-x

What Drives Digital Library User's Satisfaction Behavior? Investigating the Level and Its Determinants

Mohamad Rahimi Mohamad Rosman[1(✉)], Mohammad Azhan Abdul Aziz[2], Noor Azreen Alimin[1], Nur Arrisya Azman[1], Faten Nazirah Ramli[1], and Fatihah Balqis Shuhaimi[1,2]

[1] Universiti Teknologi MARA Cawangan Kelantan, Machang, Malaysia
rahimimr@uitm.edu.my
[2] Universiti Teknologi MARA Cawangan Kedah, Sungai Petani, Malaysia

Abstract. Novel Coronavirus 2019 or COVID-19 has shifted the traditional face-to-face interaction into digital interaction, whereby people are meeting through online; education is no longer confined within the four walls of the classroom. Subsequently, this transformation also affected the service delivery provided by the library. Recent studies show that the usage of digital libraries increases in the aftermath of COVID-19. However, there is currently scarce literature on the factors that drive the increase of usage and its impact on users' satisfaction behavior. Therefore, the purpose of this study is twofold; first, to investigate the level of digital libraries satisfaction in the aftermath of COVID-19 and second, to examine the factors that influence user' satisfaction behavior on digital libraries. A quantitative research methodology was adopted; the instrument was developed by adopting previous research studies. The instrument was pre-tested, expert reviews, pilot-tested before actual data collection. Data was collected from one of the universities on the east coast of Malaysia. A total of 358 valid responses were received; responses were performed for data cleaning and coding. The findings were then analyzed using Statistical Package for Social Sciences (SPSS) version 26 and were analyzed based on descriptive and inferential analysis. The result shows that two hypotheses were accepted; digital resources quality and digital library affinity were found to have a significant and positive relationship with user's satisfaction behavior, while digital library generic usability did not have a substantial and positive relationship with the user's satisfaction behavior.

Keywords: Digital library · User satisfaction · Information system usage · Library resources · Digital library affinity

1 Introduction

Libraries are an important social institutions that assist people in obtaining various sources of knowledge and information. According to [1], Project Gutenberg, which Michael S. Hart initiated on 4 July 1971, started the invention of eBooks. Through the

Y. Maleh et al. (Eds.): ICI2C 2021, LNNS 357, pp. 320–330, 2022.
https://doi.org/10.1007/978-3-030-91738-8_30

rapid development of information technology, there has been an increase in training centers that built digital libraries to aid training and study. Moreover, digital libraries have evolved into a digital community for collaboration, e-learning, and user search, rather than just a set of information tools. Hence, the digital library must play a vital role in providing information to their users, this is crucial to ensuring that it is the consumer's first choice for finding information that meets their needs. Through the support of the university's digital library, there will be a significant effect on user satisfaction with the use of the digital library. According to previous research, overall user satisfaction has a significant impact on expanding the usage of digital libraries [2, 3]. Researchers hypothesized that the quality of digital services influences the digital library's overall usability and affinity and that these variables and their relationships contribute to user satisfaction with digital libraries. It is expected that users' satisfaction with digital libraries can lead to user loyalty, which involves recommending and reusing them and user growth [4–10].

Some several words or terms are usually used for defining the concept of digital libraries, including "electronic library," "virtual library," "modern library," and "electronic library" [7, 8]. Digital libraries have many advantages to their users because they are easily accessible and can be accessed at any time if there is an Internet connection. Digital libraries serve the same purpose, functions, and goals as a traditional library; the "digital" part of the term indicates that the material is stored digitally and accessed over the Internet [11]. As one of the learning tools, digital libraries are widely used, especially in university libraries. Moreover, the current study also confirmed that the aftermath of COVID-19 has impacted the level of usage towards upward trends [12–14].

As a result, the digital library is the best platform for students to find eBooks, articles, and other materials relevant to their assignments. Thus, the library needs to improve its services to enhance user satisfaction. According to recent research, the affinity of digital libraries plays an important role in the form of user satisfaction [15]. For example, if a digital library is of poor quality, users may believe that it cannot provide high-quality information and students may prefer to use the search engine. Thus, digital libraries must recognize their role in gaining user satisfaction to become the user's first preference. Therefore, the purpose of this research is twofold; first, to investigate the level of digital libraries satisfaction in the aftermath of COVID-19 and second, to examine the factors that influence user' satisfaction behavior on digital libraries.

2 Research Model

The following Fig. 1 shows the research model of the study. Three independent variables were selected: namely Digital Library Generic Usability (GEU), Digital Resources Quality (DRQ), and Digital Library Affinity (DLA). On the other hand, User Satisfaction Behavior (USB) was conceptualized as the study's dependent variable.

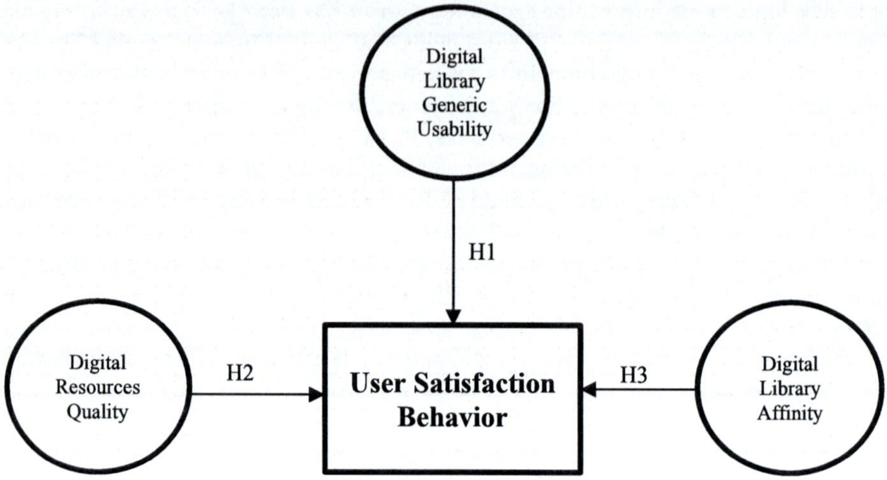

Fig. 1. Research model

Digital Library Generic Usability (GEU) is operationalized as the extent of digital library functions and features that encourage or discourage engagement with the digital library, subsequently influencing users' satisfaction behavior. Studies on human-computer interaction and information system satisfaction found out that the usability of information system influence user satisfaction and subsequent usage [15, 16]. Besides, the influence of usability is more demanded, especially in the pandemic era [17–19]. Therefore, it can be hypothesized that:

H1: Digital Library Generic Usability (GEU) has a significant and positive relationship with User Satisfaction Behavior (USB).

Digital Resources Quality (DRQ) is operationalized as the degree of digital resources quality and its capability to meet the need of the user. The quality of resources plays a vital role in determining and predicting user satisfaction behavior. As found out by previous researchers, many users, particularly students prefer to use non-DL resources due to the complexity and quality of interaction provided by those resources [14, 20–22]. On the other hand, relying on non-DL resources also negatively impacted the students – as the quality of resources provided by non-DL resources is sometimes doubtful, especially those on the list of predatory journals and publications. Therefore, it can be hypothesized that:

H2: Digital Resources Quality (DRQ) has a significant and positive relationship with User Satisfaction Behavior (USB).

On the other hand, Digital Library Affinity (DLA) is operationalized as the extent of trust and importance that users put towards using the digital library. Affinity towards the digital library will encourage or discourage engagement and usage of the digital library. Most often users particularly students are keener to use the web search engine that can be accessed anytime without restriction. On the other hand, access to digital library resources are usually restricted and user must log in before using the resources. This act can discourage engagement with the digital library resources. Thus, affinity with the digital library can encourage subsequent usage and improve user satisfaction towards the output provided by the digital library [15, 23]. Therefore, it can be hypothesized that:

H3: Digital Library Affinity (DLA) has a significant and positive relationship with User Satisfaction Behavior (USB).

3 Methodology

A quantitative research methodology was adopted for the purpose of the study. The method is deemed appropriate because: (1) in line with the previous similar studies, and (2) the study aims to establish a relationship among the variables. An instrument was developed by adapting to the previous studies of [4, 15], and [23]. The instrument was sent for expert review before an amendment was made based on the reviewers 'responses. Next, the instrument was sent for a pilot study. A total of 50 respondents was selected, in which 42 valid responses were received and indicating a response rate of 84%. Following the pilot data collection, the instrument was analyzed for reliability assessment. The following Table 1 shows the result of the reliability assessment using Cronbach's Alpha Coefficient. Reliability analysis shows that all constructs which are Digital Library Generic Usability (GEU), Digital Resources Quality (DRQ), Digital Library Affinity (DLA), and User Satisfaction Behavior (USB) exceed the minimum value of 0.7, therefore indicating a sufficient reliability of the instrument as suggested by [24].

Table 1. Reliability assessment of instrument

Construct	Items	Cronbach's alpha	Result
Digital Library Generic Usability (GEU)	2	0.836	>0.7
Digital Resources Quality (DRQ)	4	0.876	>0.7
Digital Library Affinity (DLA)	3	0.722	>0.7
User Satisfaction Behavior (USB)	5	0.881	>0.7

Next, a full-scale data collection was conducted. The study population is selected from the students enrolled in either diploma or degree level in one of the public universities in Malaysia. The non-probability sampling was selected; using convenience sampling – due to multiple factors: (1) difficulty to gain access to student details, and (2) students are not available at the campus due to the Novel Coronavirus 2019 pandemic. An invitation email was sent to all potential respondents with the help of the lecturers and students association representative. Data collection took 2 weeks in June 2021. A valid 358 responses were received, and data were imported into Statistical Package for Social Sciences (SPSS) for further analysis.

4 Findings

4.1 Demographic Analysis

The following Table 2 shows the demographic analysis of the respondents. 358 respondents have responded to the survey. The distribution of gender shows that the majority of the respondent is female (76.0% or n = 272), while 24.0% (or n = 86) are male. The faculty of the respondent is mostly from the Faculty of Information Management (n = 205 or 57.3%), Faculty of Business and Management (n = 53 or 14.8%), Faculty of Computer and Mathematical Science (n = 52 or 14.5%), Faculty of Accountancy (n = 22 or 6.1%), Faculty of Administrative Science and Policy Studies (n = 15 or 4.2%) and Faculty of Art and Design (n = 11 or 3.1%). The respondents were divided into 49.4% (or n = 177) diploma level and 50.6% (or n = 181) of degree level.

Table 2. Demographic analysis of the study.

Item		Frequency	%
Gender	Male	86	24.0
	Female	272	76.0
Faculty	Information Management	205	57.3
	Accountancy	22	6.1
	Computer and Mathematical Science	52	14.5
	Business and Management	53	14.8
	Administrative Science and Policy Studies	15	4.2
	Art and Design	11	3.1
Education	Diploma	177	49.4
	Degree	181	50.6

4.2 Assessment of Common Method Bias

Harman's Single Factor Test was conducted to determine whether the instrument is affected by the issue of common method bias. (CMB). Table 3 shows the result of the common method bias assessment. The result indicates that a single factor accounted for 42.3%, indicating that the instrument is free from common method bias.

Table 3. Assessment of common method bias

Component	Initial Eigenvalues			Extraction Sums of Squared Loadings		
	Total	% of Variance	Cumulative %	Total	% of Variance	Cumulative %
1	6.349	42.327	42.327	6.349	42.327	42.327
2	1.686	11.242	53.569			
3	1.292	8.610	62.179			
4	0.915	6.100	68.279			
5	0.880	5.870	74.149			
6	0.691	4.609	78.758			
7	0.610	4.070	82.827			
8	0.447	2.983	85.810			
9	0.392	2.615	88.426			
10	0.363	2.422	90.848			
11	0.322	2.145	92.993			
12	0.304	2.029	95.022			
13	0.268	1.784	96.806			
14	0.252	1.679	98.485			
15	0.227	1.515	100.000			

4.3 Descriptive Analysis of Instrument

Table 4 shows the descriptive analysis for Digital Library Generic Usability (GEU). The mean for GEU1 (M = 3, SD = 1.13) is 3.48 and the mean for GEU2 (M = 3, SD = 1.04) is 3.46. The overall mean for the GEU construct is 3.47, indicating a tendency toward positive responses.

Table 4. Digital Library Generic Usability (GEU)

Item		Mean	Std. dev
GEU1	How often do you refer to the digital library in completing your assignment?	3.48	1.13
GEU2	How often do you access PTA Website?	3.46	1.04

Table 5 shows the descriptive analysis for Digital Resources Quality (DRQ). The mean for DRQ1 (M = 4, SD = 0.84) is 3.85. The mean for DRQ2 (M = 4, SD = 0.74) is 4.13. The means for DRQ3 (M = 4, SD = 0.74) is 4.01. The mean for DRQ4 (M = 4, SD = 0.77) is 3.9. The overall mean for the DRQ construct is 3.97, indicating a tendency toward positive responses.

Table 5. Digital Resources Quality (DRQ)

Item		Mean	Std. dev
DRQ1	Do you think the PTA website has enough digital resources?	3.85	0.84
DRQ2	Do you feel PTA website digital resources have quality information?	4.13	0.74
DRQ3	How do you rate the quality of digital resources on our PTA website?	4.01	0.74
DRQ4	Do the digital resources in the PTA website suit your preference in doing the assignment?	3.90	0.77

Table 6 shows the descriptive analysis for Digital Libraries Affinity (DLA). The mean for DLA1 (M = 3, SD = 1.06) is 3.70. The mean for DLA2 (M = 3, SD = 0.82) is 3.91. The mean for DLA3 (M = 3, SD = 0.82) is 3.84. The overall mean for the DLA construct is 3.82, indicating a tendency toward positive responses.

Table 6. Digital Library Affinity (DLA)

Item		Mean	Std. dev
DLA1	Do you have experience in using the PTA website or any digital libraries?	3.70	1.06
DLA2	Do you think giving information on the PTA website are safe?	3.91	0.82
DLA3	Do you think that the services offered by the PTA website are sufficient?	3.84	0.82

Table 7 shows the descriptive analysis for User Satisfaction Behavior (USB). The mean for USB1 (M = 5, SD = 0.72) is 4.34. The mean for USB2 (M = 5, SD = 0.75) is 4.21. The mean for USB3 (M = 5, SD = 0.71) is 4.25. The mean for USB4 (M = 5, SD = 0.73) is 4.19. The mean for USB5 (M = 5, SD = 0.74) is 4.15. The overall mean for the USB construct is 4.23, indicating a tendency toward positive responses.

Table 7. User Satisfaction Behavior (USB)

Item		Mean	Std. dev
USB1	To what extent do you think a digital library is useful to society?	4.34	0.72
USB2	Do you think the generic usability of digital libraries influences user satisfaction?	4.21	0.75
USB3	Do you think the quality of digital resources in digital libraries influences user satisfaction?	4.25	0.71
USB4	Do you think the digital libraries affinity of digital libraries influences user satisfaction?	4.19	0.73
USB5	Do you satisfy with the PTA website?	4.15	0.74

4.4 Correlation Analysis

Figure 2 shows the Pearson's correlation coefficient of the study. Digital Resources Quality (DRQ) has the strongest correlation on the User Satisfaction Behavior (USB) with an R-value of 0.604 (n = 358). The second followed by Digital Libraries Affinity (n = 358, r = 0.551) and Digital Library Generic Usability (GEU) (n = 358, r = 0.320).

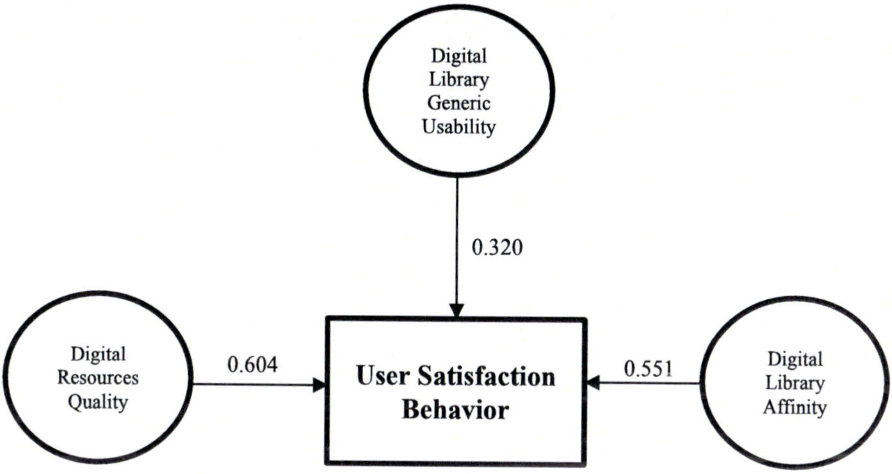

Fig. 2. Pearson's Correlation Coefficient

4.5 Multiple Regression Analysis

A linear regression analysis is conducted to determine the predictive power of the independent variable on the dependent variable. Three independent variables (Digital Library Generic Usability, Digital Resources Quality, Digital Library Affinity) were tested against a single dependent variable (User Satisfaction Behavior). The adjusted R square of 0.429 shows that model is capable to explain 43% of the variance. The following Table 8 shows the result of multiple regressions assessment.

Table 8. Multiple regression

	Hypothesis	Coefficients	t-value	p-value	Supported
H1	Digital Library Generic Usability → User Satisfaction Behavior	0.019	0.425	>0.05	No
H2	Digital Resources Quality → User Satisfaction Behavior	0.438	8.993	<0.05	Yes
H3	Digital Libraries Affinity → User Satisfaction Behavior	0.328	6.524	<0.05	Yes

5 Discussion and Conclusion

To recaps, this study aims twofold; first, to investigate the level of digital libraries satisfaction in the aftermath of COVID-19 and second, to examine the factors that influence user satisfaction behavior on digital libraries. In answering the first aim, we conducted a demographic analysis and descriptive analysis using Statistical Package for Social Sciences (SPSS) version 26. Data were analyzed based on the frequency and central tendency – mean, median, and standard deviation. The resulting mean value of 4.23 indicates that user satisfaction behavior among students at a public university in Malaysia is good. Moreover, respondents agreed that using the digital library is crucial, especially in the aftermath of the COVID-19 pandemic.

To answer the second aim, a Pearson's correlation coefficient and multiple regression analysis were conducted between the independent variables (Digital Library Generic Usability, Digital Resources Quality, Digital Library Affinity) and the dependent variable (User Satisfaction Behavior). The Pearson's correlation coefficient indicates that all variables have a positive effect on the dependent variable. However, the multiple regression analysis showed that only two hypotheses were accepted. Digital Resources Quality (DRQ) and Digital Library Affinity (DLA) were found to have a significant and positive relationship with User Satisfaction Behavior (USB). On the other hand, Digital Library Generic Usability (GEU) was found insignificant, and the hypothesis was rejected. The insignificant result of GEU can perhaps be credited due to the transformation of the educational method. The pandemic of COVID-19 has shifted traditional face-to-face education towards online learning. Many online educational tools were introduced such as Google Meet, Zoom, Webex, etc.

Exposure to these wide varieties of educational tools might enhance students' capability to engage and use varieties of software and application, thus decreasing the needs and preferences towards the usability of the software and more focusing on its usage. As asserted by [25], prior experience with information systems can negatively impact user efficiency and subsequently influence their level of satisfaction. This study is not without limitations. First, this study focuses on the generalization of theory rather than population generalization. We suggest future studies include more diverse respondents further to validate the research model in a broader audience. Second, concerning the methodology used, we adopted quantitative analysis using a questionnaire. Future studies may consider a qualitative study to gain inside into the problem. Third, we include a minimal number of variables. Future studies may consider a more complex generalization of variables into dimensions.

References

1. Stroube, B.: Literary freedom: project Gutenberg. XRDS: Crossroads ACM Mag. Students **10**(1), 3 (2003)
2. Soltani-Nejad, N., Taheri-Azad, F., et al.: Developing a model to identify the antecedents and consequences of user satisfaction with digital libraries. J. Inf. Manag. **72**(6), 979–997 (2020). https://doi.org/10.1108/ajim-04-2020-0099

3. Rosman, M.R.M., Baharuddin, N.F.S.: Are we using enough library e-services? Investigating the level of library e-services usage among students, pp. 374–378 (2021). https://ir.uitm.edu.my/id/eprint/44926/1/44926.pdf
4. O'Brien, H.L., Toms, E.G.: The development and evaluation of a survey to measure user engagement. J. Am. Soc. Inform. Sci. Technol. **61**(1), 50–69 (2010)
5. Thomas, P., O'Brien, H., Rowlands, T.: Measuring engagement with online forms. In: Proceedings of the 2016 ACM on Conference on Human Information Interaction and Retrieval, pp. 325–328 (2016)
6. Masrek, M.N., Gaskin, J.E.: Assessing user's satisfaction with web digital library: the case of Universiti Teknologi MARA. Int. J. Inf. Learn. Technol. **33**(1), 36–56 (2016). https://doi.org/10.1108/ijilt-06-2015-0019
7. Masrek, M.N., Samadi, I.: User engagement in academic web digital library. Int. J. Civil Eng. Technol. **8**(9), 789–799 (2017)
8. Mohamad Rosman, M.R., Ismail, M.N., Masrek, M.N.: How engaging are you? Asia Proc. Soc. Sci. **6**(2), 104–107 (2020)
9. Mohamad Rosman, M.R., Ismail, M.N., Masrek, M.N.: Investigating the predictors of digital library engagement: a structured literature analysis. Pak. J. Inf. Manag. Libr. **22**, 60–82 (2021)
10. Mohamad Rosman, M.R., Ismail, M.N., Masrek, M.N.: How engaging are you? Empirical evidence from Malaysian research universities. Int. J. Interact. Mob. Technol. **15**(4), 16–30 (2021)
11. Rao, K.: Digital libraries: challenges, opportunities and implications. In: Annual Conference Digital Information Exchange, pp. 236–243, January 2004
12. Ćirić, J., Ćirić, A.: The impact of the COVID-19 pandemic on digital library usage: a public library case study. J. Web Librarianship **15**(2), 1–16 (2021)
13. Goel, P., Malik, N.: A study on awareness and usage of E-resource portals among prospective teachers. Integr. J. Soc. Sci. **8**(1), 1–8 (2021)
14. Mohamad Rosman, M.R., Baharuddin, N.S.: Are we using enough library e-services? Investigating the level of library e-services usage among students, pp. 374–378 (2021). https://ir.uitm.edu.my/id/eprint/44926/1/44926.pdf
15. Zha, X., Xiao, Z., Zhang, J.: A survey of user perceptions of digital library e-quality and affinity. Ser. Rev. **40**(1), 3–11 (2014)
16. Hussaini, S.A., Jafari, S.A.S., Karim, M.A.: Effect of information systems on customer satisfaction and influence of trust: evidence from Pakistani digital banking industry. J. Independent Stud. Res.-Manag. Soc. Sci. Econ. **21**(3), 1–8
17. Shahzad, A., Hassan, R., Aremu, A.Y., Hussain, A., Lodhi, R.N.: Effects of COVID-19 in E-learning on higher education institution students: the group comparison between male and female. Qual. Quant. **55**(3), 805–826 (2020). https://doi.org/10.1007/s11135-020-01028-z
18. Mo, C.Y., Hsieh, T.H., Lin, C.L., Jin, Y.Q., Su, Y.S.: Exploring the critical factors, the online learning continuance usage during COVID-19 pandemic. Sustainability **13**(10), 5471 (2021)
19. Khattak, A., Shah, S.F., Adnan, M.: Impact of online classes during third wave of Covid-19 pandemic among university students and teachers in Pakistan. Psychol. Educ. J. **58**(5), 3951–3960 (2021)
20. Matusiak, K.K.: Perceptions of usability and usefulness of digital libraries. Int. J. Hum. Arts Comput. **6**(1–2), 133–147 (2012)
21. Sahak, M.D., Masrek, M.N.: Library usage of medical students: a comparative analysis of first year and third year students in Universiti Putra Malaysia. Procedia Soc. Behav. Sci. **129**, 127–132 (2014)

22. Sutrisno, S., Andre, H.L., Susilawati, S.: The analysis of the facilities and ICT applications usage by the university's students. In: Journal of Physics: Conference Series, vol. 1793, no. 1, p. 012050. IOP Publishing, February 2021
23. Zha, X., Cao, F., Yan, Y., Guo, J., Wang, J.: Exploring innovative information seeking: the perspectives of cognitive switching and affinity with digital libraries. J. Acad. Librarianship **45**(5), 102045 (2019)
24. Nunally, J.C., Bernstein, I.H.: Psychometric theory (1978)
25. Jing, C., Zhi, J., Yang, S., Wang, W.: Impact of driver age and experience in software usage on driving safety and usability of car-sharing software. J. Adv. Transp. **2021**, 1–14 (2021)

Managing Individual Online Learning Experience: The Roles of Perceived Engagement and Perceived Performance

Mohamad Rahimi Mohamad Rosman(✉), Izzatil Husna Arshad, Noor Rahmawati Alias, Zaila Idris, and Haslinda Abdul Hamid

Kelantan Branch, Universiti Teknologi MARA, Machang, Malaysia
rahimimr@uitm.edu.my

Abstract. The pandemic of Novel Coronavirus 2020 has shifted and transformed several industries: including educational institutions. Universities and schools were forced to turn into online education to continue the learning process. Several information systems were adopted such as a learning management system, video conferencing, google classroom, etc. However, sudden changes from face-to-face education to online distance learning raise user engagement towards the usage of the online platform, vice versa its impact on individual performance. Therefore, the purpose of this study is twofold; first, to investigate the individual's learning experience from the context of perceived engagement with the LMS and second, to investigate the impact of perceived engagement (user involvement and user participation) on perceived performance in the context of adopting a learning management system among the students from one of the local universities in Kelantan. A quantitative research instrument was developed, and data was collected. The study's findings were described in terms of descriptive and inferential analysis using Statistical Package for Social Science (SPSS) version 26. The results indicated that user involvement and user participation have a positive and significant relationship with the perceived performance.

Keywords: Perceived engagement · Perceived performance · Information management · Distance learning · online education

1 Introduction

Learning Management System (LMS) is the integration of several Information and Communication Technology (ICT) approaches in the educational context [1]. The use of LMS encourages student success in online learning [2]. LMS provides an effective mechanism in delivering and tracking learner's performance and progress [3]. LMS enables interaction and information exchange between learners and educators, especially in the Novel Coronavirus 2020 (COVID-19) pandemic; the usage of ICT helps ease the pain of connecting learners-educators and reduce the complexity of online interaction [4]. According to [5], LMS contributed to (1) ease of communication among learners and educators, (2) encourage collaborative learning, and (3) enables the learners to express themselves – depending on several criteria such as speed of Internet access, self-confidence, and the capability to use the online platform.

Y. Maleh et al. (Eds.): ICI2C 2021, LNNS 357, pp. 331–344, 2022.
https://doi.org/10.1007/978-3-030-91738-8_31

One of the main effects of the COVID-19 pandemic is learners and educators are working in social isolation [4]. Social isolation causes difficulties for educators to determine the efficiency and effectiveness of the learning process. As mentioned by [6], the widespread COVID-19 pandemic encourages the government to impose strict rules regarding social distancing, social isolation, and self-quarantine. Author [2] argues that there is a need to (1) investigate the users' perception towards the online course delivery, (2) the need to enhance online education delivery, and (3) minimize administrative issues.

Therefore, the purpose of this study is twofold; first, to investigate the individual's learning experience from the context of perceived engagement with the LMS and second, to investigate the impact of perceived engagement (user involvement and user participation) on perceived performance. The subsequent sections are structured as follows. First, we explained the conceptual model of the study, second, we discussed the methodology of the study. Third, the data were analyzed based on descriptive and inferential statistics. Fourth, the findings were discussed and discussion for the future study was outlined.

2 Research Model

The following Fig. 1 shows the proposed conceptual research model of the study. There is a total of two independent variables: namely user involvement (USI), and user participation (USP). Both independent variables were conceptualized as perceived engagement. Perceived engagement is operationalized as the extensive usage of LMS that includes multiple dimensions such as affective, cognitive, and behavioral perspectives [7–11]. On the other hand, the dependent variable of perceived performance was measured through three variables, namely task performance (TAP), task satisfaction (TAS), and task innovation (TAI).

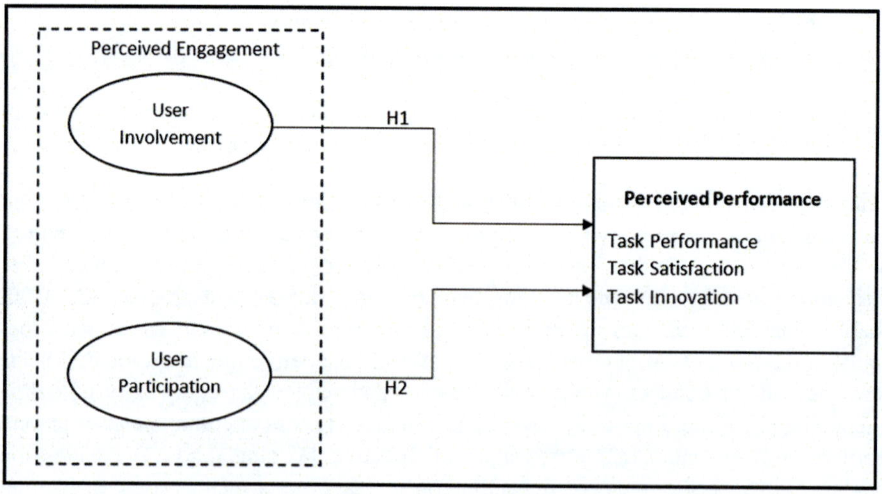

Fig. 1. Research model

2.1 Task Performance

Task performance is the user's evaluation of the degree of performance attained through engagement with the learning management system [12]. It also looks into the ability of any instructional tasks provided by the instructor for the class activity or assessment, which provide support and instruction thus assisting the students to complete their tasks. Author [13] stated that many institutions globally have heavily spent on electronic learning whereas courses that use manual methods of learning have become automated.

The work of [14] suggested eighteen instructional tasks that might be used as LMS tools, including meeting, guest speaker, probing, student questions, office, references, self-test, exam, instructor feedback, portfolio, student discussion, collaborative writing, peer feedback, blog (to share experience about courses), presentation (PowerPoint, YouTube), web lecture, and instruction. Therefore, due to the growing needs and use of LMS, determining learners' task performance when using LMS is essentially important.

Moreover, [15] highlighted that all universities should focus on the benefits associated with LMS, such as the learners' performance during a course and the learners' outcomes after course completion. Author [16] has conducted a study on a cloud-based LMS and the result has shown that the system positively affected tasks performance of the users (respondents). Meanwhile, [13] has identified the perceived usefulness of LMS and use of LMS contribute to a high percentage (70.6%) of learning and academic performance of learners.

2.2 Task Satisfaction

Task satisfaction in this study refers to the users' feeling when using the Learning Management System in terms of their knowledge, skills, and ability [17]. [18] has studied previous literature and determined that user satisfaction is the general perception of consumers regarding the process, which is also used to assess the learners' mindset. The satisfaction component evaluates interaction between user and LMS and it is regarded as among the five essential foundations of efficiency in web-based education.

The work of [19] indicated that learners' readiness towards LMS was related to their satisfaction and motivation while undertaking academic tasks. Furthermore, [20] emphasized that satisfaction is an indicator of the amount of time the user spends on the LMS. The parameter is certainly affected by the quality of learning material, but it can be assumed that users do not spend a lot of time on a LMS if they are not satisfied with the user interface.

2.3 Task Innovation

LMS enables the instructional materials to be managed through many educational activities, record, and reporting of student information, and able to facilitate the communication between educator and students. It becomes a task innovation for the student to complete their learning task. The use of LMS technologies improves students' engagement in the learning process by providing greater access to learning content and the ability for students to access the learning materials from any device. It also encourages them to take a bigger responsibility in the entire learning process and become less dependent on the lectures. Currently, the new LMS programs are becoming more student-oriented and are fostering more social and collaborative learning [21].

Moreover, LMS allows all the materials needed in one week to be visually grouped on a single page by means of contiguous placement makes it easier for students to consider all the elements as part of the week's tasks, and therefore more likely for them to access all the materials [22–24]. Author [25] has conducted a study on the availability of course materials that LMS provides, which positively impacts the ability of students to learn in and outside of the classroom. The learning skills of students have improved in courses where LMS are available [26].

2.4 User Involvement

User involvement in relation to information system (IS) development is an important mechanism in enhancing the system quality and ensuring the success of system implementation. Since the late 70s, it has been believed that user involvement in system development ensures system success [27].

User Involvement "refers to a subjective psychological state of the individual and is defined as the importance and personal relevance that users attach either to a particular system or to IS in general, depending on the user's focus" [28]. The work of [29] indicated that effective user involvement could contribute a better understanding to the system development and provide a satisfactory product as well. Furthermore, research finds that "users' involvement in design and implementation is related positively to users' perception of system usefulness" [30].

However, user involvement in relation to library and information science can be classified into group. Author [31] grouped them into specialists, students, disadvantaged majority, and the indisposed. In addition, [32] grouped users into children, pupils, students, adults, professionals, researchers, policymakers, artisans, hearing and visually handicapped, and physically handicapped. According to [31] the library user is undisputed the most important person in any library setting.

In relation to LMS, user involvement in this study refers to the students who use the platform of LMS to train and develop their skills. LMS improved students' engagement in the learning process and easier use educational resources. LMS is a software application or web-based technology used to plan, implement, and assess a specific learning process. It is used for e-learning practices and, in its most common form, consists of two elements: a server that performs the base functionality and a user interface that is operated by instructors, students and administrators [33].

2.5 User Participation

User participation refers to the observable activities of users using the LMS to collaborate in the online discussions and update their knowledge and skills in using the LMS. Students' or users' participation is considered as one of the criteria for the selection of types of LMS for education purposes, in order to increase students' contributions in collaborative interaction in learning [15, 34]. Lack of user participation in LMS can be the real challenge to the lecturers and it prevents successful online teaching and learning processes [35].

The work of [36] pointed out that user participation in using a learning management system refers to the activities of users to invite friends, post questions/answers and contribute answers. Furthermore, [37] found that users' participation in using IS/IT is influenced by the degree to which they feel engaged and involved in the training activity and they have a sense of enjoyment interacting with the system.

Authors of [38–40] agreed that in the context of information systems (IS), user participation plays an important role in the successful achievement of the system's use by the users. Users are perceived to be more active in using the information system later on when they participated in the earlier stage of the system development process [38, 40]. Active and positive user participation even increases the beneficial advantage of the information resources shared among the users who use the information systems [39].

3 Methodology

To answer both research purposes, a quantitative research instrument was developed by adopting the previous instruments from similar studies. The following Table 1 shows the sources of items adopted and adapted for this study.

Table 1. Sources of instrument

	Construct	Number of items	Reference
Perceived engagement	User Participation (USP)	3	[41, 42]
	User Involvement (USI)	5	[43]
Perceived performance	Task Performance (TAP)	3	[12]
	Task Satisfaction (TAS)	3	[17]
	Task Innovation (TAI)	3	[44]

Next, the instrument was pre-tested, and pilot tested to determine the reliability of the instrument. The Cronbach's alpha coefficient was used to determine the reliability of the instrument. The following Table 2 shows the summary of the reliability analysis. Based on Cronbach's alpha values ranging from 0.858 to 0.946, it can be assumed that there is no problem with the reliability of the instrument as it exceeds the minimum value of 0.7 as suggested by Nunnally (1994).

Table 2. Reliability analysis

Construct	Number of items	Cronbach's alpha
User Participation (USP)	3	0.858
User Involvement (USI)	5	0.937
Task Performance (TAP)	3	0.903
Task Satisfaction (TAS)	3	0.946
Task Innovation (TAI)	3	0.884

Following the success of reliability analysis, actual data collection was conducted. A convenience sampling was adopted for the study due to difficulties in obtaining students' details in the era of the pandemic of COVID-19. Convenience sampling allows the researcher to easily access the sampling frame. An online questionnaire was developed using Google Form and distributed among respondents from 6 faculties at one of the local universities in Kelantan. As a result, a valid 206 responses were received: indicating a sufficient sample size for data analysis [46]. The data was coded and imported into statistical software, Statistical Package for Social Sciences (SPSS) version 26 for further analysis. The subsequent section will discuss the findings of the study from the perspective of (1) demographic analysis, (2) descriptive analysis of variables, and (3) inferential analysis of the interrelation between variables.

4 Results and Discussion

4.1 Demographic

Table 3 shows the demographic analysis of the respondents. As expected, more than 2/3 of the respondents are female (75.2% or N = 155), while male respondents are represented by 51 individuals (24.8%). In relation to faculty, the total count indicated that most students enrolled for the Faculty of Information Management (N = 115 or 75.2%), followed by Faculty of Business and Management (N = 57 or 27.7%), Faculty of Accountancy (N = 12 or 5.8%), Faculty of Computer and Mathematical Sciences (N = 9 or 4.4%), Faculty of Administrative Science and Policy Studies (N = 7 or 3.4%), and Faculty of Art and Design (N = 6 or 2.9%). The high number of respondents from the Faculty of Information Management is due to the researcher's easy access to the respondents compared to other faculties. In terms of level of study, most of the respondents enroll for a diploma (N = 134 or 65%) and the rest enroll for a degree (N = 72 or 35%).

Table 3. Demographic analysis

	Sub-items	Frequency	%
Gender	Male	51	24.8
	Female	155	75.2
Faculty	Information Management	115	55.8
	Accountancy	12	5.8
	Computer and Mathematical Sciences	9	4.4
	Art and Design	6	2.9
	Business and Management	57	27.7
	Administrative Science and Policy Studies	7	3.4
Level of study	Diploma	134	65.0
	Degree	72	35.0

4.2 Descriptive Analysis

Table 4 shows the frequency of using the LMS among the respondents. More than half of the respondents agreed that they use the LMS on a daily basis (N = 151 or 73.3%), followed by every 2 to 3 days (N = 39 or 18.9%), weekly (N = 12 or 5.8%). Interestingly, a total of three (3) respondents indicate that they are not using the LMS, while another respondent indicates that he/she uses the LMS on a monthly basis. This scenario is perhaps contributed by several factors such as having poor access to the Internet connection, family matters, lack of motivation, etc.

Table 4. Frequency of using learning management system

	Sub-items	Frequency	%
Frequency of usage	Not applicable	3	1.5
	Daily	151	73.3
	Every 2–3 days	39	18.9
	Weekly	12	5.8
	Monthly	1	0.5

On the other hand, the following Table 5 shows the tendency of a respondent to use the online attendance provided by the LMS. The use of online attendance helps educators to manage the classroom, especially when dealing with absenteeism. The findings of the study indicated that more than half of the respondents (N = 113 or 54.9%) agreed that they usually tick the online attendance during the class, while 63 of the respondents (30.6%) tick the online attendance before the start of their class. The remaining 27 respondents (13.1%) indicate that they usually tick the online attendance after their class ends.

Table 5. The tendency of using online attendance provided by LMS

	Sub-items	Frequency	%
The tendency of using online attendance	Not applicable	3	1.5
	Before the class starts	63	30.6
	During the class	113	54.9
	After class is over	27	13.1

Table 6 shows the descriptive analysis for user participation. The mean for USP1 (M = 5, SD 1.13) is 5.23. The mean for USP2 (M = 5, SD 1.11) is 5.07, and the mean for USP3 (M = 5, SD 1.11) is 5.29. The total mean value of 5.20 indicates a tendency towards positive responses.

Table 6. User participation

Item		Mean	Std. dev.
USP1	Learning Management System enables me to share input with others	5.23	1.13
USP2	Learning Management System encourages me to speak up and participate in the digital discussion	5.07	1.11
USP3	Learning Management System encourages me to keep knowledge and skills current	5.29	1.11

Table 7 shows the descriptive analysis for user involvement. The mean for USI1 (M = 6, SD 1.11) is 5.53, followed by the mean for USI2 (M = 6, SD 1.07) is 5.58. The mean for USI3 (M = 6, SD 1.09) is 5.78, while the mean for USI4 (M = 6, SD 1.08) is 5.68. The mean for USI5 (M = 6, SD 1.14) is 5.58. The total mean value of 5.63 indicates a tendency towards positive responses.

Table 7. User involvement

Item		Mean	Std. dev.
USI1	I believe engagement with Learning Management System is essential	5.53	1.11
USI2	I believe engagement with Learning Management System is fundamental	5.58	1.07
USI3	I believe engagement with Learning Management System is needed	5.78	1.09
US14	I believe engagement with Learning Management System is relevant	5.68	1.08
USI5	I believe engagement with Learning Management System means a lot to me	5.58	1.14

Table 8 shows the descriptive analysis for task performance. The mean for TAP1 (M = 6, SD 1.11) is 5.52, followed by the mean for TAP2 (M = 6, SD 1.09) is 5.66. The mean for TAP3 (M = 6, SD 1.08) is 5.51. The total mean value of 5.56 indicates a tendency towards positive responses.

Table 8. Task performance

Item		Mean	Std. dev.
TAP1	Learning Management System enables prompt processing of my information task	5.52	1.11
TAP2	Learning Management System provides up-to-date information	5.66	1.09
TAP3	Learning Management System provides relevant output to support my information task	5.51	1.08

Table 9 shows the descriptive analysis for task satisfaction. The mean for TAS1 (M = 5, SD 1.13) is 5.29, followed by the mean for TAS2 (M = 5, SD 1.08) is 5.23. The mean for TAS3 (M = 5, SD 1.09) is 5.21. The total mean value of 5.24 indicates a tendency towards positive responses.

Table 9. Task satisfaction

Item		Mean	Std. dev.
TAS1	I am satisfied with my knowledge in using Learning Management System	5.29	1.13
TAS2	I am satisfied with my skills in using Learning Management Systems	5.23	1.08
TAS3	I am satisfied with my ability to use Learning Management System	5.21	1.09

Table 10 shows the descriptive analysis for task innovation. The mean for TAI1 (M = 5, SD 1.05) is 5.31, followed by the mean for TAI2 (M = 6, SD 1.06) is 5.51. The mean for TAI3 (M = 5, SD 1.13) is 5.30. The total mean value of 5.37 indicates a tendency towards positive responses.

Table 10. Task innovation

Item		Mean	Std. dev.
TAI1	Learning Management System encourages me to do my information task differently	5.31	1.05
TAI2	Learning Management System encourages my curiosity about how things work	5.51	1.06
TAI3	Learning Management System helps me to accomplish more work than would otherwise be possible	5.30	1.13

4.3 Correlation Analysis of Variables

The following Fig. 2 shows the results of Pearson's Correlation analysis. The test was conducted on three variables: user involvement, user participant and perceived performance. According to [47], the results of the Pearson's Correlation of this study arc interpreted as follow: (1) there is a positive and strong correlation between user involvement and perceived performance (r = 0.789, n = 206, p = 0.00), and (2) there is a positive and strong correlation between user participation and perceived performance (r = 0.703, n = 206, p = 0.00).

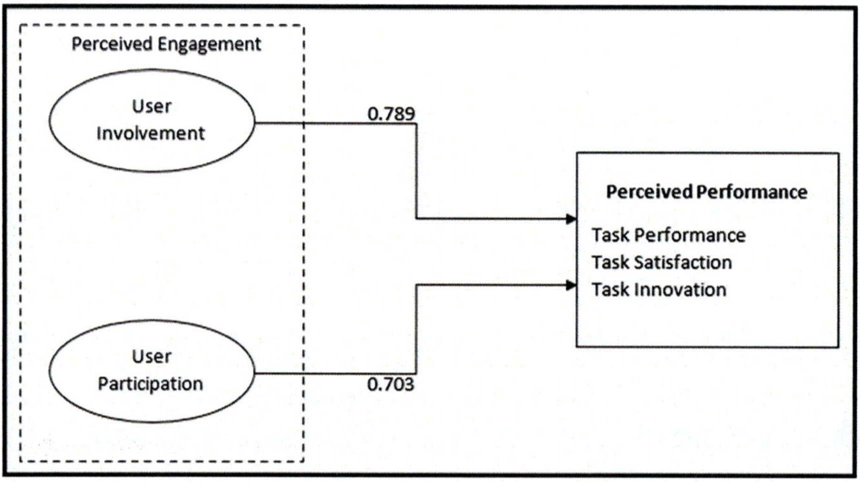

Fig. 2. Pearson's correlation coefficient

4.4 Multiple Regression

Next, a multiple regression analysis was conducted to determine the predictive power of the independent variables (user participation, user involvement) on the dependent variable (perceived performance). A multiple regression analysis was conducted using SPSS version 26. The resultant adjusted R Square of 0.683 indicates that that the model can explain 68% of the total variances in the variables. Table 11 shows the result of the analysis. Based on the result, both hypotheses of the study are supported.

Table 11. Multiple regression

Hypothesis		Coefficients	t-value	P-value	Supported
H1	User involvement → perceived performance	0.33	6.45	<0.01	Yes
H2	User participation → perceived performance	0.574	11.14	<0.01	Yes

5 Discussion and Conclusion

The study was conducted with two purposes, first, to investigate the individual's learning experience from the context of perceived engagement with the LMS and second, to investigate the impact of perceived engagement (user involvement and user participation) on perceived performance. To answer the first purpose, data were analyzed based on descriptive nature. Findings from the descriptive analysis show that (1) most student has a positive perception towards LMS, (2) majority of respondent use LMS on a daily basis, (3) most respondents prefer to tick their attendance during the class, and (4) respondents indicate a moderate level of user involvement and user participation. For the second purpose, an inferential analysis was conducted using two tests: Pearson's correlation and multiple regression. The result of Pearson's correlation coefficient indicates that both user involvement and user participation have a positive and strong correlation with perceived performance; meanings that enhancing the level of user involvement and user participation will have a positive and significant effect on the improvement of individual performance (i.e., contributed towards improving work productivity, work satisfaction, and encourage innovative behavior). The result of the study is similar to previous studies of [7, 42, 48, 49]. Author [48] asserted that user engagement with the information system contributed towards information system success and improvement of individual performance. Moreover, [42] suggests that engagement with the information system helps users to have a sense of belongings, satisfy their individual needs, and develop a better understanding of the information system. On the other hand, [7] found out that user engagement encourages subsequent usage and promotion of the usage to others.

This study is not without a limitation. First, we conceptualize perceived engagement from the perspective of user involvement and user participation. Future studies may improve the model by integrating other variables such as novelty, focus attention, felt-involvement, aesthetic, usability, etc. Second, we suggest the researcher using other quantitative analysis measurements such as Partial Least Square Structural Equation Modelling (PLS-SEM) to obtain more meaningful research findings. Third, adopting stratified random sampling against convenience sampling might give a diverse and meaningful finding. Fourth, the finding indicated that there are small issues with the frequency of using LMS and the tendency of ticking the online attendance. Embracing the issues may provide interesting new knowledge and contribute to the better utilization of LMS.

References

1. Conde, M.Á., García-Peñalvo, F.J., Rodríguez-Conde, M.J., Alier, M., Casany, M.J., Piguillem, J.: An evolving Learning Management System for new educational environments using 2.0 tools. Interact. Learn. Environ. **22**(2), 188–204 (2014)
2. Nurakun Kyzy, Z., Ismailova, R., Dündar, H.: Learning management system implementation: a case study in the Kyrgyz Republic. Interact. Learn. Environ. **26**(8), 1010–1022 (2018)
3. Kakasevski, G., Mihajlov, M., Arsenovski, S., Chungurski, S.: Evaluating usability in learning management system Moodle. In: 30th International Conference on Information Technology Interfaces, pp. 613–618. IEEE (2008)
4. Raza, S.A., Qazi, W., Khan, K.A., Salam, J.: Social isolation and acceptance of the learning management system (LMS) in the time of COVID-19 pandemic: an expansion of the UTAUT model. J. Educ. Comput. Res. **59**(2), 183–208 (2021)
5. Yunus, H.: Online learning management system (OLMS) in Indonesian Higher Education: investigating benefits and obstacles. PJEIS: Parahikma J. Educ. Integr. Sci. **1**(1), 1–8 (2021)
6. Anderson, R.M., Heesterbeek, H., Klinkenberg, D., Hollingsworth, T.D.: How will country-based mitigation measures influence the course of the COVID-19 epidemic? Lancet **395** (10228), 931–934 (2020)
7. O'Brien, H.L., Toms, E.G.: The development and evaluation of a survey to measure user engagement. J. Am. Soc. Inform. Sci. Technol. **61**(1), 50–69 (2010)
8. Masrek, M.N., Samadi, I.: User engagement in academic web digital library. Int. J. Civil Eng. Technol. **8**(9), 789–799 (2017)
9. Masrek, M.N., Razali, M.H., Ramli, I., Andromeda, T.: User engagement and satisfaction: the case of web digital library. Int. J. Eng. Technol. (UAE) **7**(4), 19–24 (2018)
10. Mohamad Rosman, M.R., Ismail, M.N., Masrek, M.N.: Investigating the determinant and impact of digital library engagement: a conceptual framework. J. Digit. Inf. Manag. **17**(4), 214–226 (2019)
11. Mohamad Rosman, M.R., Ismail, M.N., Masrek, M.N.: Investigating the predictors of digital library engagement: a structured literature analysis. Pak. J. Inf. Manag. Libr. **22**, 60–82 (2021)
12. Miller, J., Doyle, B.A.: Measuring the effectiveness of computer-based information systems in the financial services sector. MIS Q. **11**, 107–124 (1987)
13. Alam, M.M., Ahmad, N., Naveed, Q.N., Patel, A., Abohashrh, M., Khaleel, M.A.: E-learning services to achieve sustainable learning and academic performance: an empirical study. Sustainability **13**, 2653 (2021). https://doi.org/10.3390/su13052653
14. Schoonenboom, J.: Using an adapted, task-level technology acceptance model to explain why instructors in higher education intend to use some learning management system tools more than others. Comput. Educ. **71**, 247–256 (2014)

15. Khairudin, N., Khairudin, R., Hamid, M.N.A., Hancock, P., McGill, T., Zamani, Z.A.: The importance of human capital perspective in the Learning Management System (LMS) decision making process at universities. J. Psikologi Malaysia **30**(2), 102–113 (2016)
16. Cheng, Y.M.: Can tasks and learning be balanced? A dual-pathway model of cloud-based e-learning continuance intention and performance outcomes. Kybernetes (2021). https://doi.org/10.1108/K-07-2020-0440
17. Mohamad Rosman, M.R., Mohd Shukry, A.I., Alimin, N.A., Baharuddin, N.S., Nik Rosli, N.N.I., Razlan, N.M.: Level of individual performance in the context of referencing competency. In: Proceedings of the 1st International Conference on Law, Social Science, Economics, and Education (ICLSSEE 2021), pp. 606–615. EAI, Jakarta, Indonesia (2021)
18. Al-Fraihat, D., Joy, M., Sinclair, J.: Evaluating e-learning systems success: an empirical study. Comput. Hum. Behav. **102**, 67–86 (2020)
19. Yilmaz, R.: Exploring the role of e-learning readiness on student satisfaction and motivation in flipped classroom. Comput. Hum. Behav. **70**, 251–260 (2017). https://doi.org/10.1016/j.chb.2016.12.085
20. Fenu, G., Marras, M., Meles, M.: A learning analytics tool for usability assessment in moodle environments. J. E-Learn. Knowl. Soc. **13**(3), 23–34 (2017). https://doi.org/10.20368/1971-8829/1388
21. Lonn, S., Teasley, S.: Saving time or innovating practice: investigating perceptions and uses of Learning Management Systems. Comput. Educ. **53**, 686–694 (2009). https://doi.org/10.1016/j.compedu.2009.04.008
22. Clark, R.C., Mayer, R.E.: E-Learning and the Science of Instruction: Proven Guidelines for Consumers and Designers of Multimedia Learning, 2nd edn. Pfeiffer/Wiley, New York (2008)
23. Mayer, R.E.: Principles for reducing extraneous processing in multi-media learning: coherence, signaling, redundancy, spatial contiguity, and temporal contiguity. In: Mayer, R. E. (ed.) The Cambridge Handbook of Multimedia Learning, pp. 183–200. Cambridge University Press (2005)
24. Vicario, G.B.: On Wertheimer's principles of organization. Gestalt Theory **20**(4), 256–269 (1998)
25. Kim, D.: The impact of learning management systems on academic performance: virtual competency and student involvement. J. High. Educ. Theory Pract. **7**(2), 23–35 (2017)
26. Nair, S.C., Patil, R.: A study on the impact of learning management systems on students of a University College in Sultanate of Oman. Int. J. Comput. Sci. Issues **9**(2), 379–395 (2012)
27. Ives, B., Olson, M.H.: User involvement and MIS success: a review of research. Manage. Sci. **30**(5), 586–603 (1984)
28. Barki, H., Hartwick, J.: Rethinking the concept of user involvement. MIS Q. **13**, 53–63 (1989)
29. Sun, Z.: User involvement in system development process. In: 2nd International Conference on Computer Science and Electronics Engineering (ICCSEE 2013) (2013)
30. Franz, C.R., Robey, D.: Organizational context, user involvement, and the usefulness of information systems. Decis. Sci. **17**(3), 329–356 (1986). https://doi.org/10.1111/j.1540-5915.1986.tb00230.xC
31. Nwalo, K.I.N.: Fundamentals of library practice: a manual on library routines. Sterling-Horden (2003)
32. Aina, L.O.: Library and information science text for Africa. Third World Information Services (2004)
33. Brush, K.: Definition learning management system (LMS) (2019). https://searchcio.techtarget.com/definition/learning-management-system

34. Aldiab, A., Chowdhury, H., Kootsookos, A., Alam, F., Allhibi, H.: Utilization of Learning Management Systems (LMS) in higher education system: a case review for Saudi Arabia. Energy Procedia **160**, 731–737 (2019)
35. Waheed, I., Seddon, A.: Utilisation and challenges of the learning management systems in the higher education institutions in Maldives: the lecturers' and students' perspectives. Int. J. Inf. Technol. Manag. Inf. Syst. **10**(1), 1–10 (2019)
36. Lek, H.H., Poo, D.C.C.: Knowledge community: moving beyond conventional learning management systems. Paper presented at the CSEDU 2011 - Proceedings of the 3rd International Conference on Computer Supported Education, vol. 1, pp. 173–178 (2011)
37. Reychav, I., Wu, D.: Are your users actively involved? A cognitive absorption perspective in mobile training. Comput. Hum. Behav. **44**, 335–346 (2015)
38. Gao, J.: Design strategy of information construction based on user participation. MATEC Web Conf. **136**, 2009 (2017). https://doi.org/10.1051/matecconf/201713602009
39. Puspitasari, I., Fauzi, S.S.M., Ho, C.: Factors driving users' engagement in patient social network systems. Informatics **8**(1), (2021). https://doi.org/10.3390/informatics8010008
40. Vahteristo, A., Jylhä, V.: Effects of user participation in the development of health information systems on their evaluation within occupational health services. Stud. Health Technol. Inform. **275**, 207–211 (2020). https://doi.org/10.3233/SHTI200724
41. Hunton, J.E., Jesse, D.B.: Effects of user participation in systems development: a longitudinal field experiment. MIS Q. **21**, 359–388 (1997)
42. Barki, H., Hartwick, J.: Measuring user participation, user involvement, and user attitude. MIS Q. **18**, 59–82 (1994)
43. Bartel, C.A.: Social comparisons in boundary-spanning work: effects of community outreach on members' organizational identity and identification. Adm. Sci. Q. **46**, 379–413 (2001)
44. Rowan, B., Robert, J.M.: Organizational strategies for promoting instructional change: implementation dynamics in schools working with comprehensive school reform providers. Am. Educ. Res. J. **44**, 252–297 (2007)
45. Nunnally, J.C.: Psychometric Theory 3E. Tata McGraw-Hill Education (1994)
46. Raosoft: Raosoft Sample Size Calculator. Raosoft, Inc., Seattle (2004). http://www.raosoft.com/samplesize.html
47. Cohen, J.: Statistical Power Analysis for the Behavioral Sciences, 2nd edn. Lawrence Erlbaum Associates, Hillsdale (1988)
48. Baroudi, J.J., Olson, M.H., Ives, B.: An empirical study of the impact of user involvement on system usage and information satisfaction. Commun. ACM **29**(3), 232–238 (1986)
49. Khan, A., Masrek, M.N., Mahmood, K., Qutab, S.: Factors influencing the adoption of digital reference services among the university librarians in Pakistan. Electron. Libr. **36**(6), 1225–1246 (2017)

Consumer's Organic Food Buying Intention in COVID-19 Pandemic: Evidence from Vietnam

Nguyen Thi Ngan and Bui Huy Khoi(✉)

Industrial University of Ho Chi Minh City, Ho Chi Minh City, Vietnam
buihuykhoi@iuh.edu.vn

Abstract. During the ongoing Covid-19 Pandemic in the world and Vietnam, the increasing interest in health and people's health consciousness affects the use of safe food, including organic foods. The desire to promote green consumption and develop green products for people to have good health and a healthy lifestyle to fight against the great tribulation has enabled the research on organic food. The study was conducted to determine the factors affecting consumers' intention to buy organic food conducted in Ho Chi Minh City, Vietnam, during the Covid-19 Pandemic. Collected 299 survey samples through questionnaires and processed the data using R software. The results showed 6 factors: Food safety, Health Consciousness, Environment care, Paying ability, Food quality, Social lifestyle positively influencing consumers' intention to buy organic food in Ho Chi Minh City, Vietnam, during the COVID-19 Pandemic. The study also proposes some managerial implications for related businesses and future research directions based on the analyzed data. This study uses the optimal choice of the BIC for the Consumer's Organic Food Buying Intention.

Keywords: BIC · COVID-19 · Consumer's Organic Food Buying Intention · Food safety · Health consciousness · Environment care · Paying ability · Food quality · Social lifestyle

1 Introduction

Vietnam's economy has been heavily affected by the COVID-19 pandemic but has also shown considerable resilience. In 2020, Ho Chi Minh City contributed 22.2% of GDP and 27% of national budget revenue. Labor productivity equals 2.6 times the national productivity, has the densest population density, and education level, high income [1]. Organic food is considered a product of organic agriculture, especially in the ongoing pandemic caused by the new virus (COVID-19) and climate change, which are becoming a serious global threat. It is a significant threat to the life and survival of people. Only then do people understand the importance of health and the life of nature. Having good health and knowing how to protect yourself, your family, and the safety of society is a prerequisite in today's times. Consumers have been interested in organic foods because they are produced without pesticides and other agricultural-related materials harmful to human health. The paper uses the optimal choice by the BIC for the Consumer's Organic Food Buying Intention.

Y. Maleh et al. (Eds.): ICI2C 2021, LNNS 357, pp. 345–353, 2022.
https://doi.org/10.1007/978-3-030-91738-8_32

2 Literature Review

2.1 Consumer's Organic Food Buying Intention (COFI)

According to Ajzen and Fishbein [2], behavioral intention expresses each person's willingness to perform a specified behavior, and it is considered a direct premise leading to the behavior. According to the theory of rational behavior, one of the most important factors for determining human behavior is the intention to perform the behavior. The relationship between behavioral intention and human behavior is used simultaneously to study consumer buying behavior. In the author's study, using the definition of Rashid et al. [3], the intention to buy organic food is the ability. The individual will give his preference to organic food over organic food are common food in shopping considerations. In summary, the intention to buy organic food is a set of factors that motivate consumers to purchase Organic Food such as considering purchasing organic food, buying soon, willing to pay a lot for Organic Food, gender, etc., recommending buying organic food products to my acquaintances [4]. If organic foods are available at supermarkets, consumer food stores will buy those [5].

2.2 Food Safety (FS)

Food safety control is important for government image, business survival, public interest, and social stability [6]. Despite continuous optimization, technological improvement, and other aspects in recent years, global food safety incidents still occur [7]. The continued supply of safe food depends on effective food safety management processes and the behaviors of producers and operators [8]. Because of the above, organic food is a food-safe option. Because organic food is natural, food is grown without chemicals (pesticides, fertilizers, etc.), preservatives, and genetic modification. Products that want to be recognized as organic must go through rigorous testing and be certified as organic.

In summary, consumers' concern about food safety and the origin of products motivates them to buy organic products. Organic food is a safe, natural food cultivated, grown without chemicals (pesticides, fertilizers, etc.), preservatives, and genetic modification. That reduces the risk of food poisoning [9].

2.3 Health Consciousness (HS)

Health consciousness can be indicated by an individual's interest in health in their daily activities [10]. And especially when the whole world is looking for ways to fight the disease called Covid-19, people are more aware of the importance of health for themselves, their families, and society. The consequences it causes are poisoning, soil degradation, erosion, water pollution, dangers to human health such as the appearance of skin diseases, cancer, neurological disorders, genetic disorders, transmission, severe poisoning, etc., and environmental degradation [11, 12]. In summary, organic food is healthier than conventional food because no chemicals are used in the production process [9]. Therefore, using organic products can reduce the potential risk of disease to humans [9] and help improve health, promote health, and may be useful in disease prevention [13].

2.4 Environment Care (EC)

The environment is the living space of people and living things. In recent years, the interest in organic food and organic agriculture in most developed and developing countries has been stimulated along with increasing public concern about food quality. and people's health, especially natural resource degradation [14]. Unlike conventional farming, organic farming is an environmentally friendly form of farming. Organic food production is based on the biological balance of the soil-plant-animal-human system. It aims to produce food with high nutritional value, develop sustainable agriculture coupled with ecosystem conservation such as reducing pollution, saving water and resources, and improving soil fertility [15].

Finally, the cultivation of organic food also helps to protect the environment, balance the ecosystem and the life of nature [10]. Organic food is grown to be environmentally friendly such as preventing soil and water pollution [9].

2.5 Paying Ability (PA)

Paying ability refers to the consumer's ability and willingness to pay for the price of the product offered. Generally, organic food products are sold at higher prices than conventional foods [16], raising affordability problems. Organic food price is a barrier to the affordability of consumers intending to buy organic food [9]. Not only was that, during the Covid-19 epidemic season, but all sources of income in the world were also seriously affected. That also impacts individual spending, especially on organic food.

2.6 Food Quality (FQ)

Consumer preferences, their concerns about food quality, and their need for chemical-free foods will influence manufacturers' policies and production processes, including sellers. Wholesale and retail [17]. In the study by Jeyakumar Nathan et al. [18], Malaysian consumers compared conventional food with organic food based on the quality of organic food. And organic food is seen as a healthy food option. Adopting advanced farming methods that improve product quality and they accept to pay more to get higher quality. Research results by [13] show that high-quality food products are often understood and defined as delicious, fresh and natural, low in processing, and with expected ingredients.

2.7 Social Lifestyle (SL)

Lifestyles are typical features, repeated and shaped into styles and habits in the life of an individual, a social group, a nation, or an entire culture. One of the younger generations that have unique characteristics from the previous generation is the way of life. Life is developing more and more, modern lifestyle, making information and content accessible to everyone easily through media. The healthy lifestyle chosen by the younger generation is a healthy or clean diet [19]. The Covid-19 epidemic has affected people's health consciousness and everyone needs to maintain a healthy lifestyle. This trend has given rise to the rapid growth of organic products. So social lifestyle is considered the most important factor contributing to buying organic food in Malaysia [18].

3 Methodology

3.1 Sample Approach

Based on the study of Hair et al. [20] for reference to the expected sample size. Table 1 describes statistics of sample characteristics. We see that the gender is Fe-male accounted for 60.2% and Male accounted for 39.8%.

Table 1. Statistics of sample

Characteristics		Amount	Percent (%)
Sex and age	Male	119	39.8
	Female	180	60.2
	Below 18	35	11.7
	18–30	84	28.1
	31–40	140	46.8
	Above 40	40	13.4
Income/month	Below 5 million VND	20	6.7
	5–10 million VND	69	23.1
	11–15 million VND	140	46.8
	Over 15 million VND	70	23.4
Job	Student	20	6.7
	Officer	70	23.4
	State staff	66	22.1
	Freelance	86	28.8
	Homemakers	29	9.7
	Others	28	9.4

Table 1 describes statistics of sample characteristics. We see the gender is female accounted for 60.2% and Male accounted for 39.8%. This difference shows that the female gender is more interested in food safety and health awareness than the male gender. We see that the age group from 31 to 40 years old accounted for 46.8%, from 18 to 30 years old accounted for 28.1%, over 40 years old accounted for 13.4%, and finally less than 18 years old accounted for 11.7%. This difference is because middle-aged people are more concerned about food safety than other age groups. From 11 to 15 million accounted for 46.8%, from 5 to 10 million accounted for 23.3%, over 15 million for 23.4%, and finally under 5 million accounted for 6.7%. This shows that the customer's income is relatively high. The reality of such income is considered relatively average. We see Freelance accounted for 28.8%, officers accounted for 23.4%, State staff accounted for 22.1%, homemakers accounted for 9.7%, other occupations accounted for 9.2% and finally students accounted for 6.7%.

3.2 Measures

To design a scale suitable for the topic, the author has consulted, inherited, and developed a scale of intention to buy organic food synthesized from many research works, which is the intention to buy organic food of people in Ho Chi Minh City during the Covid-19 Pandemic. Factor and Item are in Table 2.

4 Results

4.1 BIC Model Selection

BIC was used to choose the best model by R software. BIC has been used in the theoretical context for model selection. As a regression model, BIC can be applied to estimate dependent variables from one or more independent variables [21]. An essential and valuable measurement for deciding a complete and straightforward model is the BIC. Based on the BIC information standard, a model with a lower BIC is selected. The best model will stop when the minimum BIC value [21–23]. R report shows every step of searching for the optimal model. BIC selects the best 2 models as Table 3.

There are six independent and one dependent variable. Food safety (FS), Health Consciousness (HS), Environment care (EC), Paying ability (PA), and Social lifestyle (SL) influence Consumer's Organic Food Buying Intention (COFI) with a Probability is 100%. Food quality (FQ) has a probability of 90.8%.

4.2 Model Evaluation

According to the results from Table 4, BIC shows model 1 is the optimal selection because BIC (−283.42493) is minimum. Food safety (FS), Health Consciousness (HS), Environment care (EC), Paying ability (PA), Food quality (FQ), Social lifestyle (SL) impact Consumer's Organic Food Buying Intention (COFI) is 65.4% in Table 4. BIC finds model 1 is the optimal choice and six variables have a Probability of 90.8%. The above analysis shows the regression equation below is statistically significant.

$$\mathbf{COFI = 0.42925 + 0.12317FS + 0.27645HS + 0.12498EC + 0.08676PA + 0.15746FQ + 0.15521SL}$$

Table 2. Factor and item

Factor	Mean	Item
COFI	3.6582	I would consider buying organic food
		I will buy organic food soon
		If organic foods were available at the supermarket or grocery store, I would buy them
		I will pay a lot of money to buy organic food for better health
		I will recommend buying organic food for my acquaintances
FS	3.4756	I care about food hygiene and safety
		Organic food is natural food
		Organic food can reduce the risk of food poisoning
		I am always interested in the product's origin
		I think organic food does not contain chemicals (pesticides, fertilizers, etc.), preservatives, and change genes
HS	3.3351	The Covid-19 pandemic has had a powerful impact on our health consciousness
		I need to know food choices to ensure a healthy, healthy lifestyle
		Health concerns are the reason to consume organic food
		Organic food is healthier than regular food
		Consuming organic food helps reduce the risk of disease
EC	3.4635	I am concerned about the state of our environment today
		To survive, humanity should protect the balance and survival of nature
		Organic food is produced using environmentally friendly methods
		Organic farming can prevent soil, water, and air pollution
		Using organic food helps to protect the environment
PA	3.7177	The price of organic foods is a barrier to buying them
		Organic food usually costs more than regular food
		My income during Covid-19 affects my ability to buy organic food
		I can afford to buy organic food
		Higher prices will cause better food quality
FQ	3.5130	Quality food is fresh, proven safe stuff
		Organic food is a healthy, high-quality food option
		Organic food is of higher quality than conventional food products
		Organic food is of better quality because of the application of advanced farming methods
		Although I may have to pay more, I get better quality organic food
SL	3.6742	The Covid-19 epidemic changes people's food choices
		Consumption of organic food ensures a healthy lifestyle for a healthier body
		Organic food is trending in society
		The change to organic food helps improve the living environment and develop agriculture in Vietnam
		The media are promoting organic food widely

Table 3. BIC model selection

COFI	Probability (%)	SD	Model 1	Model 2
Intercept	100.0	0.22584	0.42925	0.75443
FS	12.5	0.02872	0.12317	0.12022
HS	100.0	0.02431	0.27645	0.29089
EC	100.0	0.03158	0.12498	0.12360
PA	100.0	0.03606	0.08676	
FQ	90.8	0.02742	0.15746	0.16008
SL	100.0	0.03823	0.15521	0.14297

Table 4. Model test

Model	nVar	R^2	BIC	Post prob
Model 1	6	0.654	−283.42493	0.908
Model 2	5	0.642	−278.85448	0.092

5 Conclusions

The optimal choice by the BIC for the Consumer's Organic Food Buying Intention in COVID-19 Pandemic explored the factors affecting consumers' intention to buy organic food in Ho Chi Minh City during the Covid-19 Pandemic, research has shown that there are 6 influencing factors order from strongest to the weakest: Health Consciousness (0.27645), Food quality (0.15746), Social lifestyle (0.15521), Environment care (0.12498), Food safety (0.12317), and paying ability (0.08676). The factors in the model are suitable for the survey site (Ho Chi Minh City). These factors are mostly the factors that strongly influence the intention to buy organic food. The importance of health consciousness factors to the intention to buy organic food. Health consciousness is a factor that has an average influence on the intention to buy organic food in the study of Imani et al. [10] and that factor has the most significant influence on the intention to buy organic food in the study of Lian and Yoong [9]. Because the research is surveying at the time of the ongoing Covid-19 Pandemic, consumers should put their health first. Paying ability (PA) with the lowest Beta value is like the study of Lian and Yoong [9]. For Environment care (EC) is at the average level of impact on purchase intention. Factors also positively affect the intention to buy organic food in Ho Chi Minh City, Vietnam. With the results, this study has positive contributions to management practice. Businesses, retailers, and marketers can more effectively plan and execute targeting and positioning strategies to drive consumers to buy organic food. The study provides valuable information for organic food businesses, retailers, or marketers to improve sales and achieve continued business growth. Thus, managers need solutions to improve consumers' intention to buy organic food in Ho Chi Minh City during the Covid-19 Pandemic even more soon.

Acknowledgments. This research is funded by the Industrial University of Ho Chi Minh City, Vietnam.

References

1. https://bnews.vn/thanh-pho-ho-chi-minh-tu-tin-buoc-vao-nam-2021/182541.html
2. Ajzen, I., Fishbein, M.: Belief, Attitude, Intention, and Behaviour: An Introduction to Theory and Research. Addision-Wesley, Reading (1975)
3. Rashid, N.R.N.A., Jusoff, K., Kassim, K.M.: Eco-labeling perspectives amongst Malaysian consumers. Can. Soc. Sci. **5**(2), 1–10 (2009)
4. Leyva-Hernández, S.N., Toledo-López, A., Hernández-Lara, A.B.: Purchase intention for organic food products in Mexico: the mediation of consumer desire. Foods **10**(2), 245 (2021)
5. Soroka, A., Wojciechowska-Solis, J.: Consumer motivation to buy organic food depends on lifestyle. Foods **8**(11), 581 (2019)
6. Kumar, P., Mahato, D.K., Kamle, M., Mohanta, T.K., Kang, S.G.: Aflatoxins: a global concern for food safety, human health and their management. Front. Microbiol. **7**, 2170 (2017)
7. Downing, N.S., et al.: Postmarket safety events among novel therapeutics approved by the US Food and Drug Administration between 2001 and 2010. JAMA **317**(18), 1854–1863 (2017)
8. Shen, C., Wei, M., Sheng, Y.: A bibliometric analysis of food safety governance research from 1999 to 2019. Food Sci. Nutr. **9**(4), 2316–2334 (2021)
9. Lian, S.B., Yoong, L.C.: Assessing the young consumers' motives and purchase behavior for organic food: an empirical evidence from a developing nation. Int. J. Acad. Res. Bus. Soc. Sci **9**, 69–87 (2019)
10. Imani, B., Allahyari, M.S., Bondori, A., Surujlal, J., Sawicka, B.: Determinants of organic food purchases intention: the application of an extended theory of planned behaviour. Future Food J. Food Agric. Soc. **9**, 1–12 (2021)
11. Bondori, A., Bagheri, A., Allahyari, M.S., Damalas, C.A.: Pesticide waste disposal among farmers of Moghan region of Iran: current trends and determinants of behavior. Environ. Monit. Assess. **191**(1), 1–9 (2019). https://doi.org/10.1007/s10661-018-7150-0
12. Bondori, A., Bagheri, A., Damalas, C.A., Allahyari, M.S.: RETRACTED: Use of personal protective equipment towards pesticide exposure: farmers' attitudes and determinants of behavior. Sci. Total Environ. **639**, 1156–1163 (2018)
13. Kaczorowska, J., Prandota, A., Rejman, K., Halicka, E., Tul-Krzyszczuk, A.: Certification labels in shaping perception of food quality—insights from Polish and Belgian urban consumers. Sustainability **13**(2), 702 (2021)
14. Muhammad, B., Khan, M.K., Khan, M.I., Khan, S.: Impact of foreign direct investment, natural resources, renewable energy consumption, and economic growth on environmental degradation: evidence from BRICS, developing, developed and global countries. Environ. Sci. Pollut. Res. **28**(17), 21789–21798 (2021). https://doi.org/10.1007/s11356-020-12084-1
15. Chekima, B.: Consumer values and green products consumption in Malaysia: a structural equation modelling approach. In: Green Business: Concepts, Methodologies, Tools, and Applications, pp. 206–231. IGI Global (2019)
16. Singh, S., Alok, S.: Drivers of repurchase intention of organic food in India: role of perceived consumer social responsibility, price, value, and quality. J. Int. Food Agribus. Mark. 1–23 (2021)

17. Ghorbani, M., Koocheki, A., Rajabzadeh, M., Mansori, H.: The study of willingness to accept of Khorasan Razavi Province farmers to produce greenhouse organic cucumber. J. Econ. Agric. Dev. **28**, 149–156 (2014)
18. Jeyakumar Nathan, R., Victor, V., Popp, J., Fekete-Farkas, M., Oláh, J.: Food innovation adoption and organic food consumerism—a cross national study between Malaysia and Hungary. Foods **10**(2), 363 (2021)
19. Hu, F., Liu, Y., Willett, W.: Preventing chronic diseases by promoting healthy diet and lifestyle: public policy implications for China. Obes. Rev. **12**(7), 552–559 (2011)
20. Hair, J., Black, W., Babin, B., Anderson, R., Tatham, R.: Multivariate Data Analysis. Cengage (2019)
21. Raftery, A.E., Madigan, D., Hoeting, J.A.: Bayesian model averaging for linear regression models. J. Am. Stat. Assoc. **92**(437), 179–191 (1997)
22. Kaplan, D.: On the quantification of model uncertainty: a Bayesian perspective. Psychometrika **86**(1), 215–238 (2021). https://doi.org/10.1007/s11336-021-09754-5
23. Raftery, A.E.: Bayesian model selection in social research. Sociol. Methodol. **25**, 111–163 (1995)

Students' Performance and Engagement in Discrete Mathematics Online Learning During COVID-19 Pandemic

Shiau Foong Wong[1](✉), Malissa Maria Mahmud[1], and Shiau San Wong[2]

[1] Sunway University, Petaling Jaya, Selangor, Malaysia
{janicew,malissam}@sunway.edu.my
[2] Help University, Kuala Lumpur, Malaysia
wong.shiau.san@help.edu.my

Abstract. The COVID-19 pandemic has led to a paradigm shift in higher education. Online learning emerges as a new norm and is used as the core method in teaching the curriculum in helping to curb the pandemic. It provides the students the flexibility to learn anytime and anywhere. However, flexibility, on the other hand, is putting students with self-discipline difficulties at a disadvantage, affecting the students' engagement in an online learning environment. Considering discrete mathematics constitutes real-life application, it is interesting and significant for students' lifelong learning skills; thus, the purpose of this study is to examine the relationship between students' academic performance and engagement in discrete mathematics learning in a full online environment during the COVID-19 pandemic. The findings indicated that a higher level of student engagement in an online learning environment does not guarantee better academic performance. It also shows that the assignment completion has no relation with the academic performance. This study implies that different online strategies on student engagement and academic performance in designing an online course are a crucial avenue that cannot be neglected.

Keywords: Discrete mathematics · Engagement · Performance · Online · Pandemic · Higher education

1 Introduction

The COVID-19 pandemic has disrupted and impacted various sectors from economic, social to education. More than 91% of the students around the world cannot go to schools or educational institutions due to the closure of schools as a result of the pandemic since April 2020 [1]. The rapid spread of the COVID-19 pandemic has forced universities and educational institutions worldwide to suspend all face-to-face classes to curb the pandemic. To mitigate the impact of the pandemic, an overwhelming number of educational institutions have transformed the teaching to emergency remote teaching (ERT) through online platforms. Online learning emerges as the new trend and is used as the core curriculum method during the COVID-19 pandemic [2]. Higher

Y. Maleh et al. (Eds.): ICI2C 2021, LNNS 357, pp. 354–365, 2022.
https://doi.org/10.1007/978-3-030-91738-8_33

educational institutions must stay competitive in the education industry and promote students' learning experience and improve students' learning outcomes [3] through online learning. There are many factors in making online learning successful, including using the right tools and methods [4], accessibility, course content and assessment criteria. On the other hand, online learning also has its limitations, including poor internet connection quality and digital literacy challenges for both academics and students. According to Parienen [5], lecturers without technology skills must be empowered to have the ability to change the learning content, which provides students better access to knowledge.

Technology adoption has become a necessity for all types of courses, including mathematics [6], especially during this pandemic. The well-developed technology provides the mathematics instructors the opportunity to implement online learning with various strategies, thus bringing a positive impact on students' learning experiences. Sujarwo et al. [7] discovered that students have a positive perspective towards online learning. They are interested in online learning, which provides them the flexibility to learn anytime and anywhere in the midst of the pandemic. Foong and Mahmud [8] found that online learning helps students learn different levels of Mathematics in higher education, including high-level mathematics subjects.

Despite the many good observations on online learning from several researches [9, 10], some research studies have investigated the effects of online learning during the pandemic and realized that the learning is not as effective as expected. Agung et al. [11] noticed that students were unprepared for the new teaching and learning style. About 50% of the students managed the online class well and some were more favorable to offline classes. Adnan [12] stated that students' performance is not up to expectation due to the inaccessibility to the internet in Pakistan. Another research by Irawan et al. [13] showed that online learning is not effective as students were feeling bored after two weeks of home studying and were given many tasks that discouraged them from learning.

Self-discipline is very important in online learning environment as the environment is highly autonomous and the lecturers are not present physically to offer support [14]. Flexibility puts students with self-discipline difficulties at a disadvantage [15, 16] and thus affects their engagement in the class. Along with the growing number of online classes, it brings attention to the instructors concerning the students' engagement and high dropout rates [17] compared with face-to-face classes [18]. According to Rajabalee et al. [19], one of the components of the quality of online education is to ensure learner engagement. Previous research indicated that low engagement and poor self-discipline, which reflects low study persistence, contribute to poor academic performance and high attrition rate among students in online courses [18, 20]. Thus, it is necessary to understand the influence of high engagement on learning and motivation among students in online class. Considering discrete mathematics constitute real-life application, it is interesting and significant for students' lifelong learning skills; thus, in this context, this study is aimed to examine the relationship between students' academic performance and their level of engagement in discrete mathematics learning in an online classes environment during the COVID-19 pandemic.

2 Methodology

In curbing the pandemic, the university has decided to implement full online classes without face-to-face teaching and learning. Considering the limitation of the internet access and the different time zone between Malaysia and the region that the international students are staying, thus this study employed a research to deploy full online teaching and learning of discrete mathematics to examine the relationship between students' academic performance and engagement in a full online environment. Discrete Mathematics is a bachelor's degree course' subject with 14 weeks of lessons in an American Degree Program (ADP), Sunway University, Malaysia. The learning outcomes of this subject are:

a) Demonstrate the understanding of mathematical statements.
b) Explain the differences between combination and permutation, and to use them to compute the probability of an event.
c) Apply the knowledge of recurrence and iterations in writing programming languages.
d) Apply the concepts of tree and graph theory to model a real-life problem or use the shortest path algorithm to solve combinatorial problems such as traveling salesman problems.

2.1 Procedure

On the first day of the semester, a live class was conducted and students were informed about the full online mode of teaching and learning of discrete mathematics throughout the whole semester, August 2020, over 14 weeks of lessons, with a total of 21 online classes and 5 online assessments including quizzes, tests and final examination. Each online class is 1.5 h in duration. Besides, the instructor's expectations were also communicated to the students to encourage them to fulfill their potential. The pre-recorded videos demonstrating the concepts involved in each learning module and the assignment will be made available on the learning management system, eLearn on the class day (Tuesday and Thursday). Students are required to access the material by viewing the videos at their own pace and assignment needs to be completed to assess their understanding of the content learned within the same day. They are required to submit the assignment through eLearn by the end of the day. Whenever there is an issue, the students could contact the instructor through email for guidance and support. Live discussion might be arranged if necessary to ensure the students get a better understanding. Assignment submitted would be marked and feedback would be provided to students through eLearn platform. It would also be used for attendance purpose. Moreover, the material for each learning module and the fully worked solutions would be shared in eLearn on the next day.

Figure 1 shows the samples of the learning module in which clear instructions are given in the online classes through eLearn.

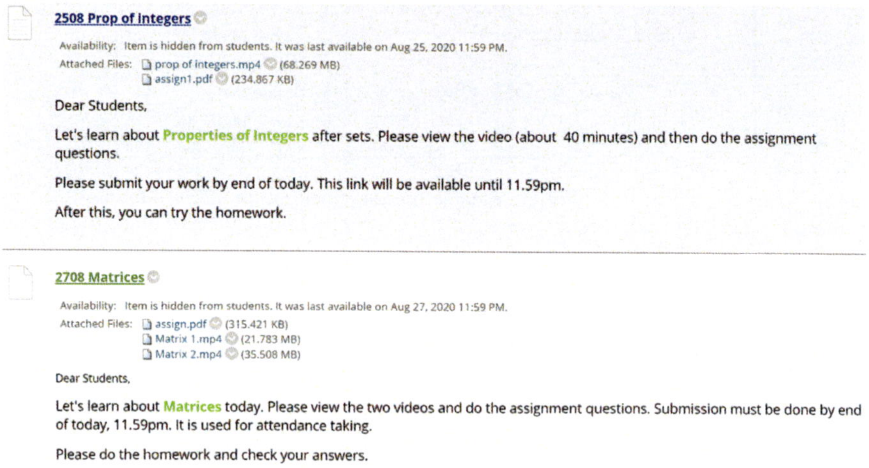

Fig. 1. Sample of each learning module

Figure 2 displays the pre-recorded video uploaded in one of the online classes.

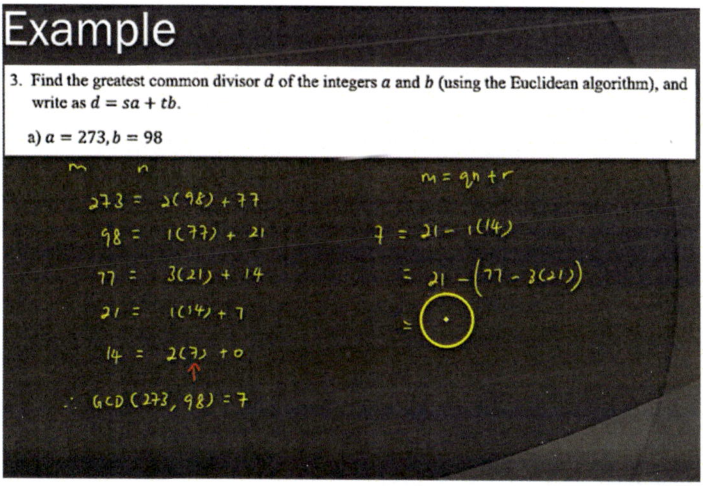

Fig. 2. Sample of pre-recording video

Figure 3 shows the sample of the assignment for one of the online classes.

MATH 1053
Properties of Integers

1. Prove the following statement.
   If $a|b$ and $b|c$, then $a|c$. [4 marks]

2. Find the GCD of $a = 540$ and $b = 504$ by using Euclidean algorithm and write as $d = sa + tb$. [4 marks]

3. Find GCD and LCM of 36 and 90 by using product of primes. [4 marks]

Fig. 3. Sample of assignment

2.2 Participants

The sample size for this class is 10 in the beginning of August 2020 semester in American Degree Program, Sunway University. There are two international students—one from Japan and another one from Bangladesh. However, in week 10 (add/drop week), three students withdrew from the course and 7 students remained in the class. The sample size is small, but it is deemed apt considering the total number of students enrolled in the program in January 2020, June 2020 and August 2020 are 50, 31 and 37, respectively and the number of students who took Discrete Mathematics in previous semesters ranges from 5 to 10 students; thus, sufficient to generalize the intended population.

3 Data Analysis

In this study, the number of students who withdrew from the course indicates the dropout rate. The comparison of the students' withdrawal between this cohort of students (August 2020) and another cohort of students from the previous semester (August 2019); in which the students learned the same content of syllabus through a mix-mode of face-to-face and online classes with more face-to-face classes, 20 (71.4%) classes are face-to-face mode and 8 (28.6%) classes are online learning mode. For face-to-face classes, students attend and participate in the class. These are viewed as engagement and they do not have direct online analogs. The comparison of the students' withdrawal between this two cohort of students allows us to look into the students' engagement in a fully online learning environment compared to a mix-mode with a more face-to-face learning environment.

The students' course activity in the eLearn platform is overviewed to examine the students' engagement in a fully online course. It includes the total time spent online per day and the total time spent online by each active student in the course in eLearn.

Besides, to study the students' academic performance, the overall mean of all the assessments, which contain two quizzes, two tests, and a final examination, were calculated. Students' performance and their course activity online that indicating students' engagement are compared. It is worth mentioning that each assessment was a closed-book test which was conducted online through Respondus lockdown browser and proctoring. Any suspected of misconduct will be penalized accordingly as a promotion of integrity, one of Sunway University's core values.

3.1 Statistics

Table 1 shows the number of students who withdrew from the course in week 10 for August 2020 and August 2019. Three students withdrew from the course in August 2020, while there was no student withdrew from the course in August 2019.

Table 1. Number of withdrawals in week 10 for August 2019 and August 2020

	August 2019	August 2020
The number of students enrolled in the course	10	10
The number of students withdrew from the course	0	3

The course overview, which indicates the total time spent online by day over the 14 weeks of the lesson is shown in Fig. 4. The total time spent on course activity on eLearn for 7 active students is 479.64 h in this semester. Students spent most of their course activity time on Tuesday and Thursday, 40.7% (195.33 h) and 39.10% (187.54 h) of the total course activity time throughout the whole semester. This means that each student spent an average of 2.33 h and 2.06 h on Tuesday and Thursday, respectively to learn the content through online material on eLearn. It is about 30 to 48 min more than the class time, which is 1.5 h. They also spent some time on Sunday, Monday and Wednesday for self-revision, averaging about 23 min per student per day.

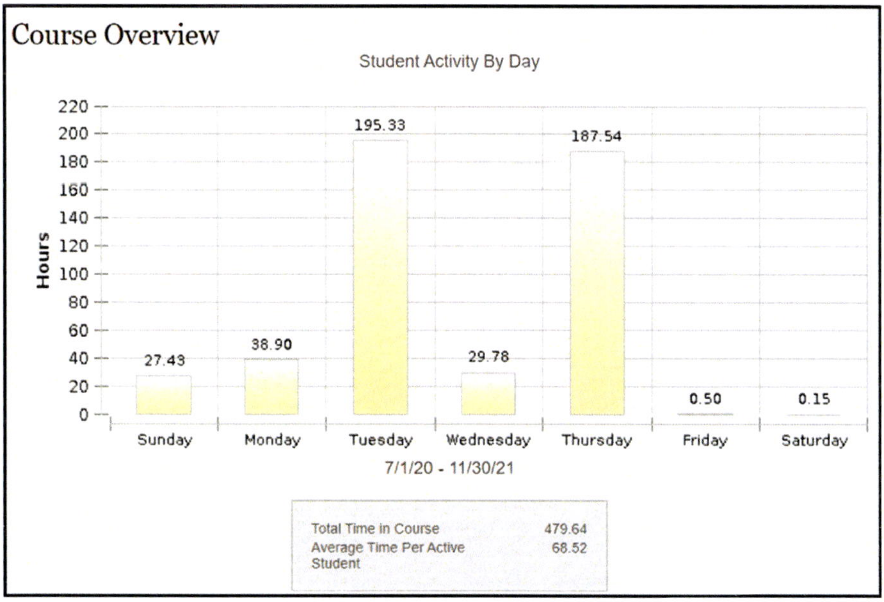

Fig. 4. Total time spent online by day over 14 weeks of lesson

The student overview displays the total time spent online by each active student over the 14 weeks of the lesson is shown in Fig. 5. Student B spent the most time on course activity on eLearn, which is about 29.1% of the total course activity time, followed by Student G, about 23.6%. Student C spent the least time on course activity online through eLearn, which is about 0.87%.

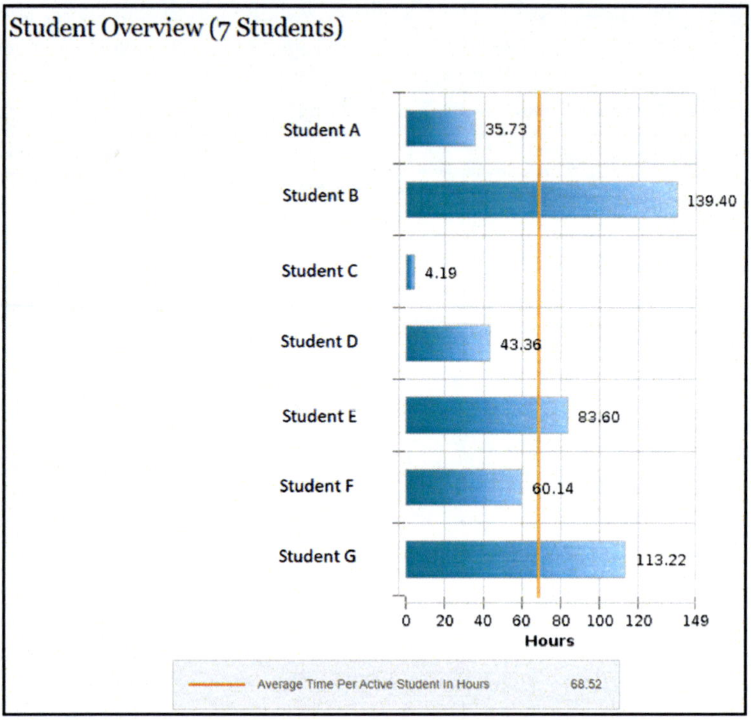

Fig. 5. Total time spent online by each student over 14 weeks of lesson

Table 2 shows the mean result from all the assessments and the percentage of assignment completion from all online classes of each student. Students D, E, F and G got a mean result of over 90%. While for the assignment completion, student A completed the least number of assignments, only at 38.9%, but with a mean result of 85.1%. Student B and C completed 94% and 100% of the assignment, respectively, but with a mean result of only 70.3% and 70.1% respectively. The rest of the students completed all the assignments.

Table 2. Mean result and assignment completion percentage

Student	Mean result	Assignment completion
Student A	85.1	38.9
Student B	70.1	94.4
Student C	70.3	100
Student D	95.8	100
Student E	95.8	100
Student F	90.8	100
Student G	90	100

4 Discussion

This study incorporated a fully online learning environment with the aim to examine the relationship between students' academic performance and engagement in discrete mathematics learning during the pandemic. The usage of eLearn is important due to the current teaching and learning trends, thus, the average time spent on eLearn platform would be able to indicate the online engagement of the students. While the analyses of the total time spent online by each student and the mean result of each student are suitable to probe the postulated aim. Consequently, the findings from the study would provide the instructors an insight and assist them in designing an effective online course and supporting the students [21].

Three students withdrew from the course in week 10, while the remaining 7 students continued the course until the end of the semester. This indicates a 30% dropout rate in the fully online learning environment compared to 0% of dropout rate in a mixed mode of teaching and learning environment. It reflects that the course dropout rate in the fully online learning environment is higher than the mixed mode learning environment. It is consistent with the findings from the researches of Kovanović et al. [17] and Otter et al. [18]. Students feel more disconnected from lecturers and peers, having less engagement with others. The students who are weak in student-content interaction, having less engagement and poor self-discipline due to flexibility provided, and cannot adjust to the online learning environment might feel being left behind and thus, decide to withdraw from the course. To be successful in learning through an online learning environment, the students must be more self-directed, but the lecturer also needs to be careful in designing the online course to engage the students.

The statistics show that, on average, the active students spent most of the total course activity time on Tuesday and Thursday as these two days are the online class day for this course. The online class time is 1.5 h per day, and they spent an average of 2.2 h on online class days to learn the content of the course. They were willing to spend more time than the class time to learn the content at their own pace and being self-discipline throughout the whole semester (14 weeks). The data also shows that each active student would spend an average of 23 min each day on Sunday, Monday, and Wednesday, respectively, for self-revision. With this data presented, active students can manage their time and be self-regulated on online learning. They are found to be well engaged in the fully online learning environment.

When we look into the average time spent by individual students, the more time spent on course activity online, the higher level the student engagement to the material. To be precise, it is online engagement. Student B spent the most time on course activity on eLearn compared to other students, he is the most online-engaged student, but he got the lowest score in the class. On the other hand, student C spent the least time on eLearn, he has the lowest level of online engagement, but he managed to pass this course. The more time the student spent on course activity, the higher level of online engagement is. However, this does not reflect better academic performance.

Conversely, less time spent on course activity online does not mean will have poor academic performance. This indicates that high or low student engagement in an online learning environment does not necessarily anticipate better or poorer academic performance, respectively. This finding contrasts with the findings from You [20], where it demonstrated that the frequency of course logins predicted their course achievement. Besides, the analysis from Lei et al. [22] revealed that method of reporting engagement, cultural value and gender are the factors that influence the relationship between student engagement and academic achievement. Therefore, we need to consider many other factors, including students' academic abilities, demographics, financials, emotions, and others, which could play an important role in the duration of the engagement.

Take offline engagement as an example. Students might spend more time having their revision offline, and this is considered offline engagement. Sutherland et al. [23] discovered that using technology and social media within students' academic community facilitates offline engagement and its influence on academic performance. It helps to reduce the student dropout rate too. Besides, we also should consider the effect of prior knowledge. Many researchers found that learning performance is strongly influenced by prior knowledge [24, 25]. Dong et al. [26] discovered the students with high prior knowledge are more likely to engage in adaptive instrumental help-seeking, which means the students might get help from peers or have better self-regulated learning in an offline environment.

Comparing the assignment completion percentage and the mean result, students C, D, E and F completed all the assignments throughout the whole semester. However, student C's mean result is 70.1%, while the rest of the three students scored over 90%. On the other hand, student A completed only 38.9% of the assignment, but his mean result was over 85%. This information shows that there is no relationship between assignment completion and academic performance. However, assignment completion is a good indicator of positive student-content engagement [27] and interaction regarding learning [28].

We should pay attention to designing an online course that can gain students' interest. Strategies are needed to enhance the students' engagement with not only the instructors, but also with the material. We can involve more activities like jigsaw method, group activities and peer discussion to promote students' engagement. Mayer et al. [29] mentioned that technology-based methods such as clickers can help to increase students' engagement in a large group and, thus, improve academic performance. Irani and Denaro [30] found that embedding interactive features into online tools such as video lectures and online discussion promotes student engagement with the material. Mahmud and Wong [31] suggested incorporating online peer assessment and a collaborative learning environment through online networking in virtual groups to facilitate peer interaction for better engagement.

5 Conclusion

This study examines the relationship between students' academic performance and engagement in discrete mathematics learning in a fully online learning environment during the COVID-19 pandemic. The findings indicated that a higher level of student engagement in an online learning environment does not guarantee better academic performance. The offline engagement within the academic community facilitated through technology and social media could be the factor in academic performance promotion. The effect of prior knowledge should also be considered as another factor that influences learning performance. High prior knowledge could lead to better academic performance. A practical implication of this study is that the instructors should pay attention to different online strategies on student engagement and academic performance in designing an online course, which is also a crucial avenue for future study.

References

1. United Nations (UN): Sustainable Development Goals (2020). https://www.un.org/sustainabledevelopment/education/. Accessed 19 June 2020
2. Mardiah, H.: The use of E-learning to teach English in the time of the Covid-19 pandemic. Engl. Teach. Linguist. J. (ETLiJ) **1**(2), 49–55 (2020)
3. Stefanovic, D., Drapsin, M., Nikolic, J., Scepanovic, D., Radjo, I., Drid, P.: Empirical study of student satisfaction in e-learning system environment. Tech. Technol. Educ. Manag. **6**(4), 1152–1164 (2011)
4. Zelkowski, J., Gleason, J., Cox, D., Bismarck, S.: Developing and validating a reliable TPACK instrument for secondary mathematics preservice teachers. J. Res. Technol. Educ. **46**(2), 173–206 (2013). https://doi.org/10.1080/15391523.2013.10782618
5. Perienen, A.: Frameworks for ICT integration in mathematics education - a teacher's perspective. Eurasia J. Math. Sci. Technol. Educ. **16**(6), em1845 (2020). https://doi.org/10.29333/ejmste/7803
6. Kaleli-Yilmaz, G.: The views of mathematics teachers on the factors affecting the integration of technology in mathematics courses. Aust. J. Teach. Educ. **40**(40), 132–148 (2015). https://doi.org/10.14221/ajte.2015v40n8.8
7. Sujarwo, S., Akhiruddin, R., Siradjuddin, S.: An analysis of university students' perspective on online learning in the midst of Covid-19 pandemic. Jurnal Pendidikan Dan Pengajaran **53** (2), 125–137 (2020)
8. Foong, W.S., Mahmud, M.M.: I can learn calculus virtually: personalizing eLearn and Screencast-O-Matic in blended learning environment. In: Proceedings of the 5th International Conference on Frontiers of Educational Technologies, pp. 13–19. ACM, June 2019
9. Macfadyen, L.P., Dawson, S.: Mining LMS data to develop an "early warning system" for educators: a proof of concept. Comput. Educ. **54**, 588–599 (2010). https://doi.org/10.1016/j.compedu.2009.09.008
10. Roby, T., Ashe, S., Singh, N., Clark, C.: Shaping the online experience: how administrators can influence student and instructor perceptions through policy and practice. Internet High. Educ. **17**, 29–37 (2013). https://doi.org/10.1016/j.iheduc.2012.09.004

11. Agung, A.S.N., Surtikanti, M.W., Quinones, C.A.: Students' perception of online learning during COVID-19 pandemic: a case study on the English students of STKIP Pamane Talino. SOSHUM: Jurnal Sosial dan Humaniora [J. Soc. Sci. Hum.] **10**(2), 225–235 (2020). https://doi.org/10.31940/soshum.v10i2.1316
12. Adnan, M.: Research Article Online learning amid the COVID-19 pandemic: students' perspectives. J. Pedagogical Sociol. Psychol. **2**(1), 45–51 (2020). https://doi.org/10.33902/JPSP.CITATION
13. Irawan, A.W., Dwisona, D., Lestari, M.: Psychological impacts of students on online learning during the pandemic COVID-19. KONSELI : Jurnal Bimbingan dan Konseling (E-Journal) **7**(1), 53–60 (2020). https://doi.org/10.24042/kons.v7i1.6389
14. Wong, J., Baars, M., Davis, D., Zee, T.V.D., Houben, G., Paas, F.: Supporting self-regulated learning in online learning environment and MOOCs: a systematic review. Int. J. Hum.-Comput. Interact. **35**, 356–373 (2019)
15. Dyrbye, L., Cumyn, A., Day, H., et al.: A qualitative study of physicians' experiences with online learning in a master's degree program: benefits, challenges, and proposed solutions. Med. Teach. **31**, 40–46 (2009)
16. Bediang, G., Stoll, B., Geissbuhler, A., et al.: Computer literacy and e-learning perception in Cameroon: the case of Yaounde Faculty of Medicine and Biomedical Sciences. BMC Med. Educ. **13**, 1–8 (2013)
17. Kovanović, V., Joksimović, S., Gašević, D., Owers, J., Scott, A.M., Woodgate, A.: Profiling MOOC course returners: how does student behavior change between two course enrollments? In: Proceedings of the Third (2016) ACM Conference on Learning@ Scale, pp. 269–272. ACM (2016)
18. Otter, R.R., et al.: Comparing student and faculty perceptions of online and traditional courses. Internet High. Educ. **19**, 27–35 (2013). https://doi.org/10.1016/j.iheduc.2013.08.001
19. Rajabalee, B.Y., Santally, M.I., Rennie, F.: A study of the relationship between students' engagement and their academic performances in an eLearning environment. E-learning Digit. Media **17**(1), 1–20 (2020)
20. You, J.W.: Identifying significant indicators using LMS data to predict course achievement in online learning. Internet High. Educ. **29**, 23–30 (2016). https://doi.org/10.1016/j.iheduc.2015.11.003
21. Robinson, C.C., Hullinger, H.: New benchmarks in higher education: Student engagement in online learning. J. Educ. Bus. **84**(2), 101–109 (2008). https://doi.org/10.3200/JOEB.84.2.101-109
22. Lei, H., Cui, Y., Zhou, W.: Relationships between student engagement and academic achievement: a meta-analysis. Soc. Behav. Personal. Int. J. **46**(3), 517–528 (2018)
23. Sutherland, K., Davis, C., Terton, U., Visser, I.: University student social media use and its influence on offline engagement in higher educational communities. Student Success **9**(2), 13–24 (2018)
24. Yang, J.C., Quadir, B.: Effects of prior knowledge on learning performance and anxiety in an English learning online role-playing game. J. Educ. Technol. Soc. **21**(3), 174–185 (2018)
25. Lee, J.Y., Donkers, J., Jarodzka, H., Van Merriënboer, J.J.: How prior knowledge affects problem-solving performance in a medical simulation game: using game-logs and eye-tracking. Comput. Hum. Behav. **99**, 268–277 (2019)
26. Dong, A.M., Jong, M.S.Y., King, R.: How does prior knowledge influence learning engagement? The mediating roles of cognitive load and help-seeking. Front. Psychol. **11**, 591203 (2020)
27. Ozan, O., Ozarslan, Y.: Video lecture watching behaviors of learners in online courses. Educ. Media Int. **53**(1), 27–41 (2016). https://doi.org/10.1080/09523987.2016.1189255

28. Ma, J., Han, X., Yang, J., Cheng, J.: Examining the necessary condition for engagement in an online learning environment based on learning analytics approach: the role of the instructor. Internet High. Educ. **24**, 26–34 (2015). https://doi.org/10.1016/j.iheduc.2014.09.005
29. Mayer, A., et al.: Using technology-based methods to foster learning in large lecture classes: evidence for the pedagogic value of clickers (n.d.)
30. Irani, S., Denaro, K.: Incorporating active learning strategies and instructor presence into an online discrete mathematics class. In: Proceedings of the 51st ACM Technical Symposium on Computer Science Education, pp. 1186–1192, February 2020
31. Mahmud, M.M., Wong, S.F.: Facebook and collaborative learning: an empirical study on online assessment. Int. J. Learn. Teach. **4**(2), 107–113 (2018)

Advances in Smat Healthcare

A Smart Health Monitoring Application for Patients to Improve Health

Paola Ariza-Colpas[1], Shariq Aziz Butt[2(✉)], Muhammad Imran Tariq[3], Roberto Cesar Morales Ortega[1], Muhammad Waqas Anjum[4], and Marlon Alberto Piñeres Melo[1]

[1] Universidad De La Costa, CUC, Barranquilla, Colombia
{parizal, rmoralesl}@cuc.edu.co, pineresm@uninorte.edu.co
[2] University of Lahore, Lahore, Pakistan
[3] The Superior University, Lahore, Pakistan
[4] Khwaja Fareed University of Engineering and Information Technology, Rahim Yar Khan, Pakistan

Abstract. Health is growing nowadays due to technological revolutions. These revolutions make health smart and easy to monitor the patient at their location. In this paper, we are suggesting and developing the smart health monitoring scenario. In our system, we are monitoring patients with the disease of frequent urine due to diabetes, stiffness of bones due to ortho disease, and heart rate monitoring due to heart disease we used some statically measured to find results from the generated data during the monitoring. Our system reveals that the scenario we created for monitoring patients at their location is more accurate and valuable during the decision-making to take appropriate action. Objective of this study is to support the elder patients for their remote health monitoring with the trustworthy scenario. We used an app development method for it. We used almost 10 patients to evaluate the results of our developed system. The study reveals that our developed system is very valuable for elder patients' health monitoring. We selected 10 patients with the mentioned diseases to monitor them and check the validity of our developed system. We find that our developed system is very efficient and useful for monitoring and generating an alarm in the case of any abnormality situations. In conclusion, there is an essential need for such systems that monitor the patient effectively and take decisions timely. In this paper, we implemented a model for elder patients to monitor the patient with different diseases such as frequent urine sugar levels, and stiffness in bones. For this, we deployed sensors in the patient's body and attached them to the mobile to transmit the data to the other ends. We also set an alarming stage in the system when there is any abnormality condition observed. The study reveals that there is a need for a more precise and decision support mechanism for disease monitoring to enhance the elder people life.

Keywords: Smart Health · Elder monitoring · Mobile application development

Y. Maleh et al. (Eds.): ICI2C 2021, LNNS 357, pp. 369–382, 2022.
https://doi.org/10.1007/978-3-030-91738-8_34

1 Introduction

In the 9th century, direct challenges were faced by older people due to fewer and low advancements in technologies. More or less, these technologies are non-affective for people to use for health purposes. The changes and advancements in the devices make hospitals, clinics, and medical services more feasible for patients. Now the health can be demonstrated through smart mobile applications and web interfaces. Success and unbelievable thoughts can appear through applications on PDAs or other splendid sensors. The accessibility and correspondence openings are crossed over these inventive applications in a useless style [1, 2].

In previous centuries, eHealth, as described by the World Health Organization, through the use of data and networking technology for health has extended dramatically. As an alternative to or substitute for treatment-as-usual, digital innovations are used in clinical practice to facilitate patient involvement and to track and control a patient's health condition. While eHealth solutions have primarily been used in young people, they can also be useful in promoting patient independence, and mental well-being, and state of health in the elderly [3]. The efficacy of technical technologies for self-management and remote monitoring has also been investigated mostly in older patients with chronic diabetes or heart disease, instead of in younger surgery patients. Telehealth after elective internal medicine is the addition since it is non - inferior to face-to-face psychotherapy in older patients. Leading to comorbid conditions and earlier hospital release after surgery, older patients who experience more complicated surgery have a higher risk of experiencing preoperative problems [4]. These issues are more likely to arise at home. As a result, eHealth approaches with objectives like an earlier diagnosis of problems may theoretically support this community; nevertheless, descriptions in the research are limited. One reason for the scarcity of research on eHealth in older patients experiencing complicated surgery could be that virtual care technologies are deemed unfeasible to such communities due to uncertainties about accessibility, enforcement, and technology availability [5].

The Internet of things (IoT) has revolutionized the technology for the coming generation with new amendments and concepts. The IoT improves the sensors and devices to monitor any environment [6]. Now, these sensors are connected in a more advanced manner and now can support device-to-device connectivity. Using this automation, the old human services systems can be made in the future. Overall prosperity services are standing up to challenges considering the quick improvement of old masses [7] accordingly a creative way is depended upon to go facing this test considering the late progress in gear and several no of gadgets have developed that can screen patients' prosperity data perseveringly similarly via the use of internet remotely. Help to patients, the medicinal services framework for the old is imperative to fulfill the wellbeing administration needs to screen day-by-day exercises of the old individuals. There is meant to approve the proficiency of coordinating various advancements in recording day-by-day exercises of old individuals through gadgets such as wearable sensors in actual time [8].

2 Background of Study

The use of wireless connectivity devices to promote public wellbeing and medical care is known as mobile health or mHealth. Smartphones, personal digital assistants, patient monitoring systems, and other cellular devices are examples of smartphones. Because of the widespread adoption of mobile devices, mHealth apps are gaining momentum. With over 6 billion smartphone subscribers, it is reported that over 80 percent of the global population is already served with a cellular wireless connection. Over and above roads and energy, digital technology has enhanced network growth in low- and middle countries. Communication among patients and health services (like call services, scheduling notifications, and medication compliance) for health tracking and monitoring (like surveys, patient monitoring devices) and accessibility of data at the time of service are all examples of remote health apps (health records, decision support).

Health awareness services distribute knowledge about disease detection and well-being promotion. Point-of-care services and diagnostics can be used to enable physicians with medical reference knowledge and medical decision assistance. Patient monitoring helps patients stick to their medication schedule. Disease and epidemic outbreak monitoring keeps a record of diseases in actual time. Deaths and injuries and disasters are signaled by emergency medical response services. Information for use in medical care is handled via healthcare systems. Mobile channels for education services to medical practitioners are provided by learning [9, 10].

Finally, apps that allow the usage of smart mobile payments are referred to as healthcare finance applications.

Several different types of mobile medical apps have been suggested for inclusion in regulations. An app used as an aspect of a device, including a remote view of information through a desk monitor, is one of these forms. It will be critical to make sure that the data presented to the physician is reliable across all displays. mHealth apps which use attachments, view displays, or sensors that are identical to those used in current health devices will also be subject to regulations. Apps that connect a smartphone to a glucose strip reader fall into this group. Once again, reliability is needed for proper diagnosis and treatment [11].

A wide range of health effects is closely related to daily decisions about sleep, diet, exercise, and socialization. The sensor-based observations of user behavior are used in the health monitoring process (e.g., sleeping, exercising, and socialization). The wellness framework analyses these behavior trends and evaluates a multidimensional wellness rating to help the user understanding the influence on health. Behavioral trends and wellness estimates are used to elicit user input that is intended to notify users of their current state of wellbeing while also identifying the behavior adjustments that are required to enhance it [12].

Although many issues about the connection between action and health remain unresolved, the need to take a holistic approach to health and develop an initial wellness model that can be expanded and refined.

In the concrete evidence design, health is measured separately for each of the three activities, yielding three basic health ratings ranging from 0 to 100. Each score attempts to analyze the effect of recent user activity on overall health for only one of the key activities tracked. A score of 100 clearly shows that the individual is meeting the agreed performance rules for those actions. Scores of 0 mean that the individual is not even following the minimum recommended behavioral trends [13].

3 Health of Patients System

3.1 Design

IoT Help to patient's medicinal services is an assorted processing framework of apps and wearable gadgets that associate patients and medicinal services with specialist organizations distantly [14]. The framework capacities by sending a wide assortment of sensors to screen safety and wellbeing of household conditions, individual wellbeing, indispensable wellbeing signs, every day, exercises, and so on that intuitively associate the patient with the specialist on a 24/7 per week premise through calling, video calling, e-prescription, and so on [15].

The gathered clinical information is put away in versatile applications and communicated to the focal database worker promptly or occasionally through the Internet. If there should be an occurrence of Complications, Help to Patients the medicinal services framework sends caution and generate an alarm specialist just like his/her family members [16]. When the alarm is set, the doctor should be prepared to provide emergency assistance to the patient, and in the meantime, the doctor will review the patient's health records using the requested health information from the patient's information base, which is currently stored in the cloud. Figure 1 illustrates the communication of the devices and sensors attached with the application and mobile device associations, with the board to help considerable reliably clinical reports like a heartbeat, diabetes level, are noted and spared in the advantageous application and focal database of the old social security framework. This aggregated clinical information is collected and is being utilized for the making of the patient's history and take any appropriate action against the inputs.

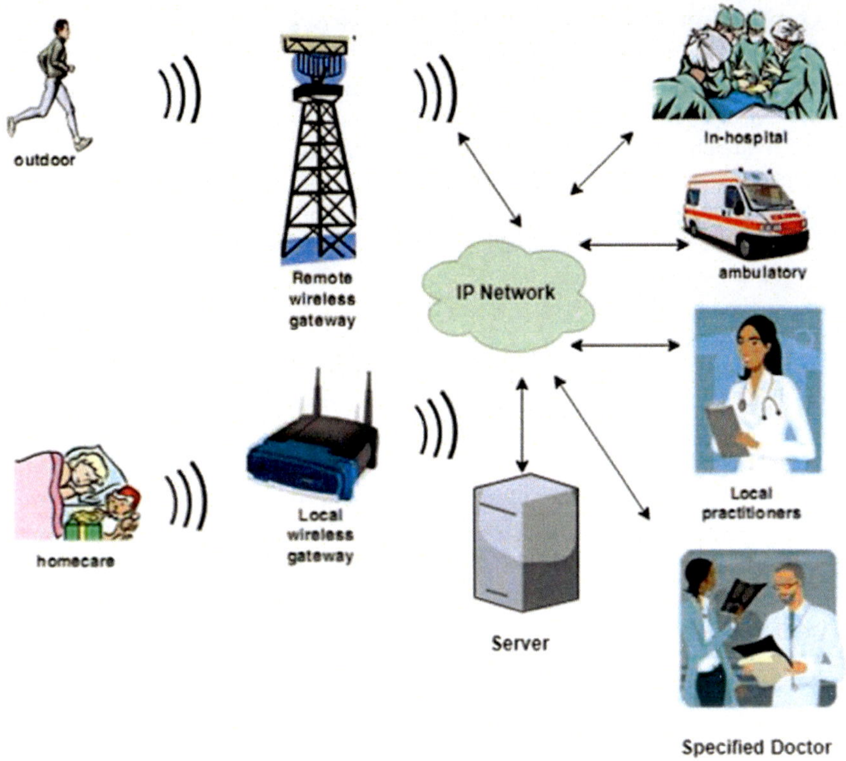

Fig. 1. Design of help to patients healthcare system.

3.2 Protection and Safety in Help with Patients Human Services Framework

Internet of things services such as gadgets and internet associations are more helpless against the security assault because of the increment of unapproved access on the Internet. Programmers cannot just focus on an individual yet can likewise harm the IoT worker and gadgets which thusly can bring about considerable misfortune. In a not-so-distant future, there will be security matters identified with the IoT. Subsequently, security can involve care in a not-so-distant future which wants to be dealt with using alert and appropriate conventions. Safety furthermore, protection, wellbeing, delicate and individual patient clinical information are significant issues and concerns while gathering tolerant clinical information from sensors to cell phone and additional presenting information to the brought together worker [17].

Currently, there is no standard show and rules are open to the screen by what method the data become assembled done in the internet of things. In the Help with Patient prosperity administration structure, each client has connected with sign-in the interface and encoded security to store the affirmation confirmation. Under the suggested framework, the system therefore can be protected after unapproved customers [18].

4 Application of the Suggested Framework

The IoT empowered wellbeing checking framework has an incredible advantage over conventional human services checking frameworks. The older patients can without much of a stretch wear the wellbeing detecting part the time, in this way permitting consistent checking. The framework serves exceptionally indispensable just as advantageous since the old residents need consistent consideration. As consistent checking of patients is not conceivable by specialists nonstop, IoT empowered Help to Patients human services end up being far superior as the basic information can be got to effectively through the Internet. This is a manner that turns out to be profoundly profitable and advantageous as a significant ailment can be identified at the correct time [19]. A remote sensor is set in conditions that are essential to make the remote sensor environment. Since IoT gives ongoing information, it will permit better understanding and knowledge of infection advancement and the impacts of medication treatment [20].

The IoT has a focal choice that can recognize a Complications and hazardous circumstances dependent on the information received by the connected sensors of the patient. In addition, older patients can be checked indoors the same as outside by utilizing IoT support for patients' wellbeing [21].

4.1 Emergency Alert

The sensors are deployed in the environment to sense the patient health and monitor their diseases. When any abnormality is sensed during the monitoring and the patient needs any medical assistance, then the alert message concept is used here. The more established nursing accepts pivotal work and plans to give a fulfilled condition and living space and help as an ensured and sound spot for the old. Help with Patients human services gives a Complications a structured interface stage to screen and help older individuals in a Complications condition. The Complications observing stage gives a top-notch assessment and judgment of the prosperity information of old. This interface permits the client to set up programmed correspondence through calls to the Complications control specialist of the emergency clinic or parental figure intermediary administration. It can likewise report new cautions for the Help of Patients wellbeing application by tapping the button [21].

This emergency stage, moreover, gives an interface to screen the prosperity condition of the old, for instance, tumbling down, cardiovascular disappointment, etc. The patient's family member mobile no or other numbers suggested by the patient are added to the patient information on the application site. Hence, in the case of any abnormality, the interface informed one of them and the specified doctor as well.

4.2 Medicine Reminder

In this phase of the application, we aimed to remind the medicine to patients because due to their elder age, illness, less hope for living, less active body, less active mind, and less remembering of medicine time. Along these lines, the more prepared person's pessimistic procedure towards developing gets unavoidable and achieves more noteworthy than their odds for the duration of regular daily existence. A developing people will by and large have a higher inescapability of relentless afflictions, physical insufficiency, and mental hazards [17]. The quality requirements and prosperity associated with complications of elderly people cannot be found in control.

This interface is supporting the real help-to–a to-patient scenario in our study. It's the main part of our study and the suggested solution. For abnormality situations, the sensor is programmed and set the medicine information in the sensor. The reason for the medicine set in the sensors is to send an alert to the patient to take medicine and send an alert to the doctor with the patient information. This interface will notify or remind the scheduled prescription of the clinical master's step-by-step instructions. When the time limit is reached, the support to the Patient medication notification application will submit a message in the same way as it will warn on the user's smartphone. The objective of this clinical update is to offer a simple to utilize and assist the senior people with keeping up a sound presence with genuine remedies.

4.3 Symptoms Checker

The symptoms checker will enroll such responses to different sicknesses that the more settled the patient has. The end clients will be permitted to pick different appearances as per their risky flourishing situation. At the point when the assurance is done, the interface will select such ailment and illustration the associated division that the patient can guidance to for extra amplification and test. We present MCDM (Multi Rules Decision Making) informative dynamic framework measure (AHP) that is a technique with hierarchal building for decision-making issues [19, 20] to decide and foresee the illnesses of older individuals dependent on various side effects endured by them. MCDM AHP causes the client to comprehend the suitable assessment of their ailments which would along these lines order anticipation through conventional methodologies. This is a superior approach to structure data and to counsel a shrewd social insurance dynamics of loathsome malady to improve personal satisfaction. The approach of MCDM AHP is surrendered in Fig. 2.

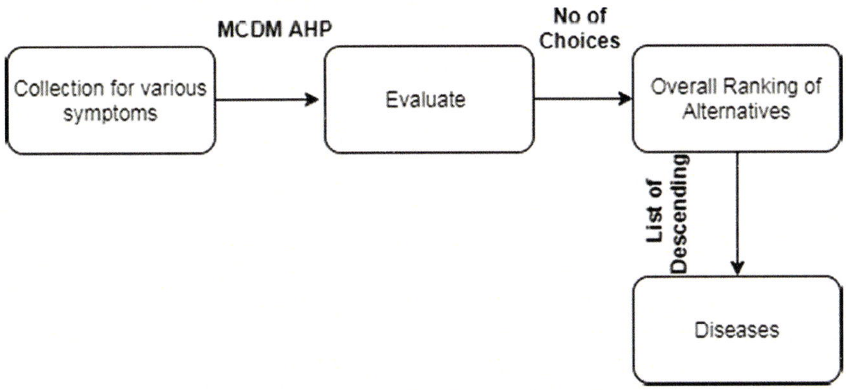

Fig. 2. MCDM AHP for symptoms checker.

Stages for MCDM AHP in the symptoms checker are presented below:

- Express the limitations of different manifestations and decided the sort of patient information.
- Syndicate different contributions to frame a united result of various illnesses.
- Build a lot of pairing for examination networks of different ailments and indications.
- The whole position of the elective ailments from the correlations with the weight needs of the side effects picked by the patients

5 Case Study

For the appropriate analysis of a patient with given different side effects, clinical information is basic to examining the patient for different medical problems. After the examination, we give medication support to patients with a kind of sort of ailments that they feel during their disease. We assumed the MCDM AHP. For the selection of illnesses, IL = {Urine, heartbeat, ortho} endured by older individuals, 3 etymological variable sizes of outright numbers are taken to portray the degree of match of patient indications S = {Numerous pee, heart consume, Stiffness in joint} and patients p_i = {p_1, p_(2), p_(3), p_4} on the choice models, execution producers are abstractly to pick between more than seven options to maintain a strategic distance from the challenges in dynamic. Table 1 presents the phonetic measures related to comparing etymological factors.

Constancy Ratio	CR
Constancy Index	CI
Constancy List	CL
Random Index	RI

Table 1. Linguistic variables scale of absolute value.

Etymological factors	Scales
Weakly and slightly	1
Moderate importance	2
Moderate plus	3
Strong rank	4
Very strong	5
Extremely important	6

From the contextual analysis, we can have 6 correlations of ailments, in this manner we have a 6u6 lattice. The components of the framework are constantly set to 1 and afterward, we have to top off the upper three-sided framework dependent on the significance of options as appeared in Table 2. To utilize the framework in these 3 illnesses, the following guidelines are utilized.

- Either the decision esteem is on the left half of 1, we mark the genuine decision esteem.
- Either the decision esteem is on the correct side of 1, we mark the complementary worth.

Table 2. Pair-wise assessment of the matrix of diseases

	Urine	Heartbeat	Ortho
Urine	1	5	4
Heartbeat		1	1
Ortho			1

To fill the lower three-sided lattice, we utilize the corresponding estimations of the upper corner to corner. If b_xy is the part of row_i and section j of the matrix, at that point to bottom, corner-to–to-corner the components are loaded up with the accompanying formula.

$$b_{xy} = \frac{1}{b_{xy}} \tag{1}$$

Hence, the lower three-sided lattice examination can be finished as appeared in Table 3.

Table 3. Complete pairwise comparison matrix diseases.

	Urine	Heartbeat	Ortho
Urine	1	5	4
Heartbeat	1/5	1	1
Ortho	1/5	1/2	1

Later a detailed examination, matrix pairing has made the standardized matrix Table 4 with the rules as mentioned:

$$y_{xy} = \frac{b_{xy}}{\sum_{x=1}^{m} b_x} \tag{2}$$

Table 4. Normalized pairwise comparison matrix of diseases

	Urine	Heartbeat	Ortho
Urine	0.37	0.37	0.23
Heartbeat	0.37	0.07	0.08
Ortho	0.01	0.03	0.04

The relative loads among the thoughts about various infections are determined and the outcomes are presented in Table 5:

$$w_i = \frac{\sum_{j=1}^{n} y_{xy}}{n} \qquad i = 1, \ldots\ldots, m. \tag{3}$$

Table 5. Relative weights of diseases

Disease attributes	Sum of column	Weight (Wi)
Urine	0.37	0.37
Heartbeat	0.37	0.07
Ortho	0.01	0.03

To decide the judgment of the constancy of a selection to be alluring, CR is the proportion of CI and RI as given:

$$CR = \frac{CI}{RI} \tag{4}$$

Saaty provided a proportion of constancy called CL utilizing the subsequent equation [4]:

$$CI = \frac{\lambda_{max} - n}{n - 1} \tag{5}$$

Where

λ_{max} = (2.58 * 0.35) + (0.38 * 0.07) + (0.25 * 0.03) + (0.67 * 0.13) + (0.57 * 0.22) + (1.23*0.17) = 5.46 and n is matrix's scope.

$$CL = \frac{5.46 - 6}{6 - 1} = 0.114 \tag{6}$$

And RI is specified in Table 6.

Table 6. Random Index RI

N	1	2	3	4	5	6
RI	0	0	0.57	0.7	1.12	1.13

The CR is determined as follows:

$$CR = \frac{0.115}{1.24} 0.092 \tag{7}$$

The estimation of CR is little or equivalent to 10%, the irregularity is worthy. When the CR is more noteworthy than 12%, then there is a need to change the disease's monitoring/evaluating decisions. For this situation, CR is 9.2%, thus the judgment is acknowledged to the illness combined examination.

More to locate the general load of various illnesses considering the appearance of the patient, we pondered pairwise reactions for arranged afflictions. The last weight for various afflictions on the extent pairwise sign assessments is shown up in Table 7.

Table 7. Relative pairwise weights for the diseases with the concern of symptoms.

	Numerous urination	Heart Burn	Rigorousness of joint
Urine	0.32	0.32	0.24
Heartbeat	0.07	0.10	0.07
Ortho	0.04	0.04	0.04

From the CR estimations of various illnesses concerning indications, pairwise correlations show that the finding of side effects pairwise examination is worthy. Plausibility weight rates of various infections are given in Table 8.

Table 8. Consistency Ration of CR in disease.

Disease attributes	CR
Urine	0.078
Heartbeat	0.068
Ortho	0.037

More pij is the weight for illness manifestations combined correlation from Table 9 and wi is the comparative load for various maladies combined examination.

$$\text{Weight} = \sum\nolimits_{i=1}^{m} p_{ij} * w_i, j = 1, \ldots\ldots, n. \tag{8}$$

Table 9. Weight Score of disease

Disease attributes	Weight %
Urine	0.253
Heartbeat	0.072
Ortho	0.034

The proposed logical examination for the appearance assessor by MCDM AHP shows affirmation of the comparative norms, weight, and surveying the ailment of, respectively, individuals considering their clinical issues as per the indication affirmation gives the likelihood of affliction that the patient may persevere. It has been shown that the suggested method is effective and persuasive in calculating the patient's risk of immorality based on the level relative to the patient's symptoms. Similarly, we can show that a patient has a greater chance to suffer urine by 36, heartbeat by 9%, and followed ortho by 5%, independently, as showed up in Fig. 3.

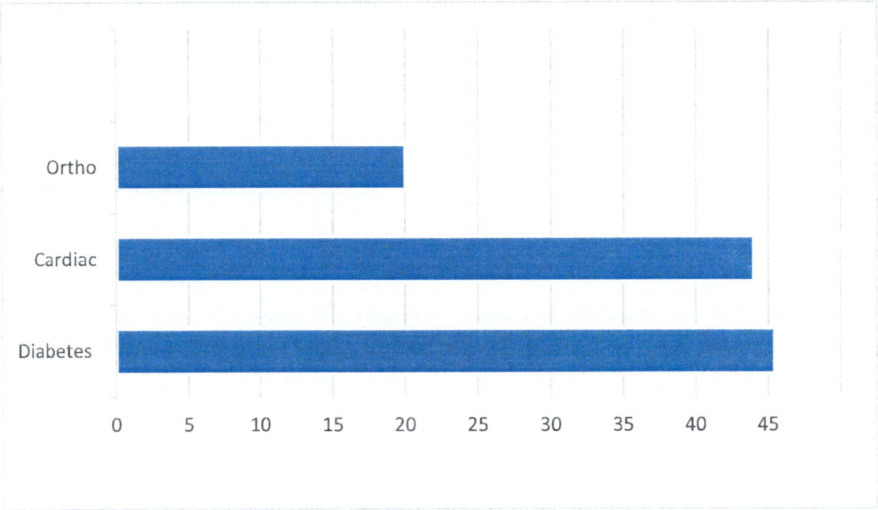

Fig. 3. Rate score of patient ailment based on the general load of manifestations

6 Conclusion

As the IoT promotes, revolutions in technology and devices are used to monitor the environment. This revolution enhances the healthcare domain and made it smarter to monitor elder patients. There is an essential need for such systems that monitor the patient effectively and take decisions timely. In this paper, we implemented a model for elder patients to monitor the patient with different diseases such as frequent urine sugar levels, and stiffness in bones. For this, we deployed sensors in the patient's body and attached them to the mobile to transmit the data to the other ends. We also set an alarming stage in the system when there is any abnormality condition observed. The study reveals that there is a need for a more precise and decision support mechanism for disease monitoring to enhance elder people's life.

References

1. Aula, A., Käki, M.: Less is more in Web search interfaces for older adults. First Monday **10** (2005)
2. Watts-Schacter, E., Kral, P.: Mobile health applications, smart medical devices, and big data analytics technologies. Am. J. Med. Res. **6**(1), 19–24 (2019)
3. Gubbi, J., Buyya, R., Marusic, S., Palaniswami, M.: Internet of Things (IoT): a vision, architectural elements, and future directions. Futur. Gener. Comput. Syst. **29**(7), 1645–1660 (2013)
4. Misra, G., Kumar, V., Agarwal, A., Agarwal, K.: Internet of things (IoT)–a technological analysis and survey on vision, concepts, challenges, innovation directions, technologies, and applications (an upcoming or future generation computer communication system technology). Am. J. Electr. Electron. Eng. **4**(1), 23–32 (2016)
5. Jonker, L.: Perioperative telemonitoring of older patients with cancer (2021)
6. Din, I.U., Guizani, M., Kim, B.S., Hassan, S., Khan, M.K.: Trust management techniques for the Internet of Things: a survey. IEEE Access **7**, 29763–29787 (2018)
7. Krishnan, D.S.R., Gupta, S.C., Choudhury, T.: An IoT based patient health monitoring system. In: 2018 International Conference on Advances in Computing and Communication Engineering (ICACCE), pp. 01–07. IEEE, June 2018
8. Butt, S.A., Gochhait, S., Andleeb, S., Adeel, M.: Games features for health disciplines for patient learning as entertainment. In: Digital Entertainment, pp. 65–86. Palgrave Macmillan, Singapore (2021)
9. Shafi, J., Waheed, A.: Role of smart wearable in healthcare: wearable Internet of Medical Things (WIoMT). In: The IoT and the Next Revolutions Automating the World, pp. 133–155. IGI Global (2019)
10. Jang-Jaccard, J., Nepal, S., Celler, B., Yan, B.: WebRTC-based video conferencing service for telehealth. Computing **98**(1–2), 169–193 (2014). https://doi.org/10.1007/s00607-014-0429-2
11. Vahedi, I.: Applications of telehealth in the practice, upgrading of knowledge, and communication of physicians with their colleagues and patients in Canada (Doctoral dissertation) (2017)
12. Tariq, M.I.: Big Data with Green Internet of Things (G-IoT) for the Development of Smart Cities and Technologies (2020)
13. Burnett, A.D.: How does leadership synergistically impact employee performance and frontline care effectiveness (Doctoral dissertation, Capella University) (2017)
14. Fatima, T., Malik, S.A., Shabbir, A.: Hospital healthcare service quality, patient satisfaction and loyalty. Int. J. Qual. Reliab. Manag. **35** (2018)
15. Pramanik, P.K.D., Upadhyaya, B.K., Pal, S., Pal, T.: Internet of Things, smart sensors, and pervasive systems: enabling connected and pervasive healthcare. In: Healthcare Data Analytics and Management, pp. 1–58. Academic Press (2019)
16. Alsina-Pagès, R.M., Navarro, J., Alías, F., Hervás, M.: Homesound: real-time audio event detection based on high performance computing for behaviour and surveillance remote monitoring. Sensors **17**(4), 854 (2017)
17. Andriopoulou, F., Orphanoudakis, T., Dagiuklas, T.: IoTA: IoT automated SIP-based emergency call triggering system for general eHealth purposes. In: 2017 IEEE 13th International Conference on Wireless and Mobile Computing, Networking and Communications (WiMob), pp. 362–369. IEEE, October 2017

18. Minaam, D.S.A., Abd-ELfattah, M.: Smart drugs: improving healthcare using smart pill box for medicine reminder and monitoring system. Future Comput. Informat. J. **3**(2), 443–456 (2018)
19. Marsh, K., et al.: Multiple criteria decision analysis for health care decision making—emerging good practices: report 2 of the ISPOR MCDA Emerging Good Practices Task Force. Value Health **19**(2), 125–137 (2016). https://doi.org/10.1016/j.jval.2015.12.016
20. Kelley, L.T., Egan, R., Stockley, D., Johnson, A.P.: Evaluating multi-criteria decision-making in health technology assessment. Health Policy Technol. **7**(3), 310–317 (2018)
21. Butt, S.A., Anjum, M.W., Hassan, S.A., Garai, A., Onyema, E.M.: Smart health application for remote tracking of ambulatory patients. In: Smart Healthcare System Design: Security and Privacy Aspects, vol. 33 (2021)

Early Detection of the Alzheimer's Disease: A Novel Cognitive Feature Selection Approach Using Machine Learning

Muhammad Irfan[1(✉)], Seyed Shahrestani[1], and Mahmoud Elkhodr[2]

[1] School of Computer, Data and Mathematical Sciences, Western Sydney University, Sydney, Australia
19918600@students.westernsydney.edu.au,
s.shahrestani@westernsydney.edu.au

[2] School of Engineering and Technology, Central Queensland University, Sydney, Australia
m.elkhodr@cqu.edu.au

Abstract. Alzheimer's Disease (AD) is a dynamic condition that affects cognitive capabilities and functioning. It is a challenging disease to detect, particularly in its early stages. Early diagnosis of AD is the key for its treatment and slowing of its progress. This paper argues and clearly shows the benefits of using cognitive tests for efficient and early AD detection. In this study, a novel approach for the early detection of AD is proposed. We refer to it as Neighborhood Component Analysis and Correlation-based Filtration (NCA-F) and is based on selecting and identifying significant cognitive features. Cognitive features are used to train four Machine Learning (ML) classifiers, including Support Vector Machine (SVM), Naïve Bayes (NB), ANN, and AdaBoost Ensemble (AdB). Our analysis shows that the proposed approach can achieve an 88% classification accuracy. In addition, the performance of various ML classifiers with varying combinations of features has been studied. The proposed feature selection approach that implements AdB is seen to provide the best performance by various metrics.

Keywords: Alzheimer's Disease · Feature selection · Machine Learning · ANN · SVM · NB · AdaBoost Ensemble

1 Introduction

The worldwide population is getting aged, posing several challenges to the already constrained healthcare systems [1]. Globally, the population of 65 years of age and above is growing faster than all other age groups [2]. The demographic transition to increasingly older populations is likely to create an outburst in age-related conditions such as cognitive impairment and Alzheimer's Disease (AD) [1]. It is estimated that by the year 2050, there will be around 152 million people globally who have AD as every three seconds, someone develops AD [3]. This increase in AD diagnosis can cause

Y. Maleh et al. (Eds.): ICI2C 2021, LNNS 357, pp. 383–392, 2022.
https://doi.org/10.1007/978-3-030-91738-8_35

stress for both the patients and their families. In addition, it impacts the caregivers psychologically, socioeconomically, and economically [4]. Early diagnosis of AD enables patients' access to the intervention resources, support, and relevant information. More importantly, an early-stage screening in patients may help identify the AD symptoms before it is fully developed [4]. This means treatment plans can be started timely to control the progression of the disease.

Cognitive tests (CT) are commonly used in the medical domain to diagnose AD in patients. They can also help in the diagnosis of AD stages as well [4]. Since significant cognitive decline has been reported as one of the essential symptoms of early-stage AD, quantifiable measurements through cognitive tests play a vital role in the early screening of this disease [5]. However, standard and state-of-the-art diagnostic procedures for AD, such as cerebrospinal fluid analysis tests and neuroimaging tests, are expensive, time-consuming, and involve many risk factors [14]. Therefore, the research explores the use of Machine Learning (ML) technologies to predict and detect the early stages of AD. For instance, the work reported in [7] proposed using an ML model to detect AD's stages using screening tests. Many works focused on selecting ML approaches rather than selecting the relevant features used in predicting AD, such as those reported in [8–11]. When employed by ML models, feature selection approaches improve performance [12, 13].

This work proposes a novel approach that includes an ensemble feature selection method. The proposed approach combines a feature filtration method with another feature ranking method to identify the top-performing features learned by the ML model using a cognitive feature vector. The top selected features are then used to train several ML models. The performance of each of the models is then investigated and analyzed. The list of selected ML models includes the Naïve Bayes (NB) model, the more complex model, the Artificial Neural Network (ANN) model, the Support Vector Machine (SVM), and an ensemble model known as AdaBoost Ensemble (AdB).

The results show an improvement in performance measures by using the ensemble feature selection method. In addition, the results show that by incorporating the proposed approach in the ML models, the models were able to better predict the early stage of AD with the help of cognitive features.

The rest of the paper is organized as follows. Section 2 presents the related work. Section 3 introduces the proposed feature selection method. The classification results and an evaluation of the model's performance are discussed in Sect. 4. Finally, concluding remarks are drawn in Sect. 5.

2 Related Works

Machine Learning (ML) is a promising and emerging technique used for the early detection of cognitive diseases such as Alzheimer's Disease (AD) [8]. For instance, the ML technique K-nearest neighbor (KNN) algorithm has been used in [9] for the prediction and classification of AD [9]. ML-based decision-making approaches that use computer-aided design (CAD) tools have also been proposed to detect AD and the work in [10]. The use of an ensemble classification model to identify patients with high and low AD risks was proposed in [11]. The study reported a classification accuracy of 94.7% when the model is trained only using paralinguistic features. However, an increase of 2.5% in the model accuracy was achieved when paralinguistic and episodic memory features were used (achieved 97.2% accuracy when trained using combined features). An ensemble of 13 machine learning models was proposed to predict Mild Cognitive Impairment (MCI) conversion to Alzheimer's disease in [14]. The study reported that the ensemble achieved an Area Under Curve (AUC) of 0.88, specificity of 79.9%, and a sensitivity of 77.7% [14]. Early AD detection using Cambridge cognitive features is proposed in [27] where several cognitive functions are measured. A new criterion to diagnose AD is proposed that use weighted combined scores from multiple tests [28]. The Principal Components Analysis (PCA) and Discriminant analysis are employed to conduct the experiments. The study in [29] aimed to evaluate the potentials of ML in quantifying the AD detection process and to optimize or reduce the quantity of cognitive tests.

Feature selection approaches commonly used in literature include the filtering approach, the wrapper, and embedded methods [6]. The features are selected based on multiple statistical test scores and the derived correlation with a target variable. Test scores are based on a correlation coefficient, which defines the statistical relation between the variables. Other approaches use the correlation coefficient as a feature selector, such as the Pearson's correlation, Linear Discriminant Analysis (LDA), Analysis of Variance (ANOVA), and the Chi-Square approach [15]. Zhou et al. proposed a novel approach for dementia diagnosis based on a three-stage deep feature learning and fusion system [16]. Yang et al. proposed a novel feature weighting method based on the nearest neighbour method, which maximized classification accuracy compared to other benchmarking techniques [17]. Other works relied on cognitive features to predict Alzheimer's disease (AD), such as the works reported in [12, 18]. Ford et al. used 18 cognitive features for the prediction of dementia. The results reported an area under the Receiver Operating Characteristic (ROC) curve of 0.74 [18]. Gill et al. used cognitive features to predict AD. The authors used only four cognitive components and achieved 81.8% accuracy and an AUC of 0.79 [12].

3 The Proposed Feature Selection Method

This research proposes a novel approach for the early-stage detection of Alzheimer's Disease (AD). It consists of two processes: a correlation-based feature filtration and a ranking based feature selection. This proposed method selects the essential primary features using a cognitive vector-based approach. The novel feature selection method is referred to as the Neighborhood Component Analysis and Correlation-based Filtration (NCA-F) approach. The NCA-F employed by an ML model selects the feature vectors from cognitive tests, which are then fed into various classifiers, including the AdB,

ANN, SVM, and the NB, to classify the relevant instances. Figure 1 shows an overview methodology used in this work. Since the focus of this work is the early detection of AD, cognitive scores have been used as sensitive features when training the models.

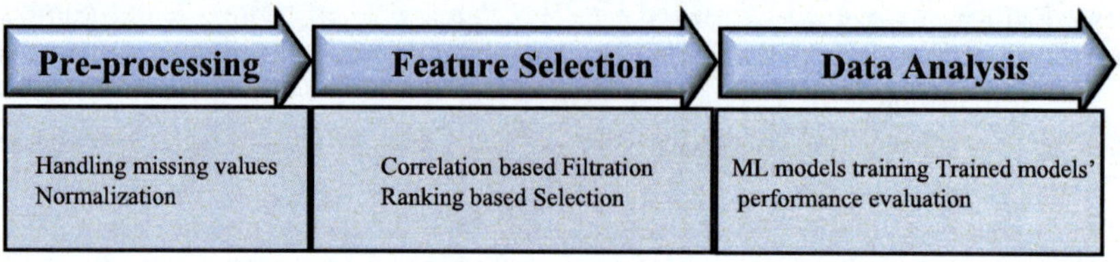

Fig. 1. Overview of the three-stage methodology.

3.1 The Dataset

The dataset used in this research is provided by the Alzheimer's Disease Neuroimaging Initiative (ADNI) [19–21], which comprises three phases. This research has focused on the data from thc first phase named ADNI1 [19]. The dataset contains cognitive test scores and values for 5013 records corresponding to 819 patients. During the trial, many patients have visited the clinic multiple times. A new cognitive test score was recorded as a new record in the dataset each time a patient visited the clinic during the trial. This is because the patients repeated the cognitive test in each of their visit. The dataset contains 1643 cognitive normal (CN) records, and 3370 Alzheimer's Disease (AD) records. The dataset has several missing values and contained many irrelevant features thus cleaning and pre-processing the data was necessary before deploying any ML models.

Therefore, a pre-processing method is first implemented to clean the data and reduce the risk of the model being underfitted and overfitted. Each record of the ADNI1 dataset is composed of 113 features. Some of these features had several missing values. The features with more than 40% missing values were discarded. The remaining was initialized using an Iterative Imputer [22]. This method imputes the missing values of the features in a round-robin technique. Also, the dataset contained features that had several values. Thus, data normalization was done on the dataset to optimise the proposed approach's performance. Thus, the dynamic range of the dataset was normalized between 0 and 1 using the min-max technique [23]. The dataset normalization's process is depicted in formula (1)

$$z_i = \frac{x_i - \min(x)}{\max(x) - \min(x)} \tag{1}$$

where x denotes the input data and z_i represents the normalized data, whereas *min* and *max* represent the minimum and maximum values, respectively.

3.2 Ensemble Feature Selection

Our proposed ensemble feature selection method, referred to as Neighborhood Component Analysis and correlation-based Filtration (NCA-F) is based on two filtration and selection processes. Initially, the features are filtered using Pearson's correlation coefficient. Then Neighborhood Component Analysis (NCA) based wrapper method [24] is applied to assign weights to the selected features.

The classification performance is improved by discarding the redundant and irrelevant features from the dataset to reduce the risks of the model being under or overfitted. The irrelevant features were identified first and then discarded from the dataset by defining a threshold. The highly correlated features with a correlated point of an absolute value greater than 0.9 were filtered. To achieve this, a correlation heat map is generated using Pearson's correlation. The correlation coefficient matrix is defined as the matrix for each pairwise variable combination, as shown in formula (2)

$$R = \begin{pmatrix} p(A,A) & p(A,B) \\ p(B,A) & p(B,B) \end{pmatrix} \tag{2}$$

where *p(A,B)* represents the correlation coefficient between features A and B and is calculated using formula (3)

$$p(A,B) = \frac{\mathrm{cov}(A,B)}{\sigma_A \sigma_B} \tag{3}$$

where σ_A and σ_B denote the mean and standard deviation, whereas *Cov*() denotes the covariance function. After the filtration process, the selected features are further processed using the NCA. NCA is a wrapper feature selection method used to assign weights to the selected features based on the classification performance of the individual and combined features. In the proposed NCA-F approach, the NCA with Stochastic Gradient Decent (SGD) method [25], as an optimizer, is used for assigning weights to the selected features. The proposed NCA-F has used the One-nearest neighbor classifier [26] for examining the performance of using different combinations of features. It then assigned weights to each feature. These weighted features were then sorted in descending order and processed by the model to detect AD stages. The resultant post-processing features are given in Table 1.

Table 1. The list of selected cognitive features after NCA-F.

S. No	Feature	Weight
1	AGE	15.288
2	EcogSPPlan	3.0912
3	PTGENDER	2.9801
4	CDRSB	2.6719
5	EcogSPOrgan	2.345
6	EcogSPLang	2.18
7	ADAS11	2.0056
8	RAVLTimmediate	1.9877
9	ADASQ4	1.7356
10	MMSE	1.565
11	EcogPtOrgan	1.5628
12	EcogSPVisspat	1.5223
13	LDELTOTAL	1.1009
14	EcogPtLang	1.0243
15	FAQ	0.91021
16	MOCA	0.79935
17	EcogPtTotal	0.75844
18	EcogPtDivatt	0.72765
19	EcogSPMem	0.33403
20	EcogPtMem	0.14011
21	TRABSCOR	0.024946
22	EcogSPDivatt	0.0033972
23	EcogPtVisspat	1.07E−07
24	RAVLTpercforgetting	1.51E−12
25	EcogPtPlan	7.03E−16
26	RAVLTlearning	8.71E−24
27	RAVLTforgetting	9.52E−47

The impact of using cognitive features on the performance of the ML model was investigated using the AdaBoost Ensemble (AdB) with a 5-fold validation set. The best performing combination of features was selected from the feature vectors. The top 18 features (from Table 1), when combined, have resulted in the overall best performance in accuracy, precision, recall, and F1-measure. The correlation matrix for the 18-top performing weighted cognitive features is provided in Fig. 2. The correlation matrix shows a strong to weak relationship amongst the features in pair variables. In addition, a high correlation can be observed amongst the features in pair variables. Figure 2 also shows a strong correlation between the features having greater absolute values.

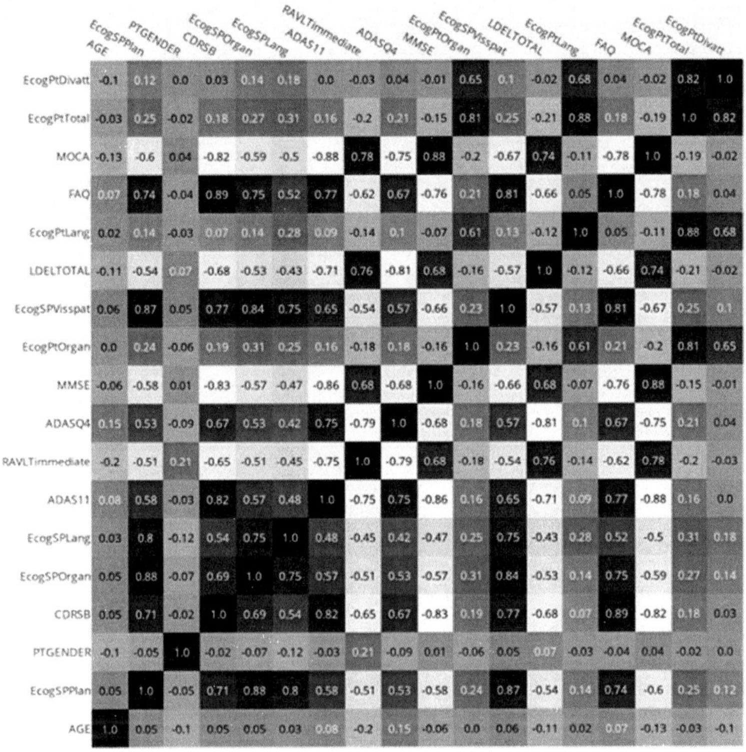

Fig. 2. Correlation matrix for selected cognitive features after NCA-F method.

4 Results and Discussions

After identifying the optimal feature combinations required for the early detection of AD, four different models were trained. These models include AdB, ANN, SVM, and NB, where the performance is evaluated for the selected features for AD detection. The experimental results are validated with a 5-folds cross-validation set. In the experiment, the 80:20 training and testing ratio of the dataset was used. Eighteen highly weighted cognitive features obtained from the proposed NCA-F were used to analyze and detect AD. Table 2 shows the experimental results of AdB, ANN, SVM and NB ML models that were trained using the 18 selected cognitive features. The results are reported in terms of Accuracy, Precision, Recall, F1-Score, and AUC.

Table 2. ML models comparison for ADNI1 dataset

Model	Accuracy	Precision	Recall	F1-Score	AUC
	Percent (%)				
AdB	**87.89**	**86.14**	86.47	**86.29**	**95.19**
ANN	86.55	84.84	84.7	84.68	94.75
SVM	85.41	85.01	81.47	82.64	92.95
NB	75.44	72.34	73.17	72.67	87.49

The AdB has outperformed the other ML models. Significant improvements are achieved with AdB. The accuracy has improved from 75% to 88%, the precision from 72% to 86%, the recall from 73% to 86%, the F1-score from 73% to 86%, and the AUC from 88% to 95%. All four ML models performed comparatively better when trained only with the selected 18 cognitive features, as shown in Table 2. The results corresponding to the four models are reported in Fig. 3, showing the Receiver Operating Characteristic (ROC) curves for the four ML models. The AUC curves for the SVM, ANN, and AdB models are above 92 to 95%, whereas the AUC of the NB model reached 87%. The AdB model outperformed the three other ML models in terms of AUC.

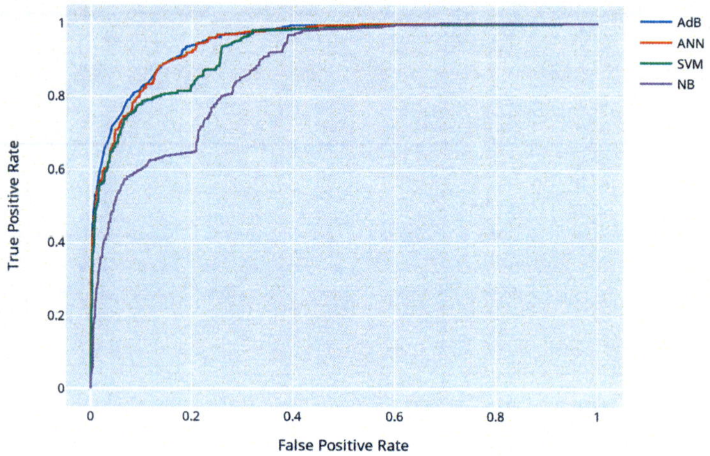

Fig. 3. ROC curve obtained for selected features on 4 ML models.

The experimental results obtained with the proposed NCA-F model and the existing filtrations methods on diverse feature configurations are presented in Table 3.

Table 3. Performance Comparison of the Proposed NCA-F against existent approach.

Method	Dataset	Selected Features	Number of records	Accuracy (%)	AUC
Gill et al. [12]	ADNI1	4	600	81.80	0.79
NCA-F	ADNI1-Updated	18	5013	**87.89**	**0.95**

From Table 3, it can be observed that the performance of the proposed NCA-F approach has achieved better performance compared to the feature filtration method used by Gill et al. [12]. The ADNI1 dataset used by Gill et al. [12] is composed of limited records that have created overfitting. On the other hand, ADNI1-Updated has a higher number of records that have resolved the overfitting problem. Table 3 indicates that the proposed NCA-F on cognitive (clinical) features achieved the best performance accuracy.

5 Conclusions

Alzheimer's disease (AD) is challenging to detect in its early stages. This work has proposed a novel NCA-F-based feature selection approach for the early detection of AD through cognitive tests. Four ML models have been trained on the ADNI dataset. They employed the proposed NCA-F, a feature selection method. AdaBoost model produced a prediction accuracy of 88%. Compared to the current state of the art feature selection methods, the proposed NCA-F feature selection method has achieved the best performance when used by the AdaBoost model. Future work will improve the AD prediction in different stages by using ML models incorporated with the feature selection method.

References

1. Tong, T., Chignell, M., Lam, P., Tierney, M.C., Lee, J.: Designing serious games for cognitive assessment of the elderly. Presented at the Proceedings of the International Symposium on Human Factors and Ergonomics in Health Care (2014)
2. Vespa, J., Armstrong, D.M., Medina, L.: Demographic turning points for the United States: Population projections for 2020 to 2060. US Department of Commerce, Economics and Statistics Administration, US (2018)
3. Prince, M.J., Wimo, A., Guerchet, M.M., Ali, G.C., Wu, Y.-T., Prina, M.: World Alzheimer Report 2015-The Global Impact of Dementia: an analysis of prevalence, incidence, cost and trends (2015)
4. Australia, D., Baker, S., Banerjee, S.: Alzheimer's Disease International. World Alzheimer Report 2019: Attitudes to dementia. Alzheimer's Disease International, London, UK (2019)
5. Jack, C.R., Jr., Holtzman, D.M.: Biomarker modeling of Alzheimer's disease. Neuron **80**, 1347–1358 (2013)
6. Guyon, I., Gunn, S., Nikravesh, M., Zadeh, L.A.: Feature Extraction: Foundations and Applications. Springer, Heidelberg (2008)
7. Youn, Y.C., et al.: Detection of cognitive impairment using a machine-learning algorithm. Neuropsychiatr. Dis. Treat. **14**, 2939 (2018)
8. Bratić, B., Kurbalija, V., Ivanović, M., Oder, I., Bosnić, Z.: Machine learning for predicting cognitive diseases: methods, data sources and risk factors. J. Med. Syst. **42**, 1–15 (2018)
9. Kruthika, K., Maheshappa, H.: Alzheimer's disease neuroimaging initiative: multistage classifier-based approach for Alzheimer's disease prediction and retrieval. Inform. Med. Unlocked. **14**, 34–42 (2019)
10. Veeramuthu, A., Meenakshi, S., Manjusha, P.: A new approach for Alzheimer's disease diagnosis by using association rule over pet images. Int. J. Comput. Appl. **91** (2014)
11. You, Y., Ahmed, B., Barr, P., Ballard, K., Valenzuela, M.: Predicting dementia risk using paralinguistic and memory test features with machine learning models. Presented at the 2019 IEEE Healthcare Innovations and Point of Care Technologies, (HI-POCT) (2019)

12. Gill, S., et al.: Alzheimer's disease neuroimaging initiative: using machine learning to predict dementia from neuropsychiatric symptom and neuroimaging data. J. Alzheimers Dis. **75**, 277–288 (2020)
13. Tang, J., Alelyani, S., Liu, H.: Feature selection for classification: a review. Data Classif. Algorithms Appl. **37** (2014)
14. Grassi, M., et al.: Alzheimer's disease neuroimaging initiative: a novel ensemble-based machine learning algorithm to predict the conversion from mild cognitive impairment to Alzheimer's disease using socio-demographic characteristics, clinical information, and neuropsychological measures. Front. Neurol. **10**, 756 (2019)
15. Fisher, R.A.: Statistical methods for research workers. In: Kotz, S., Johnson, N.L. (eds.) Breakthroughs in Statistics, pp. 66–70. Springer, New York (1992). https://doi.org/10.1007/978-1-4612-4380-9_6
16. Zhou, T., Thung, K., Zhu, X., Shen, D.: Effective feature learning and fusion of multimodality data using stage-wise deep neural network for dementia diagnosis. Hum. Brain Mapp. **40**, 1001–1016 (2019)
17. Yang, W., Wang, K., Zuo, W.: Neighborhood component feature selection for high-dimensional data. J Comput. **7**, 161–168 (2012)
18. Ford, E., et al.: Identifying undetected dementia in UK primary care patients: a retrospective case-control study comparing machine-learning and standard epidemiological approaches. BMC Med. Inform. Decis. Mak. **19**, 1–9 (2019)
19. Weiner, M.W., et al.: The Alzheimer's disease neuroimaging initiative: progress report and future plans. Alzheimers Dement. **6**, 202–211 (2010)
20. Protocol Consultants Coordinating Center, Neuropathology Biomarker Core: Alzheimer's Disease Neuroimaging Initiative 2 (ADNI2) Protocol (ADC-039) (2014)
21. Weiner, M.W., et al.: The Alzheimer's disease neuroimaging initiative 3: continued innovation for clinical trial improvement. Alzheimers Dement. **13**, 561–571 (2017)
22. Van Buuren, S., Groothuis-Oudshoorn, K.: Multivariate imputation by chained equations in RJ Stat
23. Zang, I.: A smoothing-out technique for min—max optimization. Math. Program. **19**, 61–77 (1980)
24. Yang, W., Wang, K., Zuo, W.: Fast neighborhood component analysis. Neurocomputing **83**, 31–37 (2012)
25. Ketkar, N.: Stochastic gradient descent. In: Ketkar, N. (ed.) Deep Learning with Python, pp. 111–130. Apress, Berkeley (2017). https://doi.org/10.1007/978-1-4842-2766-4_8
26. Kontorovich, A., Weiss, R.: A Bayes consistent 1-NN classifier. Presented at the Artificial Intelligence and Statistics (2015)
27. Schmand, B., Walstra, G., Lindeboom, J., Teunisse, S., Jonker, C.: Early detection of Alzheimer's disease using the Cambridge Cognitive Examination (CAMCOG). Psychol. Med. **30**(3), 619–662 (2000)
28. Chapman, R.M., et al.: Diagnosis of Alzheimer's disease using neuropsychological testing improved by multivariate analyses. J. Clin. Exp. Neuropsychol. **32**(8), 793–808 (2010)
29. Battista, P., Salvatore, C., Castiglioni, I.: Optimizing neuropsychological assessments for cognitive, behavioral, and functional impairment classification: a machine learning study. Behav. Neurol. **2017** (2017)

An Artificial Neural Network-Based System to Predict Cardiovascular Disease

Hayat Bihri(✉), Rachid Nejjari, Salma Azzouzi, and My El Hassan Charaf

Faculty of Sciences, Ibn Tofail University, Kenitra, Morocco

Abstract. In recent years, heart disease has become one of the leading causes of death worldwide. In this context, numerous risk factors such as smoking, high blood pressure, high cholesterol, and diabetes often help early predict heart disease, decreasing the cost of diagnosis and treatment. Today, through medical and technology innovation, the medical field promises to change healthcare services and reach full potential advancement. In fact, the wide adoption of electronic medical data systems makes more and more health-related data available. This paper aims to predict people likely to have cardiovascular disease by examining available data and analyzing some attributes recognized as high-risk factors. To achieve this objective, we suggest a neural network-based approach to predict cardiac patients using patient data measurements and the analysis of some high-risk factors in this study. The simulation results of our model show that the accuracy achieved is 71.86%, with a precision of 75.20% and a specificity of 78.40%.

Keywords: Predictive analytics · Artificial neural network · CVD · Risk factors

1 Introduction

Cardiovascular disease is one of the most dangerous diseases globally as it is not easily detected in the early stages of the disease. Yet, early diagnosis is important for effective treatment. For this purpose, predicting cardiovascular disease early can identify patients with a high risk of hospital readmission and lead to taking timely measures to avoid cardiac arrest. Thus, many risk factors such as smoking, high blood pressure, high cholesterol and diabetes can often predict heart disease early.

In general, medical experts make diagnoses based on the results of electrocardiography or blood tests and their personal experiences and understanding of the disease, which can increase the risk of error and delay the appropriate treatment. To solve these problems, the adoption of new technologies in health care has been presented to reduce costs and improve quality. In this context, many studies have been conducted on innovative techniques to predict chronic diseases as the traditional prediction methods have insufficient prediction accuracy, so they cannot provide effective decision support for medical experts. As a result, more advanced computational techniques, such as machine learning, become more convenient to be used and explored in disease prediction. As for machine learning algorithms (e.g., logistic regression, support vector machine and artificial neural network), especially - supervised learning algorithms, the

Y. Maleh et al. (Eds.): ICI2C 2021, LNNS 357, pp. 393–402, 2022.
https://doi.org/10.1007/978-3-030-91738-8_36

disease risk prediction models based on these algorithms use for training: labelled training data of patients whereas, for the test set, patients are classified into low-risk and high-risk groups for example.

Furthermore, the availability of electronic health data and the widespread adoption of electronic medical records (EMR) has transformed the traditional way of managing medical records. The digital method of storing patient medical records has revolutionized the healthcare field by making it easier to share records. In fact, electronic medical data systems make more and more health-related data available. These electronic data are being used across a broad range of healthcare-related areas such as measuring the performance of a hospital care network, developing a disease risk prediction model or monitoring chronic diseases. For this purpose, we suggest predicting people likely to have cardiovascular diseases by examining available data and analyzing some attributes recognized as high-risk factors. Furthermore, we use a neural network-based approach to improve performance accuracy as the NN based methods are suitable to discover new patterns and information related to cardiovascular disease but also at generalizing data without domain knowledge before training.

The organization of the paper is as follows. The second section describes some basic concepts. Then, we introduce the problem statement and some related works, respectively, in Sects. 3 and 4. Afterward, we present in Sect. 5 the prototype and the methods used to implement our Cardiovascular Disease Prediction System. Then simulation results and discussion are given in Sect. 6. Finally, the conclusion and future works are explained.

2 Preliminaries

2.1 CVD High-Risk Factors

Hypertension and diabetes have commonly been highlighted as well-established predictors of cardiovascular disease.

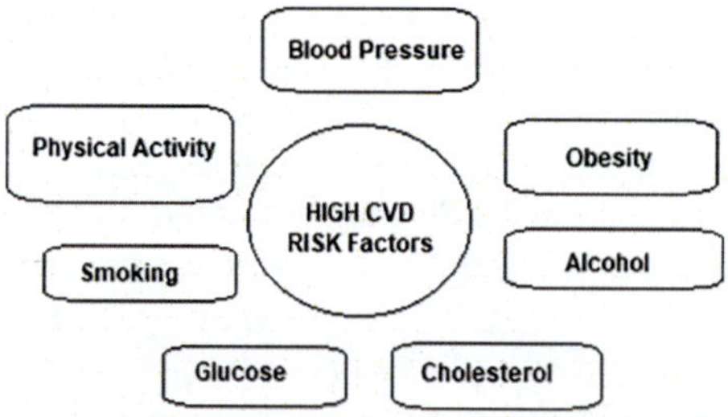

Fig. 1. CVD high-risk factors

These risk factors have become prime targets for assessment of cardiovascular risk as well as monitoring, and treatment. In addition, lifestyle risk factors including dietary habits, lack of physical activity and smoking significantly influence cardiovascular risk factors [1]. Figure 1 summarizes the main risk factors related to cardiovascular disease.

Thus, clinical assessment and treatment of dietary habits, physical activity, and smoking should become as routine and familiar as assessing blood pressure, cholesterol, and blood glucose.

2.2 ANN vs. Machine Learning Algorithms

An artificial neural network (ANN) is a computational model that imitates the behavior of the human brain. Artificial neural networks can make adjustments independently or learn when they receive new data [2].

On the other hand, machine learning algorithms use various statistical, probabilistic, and optimization methods to learn from experience and detect valuable patterns in datasets [3]. These algorithms can be grouped into three categories based on their goals: Supervised, Unsupervised and Semi-supervised machine learning algorithms. The principal difference between supervised and unsupervised methods is the use of labeled data sets. Within supervised learning, the algorithm "learns" from the training dataset through iterative predictions about the data and makes adjustments to get the appropriate answer. In contrast, unsupervised learning models operate autonomously to discover the inherent pattern of the unlabeled data. Finally, semi-supervised learning uses a training data set containing both labeled and unlabeled data. It is most useful in cases of difficulty in extracting relevant features from data.

Machine learning algorithms that use artificial neural networks generally do not need to be programmed with specific rules that define what to expect from the input. Once a sufficient number of examples have been processed, the neural network can process new, unseen inputs and successfully return accurate results [4].

3 Problem Statement

Identifying patients having cardiovascular disorders can prevent premature death and thus potentially help to save lives. The objective of the present research study is to provide medical experts with a feature that may assist them in performing a preliminary diagnosis of cardiovascular disease based on high-risk factors abnormalities and to warn about the state of patients with probable heart disorder.

To this end, we suggest using a machine learning technique by considering specific attributes known as high-risk factors for cardiovascular disease to predict people at risk accurately. More precisely, we propose using neural network technics to predict cardiac patients based on available measurements of principal high-risk factors of cardiovascular disease stored in the medical information system. Hence, the patients identified as high-risk individuals are advised to undergo treatment to control elevated features.

4 Related Works

Today, many research studies have been undertaken to investigate disease prediction and suggest proposals to predict the outcomes of some diseases based on the observed data. In this context, the authors in [5] propose a novel model, namely "hybrid random forest with a linear model" (HRFLM), that combines two existing models for heart disease prediction, namely Random Forest (RF) and Linear Method. Thus, the proposed method significantly enhances the prediction accuracy and decreases the healthcare industry's classification error, especially for cardiovascular disease.

Moreover, the work [6] provides a predictive model for potential cardiovascular patients based on their present features. Therefore, the authors use several machine learning techniques such as Gradient Boosting Classifier, Logistic Regression, Multi-Layer Perceptron (MLP), Random Forest Classifier, Support Vector Machine and Extremely Randomized Trees Classifier to ensure the performance of the proposed diagnosis model.

Furthermore, the authors in [7] suggest a hybrid model predicts probable patients to develop type 2 diabetes and highlight techniques used for cardiovascular disease diagnosis. They use the Improved Fuzzy C Means (IFCM) algorithm for clustering and the Support Vector Machine (SVM) algorithm to build the final model through the k-fold cross-validation method. In addition, the authors in [8] combine carotid ultrasound imaging and conventional risk factors to assess the predictive risk of CVD/CVA. For this purpose, they experimented on 500 contributors using algorithms based on multiclass machine learning (MCML) and conventional CVD risk calculators (CCVRC). The proposed method performed significantly in terms of reliability, accuracy, and CVD/CVA risk prediction.

Another work [9] conducts a study to compare the standard logistic regression method with five machine learning (ML) methods, including: single-hidden-layer, random forest, neural network, gradient boosting machine, support vector machine, and k-nearest neighbors method using simple clinical predictors. The objective is to evaluate the performance of machine learning (ML) used for risk prediction of cardiovascular diseases (CVDs) and other chronic diseases. The analyses were performed on Asian adult participants and the best result was found using logistic regression, gradient boosting machine and neural network.

The study [10] presents an overview of machine learning techniques used for Artificial Neural Networks (ANNs) and Bayesian Networks (BNs) to classify cardiovascular and diabetes diseases. To this end, the authors conduct a comparative study on selected articles published between 2008 and 2017 and conclude that the most used learning algorithm for ANN and BN are Levenberg-Marquardt and the Naive Bayesian network. The study also highlighted that the best accuracy performance is obtained by applying ANN for diabetes and cardiovascular diseases prediction.

Finally, the authors in [11] show that the predictive ability of ML algorithms for cardiovascular diseases is promising, particularly SVM and boosting algorithms. They also affirm the heterogeneity among ML algorithms in terms of multiple parameters and how such information may assist clinicians in interpreting data and implementing optimal algorithms for their datasets.

In this paper, we suggest an ANN-based model predict CVD disease. Experimentally, we run our model on a medical dataset freely available online containing the main CVD risk factors measurements. For performance purposes, we aim to analyze more the obtained prediction metrics in order to enhance our model accuracy compared to previous works in the literature.

5 ANN-Based Model for CVD Prediction

5.1 Data Preparation

In this study, we used a dataset containing information related to cardiovascular disease as described in [12].

Data Collection. The dataset values were collected at the moment of medical examination and stored according to the following structure (Table 1):

Table 1. CVD dataset overview

Input feature	Feature type	Column	Data type
Age	Objective	Age	int (days)
Height	Objective	Height	int (cm)
Weight	Objective	Weight	float (kg)
Gender	Objective	Gender	Categorical code
Systolic blood pressure	Examination	Ap_Hi	int
Diastolic blood pressure	Examination	Ap_Lo	int
Cholesterol	Examination	Cholesterol	1: normal, 2: above normal, 3: well above normal
Glucose	Examination	Gluc	1: normal, 2: above normal, 3: well above normal
Smoking	Subjective	Smoke	Binary
Alcohol intake	Subjective	Alco	Binary
Physical activity	Subjective	Active	Binary
Cardiovascular disease (Presence/Absence)	Target Variable	Cardio	Binary: (0: absent 1: present)

Therefore, we distinguish three types of features:

- Objective features related to factual information;
- Examination features comprises medical examination results;
- Subjective features related to information provided by the patient.

Data Analysis. In this step, we prepare collected data for processing to extract relevant information for decision making. Data preparation consists of examining data and performing a set of operations, including eliminating redundant values and the appropriate treatment of missing values.

In addition, we checked the data distribution, which is useful for classification but also for prediction purposes. Indeed, the data set is quite well distributed: (49.97% the number of cardio patients while the number of non-cardio patients is about 50.03%).

5.2 Model Description

The proposed model uses an ANN network to predict the risk of cardiovascular diseases by considering some features related to high-risk factors. Hence, the original dataset is equitably divided into two parts: training and testing datasets for performance and model accuracy.

To carry out this study, we implement our model using the open-source tool PyCharm (2021.1.2) [13] as an integrated development environment of the Python language. The software was used in both the training and the evaluation stages. Therefore, we use the sequential model to train the proposed ANN network and get suitable weights to map inputs to outputs. The model is used to manage different layers and to implement the activation operation on the input. Furthermore, we used for the compilation purpose. Adam [14] as an optimizer and the cross-entropy [15] as the loss argument since it's a binary classification problem.

The model, as designed uses eleven features for input entries and two hidden layers with a number of nodes (perceptrons). The number of nodes in each layer is identified experimentally. Therefore, the output results of our proposal will predict the presence or absence of cardiovascular disease. In the next section, we overview some simulation results in order to assess the model performance and accuracy.

6 Simulation Results

6.1 Experimental Procedure

First Layer. The first hidden layer uses all our dataset features as input in order to determine the best number of nodes. Therefore, we will use 17 nodes for better performance, as described in the table below (Table 2).

Table 2. First hidden layer: Accuracy according to the number of nodes.

Number of nodes	Accuracy (%)
12	61.75
13	68.35
14	70.84
15	50.13
16	71.25
17	71.51
18	71.21
19	69.68

Second Layer. The second hidden layer uses the outputs of the first hidden layer. For better performance, the layer will comprise 16 nodes as given below (Table 3):

Table 3. Second hidden layer: Accuracy according to the number of nodes.

Number of nodes	Accuracy (%)
11	71.16
12	71.35
13	71.51
14	71.35
15	71.32
16	71.86
17	70.83
18	71.26
19	70.74

Moreover, we suggest using the Rectified Linear Unit Activation (ReLU) as a linear activation function for our hidden layers. The function returns the input directly if it is positive. Otherwise, it returns zero (1):

$$f(x) = \begin{cases} x & x \geq 0 \\ 0 & x < 0 \end{cases} \tag{1}$$

Third Layer. The last layer (output layer) comprises one node using for activation the following sigmoid function (2) to ensure that the output result is between 0 and 1:

$$f(x) = \frac{1}{1 + e - x} \tag{2}$$

After that, we illustrate the overall of our architecture for an ANN-based model to predict cardiovascular disease (Fig. 2).

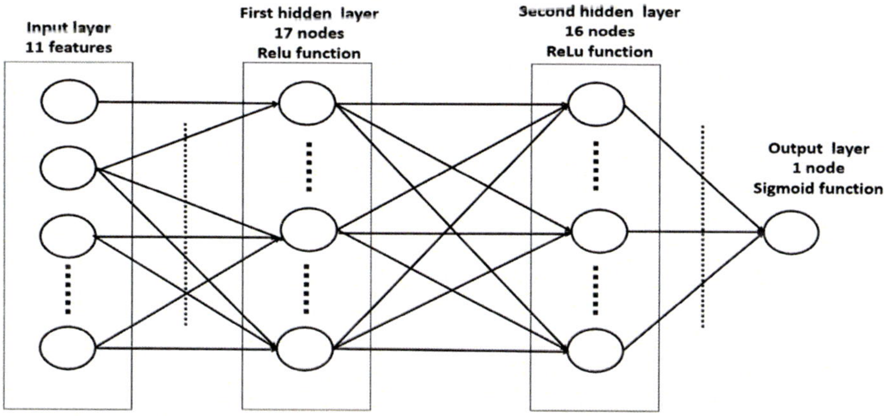

Fig. 2. The architecture of our ANN-based model for CVD prediction

6.2 Evaluation Metrics

In the current research, we focus on using an ANN model to predict individuals with cardiovascular disease risk. To assess the performance of our model, we suggest using the confusion matrix with the four base performance measures, namely: True Positive (TP), False Positive (FP), True Negative (TN), and False Negative as given in Table 4:

Table 4. Confusion Matrix.

		Actual values	
		Positive	Negative
Predicted values	Positive	TP	FP
	Negative	FN	TN

In fact, serval metrics are available [16] such as:

- Accuracy score defined as the ratio of correctly predicted values and all the values of the output:

$$f(x) = \frac{TP + TN}{TP + TN + FP + FN} \tag{3}$$

- Specificity (True Negative Rate) is defined as true negative values and the actual negative values

$$f(x) = \frac{TN}{TN + FP} \tag{4}$$

In summary, the model as designed shows that the accuracy achieved is 71.86% with a specificity rate of about: 78.40%. The result obtained means that 71.86% of the available instances are correctly predicted. This type of prediction will help practitioners predict the "presence" or "absence" of CVD at an early stage based not on operating an Electrocardiogram ECG but on risk factors measurements. These attributes can then be uploaded to a website by the patient or collected at the nearest medical unit.

As early diagnosis of heart disease is crucial to reduce health risks and prevent cardiac arrest, the physicians at the central medical unit then analyze the prediction results upon receipt and contact - if necessary - the target patients (presence of the disease) to take preventive measures.

7 Conclusion

Machine learning is a widely used technique to predict diseases in the healthcare domain. The goal is to predict outcomes of some diseases based on observed data and to help medical experts make a decision about the patients' situation according to some high-risk factors.

In this paper, we use an artificial neural network with two layers and 33 neurons. The activation function for the two hidden layers is ReLU, while we used the sigmoid function for the output layer. The simulation results show that the accuracy achieved is 71.86%, with a precision of 75.20% and a specificity of 78.40%. The results are promising with higher prediction performance. In this context, we suggest investigating more other machine learning technics to predict cardiovascular disease.

As prospects, we plan to extend our model to predict the readmission rate of cardiovascular patients at the hospital to prevent their early mortality and do preventive care accordingly.

References

1. O'Connor, E.A., Evans, C.V., Rushkin, M.C., Redmond, N., Lin, J.S.: Behavioral Counseling to Promote a Healthy Diet and Physical Activity for Cardiovascular Disease Prevention in Adults with Cardiovascular Risk Factors: Updated Evidence Report and Systematic Review for the US Preventive Services Task Force (2020). https://pubmed.ncbi.nlm.nih.gov/33231669/
2. Panambur, A.H., Manoj, M.V., Mathur, A.: Theory, concepts, and applications of artificial neural network. In: Applied Soft Computing Algorithms and Applications. Apple Academic Press (2021)
3. Dineva, K., Atanasova, T.: Systematic look at machine learning algorithms - advantages, disadvantages and practical applications. In: International Multidisciplinary Scientific GeoConference Surveying Geology and Mining Ecology Management, SGEM, pp. 317–324. International Multidisciplinary Scientific Geoconference (2020)
4. Sahoo, A.K., Pradhan, C., Das, H.: Performance evaluation of different machine learning methods and deep-learning based convolutional neural network for health decision making. In: Rout, M., Rout, J.K., Das, H. (eds.) Nature Inspired Computing for Data Science. SCI, vol. 871, pp. 201–212. Springer, Cham (2020). https://doi.org/10.1007/978-3-030-33820-6_8
5. Mohan, S., Thirumalai, C., Srivastava, G.: Effective heart disease prediction using hybrid machine learning techniques. IEEE Access **7**, 81542–81554 (2019). https://doi.org/10.1109/ACCESS.2019.2923707
6. Arunachalam, S.: Cardiovascular disease prediction model using machine learning algorithms. Int. J. Res. Appl. Sci. Eng. Technol. **8**, 1006–1019 (2020). https://doi.org/10.22214/ijraset.2020.6164
7. Radha, P., Srinivasan, B.: Diagnosing heart diseases for type 2 diabetic patients by cascading the data mining techniques. Int. J. IJRITCC **2**, 2503–2509 (2014)
8. Jamthikar, A.D., et al.: Multiclass machine learning vs. conventional calculators for stroke/CVD risk assessment using carotid plaque predictors with coronary angiography scores as gold standard: a 500 participants study. Int. J. Cardiovasc. Imaging **37**(4), 1171–1187 (2020). https://doi.org/10.1007/s10554-020-02099-7
9. Nusinovici, S., et al.: Logistic regression was as good as machine learning for predicting major chronic diseases. J. Clin. Epidemiol. **122**, 56–69 (2020). https://doi.org/10.1016/j.jclinepi.2020.03.002
10. Alić, B., Gurbeta, L., Badnjević, A.: Machine learning techniques for classification of diabetes and cardiovascular diseases. In: 2017 6th Mediterranean Conference on Embedded Computing, MECO 2017 - Including ECYPS 2017, Proceedings. Institute of Electrical and Electronics Engineers Inc. (2017)

11. Krittanawong, C., Virk, H.U.H., Bangalore, S., et al.: Machine learning prediction in cardiovascular diseases: a meta-analysis. Sci. Rep. **10**, 16057 (2020). https://doi.org/10.1038/s41598-020-72685-1
12. Cardiovascular Disease dataset—Kaggle. https://www.kaggle.com/sulianova/cardiovascular-disease-dataset
13. Nguyen, Q.: Hands-On Application Development with PyCharm: Accelerate Your Python Applications Using Practical Coding Techniques in PyCharm. Packt Publishing (2019)
14. Kingma, D.P., Ba, J.: Adam: a method for stochastic optimization. arXiv preprint arXiv:1412.6980 (2014)
15. Mannor, S., Peleg, D., Rubinstein, R.: The cross entropy method for classification. In: Proceedings of the 22nd International Conference on Machine Learning (ICML 2005), pp. 561–568. Association for Computing Machinery, New York (2005). https://doi.org/10.1145/1102351.1102422
16. Canbek, G., Temizel, T.T., Sagiroglu, S., Baykal, N.: Binary classification performance measures/metrics: a comprehensive visualized roadmap to gain new insights. In: 2nd International Conference on Computer Science and Engineering, UBMK 2017, pp. 821–826. Institute of Electrical and Electronics Engineers Inc. (2017)

Hybrid Machine and Deep Transfer Learning Based Classification Models for Covid 19 and Pneumonia Diagnosis Using X-ray Images

Alassane Bonkano Abdoul-Razak[1(✉)], Mounia Mikram[2], Maryem Rhanoui[2], and Sanaa Ghouzali[3]

[1] Laboratory of Mathematics, Computer Science, Applications and Information Security, Mohammed V University in Rabat, RP, Rabat, Morocco

[2] Meridian Team, LYRICA Laboratory, School of Information Sciences, Rabat, Morocco

[3] Information Technology Departement, College of Computer and Information Sciences, King Saud University, Riyadh, Saudi Arabia

Abstract. The novel disease COVID-19 is due to infection by a new virus named SARS CoV-2 which is very different from other corona-viruses such as MERS or SARS and has a higher infection rate. The infection rate could be kept low if COVID-19 cases were detected and isolated sooner. Even though many vaccines were made available, SARS CoV-2 has multiple variants such as Alpha, Beta, Gamma, and Delta which are more contagious and resistant to vaccines. Moreover, only 10% of the worldwide population is fully vaccinated up to this date due to vaccines shortage and anti-vaxxers campaigns. Thus, according to these reasons, more measures need to be considered to contain this pandemic. This study proposed to perform COVID-19 cases detection, based on X-ray images, using machine learning and deep learning algorithms. Some hybrid approaches are investigated such as performing feature extraction using deep learning algorithms and classification using machine learning methods. The best model on four class data sets that achieved 0.9109 accuracy was a hybrid model where features were extracted using VGG16 deep neural network and the classification was done by VotingClassifier.

Keywords: Classification · Medical images · Deep learning · Transfer learning · Covid19

1 Introduction

In late January 2020, a new virus named SARS-CoV-2 (Syndrome Respiratory Acute Severe-coronavirus) has been identified in Wuhan, China. SARS-CoV-2 belongs corona-viruses family that infects animals especially bats which then infect humans who infect other humans. SARS-CoV-2 and MERS-CoV (Middle

Y. Maleh et al. (Eds.): ICI2C 2021, LNNS 357, pp. 403–413, 2022.
https://doi.org/10.1007/978-3-030-91738-8_37

East Respiratory Syndrome-Coronavirus) are the most contagious viruses of this family. SARS-CoV-2 is very contagious and can be transmitted from one person to another while being asymptomatic; which makes it difficult to control the virus propagation. In addition, multiple SARS-CoV-2 variants have impacted human health worldwide through accrue transmission, acute infection, immune escapement. These variants known as Alpha, Beta, Gamma, and Delta have initially emerged in distinct areas such as United Kingdom, South Africa, Brazil, and India. A number of vaccines were developed to protect against this virus. However, the COVID-19 pandemic due to infection caused by SARS-CoV-2 is still out of control since only 10% of the worldwide population is fully vaccinated up to this date. This low vaccination rate is due to vaccines shortage and anti-vaxxers campaigns. Henceforth, COVID-19 control mechanisms will continue to rely on screening and detection infections. Since the SARS-CoV-2 apparition, various measures have been provided to identify infected people and isolate them in order to avoid further spread of the virus including highly specific and sensitive diagnostic such Reverse Transcription Polymerase Chain Reaction (RT-PCR), Antibody, Molecularly imprinted polymer, and so forth [12].

Real-time RT-PCR is the major workhorse in the field of molecular diagnostics. It has been extensively used for high-throughput screening and early diagnosis of infectious diseases in the past. At the moment, a vast majority of the commercially available technologies (tests) for early diagnosis of COVID-19 are based on real-time RT-PCR [1].

Nevertheless, the limited availability of the test kits makes it challenging to detect every individual affected by the virus. Furthermore, these tests take from few hours to a day or two to produce the output, which becomes tedious, time-consuming, and most of the time error-prone in the current state of emergency. Therefore, faster and reliable screening techniques that could be further confirmed by the RT-PCR test are urgently required [5].

In this study, we will investigate the effectiveness of machine learning and deep learning to improve detection algorithms using X-ray images of COVID-19 and PNEUMONIA from different data set. The paper is structured as follows. Section 2 provides basic knowledge on Convolutional Neural Network (CNN), classification model and performance evaluation metrics, ensemble learning concept and an overview of related works. The data-sets used to conduct the study and comparisons results of hybrid machine learning and deep learning models are given in Sect. 3. Conclusion is drawn in Sect. 4.

2 Materials and Methods

2.1 Convolutional Neural Network (CNN)

The growth of deep learning over the past two decades has spurred research into solutions for improving the diagnosis of several complex and chronic diseases. Deep learning models have proven useful for the diagnosis of such diseases as cancer [3,4,23,24], cardiovascular disease [7,25,26], lung disease [2,8,10], Ophthalmologic [16,19], neurodegenerative diseases [14,22] and orthopedics [17,18].

As a result, a wide range of methods have been utilized, but over the last decade, a significant trend toward the usage of convolutional neural networks (CNN) has been seen.

For multi-dimensional resources as images and videos, the CNN is known for its automatic feature maps extractor. As it is named, convolution is the crux of CNN process. Instead of having an input vector for most algorithms, in this case, the input tensor has new dimensions (height, width, channels). According to the task we have to do, the nature of the input data, depends on the depth of the architecture; however, in most cases, we have these layers: convolution, pooling, dropout, dense, and flatten [21].

Convolution is the mathematical operation, which can be seen as the process by which the kernel slides over a target array from left to right, and from top to bottom; obtaining the sum of element-wise multiplication between the kernel and a subset of equal size in the whole image. The reduced information learned by the kernel from the input array is mapped into a feature; according to some arguments such as the number of kernels, the size of each kernel, the shape of the input array, or even the activation function. Then this feature is transformed into another one according to the next layer until the top layers [6].

Max pooling can be seen as the process of summarizing a group of values with the maximum value within that group. Similarly, if you compute the average, it would be average pooling. Pooling operations are usually performed on the generated feature maps after convolution to reduce the number of parameters. Then each pixel of the feature is mapped to an element in the vector: it is the flattening process. Each element in the vector represents a neuron in the layer. For example, if the feature map has $3 \times 3 = 9$ pixels, then there will be nine neurons in the flatten layer. According to over-fitting and the attending output we can add dropout and dense layers.

2.2 Performance Evaluation and Model Improvement

To compare the model according to metrics, we have intra-evaluation and inter-evaluation; both are based on some metrics such as matrix confusion through some scores among which the most used are accuracy score, precision score, recall score, specificity score, and F1 score.

A classifier in binary decision classification identifies examples as positive or negative. The classifier's response can be expressed in a structure called a confusion matrix, which contains four groups: True positives (TP) are positive examples that have been accurately identified as such. False positives (FP) are negative cases that have been mistakenly classified as positives. True negatives (TN) are negatives that have been accurately classified as such. Finally, false negatives (FN) are positive examples that have been mistakenly classified as negative.

$$Precision = \frac{TP}{TP + FP} \tag{1}$$

$$Recall = \frac{TP}{TP + FN} \quad (2)$$

$$Specificity = \frac{TN}{FP + TN} \quad (3)$$

$$F1 = \frac{2 * Precision * Recall}{Precision + Recall} \quad (4)$$

$$Accuracy = \frac{TP + TN}{TP + TN + FP + FN} \quad (5)$$

Inter-evaluation consists of comparing two or more models through these previous scores while intra-evaluation (improvement) concerns the same model but with different parameters like the optimizer, the number of hidden layers, fine-tuning, semi fine-tuning, leave-one-out, k-fold validation, and so forth. Fine-tuning is the process by which, we use transfer learning models while leaving weights trainable; however, frozen consists of using the model with related task weight.

2.3 Ensemble Learning

Many recent approaches to machine learning have benefited from the idea that predictions of an ensemble of models will usually be better than predictions of a single model. The errors of one model can be counteracted by the hits (in the context of incorrect predictions) of other models [13].

The premise of using ensemble-based decision systems in computational intelligence is fundamentally not different from their use in our daily lives. We consult with others before making a decision often because of the variability in the previous record and accuracy of any of the individual decision makers [27].

For data analysis, ensemble techniques integrate a group of learners. Learning multiple classifier systems is another name for ensemble learning. Ensemble techniques integrate several models, whereas several machine learning algorithms create a single model (e.g., a decision tree or neural network). Ensemble techniques are appealing because they may help weak learners make correct predictions. Ensemble learners are machine learning algorithms that use the power of several models to achieve more accuracy than any one model could accomplish on its own. The most important ensemble techniques are bagging, voting, boosting and stacking.

2.4 Related Works

Simonyan et al. [21] presented how a model's accuracy based on CNN can be improved through layers depth increasing.

Horry et al. [9] presented a new approach to confirm COVID-19 disease using pre-trained deep learning models (Transfer Learning) in the name of VGG 16 and VGG 19, ResNet50, Inception V3 and Xception through X-ray images. Similar works have recently been proposed [11]. In their study, VGG16 achieves the best metrics performance, then VGG19, and the metrics performance decrease where the depth layers increase.

Also in the work of [20], ResNet50 achieves 0.9934 accuracy counter 0.9678 for VGG 19. Through [28], the authors compare a model named COVID19XrayNet made from scratch according to the pre-trained deep transfer learning model ResNet34. In their study, they show that the COVID19XrayNet model outperformed the ResNet34 one.

A hybrid approach for skin lesion classification is proposed in [15]. The study was based on feature extraction using pre-trained models, then was given a classification using an SVM classifier.

In Table 1, we summarize the main properties and performances of the related works on Covid 19 diagnosis.

3 Experimental Results and Discussions

In this section, we will use some ML and DL algorithms on a given data-sets. According to the related works, we did an important observation which is, bigger networks do not necessarily translate to better performance. In this study the best train and test accuracy have been obtained using ReLu function, SGD as an optimizer, Dropout(0.4) regulizer, and 512 units for layers 1 to 3.

3.1 Data Set

In this study we used different chest X-ray datasets obtained from Kaggle namely QaTa-COV19 and chest X-ray (PNEUMONIA). The QaTa-COV19 data set, consists of 4603 COVID-19 chest X-rays images. Among 4603 images, 2951 have their corresponding ground-truth segmentation masks. Early-QaTa-COV19 is a subset of the QaTa-COV19 data set, which consists of 1065 chest X-rays including no or limited signs of COVID-19 pneumonia cases for early COVID-19 detection. In this study, we used COVID-19 and normal X-ray images; some of these images are visualized respectively in Fig. 1.

The second data-set of chest X-ray images is composed of 5863 images divided into two categories (normal and pneumonia); 5216 images are used for training, 624 images are used for validation. Visualization of normal and pneumonia X-ray images are shown in Fig. 2.

Table 1. Models setup through many data sets

Algorithm	Layers	Weight state	Metrics	Configuration	Data set	Data augmentation
Model [20]	12 (conv, max pool, batch normalization, dropout, relu)	All trainable (From scratch)	Accuracy = 0.99 Precision = 0.99 Recall = 0.99 F1 = 0.99	Batch size = 32 epochs = 100 5-fold validation	253 covid vs500 normal 100 hold out fortest	Add up from 753 to 1000
VGG16 [9]	16 (includetop = False)	Frozen (non trainable except the top layer)	Precision = 82 Recall = 80 F1 = 80	Batch size = 8 epochs = 20 cross validation	Covid 100pneumonia 100and normal 200	No
VGG19 [9]	19 (includetop = False)	Frozen (non trainable except the top layer)	Precision = 83 Recall = 80 Score = 80	Batch size = 8 epochs = 20 cross validation	3 classe(100,100,200)	No
ResNet50 [9]	50 (includetop = False)	Frozen (Non trainable)	Precision = 70 Recall = 67 F1 = 66	Batch size = 8 epochs = 20 cross validation	3 classe(100,100,200)	No
ResNet50 [20]	50 (includetop = False)	Frozen (non trainable)	Accuracy = 0.98 Precision = 1 Recall = 0.96 F1 = 0.98	Batch size = 2 epochs = 303 fold validation	50 covid 50 normal 20 for test	No
Inception-V3 [20]	159 (includetop = False)	Frozen (non trainable)	Accuracy = 0.97 Precision = 1 Recall = 0.94 F1 = 0.96	Batch size = 2 epochs = 303 fold validation	50 covid 50 normal 20 for test	No
Inception-V3 [9]	153 (includetop = False)	Frozen (non trainable)	Precision = 68 Recall = 61 F1 = 61	Batch size = 8 epochs = 20 cross validation	Covid 100 pneumonia 100 normal 200	No
Inception ResNetV2 [11]	572 (includetop = False)	Frozen (non trainable)	Accuracy = 0.87 Precision = 0.91 Recall = 0.84 F1 = 0.86	Batch size = 2 epochs = 303 fold validation	50 covid 50 normal 20 for test	No
Xception [9]	126 (includetop = False)	Frozen (non trainable)	Precision = 69 Recall = 57 F1 = 48	Batch size = 8 epochs = 20 cross validation	Covid 100 pneumonia 100 normal 200	No
COVID19XrayNet [28]	Modified resnet34 by adding 3 layers	Fine tuning	Precision = 0.92 vs 0.95 Specificity = 0.9017 Sensitivity = 0.9744	Batch size = 16 and 32 epochs = 10 and 20 train_test split	Data set released 5860 chest X-ray	No
Fusion of all networks [15]	AlexNet + VGG16+ ResNet-18 fusion	All FC	MM AUC = 83.83 SK AUC = 97.55 avg. AUC = 90.69	Feature extractor uses 3 pre-trained networks and SVM classifiers	Validation and test images of the ISIC 2016 competition	Not mentioned

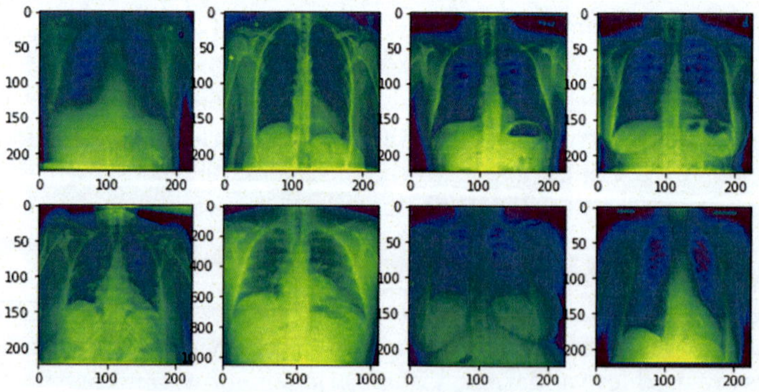

Fig. 1. Respectively 4 covid and 4 normal samples of Qata-cov data set

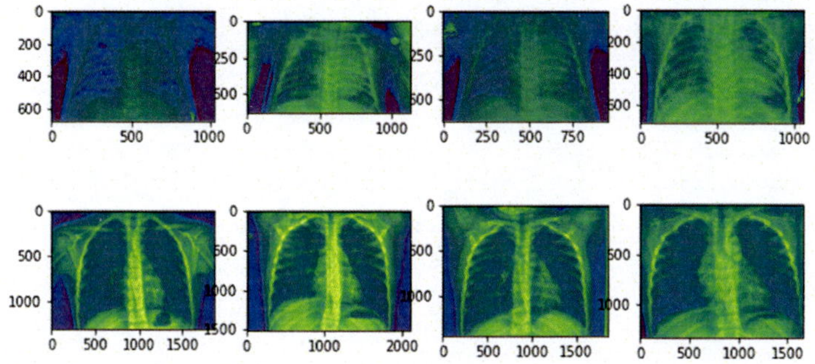

Fig. 2. Respectively 4 normal and 4 pneumonia samples of chest xray data set

3.2 Machine Learning Models

Here we add the COVID-19 images to the chest X-ray dataset and then train some ML classifiers on the three classes datasets. Counter to the deep learning approach where vector descriptor of each images extracted automatically; with machine learning it was extracted manually according to some characteristics of the image as the form, texture, histogram, and color. The results are summarized in the (Table 2) according to the accuracy score and different data partition such 10-fold, leave-one-out, and standard train-test split validation. As it is shown in the table, the KNN classifier achieved the highest accuracy score of 0.87 for Qata-cov19 dataset where SVM and Logistic regression achieved best performance on the chest X-ray dataset.

3.3 Deep Transfer Learning Models

In this experiment, deep transfer learning models based on CNN are trained using the chest X-ray dataset. Some of these models are made from scratch where others are transferred from older tasks through different depth (above 9 layers). Table 3 summarised the results obtained from these models. Basic_CNN1 that was made from scratch with 9 layers without regularizer achieved 0.98 as training accuracy against 0.76 validation accuracy. This result proves that the model is over-fitted. To fix this problem, we increased the layers depth of the layers from 9 to 14 (2 of these 14 are Dropout layers rated to 0.3, Dropout is a regularizer which is known to decrease the over-fitting). The model (Basic_CNN2) with image DataGenerator achieved better results than the previous one with a 0.85 validation accuracy.

Table 2. Machine learning models evaluation

Classifier	Decision-tree	KNN	SVM	Logistic regression
Model	10-fold loo train-test split	10-fold loo train-test split	10-fold loo train-test split	10-fold loo train-test split
Acc_score (Qata-cov19 Dataset)	0.8363 0.8417 0.8338	0.8709 0.8735 0.8696	0.8641 0.8580 0.8497	0.8183 0.8097 0.8084
Acc_score (Chest xray Dataset)	0.9334 0.9348 0.6698	0.9200 0.9233 0.6955	0.9641 0.9641 0.6778	0.9633 0.9629 0.6746

Table 3. Transfer learning models evaluation

Model	Options	Train acc	Val acc	Data-set
Basic_CNN1	9 layers (from scratch)	0.98	0.76	chest_X-ray
Basic_CNN2	14 layers (from scratch)	0.98	0.85	chest_X-ray
VGG16	include_top = False weights = 'imagenet' input_shape = (150, 150, 3)	0.93	0.89	chest_X-ray
InceptionV3	include_top = False weights = 'Imagenet' input_shape = (150, 150, 3)	0.93	0.85	chest X-ray
ResNet50	include_top = False weights = imagenet input_shape = (150, 150, 3)	0.84	0.75	chest X-ray
VGG 16	include_top = False weights = imagenet input_shape = (150, 150, 3)	0.97	0.92	Qata-Cov

Comparison is also made with deep transfer models such as VGG16, InceptionV3 and ResNet50. Even though the VGG16 has the lowest layers depth, the obtained result are still the best.

3.4 Hybrid Models

In this experiment, we investigated the effectiveness of hybrid models based on VGG 16 and machine learning classifiers to detect Covid-19 cases. Thus the hybrid models were trained on data that represent features which were extracted from images using VGG16 and a classification was made with machine learning classifiers; this approach gave us different accuracy scores where the highest score was obtained by VotingClassifier (0.9109); as it can be seen in Table 4.

Table 4. Hybrid models summary

Name	Features extractor	Classifier	Data used	Accuracy score
Model_1	VGG16	RandomForest	chest xray	0.6185
		SVM		0.6651
		DecisionTree		0.6170
		LogisticRegression		0.6362
Model_2	VGG 16	RandomForest	Qata-Cov	0.8847
		SVM		0.5864
		DecisionTree		0.7968
		LogisticRegression		0.9086
Model_3	VGG 16	VotingClassifier	chest xray	0.6730
Model_4	VGG 16	VotingClassifier	Qata-Cov	0.9109

This show the power of deep learning models for accurate features extraction over machine learning models; especially in ensemble learning mode.

4 Conclusion

Traditional methods of diagnosing PCR Covid-19 are time-consuming, expensive, and have a high rate of misdiagnosis. This study addresses such limitations by proposing a hybrid Machine and Deep Transfer Learning Based Classification models for Covid-19 and Pneumonia diagnosis using X-ray Images. The proposed approach uses Deep learning models known for their discriminatory power to extract relevant features combined with Machine Learning models to achieve an accurate classification. The VotingClassifier for the diagnosis of covid-19 achieve almost 92% accuracy, which coincide to the VGG 16 validation accuracy; whatever the best accuracy on pneumonia data is secured by a machine learning classifier with leave one out validation, which achieve more than 0.95 as accuracy score.

References

1. Afzal, A.: Molecular diagnostic technologies for Covid-19: limitations and challenges. J. Adv. Res. (2020)
2. Anthimopoulos, M., Christodoulidis, S., Ebner, L., Christe, A., Mougiakakou, S.: Lung pattern classification for interstitial lung diseases using a deep convolutional neural network. IEEE Trans. Med. Imaging **35**(5), 1207–1216 (2016)
3. Bardou, D., Zhang, K., Ahmad, S.M.: Classification of breast cancer based on histology images using convolutional neural networks. IEEE Access **6**, 24680–24693 (2018)
4. Dorj, U.O., Lee, K.K., Choi, J.Y., Lee, M.: The skin cancer classification using deep convolutional neural network. Multimed. Tools Appl. **77**(8), 9909–9924 (2018)
5. Farooq, M., Hafeez, A.: Covid-ResNet: a deep learning framework for screening of covid19 from radiographs. arXiv preprint arXiv:2003.14395 (2020)

6. Goodfellow, I., Bengio, Y., Courville, A.: Deep Learning. MIT Press, Cambridge (2016)
7. Hasan, N.I., Bhattacharjee, A.: Deep learning approach to cardiovascular disease classification employing modified ECG signal from empirical mode decomposition. Biomed. Signal Process. Control **52**, 128–140 (2019)
8. Hasenstab, K.A., et al.: Automated CT staging of chronic obstructive pulmonary disease severity for predicting disease progression and mortality with a deep learning convolutional neural network. Radiol. Cardiothorac. Imaging **3**(2), e200477 (2021)
9. Horry, M.J., et al.: X-ray image based Covid-19 detection using pre-trained deep learning models (2020)
10. Huang, S., Lee, F., Miao, R., Si, Q., Lu, C., Chen, Q.: A deep convolutional neural network architecture for interstitial lung disease pattern classification. Med. Biol. Eng. Comput. 1–13 (2020)
11. Islam, M.M., Karray, F., Alhajj, R., Zeng, J.: A review on deep learning techniques for the diagnosis of novel coronavirus (COVID-19). arXiv, vol. 1, pp. 1–18 (2020)
12. Jalandra, R., et al.: Strategies and perspectives to develop SARS-CoV-2 detection methods and diagnostics. Biomed. Pharmacother. **129**, 110446 (2020)
13. Kozak, J.: Decision Tree and Ensemble Learning Based on Ant Colony Optimization. Springer, Heidelberg (2019). https://doi.org/10.1007/978-3-319-93752-6
14. Lahrichi, S., Rhanoui, M., Mikram, M., El Asri, B.: Toward a multimodal multitask model for neurodegenerative diseases diagnosis and progression prediction. In: Proceedings of the 10th International Conference on Data Science, Technology and Applications - DATA, pp. 322–328. INSTICC, SciTePress (2021)
15. Mahbod, A., Schaefer, G., Wang, C., Ecker, R., Ellinge, I.: Skin lesion classification using hybrid deep neural networks. In: 2019 IEEE International Conference on Acoustics, Speech and Signal Processing (ICASSP), ICASSP 2019, pp. 1229–1233. IEEE (2019)
16. Mikram, M., Moujahdi, C., Rhanoui, M., Meddad, M., Khallout, A.: Hybrid deep learning models for diabetic retinopathy classification. In: Proceedings of the 5th International Conference on Big Data and Internet of Things (BDIOT2021) (2021)
17. Norman, B., Pedoia, V., Majumdar, S.: Use of 2D U-Net convolutional neural networks for automated cartilage and meniscus segmentation of knee MR imaging data to determine relaxometry and morphometry. Radiology **288**(1), 177–185 (2018)
18. Ounasser, N., Rhanoui, M., Mikram, M., El Asri, B.: Anomaly detection in orthopedic musculoskeletal radiographs using deep learning. In: Proceedings of the International Conference on Computing and Communication Networks, Manchester, UK (2021)
19. Rahimy, E.: Deep learning applications in ophthalmology. Curr. Opin. Ophthalmol. **29**(3), 254–260 (2018)
20. Sarv Ahrabi, S., Scarpiniti, M., Baccarelli, E., Momenzadeh, A.: An accuracy vs. complexity comparison of deep learning architectures for the detection of Covid-19 disease. Computation **9**(1), 3 (2021)
21. Simonyan, K., Zisserman, A.: Very deep convolutional networks for large-scale image recognition. In: 3rd International Conference on Learning Representations, ICLR 2015 - Conference Track Proceedings, pp. 1–14 (2015)
22. Sivaranjini, S., Sujatha, C.: Deep learning based diagnosis of Parkinson's disease using convolutional neural network. Multimed. Tools Appl. **79**(21), 15467–15479 (2020)

23. Sun, W., Tseng, T.L.B., Zhang, J., Qian, W.: Enhancing deep convolutional neural network scheme for breast cancer diagnosis with unlabeled data. Comput. Med. Imaging Graph. **57**, 4–9 (2017)
24. Ting, F.F., Tan, Y.J., Sim, K.S.: Convolutional neural network improvement for breast cancer classification. Expert Syst. Appl. **120**, 103–115 (2019)
25. Wang, J., et al.: Detecting cardiovascular disease from mammograms with deep learning. IEEE Trans. Med. Imaging **36**(5), 1172–1181 (2017)
26. Yıldırım, Ö., Pławiak, P., Tan, R.S., Acharya, U.R.: Arrhythmia detection using deep convolutional neural network with long duration ECG signals. Comput. Biol. Med. **102**, 411–420 (2018)
27. Zhang, C., Ma, Y.: Ensemble Machine Learning: Methods and Applications. Springer, Heidelberg (2012). https://doi.org/10.1007/978-1-4419-9326-7
28. Zhang, R., et al.: Covid19XrayNet: a two-step transfer learning model for the Covid-19 detecting problem based on a limited number of chest X-ray images. Interdisc. Sci. Comput. Life Sci. **12**(4), 555–565 (2020)

Applying Advanced IoT Network Topologies to Optimize COVID-19 Sanitary Passport Platform Based on CoAP Protocol

Fatima Zahra Hamza[1(✉)], Sanaa El Aidi[1], Siham Beloualid[1], Abdelhadi El Allali[1], Abderrahim Bajit[1], Habiba Chaoui[1], and Ahmed Tamtaoui[2]

[1] Laboratory of Advanced Systems Engineering (ISA), National School of Applied Sciences, Ibn Tofail University, Kenitra, Morocco
{fatimazahra.hamza, sanaa.elaidi, siham.beloualid, abderrahim.bajit, habiba.chaoui}@uit.ac.ma

[2] National Institute of Posts and Telecommunications (INPT-Rabat), SC Department, Mohamed V University, Rabat, Morocco
tamtaoui@inpt.ac.ma

Abstract. The product of this work is based on the creation of a medical IoT platform that serves to control citizens to access public or private space to prevent the spread of the covid-19 pandemic. Our platform discusses 3 scenarios in which a citizen can be allowed to: the first scenario is used to verify citizens who are vaccinated by their QR Code and facial authentication, the second scenario is used to verify citizens who have done a PCR test (which is only valid for 48 h) and the facial recognition test as well, and finally we have the third scenario which starts by measuring the body temperature and then verifying the identity by presenting an RFID number and passing the facial recognition test. This platform is implemented on 4 types of network topologies (Star, Tree, Mesh, Cluster) to choose which one best matches our platform's needs in terms of execution time, memory occupation, and energy consumption. Since this platform is secure and intelligent, we opted for using the CoAP communication protocols to ensure communication between the nodes. We used the AES/RSA-SHA256 encryption algorithm to ensure data security and protection from any attacks.

Keywords: CoAP IoT protocol · IoT Star/Tree/Cluster/Mesh topologies · AES · RSA · AI · Sanitary passport PCR/QR/Barre/RFID code

1 Introduction

The Internet of Things allows an interaction between the physical and digital worlds. The digital world interacts with the physical world through sensors and actuators. These sensors collect information that must be stored and processed. Data processing can occur at the edge of the network, on a remote server, or in the cloud [1].

The covid-19 pandemic has changed our lives, and we have to get used to it. After several waves of pandemics, we have reached a point where respecting the sanitary

Y. Maleh et al. (Eds.): ICI2C 2021, LNNS 357, pp. 414–424, 2022.
https://doi.org/10.1007/978-3-030-91738-8_38

rules is mandatory for reopening public or private spaces to reduce the spread of the virus. We then proposed an IoT platform that allows managing access to these places. The access requires that people have a QR code obtained after the second dose of vaccination, or a negative PCR test dated less than 48 h, or a valid UID and body temperature. Without these conditions, the visitor is not allowed to access.

To further explain, the platform is divided into 3 tests: The first test consists of validating the QR code of the vaccine and the facial recognition of its property to verify if it is the same person. If the visitor is not vaccinated, we check in a second test, a negative PCR test certificate that must not exceed 48 h, and then we proceed to facial recognition. If the visitor has neither his vaccine nor his PCR test, we check his body temperature and if his/her UID already registers him/her in the database. If so, we proceed to facial recognition. The access is then refused in the negative case of these tests.

We have also used artificial intelligence, specifically facial recognition, for image detection and processing in our platform. The human face provides several social signals essential for our public life [1]. Face recognition is the detection of a person by their facial characteristics. The face mediates person identification, attractiveness, and facial communication. This approach is very important to provide security and user access control [2]. IoT needs artificial intelligence AI and vice versa, and AI is beneficial for real-time processing [3, 4]. In our platform, IoT has been used to connect medical devices to the internet and collect data about the citizen to process its analysis and act the citizen's access right appropriately. The system can analyze the facial structure by comparing it to the information in the database and then identifying the detected person. We devoted ourselves to the implementation of a CoAP protocol with an authentication server in the IoT. The intervention of encryption algorithms to solve the problems mentioned above by encrypting/decrypting messages. Therefore, we have implemented a secure version of the CoAP protocol using the AES encryption algorithm [5].

To manage citizens' access to public spaces, an intelligent, secure and synchronized platform was created using 4 nodes: a PIR sensor to detect the presence of a person at the entrance of the space. An RFID tag for data identification, a temperature sensor to measure the temperature of the citizen, a camera to scan the QR code of the vaccine pass and the barcode of PCR and for the identification of the person. To secure the data, we used the RSA-SHA256 and AES-SHA256 encryption algorithms and the CoAP communication protocol to see its efficiency in real-time (Fig. 1).

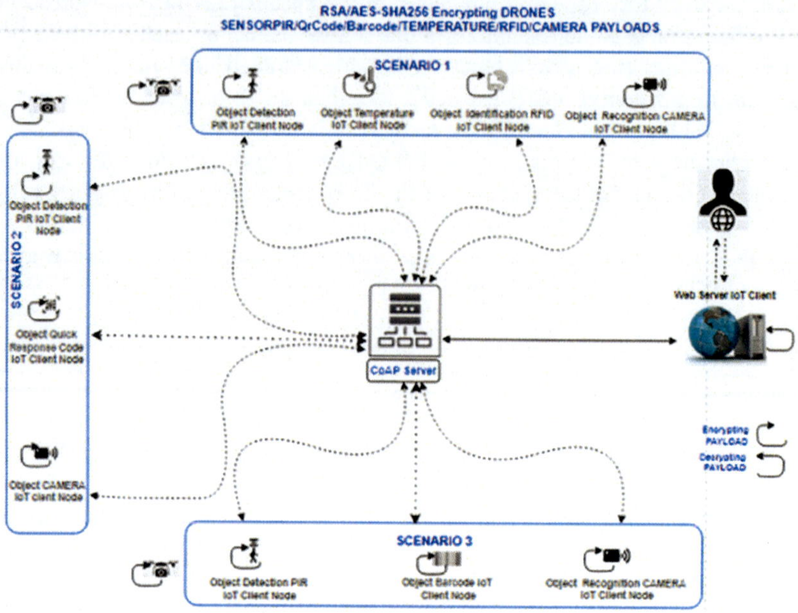

Fig. 1. Medical IoT architecture

2 Functional Medical IoT Scenarios

Our platform is composed of three major scenarios, on which our algorithm is based. The first scenario checks the vaccinated people against covid-19 by presenting their vaccine pass to scan the QR Code, which contains a unique RFID number. If this number is valid, we proceed by applying the facial recognition test, which is based on artificial intelligence, to confirm the individual's identity. The second scenario is used when the citizen does not have a vaccine pass, so we check if he/she has a negative PCR test that must not exceed 48 h. Then we apply the facial recognition test to ensure that this person is indeed the owner of the test. Last but not least, our third scenario is performed when the individual does not possess a vaccine pass nor a PCR test, so we verify his/her body temperature, which should not exceed 37°, as well as their RFID's UID to check if it's a valid one, and then apply facial recognition test (Fig. 2).

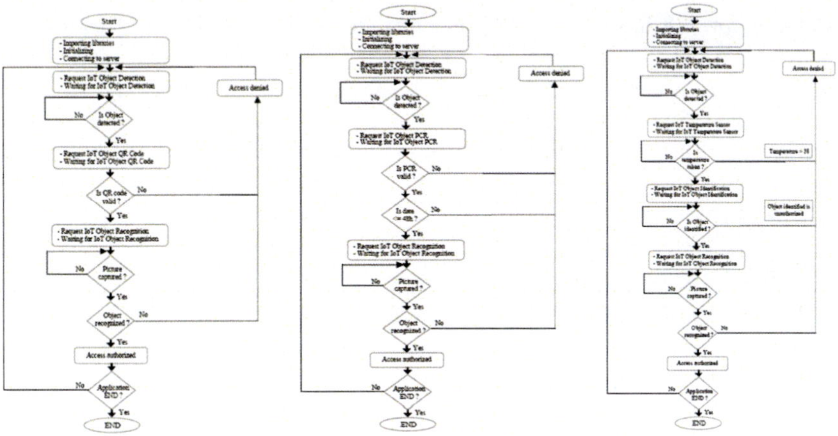

Fig. 2. Functional algorithm

In our platform, we used the CoAP protocol to ensure the communication between the nodes, and based it on four different topologies (Star, Tree, Mesh, Cluster). And to secure the payloads, we used AES-SHA256 and RSA-SHA256 encryption algorithms.

2.1 CoAP Protocol

CoAP protocol is used to address HTTP's limitations in a constrained environment. It is based on a client-server model and supports HTTP's REST methods to retrieve information from a sensor. CoAP messages are by default transported over UDP datagrams that offer additional security guarantees [10]. CoAP consists of a messages layer That manages the UDP and the messages, and the Request/Response layer controls the request/response interaction based on the messages. CoAP has 4 types of messages: confirmable/unconfirmable message, acknowledgment, and reset. The confirmable message is a reliable message that the client is certain will be sent to the receiver. Once the server receives the data, it sends a confirmation message with the same identification number of the confirmable message. In an unreliable transmission, a non-confirmable message is sent, and it does not need to be acknowledged by the server [11, 12]. IoT communication relies on the CoAP protocol to transmit data in a restricted network and supports connection-oriented transmission.

2.2 RSA and AES Algorithms

The Advanced Encryption Standard (AES) is used to protect data from unauthorized access and encrypt it. It provides a means of proving that information is authentic and comes from the point of origin it claims to be. It also provides a way to confirm that the information has not been altered during transmission. AES is the most common security algorithm, used for a wide variety of applications such as wireless communication, financial transactions, e-commerce, etc. It uses larger key sizes such as 128, 192 and 256 bits for encryption. This makes the AES algorithm more robust against hacking, as for 128 bits, about 2128 attempts are needed to break it. Therefore, it is complicated to hack it, which makes it a very secure protocol [6–8].

RSA is an asymmetric cryptography algorithm; that works on two keys (public/private key). The public key is given to everyone to encrypt and the private key is kept private. The RSA algorithm is hard to crack since it involves factorizing prime numbers, which are difficult to factorize. However, the problem with RSA is that it is very slow in cases where large data needs to be encrypted by the same server. Intermediaries who could temper the public key system can compromise the data transferred by the RSA algorithm. Both symmetric encryption techniques and asymmetric encryption techniques are important for encrypting sensitive data [9].

2.3 IoT Network Topologies

There are different types of IoT topologies for networking. To decide which network is best for our IoT application, we need to know the advantages and disadvantages of each. As its name suggests, the tree topology is a topology structured in such a way that the connected elements are in the format of tree branches. Tree topology brings together two other topologies: Bus topology, where the nodes are connected to a single cable and Star topology where the nodes are connected to a central hub device. This topology consists of parent and child nodes. Two child nodes can be connected to a single parent node. The advantage of this topology is that it allows to search and sort a large amount of data in the network, it requires less cable and if a node gets disconnected from the network, others work regularly. However, when the main bus is disconnected or has a problem, the network stops working. This topology is also complicated to configure, and the cable size is limited and the number of nodes used [13].

We speak of a star topology when all nodes are independently connected to the central node. Therefore, it can be said that every computer is indirectly connected to every other node. The advantage of this topology is that it's easier to install, and very simple to detect the error because the links are easily identifiable. But if a network connection failed or the central node goes down, all the nodes attached are disabled [14].

The mesh topology is used to ensure the reliability of the network communication. In this topology, there is no central connection point, we find that each node is connected to several nodes. The nodes act as relays, transmitting a message to its final destination, the strength of this topology is the speed of sending a message, there are multiple connections which means that each node can transmit and receive from several nodes at the same time, and also the addition of new nodes can be without interruption or interference with other nodes. The general scope of mesh networking can be applied physically at different physical layers and logically at different logical layers. For example, physically applicable in optical backbone networks, MHN's, networks for accessing, fixed and Wireless networks, or cognitive radio networks in peer-to-peer networks, dynamic virtual circuit networks, and overlay networks using MPLs, Mesh can be applied logically [15, 16] (Fig. 3).

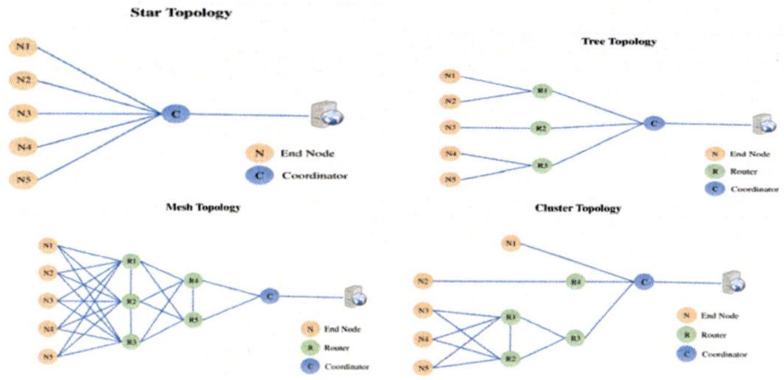

Fig. 3. IoT network topologies

3 The Proposed Approach

Our approach is based on implementing an IoT intelligent medical platform that is used to control citizens' access to public/private space. We have implemented three scenarios in the platform: the 1st one allows us to check vaccinated people by applying two tests: scanning the QR code and using facial recognition technology. The access is denied if one of these tests is invalid. The 2nd scenario is for the citizens who have the PCR test, which is checked for 48 h. If it does, we deny the access, otherwise we verify his/her identity using a facial recognition test to authorize access to the citizen. The 3rd scenario checks citizens registered in the database through their UIDs using RFID tag, body temperature facial recognition test. Our constructed platform is built based on four kinds of topologies: Star, Tree, Mesh and cluster, to settle on the most effective one in terms of three parameters: execution time, memory occupation and energy consumption, additionally to protect and guarantee the security of the data by encrypting it thanks to RSA-SHA256 and AES-SHA256 encryption algorithms to choose the best one in terms of the parameters previously mentioned.

The asymmetric encryption algorithm RSA generates a key pair: a public key used to cipher the data and a secret key to decipher it. The secret key is kept private for the receiver to decode, but the public key is passed around for the senders to encode. RSA is incredibly onerous to crack; however, it is slow since it uses large keys to cipher and decipher. Whereas AES encryption is also robust against hacking as it also uses large key sizes for coding, and it is a symmetric algorithm that only needs one key to encrypt and decrypt data. This feature makes AES encryption more practical to use in resource-constrained environments than RSA. Based on the analyses created in [1, 3], it was also concluded that AES does not have a large impact on the platform when it comes to execution time and memory occupation. In other words, AES is much less time-consuming and memory-consuming than RSA. It also provides the confidentiality, integrity, and authentication of the exchanged information, ensuring efficient communication between the IoT nodes, which is precisely what we would like to have in our platform. The choice of the CoAP communication protocol comes from the fact that, as the name suggests, it is designed for resource-constrained environments, because it runs on UDP, which requires minimal transmission overhead, and also smaller packet size, resulting in faster communication cycles, and enables the batterie life of IoT devices to last longer compared to MQTT protocol. The results of CoAP efficiency were also shown in [1, 3], where we can note that the latter protocol is practical than MQTT in terms of execution time, memory occupation and power consumption.

The network topology describes how the different units in the IoT network communicate with one another. Topologies can differ significantly in security, power consumption, time execution, cost, and complexity. Therefore, it is necessary first to understand which topology best matches our IoT application and necessities. In this work, we have compared mesh, star, tree and cluster topology. Cluster-Based topologies have been widely utilized in wireless sensor networks for different types of protocols. Clustering is especially useful for applications that require scalability to numerous amounts nodes. It is a simple hierarchical topology based on a relationship between father and children. We can find multiple clusters that are controlled with a coordinator. Each one can have its cluster head (parent) and can communicate with the coordinator. Having time, memory, and power as important parameters to consider in wireless sensor networks, the cluster topology has proved its efficiency in terms of execution time, energy and memory consumption.

4 Discussion and Results

The goal of this article is to apply the AES-SHA256 and RSA-SHA256 encryption methods on our platform, as well as to develop it according to different topologies to analyze the impact of each one on the platform, keeping in mind that CoAP is a high constraint protocol, thus it is important to preserve the power of the IoT nodes, in addition, to shrink their memory occupation and execution time. The following figures show the comparative results of the three scenarios suggested by our platform. The first scenario checks the vaccine pass, the second scenario checks the PCR test plus facial recognition, and the third scenario checks RFID's UID and body temperature.

Choosing CoAP over MQTT as a communication protocol was based on the fact that this latter takes a little more execution time and occupies more memory compared to CoAP, which is also noticed even when we add a security layer to the payloads (see Fig. 4). This big consumption comes from CoAP having a small fixed-length binary header while MQTT's packet is loaded. As for security we have also opted for CoAP since it is built on UDP and not on TCP like MQTT.

When analyzing the security aspect of the three scenarios in the results (see Fig. 5, 6, 7), it is clear that RSA is significantly expensive in terms of memory occupation and execution time, since it takes a long time to be executed and consumes more energy and memory, compared to AES which is more practical in our case as it has almost no effect on the performance of the platform. This comes as a result that RSA uses two large keys for encryption and decryption, which makes it extremely strong against attacks, but still, it is not suitable for our resource-constrained platform. As for AES is also a powerful encryption method that does not require a lot of energy, time, and memory. It is fast in processing, offers high performance, can encrypt and decrypt a

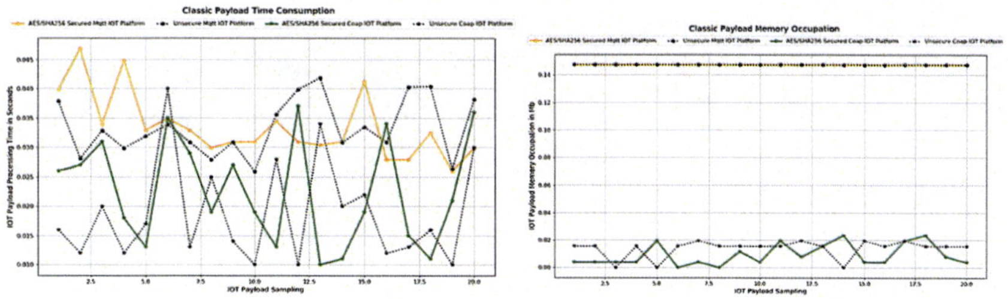

Fig. 4. Time consumption and memory occupation of CoAP/MQTT classic payload combined with AES/SHA256

large amount of data quickly, and has low resource and memory requirements, making it perfect for this platform.

By comparing the results of each topology, it is noticeable that star topology has the best results in memory and time, compared to the other topologies. Star may be a reliable, efficient and easy to implement and manage solution. However, this is only suitable for components that are near to one another. If the communication link between the coordinator and the end node is long, more power should be spent to deliver messages. Which is the contrary of what we are searching for in our platform. A tree topology is a hierarchical structure of network nodes, where every layer of nodes can form a star network with the nodes it provides services to. In this case, the structure of tree topology inherits the disadvantages of star topology. Although these two topologies (star and tree) do not consume a lot of memory space and processing time, they are not recommended for IoT platforms, since in this field, we need the response to be in real-time and fast without losing the data. So, if we had to choose between mesh and cluster-tree, we will opt for cluster-tree because it is based on metrics such as the residual energy of a node, the number of connections, and the distance between the devices of the nodes or between a node and the target. The advantage of this topology is that it gives all the advantages of the beacon mode (synchronization, QoS support by guaranteed time slots) and, at the same time, allows the construction of large networks to cover large areas, on the other hand, the network being hierarchical, the data transmission delay can be predicted which is necessary to know in this platform. The router always follows the state of its child nodes, and the connection state of the network is aggregated along the root of the network tree. Therefore, the coordinator can easily understand the connection state of the entire network. These results allow us to conclude that the adoption of a cluster-tree topology for this platform is better, in terms of efficiency and reliability (since the interconnectivity of the nodes makes them extremely resistant to failures, and there could several paths between two nodes in the network), security (the configuration is safe from any compromise), execution time and memory consumption, based on the results presented in the figures.

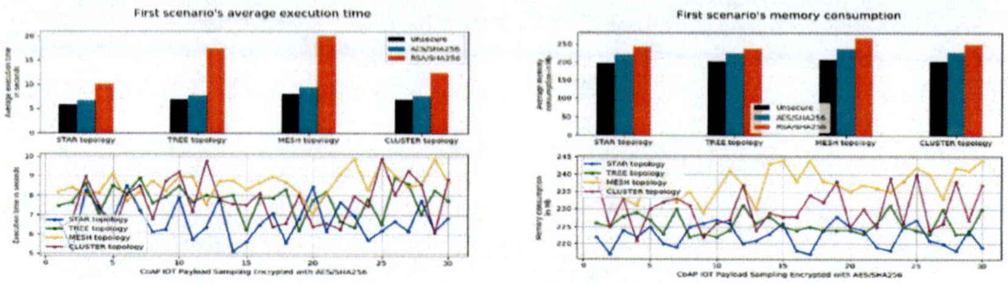

Fig. 5. First scenario's average execution time and memory consumption

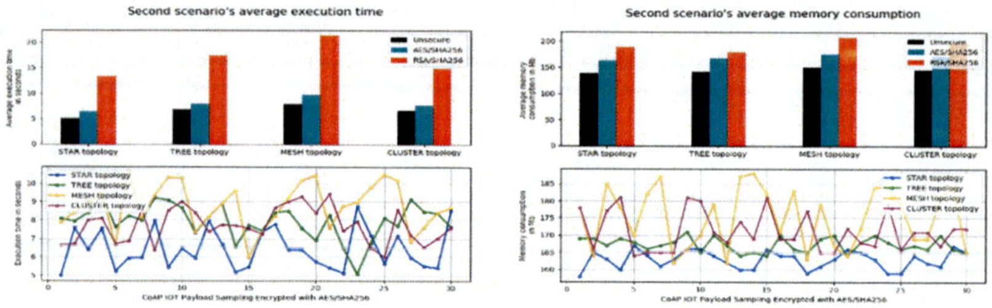

Fig. 6. Second scenario's average execution time and memory consumption

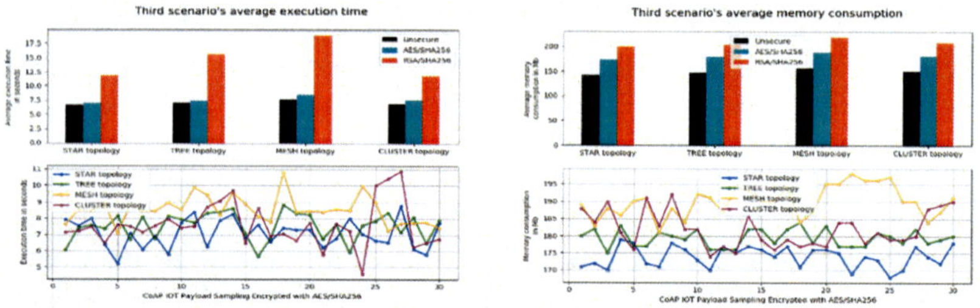

Fig. 7. Third scenario's average execution time and memory consumption

5 Conclusion and Perspectives

In this work, we created a medical platform that monitors citizens and minimizes the spread of the Covid-19 virus. And that is by taking into account that a citizen may have a vaccine pass, a PCR, or just an RFID tag.

For the development, we chose the CoAP to ensure communication between the nodes. We also based our platform on four different topologies: star, tree, mesh, cluster, and two different encryption algorithms, namely RSA-SHA256 and AES-SHA256. Eventually, we opted for the cluster topology combined with AES-SHA256 to ensure our platform's reliability, efficiency, and security.

We will implement the 6LoWpan protocol in future works, which provides the upper layer system for use with low-power wireless communications for IoT [17]. We will also deploy another security method, namely ECC (Elliptic curve cryptography), that provides a secure means of exchanging keys among communicating hosts using the DH key exchange algorithm. We will also propose a blockchain architecture for the platform. This database stores all processed transactions in chronological order, using computer memories that are tamperproof to adversaries [18].

References

1. El Aidi, S., Bajit, A., Barodi, A., Chaoui, H., Tamtaoui, A.: An advanced encryption cryptographically-based securing applicative protocols MQTT and CoAP to optimize medical-IOT supervising platforms. In: Saeed, F., Mohammed, F., Al-Nahari, A. (eds.) IRICT 2020. LNDECT, vol. 72, pp. 111–121. Springer, Cham (2020). https://doi.org/10.1007/978-3-030-70713-2_12
2. Li, L., Mu, X., Li, S., Peng, H.: A review of face recognition technology. IEEE Access **8**, 139110–139120 (2020)
3. El Aidi, S., Bajit, A., Barodi, A., Chaoui, H., Tamtaoui, A.: An optimized security vehicular Internet of Things -IoT-application layer protocols MQTT and COAP based on cryptographic elliptic-curve. In: 2020 IEEE 2nd International Conference on Electronics, Control, Optimization and Computer Science, ICECOCS 2020, p. 9314579 (2020)
4. Pantic, M., Patras, I.: Dynamics of facial expression: recognition of facial actions and their temporal segments from face profile image sequences. IEEE Trans. Syst. Man Cybern. Part B (Cybern.) **36**(2), 433–449 (2006). https://doi.org/10.1109/TSMCB.2005.859075
5. Laurière, J.-L.: Intelligence Artificielle: résolution de problèmes par l'homme et la machine. https://ulysse.univ-lorraine.fr/discovery/fulldiplay?vid=33UDL_INST:UDL&docid=alma991003670679705596&lang=fr&context=L&adaptor=Local%20Search%20Engine. Accessed 10 Aug 2021
6. Barodi, A., Bajit, A., Benbrahim, M., Tamtaoui, A.: Improving the transfer learning performances in the classification of the automotive traffic roads signs. In: E3S Web Conference Proceeding (2020)
7. Zhou, J., Wang, Y., Ota, K., Dong, M.: AAIoT: accelerating artificial intelligence in IoT systems. IEEE Wirel. Commun. Lett. **8**(3), 825–828 (2019)
8. Lu, C.-C., Tseng, S.-Y.: Integrated design of AES (Advanced Encryption Standard) encrypter and decrypter. In: Proceedings IEEE International Conference on Application - Specific Systems, Architectures, and Processors, San Jose, CA, USA, pp. 277–285 (2002). https://doi.org/10.1109/ASAP.2002.1030726
9. Kothmayr, T.: A security architecture for wireless sensor networks based on DTLS. Master's thesis in the Software Engineering Elite Graduate Program at the University of Augsburg (2011)
10. Giuseppe, N., Maria Carla, C.: Security of IoT application layer protocols: challenges and findings. MDPI (2020)
11. Chanwit, S., Chatchai, K.: Congestion control in CoAP observe group communication. MDPI (2019)
12. Naik, N.: Choice of effective messaging protocols for IoT systems: MQTT, CoAP, AMQP and HTTP. In: 2017 IEEE International Systems Engineering Symposium (ISSE), pp. 1–7 (2017). https://doi.org/10.1109/SysEng.2017.8088251
13. Estrada, R., Tomasi, C., Schmidler, S.C., Farsiu, S.: Tree topology estimation. IEEE Trans. Pattern Anal. Mach. Intell. **37**(8), 1688–1701 (2015)
14. Star Topology: Advantages and Disadvantages, Penna Sparrow. https://www.ianswer4u.com/2011/05/star-topology-advantages-and.html. Accessed 31 July 2021

15. Network Topologies, Protocols and Layers. https://www.bbc.co.uk/bitesize/guides/zr3yb82/revision/1. Accessed 31 July 2021
16. Plageras, A.P., Psannis, K.E., Ishibashi, Y., Kim, B.G.: IoT-based surveillance system for ubiquitous healthcare. In: 42nd Annual Conference of the IEEE Industrial Electronics Society, IECON 2016, pp. 6226–6230 (2016). https://doi.org/10.1109/IECON.2016.7793281
17. What is 6LoWPAN - The Basics. www.electronics-notes.com/articles/connectivity/ieee-802-15-4-wireless/6lowpan.php. Accessed 30 July 2021
18. Kodali, R.K., Narasimha Sarma, N.V.S.: Energy efficient ECC encryption using ECDH. In: Sridhar, V., Sheshadri, H., Padma, M. (eds.) Emerging Research in Electronics, Computer Science and Technology, vol. 248, pp. 471–478. Springer, New Delhi (2014). https://doi.org/10.1007/978-81-322-1157-0_48

An Approach Based on Multi-agent and Artificial Immune Algorithm for the Vehicle Routing Problem in Home-Health Care

Ettazi Haitam(✉), Amrani Ayoub, Rafalia Najat, and Abouchabaka Jaafar

University Ibn Tofail, Kenitra, Morocco
{haitam.ettazi,amrani.ayoub,najat.rafalia, jaafar.abouchabaka}@uit.ac.ma

Abstract. In this period of the COVID-19 pandemic, it is critical to adjust the resources found in multiple fields regarding logistical challenges to ensure the minimization of transportation costs and maximization of patient's comforts and preferences in-home health care. In this article, we will address the vehicle routing problem with time windows, priority, synchronization and lunch break constraints, all combined with scheduling and planning. Often, patients who no longer need to stay in the hospital must give way to another urgent case, but need more treatment than they can receive at home. Caregivers are assigned to provide special care to needy patients based on their requests within time constraints such as time windows, patient preferences, precedence, and synchronization. This paper aims to optimize the planning and scheduling of home care efficiently with consideration and respect to the criterion such as qualification of caregivers, availability and preferences of the patients. To resolve this problem, an Artificial Immune Algorithm (AIS) is proposed as a generator of routes, and a multi-agent approach is established to guarantee the upmost coordination and communication among the overall actors.

Keywords: Home health care · Vehicle routing problem · Time windows · Preferences · Synchronization · Multi-agent system · Artificial immune algorithm

1 Introduction

Over the years, the number of diseases has increased significantly, meaning the average number of patients per one hospital bed has ascended; thus, the need to give care to sick people in their homes has become crucially needed. It is very important to cite that pandemic (SAR-COV-19) [1], has led to prioritizing the necessity of moving patients from their homes or vice-versa much needed. Due to these circumstances, challenges have arisen a lot for conducting researches in the field of operations researches and decision support in this field. Coordinating either human and resources is the core to provide optimized planning to reduce/minimize the costs while maximizing the quality of home health care services for both working staff (Lunch Breaks, Preferences…) and patients (Time Windows, Preferences, Precedence…).

Y. Maleh et al. (Eds.): ICI2C 2021, LNNS 357, pp. 425–436, 2022.
https://doi.org/10.1007/978-3-030-91738-8_39

In this paper, we propose adapting the vehicle routing problem with time windows, synchronization, precedence and lunch breaks constraints. This problem consists of looking for a set of routes performed by a fleet of vehicles (regarded as caregivers) to serve a set of customers (patients) in a predetermined geographical area, respecting a specific availability most effectively and efficiently. A service that the home health care facility provides must occur within a determined and accurate time and can require the intervention of one or many caseworkers either at the same time (simultaneously) or in a specific order (precedence), which means that a customer (patient) must be ensured two or more services in a row (not at the same time). On the other hand, after the caregiver's working time, it is mandatory to have a break time to eat and rest. The lunch break generally has a time window (e.g., 13:00–14:00) and a period of time (e.g., 45 min). In our study, all those prerequisites enforce the collaboration and coordination of many caregivers to achieve the desired objectives, which lead us to consider the problem as an alternative to the vehicle routing problem with time windows called VRPTW-SPLB for the Vehicle Routing Problem with Time Windows, Synchronization, Precedence and Lunch Break constraints. This paper aims to optimize most efficiently the planning and scheduling of home care to minimize the travel costs and maximize the patient's preferences and satisfaction according to constraints such as qualification of caregivers, lunch break constraints, availability, and preferences of the patients.

The structure of this paper consists of a section highlighting concepts related to the problem considered. The following section describes the literature review, then a section prescribing the approach used to solve the problem using the AIS algorithm. Finally, a conclusion and perspective are stated.

2 Concepts

2.1 Metaheuristics

Metaheuristics are, like greedy heuristics, methods that do not guarantee to meet the optimum. However, metaheuristics are generally better performing methods than greedy heuristics. Indeed, they are equipped with devices allowing them to leave local optimums in order to push the search space further into the solution space. Of course, this increases the time needed to obtain the final solution compared to a greedy heuristic, but they are nevertheless faster than the mathematical programming methods. There are different metaheuristics, among which three are dominant in the literature. We cite: simulated annealing, genetic algorithms and tabu search. Genetic algorithms (GA) were proposed by [2]. They are based on the analogy with the evolution of species, as the name indicates. Even if technically these algorithms are sometimes challenging to implement, conceptually, their operation is simple: for each iteration, two or more solutions are chosen, evaluated as being suitable to serve as parents, and one or more derived solutions are created by combining some aspects of each of the two-parent solutions. To this is added a random mutation factor, often with a very low probability of the order of 1 in 100 or 1 in 1000 iterations.

Multiple metaheuristics can solve approximately NP-hard problems, which is the case in this paper, such as Tabu Search (TS), Simulated Annealing (SA), Particle Swarm (PS), Genetic Algorithm (GA), and the Artificial Immune System Algorithm (AIS). The main comparison between GA and AIS, according to [3], is that the second one can separate cells into good ones and bad ones. This ability ensures reasonable and flexible information proceeding that the GA could not afford. In a biological sense, the antigens, described as harmful cells, obligate the human body to construct and defend itself with its immune system [4]. Thus, the AIS is inspired by the body's natural immune system, leading to better protection against those antigens. Many authors have conducted researches based on the AIS in different fields such as [5, 6], and many others. The AIS is based upon many steps, which this paper will elaborate on in detail in the following sections.

2.2 Multi-agent System

Autonomous agents and MASs represent a new way of analyzing, designing and implementing complex computer systems. Agents are used in an ever-increasing variety of applications that are constantly growing. MASs are distributed computer systems from the field of distributed artificial intelligence (DAI). They consist of a set of "autonomous" and "intelligent" computing entities that interact with each other. In the following paragraphs, we describe the concepts and possibilities that arise from the research field of multi-agent systems. The advantages of multi-agent systems (MAS) as a tool for modeling care production systems are cited in several works. For example, the works of [7–9] are mentioned. The descriptive quality of MASs makes them so widespread for the health sector where the need for means and tools to facilitate communication and interaction (between disciplines and personnel) is very strong. The pros of the MAS is to ensure parallel proceedings, ensure good management and good cooperation. Those features suit very well considering this study as the planning and scheduling problem need a good collaboration between the overall actors engaging in the system.

3 Literature Review

This section illustrates the state of the art of the vehicle routing problem (VRP) encountered in the home health care sector (HHCS) viewed as a combinatorial optimization problem in the field of Operation Research. The philosophy derived from this type of problem is that the transportation and carriage of people (elderly/disabled/critical state), and medical devices delivery to customers' (patients') homes must occur with a time interval mostly known as time windows (VRPTW) is the essential and classical form to initiate the solution. In this study, two big clusters arise as a predominant aspect of the problem:

3.1 Planning and Scheduling Problems

This cluster tackles allocating missions to a set of caregivers according to specific planning and visit schedules. A lot of works have been conducted on this area, [10] who considered the allocation of nurses whose professional qualifications and skills vary along with [11–14] who have opted to give a set of skills to each of the staff (caregivers) such as sex gender, medical expertise, spoken languages… From what was said, restrictions have been set to decide objectively according to the caregiver skills the actual and profitable assignments to each patients' needs and demands. To illustrate and give a clear idea, each caregiver from their initial depot (start point) must visit patients according to the type of care suitable for the patient within a specific time window. Each care type has a certain amount of time (duration) not to exceed and a kind of skill.

Before moving on, taking caregivers' desires and needs into consideration is essential for the service quality of tasks performed. Hunger, thirst, fatigue, day-off, vacation, and much more needs cannot be neglected. Likewise, care is given by a caregiver to a patient within a particular time window. Lunch breaks also have a specific time window with a limited duration. In [15], the authors have considered lunch break as a stochastic service time; meanwhile, [16] included it as a requirement (constraint). Many recent works dealing with a lunch break in the Home Health Care Services have allowed it because shift length is longer than a predetermined time either without considering time windows (patients are visited at any time) or at a fixed time. In [17], the authors allowed taking a break because the successive working time prescribed for the caregiver is surpassed.

After reviewing some of the past works on this theme, we observed that the most used method for solving such problems is Mixed Integer Programming which is considered an exact method such as the branch-and-cut-and-price used by [12] and [14] without including lunch break requirement for the working staff (caregivers), meanwhile [15] and [18] included lunch break constraints. Approached methods, commonly known as metaheuristics, are most suitable for dealing with this type of problem and are mainly based on Large Neighborhood Search and Local Search as seen in [16] and [19–23].

3.2 Pick-Up and Delivery Problems

This cluster tackles the matter of delivering/picking up patients or goods from site A to a site B as explained in [24] and [25], in the same line [16, 26] and [27] were interested in solving problems of transporting medical devices and medicines to customers' (patients') homes and collecting drugs, samples that patients no longer use. Others such as [28] and [29] looked into this problem as a variation which considered heterogeneity of a fleet of vehicles which means that vehicles can deliver or collect on the same path, some customers may require several visits in the same time windows and several depots can be found, thus dividing deliveries between vehicles can occur. Another interesting case was introduced by [30], who work on a Dial-a-Ride problem (DAR) where customers (patients) demand one or more particular services before the transportation can be done.

4 Problem Description

This paper addresses a variant of the vehicle routing problem known as VRP with time windows, precedence and synchronization constraints, and staff' lunch break requirement in the Home Health Care System.

The VRPTW-SPLB can be considered as a graph G = (V, E) where V is a node-set V = N U D where N = {1… n} corresponds to a group of customers (patients) and D = {c, f} are nodes where c is the initial depot (center) and f the final depot. M denotes a margin (edge) set where each margin m = [i, j] is linked to a travel cost C_{ij} and a travel duration T_{ij}. S denotes as a set of services where the initial depot (center) c offers a choice of multiple services S = {1…s} to customers (patients). A customer i demands a subset of services ensured by a set $S_i = \{s \in S : e_{is} = 1\}$ where e_{is} matches 1 if a patient I requires service s ∈ S, and 0 if not. For each service s provided to a patient i has associated a time duration d_{is} and a time window $[a_i, b_i]$; where a_i is the earliest service start time and b_i The latest service start time of patient i. K is the set of vehicles available at the initial depot (center) c and for each k ∈ K an availability time window $[\alpha_k, \beta_k]$ where a vehicle k can only leave the depot c at α_k and must return to the final depot f at β_k. To assure the preference of each vehicle k, a number Prf_{ik} is associated with the vehicle k and denotes the non-preference of customer (patient) i to the vehicle (caregiver) k.

For the sake of this study, we suppose that k ∈ K accomplishes a unique service s. y_{ks} equals 1 if vehicle k offers the service s, 0 otherwise $(\forall k \in K, \forall s \in S)$.

P denotes the lunch break and has a duration L and a time window $[w_p, z_p]$ as put by [19] g_{ik} Equals 1 if the caregiver uses his break at patient i before service completion and 0 otherwise, g'_{ik} equals 1 if the caregiver uses his break at patient i after service completion, 0 otherwise. st_{ik} is the service start time of caregiver k at patient i and st_{pk} is the lunch break time start of the caregiver k. For the sake of simplicity, we write $k \in K_s$ if $y_{ks} = 1$ and $s \in S_i$ if $e_{is} = 1$.

For customers (patients) demanding more than one service either in a specific order or simultaneously, we define for each patient i ∈ N a $slot_{isr}$ as the time separating the time start of service s and service r demanded by patient i where services s and r have to be provided in a given order (service s before service r).

This problem formulation considers the minimization of the sum on non-preference and total traveling time.

[31] and [32] have conducted similar studies but did not consider caregivers' lunch break requirements. This kind of problem is known to be an NP-hard problem. For the first resolution, we will use the Mixed Integer Linear Program (MILP) considering the binary and real variable decision respectively as follows:

- $x_{ijk} = 1$, if k ∈ K travel from i to j, 0 otherwise
- st_{ik} Infers to the start time of service at node i if customer (patient) i is visited by vehicle (caregiver) k.

To generalize the constraints of time in this formulation, we presume that patients' requested services are asked independently from the initial depot (center) c and their duration is 0 ($D_{cs} = 0$, $\forall s \in S$).

5 Mathematical Formulation

The MILP corresponding to our formulation is as follows:

$$min \sum_{i \in V\{f\}} \sum_{j \in V \setminus \{c\}} \sum_{k \in K} C_{ij}\, x_{ijk} + \sum_{i \in N} \sum_{j \in V \setminus \{c\}} \sum_{k \in K} Prf_{ik}\, x_{ijk} \quad (1)$$

Subject to:

$$\forall k \in K, \sum\nolimits_{j \in N} x_{cjk} = 1 \sum\nolimits_{j \in N} x_{ifk} = 1 \quad (2)$$

$$\forall h \in N, \forall k \in K, \sum\nolimits_{i \in V \setminus \{f\}} x_{ihk} = \sum\nolimits_{j \in V \setminus \{c\}} x_{hjk} \quad (3)$$

$$\forall i \in N, \forall s \in S, \sum\nolimits_{j \in V \setminus \{c\}} \sum\nolimits_{k \in K_s} x_{ijk} = e_{is} \quad (4)$$

$$\forall k \in K, \sum\nolimits_{i \in V} g_{ik} + \sum\nolimits_{i \in V} g'_{ik} = 1 \quad (5)$$

$$\forall k \in K, i \in V, g_{ik} + g'_{ik} \le \sum\nolimits_{i \in V} x_{ijk} \quad (6)$$

$$\forall k \in K, j \in V, st_{pk} + L.g_{jk} \le ts_{jk} + (1 - g_{jk}) z_p \quad (7)$$

$$\forall k \in K, i,j \in V, st_{ik} + \left(T_{ij} + d_{is}\right)\left(x_{ijk} + g_{jk} - 1\right) \le st_{pk} + (2 - x_{ijk} - g_{jk}) b_i \quad (8)$$

$$\forall k \in K, i,j \in V, st_{pk} + \left(L + T_{ij}\right)\left(x_{ijk} + g'_{ik} - 1\right) \le st_{jk} + (2 - x_{ijk} - g'_{ik}) z_p \quad (9)$$

$$\forall k \in K, i \in V, st_{ik} + d_{is} g'_{ik} \le st_{pk} + (1 - g'_{ik}) b_i \quad (10)$$

$$\forall i,j \in V, \forall s \in S : s \in S_i \cup S_j, \forall k \in K_s, st_{ik} + (T_{ij} + D_{is}) x_{ijk} \le st_{jk} + b_i (1 - x_{ijk}) \quad (11)$$

$$\forall i \in N, \forall s \in S_i, \forall k \in K_s, a_i \sum\nolimits_{j \in N} x_{ijk} \le st_{ik} \le b_i \sum\nolimits_{j \in N} x_{ijk} \quad (12)$$

$$\forall k \in K, \alpha_k \le st_{ck} \le \beta_k, \alpha_k \le st_{fk} \le \beta_k, w_p \le st_{pk} \le z_p \quad (13)$$

$$\forall i,j \in V, \forall k \in K, x_{ijk} \in \{0,1\}, g_{ik}, g'_{ik} \in \{0,1\}, st_{ik} \in \{0,1\} \quad (14)$$

The objective function (1) is meant minimize the total travelling time and the sum of non-preferences. Constraint (2) ensures that each caregiver (vehicle) has to leave the central depot and return to it after the service. Constraint (3) secures the continuity of the routes and its constancy, while constraint (4) ensures that the demands of customers (patients) are served and executed appropriately. Constraint (5) assures that every caregiver has to take a break in the route. Meanwhile, constraints (6) affirm that the caregiver k uses his break at patient i on the condition that the patient i is visited by this caregiver. Constraints (7), (8), (9), (10) assert the accurate start times of breaks used by caregivers before and after services in customers' (patients') homes. Scheduling permitting the consistency between the durations of visits is maintained by constraint (11). Constraints (12), (13) guarantee the respect of patients, caregivers and breaks time windows. Constraint (14) defines the nature of the decision variables.

Just to note that lunch breaks do not depend entirely on the route's time duration, and a lunch break can be taken and utilized if the duration of the route outpaces and exceeds α_p. It is also permissible to take a lunch break at the end of a route even if the time window of the lunch break is not yet covered.

6 Proposed Approach

This approach consists of identifying two things to assimilate and to grasp the core concept of combining MAS and AIS:

- Identify the actors involved in the overall system and characterize their behaviors either as passive or active.
- Define the AIS algorithm used.

We employ MAS to assist the scheduler in making appropriate decisions during the scheduling process in a dynamic environment and guarantee effective communication between the various contributors. When adopting the MAS method, the first step is to assess each agent's level of reasoning. The multi-agent system comprises two types of agent architectures: cognitive and reactive agents [33].

The steps proceeded in this study suggests an action of a few steps in order to ensure a good quality solution. These steps are as follows:

- Step 1: The patients were randomly chosen from the health structure database, and their locations were kept constant throughout the test. On the other hand, either choose to simulate with perturbation (which is tested after specific ticks) or not and which option to test.
- Step 2: The scheduling is done by the scheduler agent, who receives the overall routes of the AIS_compute_agents for vehicle planning and chooses the best feasible solution among the routes received.
- Step 3: The vehicle routes sequence is presented, and if there is a perturbation, an update the dynamic rescheduling by agents acting autonomously following the selected alternative.

Table 1 describes the type of actors and Fig. 1 describes the overall overflow of the system proposed.

Table 1. Actors' identification

Actors	Type of agent	Description
Admin	Reactive agent	Its job is to monitor care staff at work and record patient information for each request
Scheduler	Cognitive agent	With the change of circumstances, the scheduler adapts the route dynamically. It also allocates and chooses care staff based on their skills
AIS compute	Cognitive agent	Execute the AIS algorithm
Caregiver	Reactive agent	He visits patients according to the planned route and calculates treatment time, travel time, and patient waiting time. He sends and gets information about patients' current state
Patient	Reactive agent	They are given their cares and are informed about their availability

To fulfill all the requirements for this model to work at full capacity, this study proposes an alternative of the AIS algorithm called the Clonal Selection Algorithm (CSA). This variant was firstly proposed by [34] for planning routes in an optimal way. To put this algorithm to test, a mapping is required between the aspect of routing in Home Health Care and the parameters of the CSA as illustrated in Table 2. The ultimate goal of this algorithm is to eliminate the cells that could not identify the harmful cells only known as antigens and multiply those capable of identifying those antigens (Fig. 2).

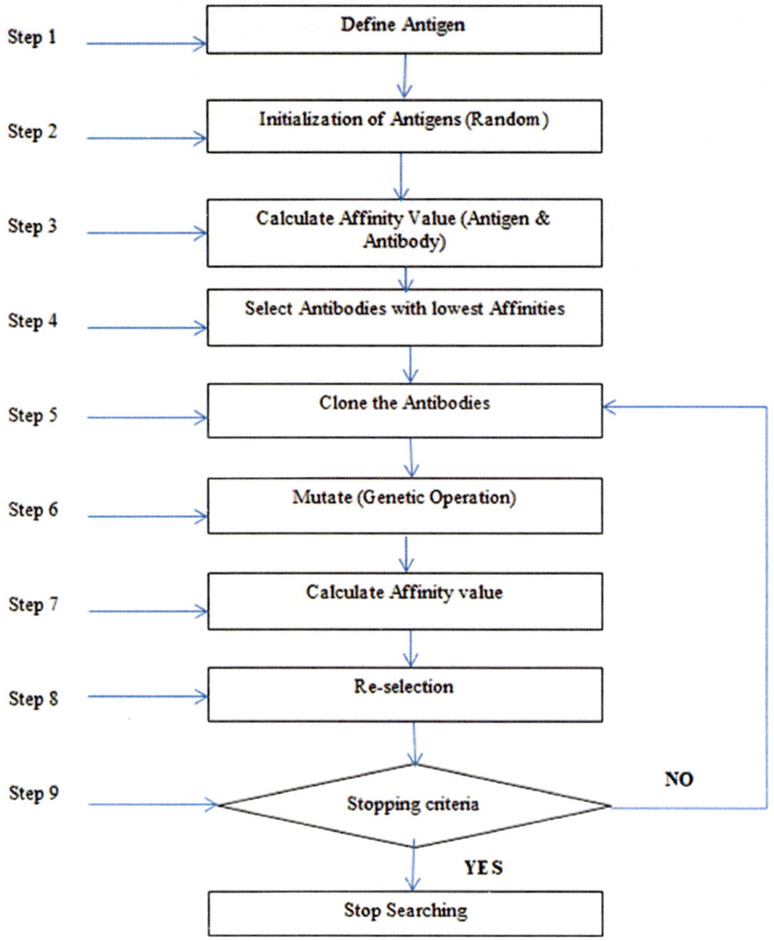

Fig. 1. Flowchart of CSA algorithm

Table 2. Interdependence between HHC and the Immune System

HHC planning & scheduling problem	Immune system
Antibodies quantity	Population
Location of patients	Gene
Initial solution (route)	Antigen
Feasible solution (route)	Antibody

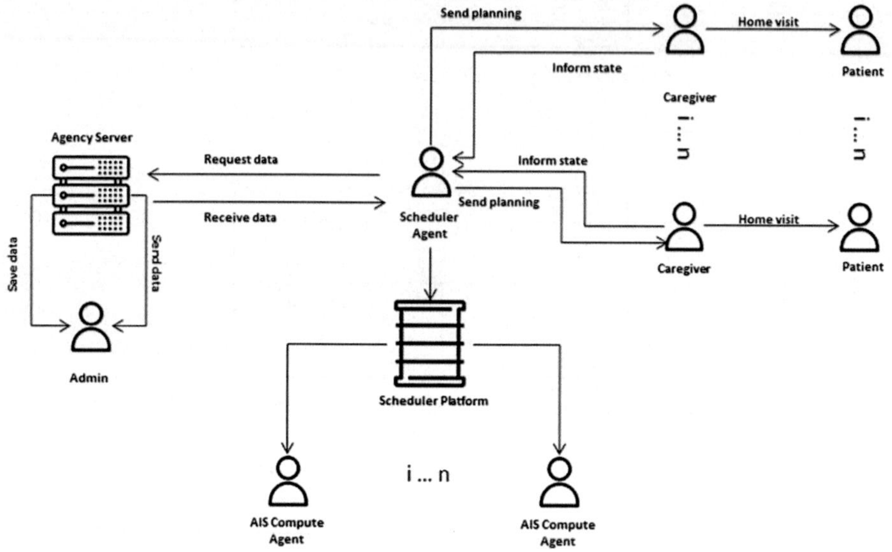

Fig. 2. Multi-agent system model for the VRPTW-SPLB

7 Conclusion and Perspectives

The problem of planning and scheduling in Home Health Care has provided much attention in recent years, especially with the spread of the COVID-19 pandemic throughout the world. In this paper, an approach based on the multi-agent model has been proposed to tackle giving care into patient's homes using an extension of the VRP with time windows, synchronization, precedence, and lunch break constraints. The goal pursued in this research is to minimize the total travelling distance for cost reduction and to maximize the patient's preference for better quality service. The algorithm used to explore the search space is a variant of the Artificial Immune System called the Cloning Selection Algorithm. As a future direction to this study, an implementation of this approach is highly needed to compare it with other metaheuristics and assess the real efficiency compared to others. Also, as an expansion, an investigation of the execution time as this approach is based on a multi-agent system since the parallel aspect is infused within the MAS will be a greater reward. To finish, an analysis of the uncertainties and unavailability of one or more of the problem's parameters can be an addition to refine the solution.

References

1. Euchi, J.: Do drones have a realistic place in a pandemic fight for delivering medical supplies in healthcare systems problems? Chin. J. Aeronaut. (2020)
2. Holland, J.H.: Adaptation in Natural and Artificial Systems. University of Michigan Press, Ann Arbor (1975)

3. Aickelin, U., Dasgupta, D.: Artificial immune systems tutorial (Chapter 7). In: Burke, E., Kendall, G. (eds.) Introductory Tutorials in Optimisation, Decision Support and Search Methodology. Kluwer (2005)
4. Tan, K.C., Goh, C.K., Mamun, A.A., Ei, E.Z.: An evolutionary artificial immune system for multi-objective optimization. Eur. J. Oper. Res. **187**(2), 371–392 (2008)
5. Coello, C.A.C., Cortés, N.C.: Solving multiobjective optimization problems using an artificial immune system. Genet. Program. Evol. Mach. **6**(2), 163–190 (2005)
6. Toma, N., Endo, S., Yamada, K., Miyagi, H.: Evolutionary optimization algorithm using MHC and immune network. In: Proceedings of the 26th IEEE Annual Conference on Industrial Electronics Society, pp. 2849–2854 (2000)
7. Paulussen, T.O., Jennings, N.R., Decker, K.S., Heinzl, A.: Distributed patient scheduling in hospitals. In: International Joint Conference on Artificial Intelligence (2003)
8. Daknou, A., Zgaya, H., Hammadi, S., Hubert, H.: Toward a multiagent model for the care of patients at the emergency department. In: MAMECTIS 2008 - Proceedings of the 10th WSEAS International Conference on Mathematical Methods, Computational Techniques and Intelligent Systems, Corfu, Greece, pp. 264–269 (2008)
9. Laleci, G.B., Dogac, A., Olduz, M., Tasyurt, I., Yuksel, M., Okcan, A.: SAPHIRE: a multi-agent system for remote healthcare monitoring through computerized clinical guidelines. In: Annicchiarico, R., Cortés, U., Urdiales, C. (eds.) Agent Technology and e-Health. Whitestein Series in Software Agent Technologies and Autonomic Computing, pp. 25–44. Springer – Birkhäuser, Basel (2008). https://doi.org/10.1007/978-3-7643-8547-7_3
10. Bertels, S., Stefan, T.: A hybrid setup for a hybrid scenario: combining heuristics for the home health care problem. Comput. Oper. Res. **33**(10), 2866–2890 (2006)
11. Bredström, D., Rönnqvist, M.: A branch and price algorithm for the combined vehicle routing and scheduling problem with synchronization constraints. NHH Department of Finance Management Science Discussion Paper (2007)
12. Bredström, D., Rönnqvist, M.: Combined vehicle routing and scheduling with temporal precedence and synchronization constraints. Eur. J. Oper. Res. **191**(1), 19–31 (2008)
13. Bräysi, O., Dullaert, W., Nakari, P.: The potential of optimization in communal routing problems: case studies from finland. J. Transp. Geogr. **17**(6), 484–490 (2009)
14. Rasmussen, M.S., Justesen, T., Dohn, A., Larsen, J.: The home care crew scheduling problem: preference-based visit clustering and temporal dependencies. Eur. J. Oper. Res. **2019**(3), 598–610 (2012)
15. Yuan, B., Liu, R., Jiang, Z.: A branch-and-price algorithm for the home health care scheduling and routing problem with stochastic service times and skill requirement. Int. J. Prod. Res. **53**(24), 7450–7464 (2015)
16. Liu, R., Xie, X., Augusto, V., Rodriguez, C.: Heuristic algorithms for a vehicle routing problem with simultaneous delivery and pickup and time windows in home health care. Eur. J. Oper. Res. **230**(3), 475–486 (2013)
17. Trautsamwieser, A., Hirsch, P.: A branch-price-and-cut approach for solving the medium-term home health care planning problem. Networks **64**(3), 143–159 (2014)
18. Liu, S., Yingfeng, Z., Yang, L., Lihui, W., Xi, V.W.: An internet of things enabled dynamic optimization method for smart vehicles and logistic tasks. J. Clean. Prod. **215**, 806–820 (2016)
19. Cheng, E., Rich, J.L.: A home health care routing and scheduling problem. Technical report CAAM TR98-04, Rice University (1998)
20. Rousseau, L.M., Gendreau, M., Pesant, G.: The synchronized dynamic vehicle dispatching problem. INFOR: Information Systems and Operational Research **51**(2), 76–83 (2013)

21. Ettazi, H., Rafalia, N., Abouchabaka, J: A vehicle routing problem for the collection of medical samples at home: case study of Morocco. Int. J. Adv. Comput. Sci. Appl. **12**(4) (2021)
22. Ettazi, H., Rafalia, N., Abouchabaka, J.: A metaheuristics methods for the VRP in home health care by minimizing fuel consumption for environmental gain. In: E3S Web Conference, vol. 234, p. 00094 (2021). https://doi.org/10.1051/e3sconf/202123400094
23. Ettazi, H., Rafalia, N., Abouchabaka, J.: GRASP combined with ILS for the vehicle routing problem with time windows, precedence, synchronization and lunch break constraints. Int. J. Adv. Comput. Sci. Appl. **12**(5) (2021)
24. Diabat, A., Abdallah, T., Le, T.: A hybrid tabu search based heuristicfor the periodic distribution inventory problemwith perishable goods. Ann. Oper. Res. **242**(2), 373–398 (2016)
25. Zhang, Y., Alshraideh, H., Diabat, A.: A two-stage multi-echelon stochastic blood supply chain problem. Comput. Oper. Res. **101**, 130–143 (2019)
26. Coppi, A., Detti, P., Rafaelli, J.: A planning and routing model for patient transportation in health care. Electron. Notes Discrete Math. **41**, 125–132 (2013)
27. Ceselli, A., Righini, G., Tresoldi, E.: Combined location and routing problems for the drug distribution. Discret. Appl. Math. **165**, 130–145 (2014)
28. Shuang, Y., Diabat, A., Liao, Y.: A stochastic reverse logistics production routing model with emissions control policy selection. Int. J. Prod. Econ. **213**, 201–216 (2019)
29. Alkaabneh, F., Diabat, A., Gao, H.O.: Benders decomposition for the inventory vehicle routing problem with perishable products and environmental costs. Comput. Oper. Res. **113**, 104751 (2020)
30. Rousseau, L.M., Gendreau, M., Pesant, G.: The synchronized vehicle dispatching problem. Citeseer (2003)
31. Labadie, N., Prins, C., Yang, Y.: Iterated local search for a vehicle routing problem with synchronization constraints. In: ICORES 2014-Proceedings of the 3rd International Conference on Operations Research and Enterprise Systems, Angers, Loire Valley, France, pp. 257–263 (2014)
32. Ait Haddadene, S.R., Labadie, N., Prodhon, C.: A GRASP × ILS for the vehicle routing problem with time windows, synchronization and precedence constraints. Expert Syst. Appl. **66**, 274–294 (2016)
33. Wooldridge, M.: An Introduction to Multiagent Systems, Chapter Preface. Wiley, Hoboken (2009)
34. De Castro, L.N., Von Zuben, F.J.: The clonal selection algorithm with engineering applications. In: Proceedings of GECCO, vol. 2000, pp. 36–39 (2000)

Ontology-Based Machine Learning to Predict Diabetes Patients

Hakim El Massari(✉), Sajida Mhammedi, Zineb Sabouri, and Noreddine Gherabi

National School of Applied Sciences, Lasti Laboratory, Sultan Moulay Slimane University, Khouribga, Morocco
{h.elmassari,n.gherabi}@usms.ma

Abstract. Diabetes is a chronic disease that increases with time and causes various other complications. Diverse traditional methods, based on physical and chemical tests, are available for diagnosing diabetes. However, early detection and prediction of diabetes is a crucial and quite challenging task for medical practitioners. So, in our work, we are focused on building an ontology-based intelligent system for preventing the diagnosis of diabetes patients based on the results of the decision tree model. This paper presents a new approach applied to medicine, that uses semantic web ontology language and machine learning to develop an ontology-based model for diabetic patients, to reach an efficient result by applying Semantic Web Rule Language. We used the decision tree algorithm to extract rules that distinguish a positive from a negative patient. Then we implemented these rules in the ontology using Semantic Web Rule Language. Thus, the article validated the result generated by the ontology model with the results of the decision tree model in terms of prediction of diabetes. Therefore, this study aims to create an ontology-based model capable of predicting diabetes with high performance.

Keywords: Machine learning · Decision tree · Ontology · SWRL

1 Introduction

Diabetes is one of the most significant global health issues of the twenty-first century. It has a direct signal of high blood sugar, and has affected hundreds of millions of people. Early detection of diabetes is crucial, because of the critical complications that it can cause. There have been various studies on the detection of diabetes, many of which have recently been developed to predict diabetes patients using machine learning algorithms. Research has concentrated on diabetes identification through Logistic regression, DNN, SVM, Decision Tree, and Naive Bayes [1]. They compare the accuracy of each classifier by obtaining some inspiring results.

This paper reports on integrating two approaches, semantic web and machine learning algorithms, to develop an ontology-based model capable of predicting diabetes with high accuracy. The semantic web offers different formats for presenting data and ontological background knowledge. Within a variety of possible applications, machine

Y. Maleh et al. (Eds.): ICI2C 2021, LNNS 357, pp. 437–445, 2022.
https://doi.org/10.1007/978-3-030-91738-8_40

learning will play an important role. So, we used the decision trees algorithm to learn the rules that differentiated diabetes patients from others. Then, we implemented those rules by using the Semantic Web Rule Language (SWRL). This integration combines the advantages of data-controlled machine learning with the expressive knowledge representation of the Semantic Web paradigm, resulting in the simplicity of explanation and maintenance.

The remainder of this paper is structured as follows: we start in Sect. 2 by giving related work drawing from prior researches in the associated areas. We introduce in Sect. 3 the methodology adopted to accomplish the our approach. In Sect. 4, we present results and discussion. Finally, Sect. 5 presents future work and conclusions.

2 Related Work

Several ways to integrate machine learning techniques and semantic web technology have been presented. In [2] the authors create an artificially intelligent prediction model for a manufacturing network that can help a project manager allocate a freshly received order among its providers. The semantic web is based on the concepts of similarity. This concept of similarity is used in several fields of research. For example, in the papers [3, 4] the authors propose two methods to compute the semantic similarity between the concepts defined in the same ontology.

According to [5] they build an efficient model to detect and classify fake Twitter accounts using Ontological Engineering and Semantic Web Rule Language. Using Semantic Web Rule Language [6, 7] this research develops an ontology-based model for diabetic patients to assist clinicians in making an effective diagnostic judgment regarding the status of diabetes. Using ontology and machine learning techniques, the authors presented a method for identifying patients with diabetes disease risk levels [8]. [9] Present a comparative analysis using different ML techniques to predict diabetes. In addition, [10] present a systematic review of the applications of machine learning, data mining techniques and tools in diabetes research.

3 Methodology

The approach proposed in this article is divided into three parts, as illustrated in Fig. 1. The following subsections present all the steps and technologies used.

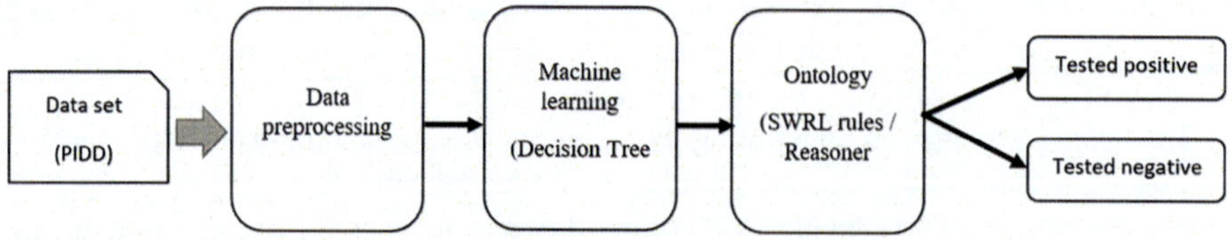

Fig. 1. Diagram of the proposed approach.

3.1 Data Preprocessing

The data set used in this article is Pima Indians Diabetes Database (PIDD)[1], having 768 instances. This data came from the National Institute of Diabetes and Digestive and Kidney Diseases. The dataset's goal is to diagnose whether or not a patient has diabetes based on certain diagnostic measurements included in the dataset. Several restrictions were imposed on the selection of these instances from a larger database. All patients here, in particular, are Pima Indian women over the age of 21. Table 1 describes the content of PIDD data set.

Table 1. Description of the PIDD attributes

Attribute	Description
1- preg	Number of times pregnant
2- plas	Plasma glucose concentration a 2 h in an oral glucose tolerance test
3- pres	Diastolic blood pressure (mm Hg)
4- skin	Triceps skin fold thickness (mm)
5- insu	2-h serum insulin (mu U/ml)
6- mass	Body mass index (weight in kg/(height in m)^2)
7- pedi	Diabetes pedigree function
8- age	Age (years)
9- Class	Variable (0 or 1). value 1 is interpreted as "tested positive for diabetes"

3.2 Decision Tree Algorithm

The decision tree algorithm is a predictive modeling technique used in statistics, data mining, and machine learning. It uses a predictive model to go from observations about an item (represented by branches) to inferences about the item's target value (represented in the leaves). We have chosen the decision tree algorithm for many reasons. Data preparation is easier, and resultant classification tree is easier to understand and interpret, and multiple data types such as numeric, nominal, categorical are supported, etc.

A predictive decision tree was created using WEKA Software [11], using J4.8 classifier algorithm. J4.8 evaluates the data using its gain ratio attribute evaluator [12]. The 10-fold cross-validation was used to validate the training set [13]. Moreover, all J4.8 parameters were left as default. We got 73% correctly classified of the data. Figure 2 shows the classifier output of the decision tree applied, and Fig. 3 illustrates the tree visualizer of the decision tree. While Fig. 4 shows a snippet of the decision tree rules, which we will use to create the SWRL rules for our ontology model, more details are in Sect. 3.4.

[1] PIID, https://www.kaggle.com/uciml/pima-indians-diabetes-database.

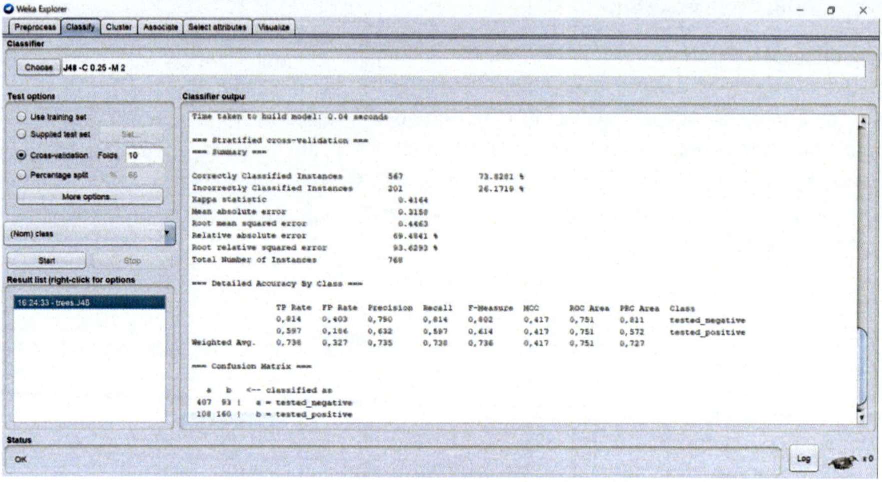

Fig. 2. Classifier output of the decision tree.

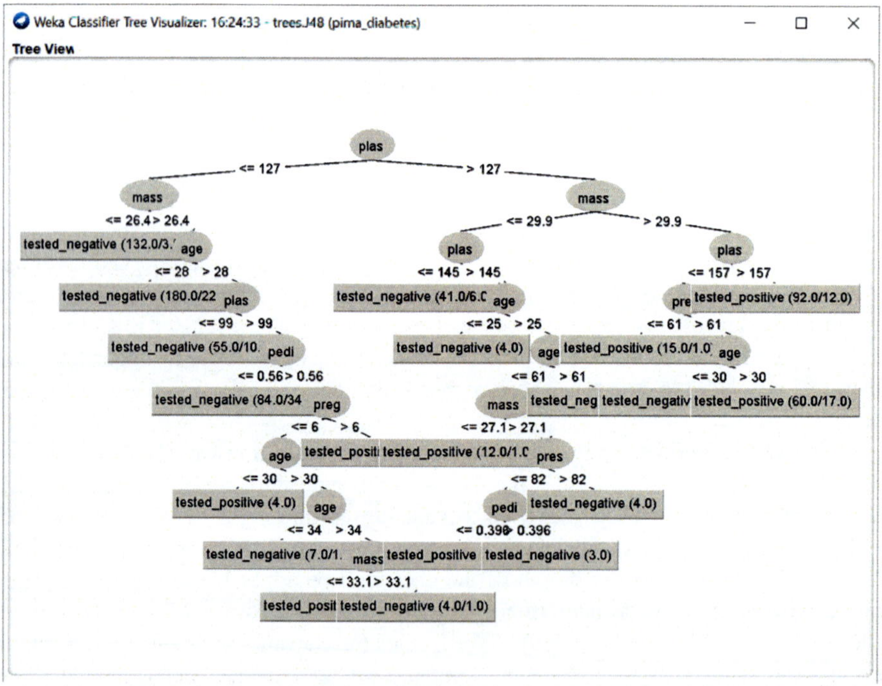

Fig. 3. Tree visualizer of the decision tree.

```
plas > 127
|   mass <= 29.9
|   |   plas <= 145: tested_negative (41.0/6.0)
|   |   plas > 145
|   |   |   age <= 25: tested_negative (4.0)
|   |   |   age > 25
|   |   |   |   age <= 61
|   |   |   |   |   mass <= 27.1: tested_positive (12.0/1.0)
|   |   |   |   |   mass > 27.1
|   |   |   |   |   |   pres <= 82
|   |   |   |   |   |   |   pedi <= 0.396: tested_positive (8.0/1.0)
|   |   |   |   |   |   |   pedi > 0.396: tested_negative (3.0)
|   |   |   |   |   |   pres > 82: tested_negative (4.0)
|   |   |   |   age > 61: tested_negative (4.0)
|   mass > 29.9
|   |   plas <= 157
|   |   |   pres <= 61: tested_positive (15.0/1.0)
|   |   |   pres > 61
|   |   |   |   age <= 30: tested_negative (40.0/13.0)
|   |   |   |   age > 30: tested_positive (60.0/17.0)
|   |   plas > 157: tested_positive (92.0/12.0)

Number of Leaves  :     20

Size of the tree :      39
```

Fig. 4. A snippet of the decision tree rules.

3.3 Ontology Building

We create the ontology using Protégé [14]. This open-source platform provides a growing user community with a suite of tools to construct domain models and knowledge-based applications with ontologies. We only used two classes (Patient and Diagnostic) and two subclasses (tested_negative and tested_positive) of the diagnostic class to keep things simple. Furthermore, the data properties are the same attributes of the data set used in Table 1. Figure 5 represents the graph of the ontology.

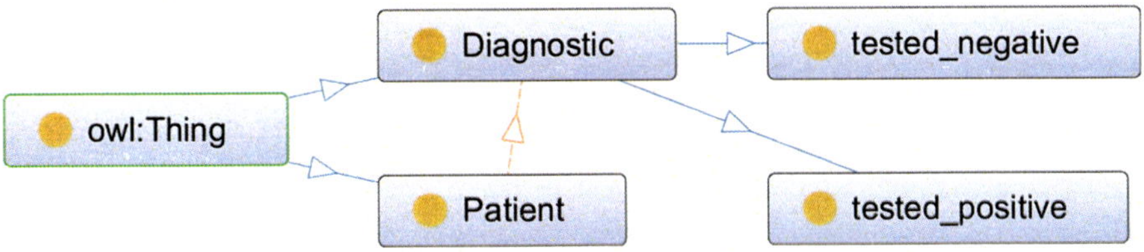

Fig. 5. Graphical representation of the ontology.

The same data set is used to import instances to our ontology with the help of Cellfie plugin, which serves to import axioms from XLS and CSV files to protégé.

3.4 SWRL Rules

After creating our ontology and importing the data set, we need to define the SWRL rules for reasoning. We used java programing language to extract these rules from the decision tree, as shown in Fig. 4. The program java extracted each leaf of the tree as a single SWRL rule. For example, the SWRL rule extracted from the first line of Fig. 4:

A leaf of decision tree

If `plas` >127 *&&* `mass` ≤ 29.9 *&&* `plas` ≤ 145 *THEN* put the patient in `tested_negative`

The SWRL rule extracted from the leaf above

`Patient`(?pt) ^ `plas`(?pt, ?P) ^ *swrlb:greaterThan*(?P, '127'^^*xsd:decimal*) ^ `mass`(?pt, ?M) ^ *swrlb:lessThanOrEqual*(?M, '29.9'^^*xsd:decimal*) ^ `plas`(?pt, ?p) ^ *swrlb:lessThanOrEqual*(?p, '145'^^*xsd:decimal*) → `tested_negative`

We generated a total of 20 SWRL rules with the same number of leaves. These rules were imported to Protégé using the SWRL tab plugin. And now it's time to use a reasoner. A Pellet reasoner [15] is applied to infer whether the patient is tested positive or negative. The reasoner uses patient diabetes data and rules to induce inference and provides the final decision to show that the patient is tested negative or positive. Figure 6 provide the statistics of inferred concepts using SWRL rules.

Fig. 6. Statistics of inferred concepts.

4 Results and Discussion

This paper aims to predict diabetes patients using the integration of machine learning (Decision Tree) and engineering ontology and SWRL rules. We use a Pima Indians Diabetes Database dataset. The decision tree is used as a classifier to classify a positive patient from a negative patient.

Table 2. Statistic of classified instances.

	Correctly classified	Incorrectly classified
Decision Tree	567 (~73.82%)	201 (~26.17%)
Ontology	571 (~74.34%)	197 (~25.65%)

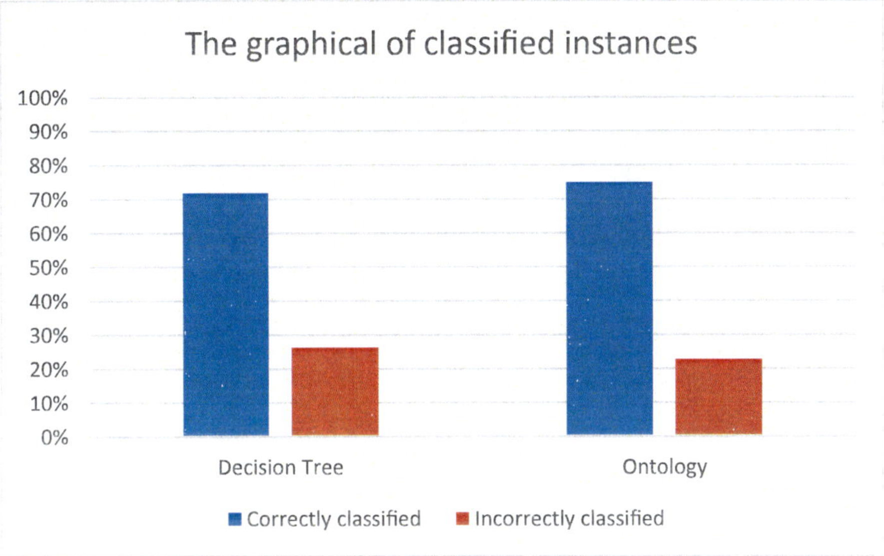

Fig. 7. The graphical of classified instances

Based on Table 2 and Fig. 7, the ontology results are seen to be the best since the decision is comprehensible and easily stated. Machine learning is an artificial intelligence approach in which machines learn from data using mathematical algorithms. Machine learning models are regarded as ambiguous black boxes. Because BlackBox classifiers do not provide human and straightforward interpretable decision rules [16], comprehending every element of the model or developing a set of rules for making choices is not always easy. On the other hand, the ontology classifier is an interpretable model that may give insights into how the model makes a decision.

5 Conclusion and Future Work

The research aims to create an artificially intelligent predictive model for solutions in the health domain, which can help in identifying patients with diabetes. This purpose was attained by developing an ontology model based on a decision tree algorithm, which was mapped into ontology using SWRL rules. The SWRL rules were extracted from the decision tree model with the help of a java program.

As future works, we plan to automate the approach by developing the whole process using java APIs of Protégé and Weka. In addition, we will test the proposed approach using different machine learning algorithms and compare the results.

References

1. Wei, S., Zhao, X., Miao, C.: A comprehensive exploration to the machine learning techniques for diabetes identification. In: 2018 IEEE 4th World Forum on Internet of Things (WF-IoT), pp. 291–295 (2018)
2. Khan, Z.M.A., Saeidlou, S., Saadat, M.: Ontology-based decision tree model for prediction in a manufacturing network. Prod. Manuf. Res. **7**, 335–349 (2019). https://doi.org/10.1080/21693277.2019.1621228
3. Daoui, A., Gherabi, N., Marzouk, A.: A new approach for measuring semantic similarity of ontology concepts using dynamic programming. J. Theoret. Appl. Inf. Technol. **95**(17), 4132–4139 (2017)
4. Daoui, A., Gherabi, N., Marzouk, A.: An enhanced method to compute the similarity between concepts of the ontology. In: Noreddine, G., Kacprzyk, J. (eds.) International Conference on Information Technology and Communication Systems. Advances in Intelligent Systems and Computing, vol. 640, pp. 95–107. Springer, Heidelberg (2018). https://doi.org/10.1007/978-3-319-64719-7_9
5. Jabardi, M., Hadi, A.: Twitter fake account detection and classification using ontological engineering and semantic web rule language. Karbala International Journal of Modern Science. **6**, 8 (2020). https://doi.org/10.33640/2405-609X.2285
6. Gherabi, N., Addakiri, K., Bahaj, M.: Discovering new technique for mapping relational database based on semantic web technology. In: International Conference on Communications and Information Technology - Proceedings, pp. 27–32, p. 6285808 (2012)
7. Chen, L., et al.: OMDP: An ontology-based model for diagnosis and treatment of diabetes patients in remote healthcare systems. Int. J. Distrib. Sens. Netw. **15**, 1550147719847112 (2019). https://doi.org/10.1177/1550147719847112
8. Lakshmi, V.S., Nithya, V., Sripriya, K., Preethi, C., Logeshwari, K.: Prediction of diabetes patient stage using ontology based machine learning system. In: 2019 IEEE International Conference on System, Computation, Automation and Networking (ICSCAN), pp. 1–4 (2019)
9. Pavani, K., Anjaiah, P., Krishna Rao, N.V., Deepthi, Y., Noel, D., Lokesh, V.: Diabetes prediction using machine learning techniques: a comparative analysis. In: Sikander, A., Acharjee, D., Chanda, C.K., Mondal, P.K., Verma, P. (eds.) Energy Systems, Drives and Automations. LNEE, vol. 664, pp. 419–428. Springer, Singapore (2020). https://doi.org/10.1007/978-981-15-5089-8_41
10. Kavakiotis, I., Tsave, O., Salifoglou, A., Maglaveras, N., Vlahavas, I., Chouvarda, I.: Machine learning and data mining methods in diabetes research. Comput. Struct. Biotechnol. J. **15**, 104–116 (2017). https://doi.org/10.1016/j.csbj.2016.12.005
11. Witten, I.H., Frank, E., Hall, M.A., Pal, C.J.: Data Mining: Practical Machine Learning Tools and Techniques. Morgan Kaufmann, Burlington (2016)
12. Osisanwo, F.Y., Akinsola, J.E.T., Awodele, O., Hinmikaiye, J.O., Olakanmi, O., Akinjobi, J.: Supervised machine learning algorithms: classification and comparison. Int. J. Comput. Trends Technol. **48**, 128–138 (2017)

13. Sahu, S., Mehtre, B.M.: Network intrusion detection system using J48 decision tree. In: 2015 International Conference on Advances in Computing, Communications and Informatics (ICACCI), pp. 2023–2026 (2015)
14. Musen, M.A.: The protégé project: a look back and a look forward. AI Matters. **1**, 4–12 (2015). https://doi.org/10.1145/2757001.2757003
15. Khamparia, A., Pandey, B.: Comprehensive analysis of semantic web reasoners and tools: a survey. Educ. Inf. Technol. **22**(6), 3121–3145 (2017). https://doi.org/10.1007/s10639-017-9574-5
16. Guidotti, R., Monreale, A., Ruggieri, S., Turini, F., Giannotti, F., Pedreschi, D.: A survey of methods for explaining black box models. ACM Comput. Surv. **51**(5), 1–42 (2019). https://doi.org/10.1145/3236009

Advances in Cybersecurity and Information Assurance

Automatic Static Vulnerability Detection Approaches and Tools: State of the Art

Oualid Zaazaa(✉) and Hanan El Bakkali

Smart Systems Laboratory, Rabat-IT Center ENSIAS, Mohammed V University in Rabat, Rabat, Morocco
oualid.zaazaa@um5s.net.ma, hanan.elbakkali@um5.ac.ma

Abstract. Software vulnerabilities are behind many cybersecurity threats. In fact, if they are exploited, they could cause severe damage to an organization. Thus, finding vulnerabilities in software is a crucial process. However, it often needs both time and expertise. Fortunately, many researchers have focused on automating this process to accelerate the finding of vulnerabilities. Some have worked on making static vulnerability discovery more effective; others have concentrated their research on developing dynamic analysis techniques. This paper presents some of the research that focuses on static analysis approaches and discusses the main differences between them. At the end of this paper, we experiment and analyze the most popular tools that have implemented these approaches and their main weaknesses.

Keywords: Static analysis · Vulnerability detection · Taint analysis · Machine learning

1 Introduction

With the rapid growth of the number of connected devices, more applications are being developed to connect people with each other. Unfortunately, a very small proportion of these applications were being developed with the best security practices. Most of the time, the developers of these applications have neither the time nor the necessary knowledge to develop such secure applications. According to the statistics published by the Common Vulnerabilities and Exposures (CVE) organization, the number of software vulnerabilities discovered in 1999 was less than 1000 while the number of vulnerabilities published each year from 2017 to 2019 in the CVE organization is more than 10000 vulnerabilities [1]. Even for cyber security professionals discovering a vulnerability in an application takes time and requires expertise. Thus, it is necessary to automate the process of finding, exploiting, and patching vulnerabilities.

The dynamic vulnerability detection techniques need the application to be fully operational to be tested, most companies still use the static vulnerability detection tools. For this reason, in this paper, we address vulnerable web applications using static source code analysis as most of the old, and recent applications are still based on the PHP language. Therefore, all the approaches and tools discussed in this paper will be working on this one language.

Y. Maleh et al. (Eds.): ICI2C 2021, LNNS 357, pp. 449–459, 2022.
https://doi.org/10.1007/978-3-030-91738-8_41

In the first section, we explain the most dangerous vulnerabilities that have been targeted in most researches. The second section presents the most relevant static analysis approaches proposed over time and discusses their limitations. In section III, we compare the most relevant and popular automatic static analysis tools. The last section presents the results of the experimentation of two of the industry's most popular static vulnerability detection tools.

2 Background

Before we start digging deeper into the static analysis approaches, in this section, we present some of the most common vulnerabilities that have been discussed recently in the literature. If they get successfully exploited, most of these vulnerabilities could lead, in some cases, to full control over the server that hosts the vulnerable application, which puts other applications and data at risk.

SQL injection is one of the most dangerous and popular vulnerabilities in web applications. It occurs when developers allow the injection of user's inputs into legitimate SQL requests without filtering them. The code below shows an example of a vulnerable PHP code to an SQL injection. The impact could go from manipulating the data to controlling the server in some conditions (Fig. 1).

```
$sql = "SELECT username FROM users WHERE id =". $_GET['ID'];
if ($result = $mysqli->query($sql))
{
    while($obj = $result->fetch_object()){
    print($obj->username);
}
```

Fig. 1. Vulnerable code to SQL injection

Cross-site scripting (XSS) is also one of the most popular vulnerabilities on the web. The idea behind this vulnerability is to inject a Javascript code to execute malicious operations within the navigator of the victim. Figure 2 shows an example of this vulnerability in a PHP source code. Most developers underestimated the power and the impact of such vulnerabilities. Still, when these vulnerabilities are combined with others, they could cause a higher impact on the attacked system and be the entry point to the whole information system.

```
//… more code
If(isset($_GET["name"]))
{
    echo "Hello ".$_GET["name"];
}
//… more code
```

Fig. 2. Vulnerable code to XSS

Local file inclusion (LFI) is a well-known vulnerability, especially in the web application, due to its critical impact on the system. This vulnerability belongs to the family of Injections, so it happens when non-filtered user data is directly injected into a file include function. Figure 3 shows an example of this vulnerability in PHP source code.

```
if(isset($_GET['page']))
{
    include("../pages/".$_GET['page']);
}
```

Fig. 3. Vulnerable code to LFI

3 Static Vulnerability Detection Approaches

Static analysis is the process of evaluating a system, based on its form, structure, content, or documentation, which does not require program execution. It is based on the idea of collecting the program source code components to analyze it. The goal of this analysis is to find defects that lead to vulnerabilities [2].

Static analysis is specially used when accessing source code, but it could also be used against binary applications by analyzing their assembly languages. Mokhov and Serguei [3] present a fast machine learning approach to static code analysis and fingerprinting for security and software engineering weaknesses. They use the N-Gram algorithm to analyze the binary code as a signal. This paper discusses some of the most important Static analysis techniques currently in use or under development to facilitate their effectiveness.

3.1 Lexical and Pattern Matching-Based Analysis

Before starting to discuss some of the most relevant static analysis techniques, we need to discuss some old ones that are still being used, like lexical and pattern-based analysis techniques. Pattern matching is the process of scanning the source code of a program to find matches between some instructions in the source code and the pre-defined list of patterns. This process could be performed with just the grep tool. The fact that we only perform a simple string matching in this technique, its false positives rate can be quite high. Hyunha Kim et al. [4], have used the pattern matching technique combined with others to find simple patterns described in the pattern specifications. They tried to look for dangerous functions call the lack of necessary methods, and so on.

To improve pattern matching analysis's accuracy, lexical analysis tries to extract the variables and the parameters of functions to build a token stream of the code. These tokens are then matched with existing vulnerabilities. Lexical analysis techniques are fast and straightforward, but their power is very limited since they do not consider the program's syntax or semantics [5]. Even with all the problems that suffer from, lexical and pattern matching techniques, are still being used in many modern methods to ameliorate their effectiveness and gain in speed.

3.2 Dataflow Based Analysis

Data-flow analysis is the process of collecting information about the way the variables are used and defined in the program. This technique is historically used in software analysis to find bugs in programs and generate efficient code by compilers. Nowadays the data-flow analysis is used to discover vulnerabilities. Hyunha Kim et al. have used this technique to find program parts that match a semantic flow pattern. Their idea is based on extraction control-flow and data-flow graphs whose nodes are program points from the source code. Then the flow analyzer starts to search for the pre-defined flow patterns in the source code [5].

Nenad Jovanovic et al. have implemented the dataflow analysis technique in their tool called *Pixy* to determine whether tainted data may reach sensitive sinks without being properly sanitized [6], which could lead to some serious vulnerabilities. Jiongyi Chen et al. have also used the data flow analysis to increase the effectiveness of the fuzzing technique. They start their approach by extracting the necessary elements from the source code (e.g., string constant, input from system APIs, etc.) whose values are related to the content of the message to be sent to the IoT device. The selected elements data-flow is tracked to sec which ones affect some message fields. Then they use those elements to mutate the content of the fields to assist the fuzzing technique. Those program elements are then used to mutate the content of the fields for fuzzing the device [7].Jorrit Kronjee et al. have combined data-flow analysis with machine learning to detect SQL injection (SQLi) and Cross-Site Scripting (XSS) vulnerabilities in PHP applications [9].

3.3 Control-Flow Based Analysis

Control flow analysis generates and analyzes the Control Flow Graph (CFG), an abstract graph representation of software where the nodes represent basic blocks. A graph characterizes the blocks, and directed edges define the paths from one block to another. If a node only has an exit edge, this is known as an 'entry' block. If a node only has an entry edge, this is known as an 'exit' block [8].

One of the most important elements in any global static analysis is the control flow of the program. The static analyzers use this graph to know the relationships between expressions and the data [8]. In cyber security, this technique is used to extract useful information to drive the vulnerability detection process. To train their machine learning probabilistic classifiers, Jorrit et al. have used this technique to generate control-flow graphs (CFGs) for the code in each file and extract features from these CFGs that relate to data flow [9].

3.4 Taint Based Analysis

The main idea of the taint-based analysis is to find the variables controlled by users and trace their use in the program source code to see if they reach any dangerous functions known as "sink". If the variable traced gets to the sink without being sanitized, the analyzer flag this as vulnerability [10]. *Pixy* [6], use this technique to detect taint style vulnerabilities. They try to taint the data that come from the users to see if it is reaching a vulnerable point in the program (sinks) without being filtered. This technique could be implemented as a static approach and a dynamic approach.

3.5 Limitations of Static Analysis Techniques

Even with all the improvements of the static analysis techniques, they still suffer from many limitations and weaknesses. One of its most critical weaknesses is the number of false positives.

False positives result from considering a line or a part of code vulnerable when it is not. This happens because static analysis generally could not be sure about the integrity of the data while it flows through the app from the user input to the sink. False positives could also be reported because the program communicates with a closed source code component, making it impossible to trace the data flow and ensure its integrity [10].

Table 1. Static vulnerability discovery tools literary comparison

Tool	Technique used	Vulnerabilities	Language	Advantages	Limitations
Pixy [6]	• Taint based analysis	• XSS	PHP	Automating the process of vulnerability detection	Not supporting object-oriented features of PHP Not allowing automatic scan of the Files included
RIPS [13]	• Taint based analysis	• XSS • SQL injection	PHP	• Fast source code audit • Easy to use (user interface) • Multiple vulnerabilities	No processing for objects
TAP [14]	• Deep Learning Data flow analysis	• XSS • SQL injection	PHP	• Multiple vulnerabilities • Automated vulnerability knowledge update	No processing for objects
Sonarqube [15]	• Lexical and pattern matching	• XSS • SQL injection	PHP	• Fast source code audit • Cover all top 10 OWASP vulnerabilities	Not easy to install Very low discovery rate

Static analysis is a time-consuming technique, especially when dealing with complex source code. This problem is very critical for software development companies as it affects their deadlines. Many types of security vulnerabilities are complicated to find using static vulnerability detections, such as authentication problems, access control issues, insecure use of cryptography, etc. [10–12]. These types of vulnerabilities still do not have a specific pattern or a structure that can be extracted from a source code, making it difficult to find them. Business logic vulnerabilities are part of these vulnerabilities. To find such vulnerabilities, it is necessary to understand the business rules that change from one company to another. These problems discover these vulnerabilities much harder and keep until now their detection related to the skills level of the security experts in manual verifications.

4 Current Implementations

In this section, we discuss the tools and frameworks developed to implement the techniques and approaches described above.

To solve some problems, every tool uses specific techniques or focuses on one or more vulnerabilities. Table 1 shows the differences between these tools based on the used techniques, the supported languages, the advantages, and the limitations. *ITs* and *Pixy* represent one of the older tools developed to automate the process of vulnerability detection. These tools use some old techniques that are still being implemented in modern ones, making *ITs* and *Pixy*, the base of the new automatic tools. *RIPS* is one of the most used tools in the market, and this is due to its low false-positive rate, fast audit, and the number of vulnerabilities that it can detect compared to the other tools. Compared to *Pixy*, one of the disadvantages of this tool is that it does not support alias analysis which could make its analysis more precise.

5 The Experimentation Study

Our experimental study consisted of four steps:

- **Preparation:** Select the static code analyzers to experiment with, the source code to be considered, the studied vulnerabilities, and the comparison parameters.
- **Execution:** use the tools to identify potential vulnerabilities in those sources.
- **Verification:** perform manual testing to differentiate the false positive from zero days in the vulnerabilities identified by the tools.
- **Analysis:** analyze the results obtained and systematize the lessons learned.

The vulnerabilities resulting from this experimentation fall in one of two categories, a false positive or a zero-day vulnerability. On the one hand, a zero-day is a vulnerability that the plugin's vendor does not yet know and has no patches yet. On the other hand, a false positive vulnerability is a portion of code that was categorized by the tool as a vulnerability while being denied by manual testing.

5.1 Preparation Step

Static Code Analyzer. In this experimentation, we choose to focus only on RIPS and Sonarqube. This choice is because Sonarqube is one of the most popular tools in the market used for quality and security code checking in the current software engineering practices (Devsecops). RIPS is also one of the most popular static code vulnerability discovery tools used by professional penetration testers to discover PHP code vulnerabilities due to its effectiveness and ease of use. The two tools give a community version, which was used in this experimentation.

Source Code. These tools were run on a group of 10 most used WordPress plugins with at least 100000–800000 active use. Due to technical limitations, the scan was performed directly on the plugins without connecting with WordPress core code.

Studied Vulnerabilities. SQL injection, LFI and XSS are the most critical and the most common on the internet. The impact of correct exploitation of an SQL injection can go from a data leak to complete control over the server, and that's the case for LFI too. XSS does not directly impact the hosting machine, but it might also be exploited to penetrate the system. All of these vulnerabilities are common on the internet [16]. For these reasons, we have focused on the discovery of only these 3 vulnerabilities.

The Comparison Parameters. Three parameters were chosen to compare these two tools. The zero-day (ZD) rate, the False Positive (FP) rate, and the test speed. The ZD rate is the number of Zero-day vulnerabilities discovered $(ZD)_x$ by the tool X, divided by the number of Zero-day vulnerabilities plus the number of existing vulnerabilities that were not discovered by the tool X $(FN)_x$:

$$((ZD)x)rate = \frac{(ZD)x}{(ZD)x + (FN)x}$$

In this experimentation, we have supposed that the source code has only the vulnerabilities discovered by both tools. So based on this supposition, and the fact that each tool has discovered a different set of vulnerabilities, the number of FN of the tool X is just equal to the number of zero-days discovered by the other one:

$$(FN)x = (ZD)y$$

The FP rate is calculated by dividing the number of false positives of the tool X $(FP)_x$ by the number of $(FP)_x$ plus the number of True negative $(TN)_x$ of the same tool.

$$((FP)x)rate = \frac{(FP)x}{(FP)x + (TN)x}$$

In the vulnerability discovery process, a portion of code that uses a sensitive sink (SS) is considered vulnerable or safe. So the total number of sinks is the sum of the Total number of True negative $(TTN)_x$ of the tool X and the total number of $(TZD)_x$ for the same tool. This means that:

$$(TTN)x = SS - (TZD)x$$

The total number of true negatives (TTN) is also equal to the number of true negatives of a tool X $(TN)_x$ plus the number of False positives of the same tool $(FP)_x$.

Taking this into consideration, the number of True negatives for a tool X could be found by the following equation:

$$(TN)x = SS - (TZD)x - (FP)x$$

The fact that we have supposed in the first place that the only vulnerabilities that exist in the source code are those that were discovered by the tools. Moreover, considering that vulnerabilities discovered by the tools are 100% different, then the total number of zero-days in the source code is the sum of the zero-day vulnerabilities discovered by each tool. So, the last equation becomes:

$$(TN)x = SS - (ZD)x - (ZD)y - (FP)x$$

Table 2. Experimentation results of both tools

WordPress Plugins	RIPS						Sonarqube					
	False positive			Zero day			False positive			Zero day		
Plugin	XSS	SQLi	LFI	XSS	SQLi	LFI	XSS	SQLi	LFI	XSS	SQLi	LFI
Affiliate manager 2.7.9	5	0	0	2	0	0	0	0	0	3	0	0
admin-and-client-message-after-order-for-woocommerce.6.9	2	0	0	4	0	0	0	0	0	0	0	0
autodescription.4.1.2	0	0	0	0	0	0	0	0	0	0	0	0
bbpress.2.6.6	14	0	0	0	0	0	0	0	0	3	0	0
complianz-gdpr.4.8.2	10	0	0	0	0	0	0	0	0	0	0	0
seo-by-rank-math.1.0.54.3	3	0	0	0	0	0	0	0	0	0	0	0
woocommerce-pdf-invoices-packing-slips.2.7.3	0	0	0	1	0	0	0	0	0	0	0	0
wp-file-manager	0	0	0	2	0	0	0	4	0	1	0	0
yikes-inc-easy-mailchimp-extender	85	0	0	25	0	2	0	0	0	0	0	0
Total	119	0	0	34	0	2	0	4	0	7	0	0

5.2 Results and Discussion

This section presents the experimental results. As the first step, we have launched both tools on the defined plugins with a default configuration in the same environment to calculate the scan speed of both of them. The environment has 8 CPUs, 30 GB RAM, and an SSD Disk. The results have shown that the Sonarqube is way faster than RIPS and this can be explained by the nature of the algorithm used by this tool (pattern matching). The scan of Sonarqube was performed in 5 min and 22.986 s, while RIPS finished his scan after 47.694 s only.

After the end of the scan, manual verification of the discovered vulnerabilities was performed to eliminate all the false positives. Table 2 present the results of this experimentation.

RIPS was able to find 155 vulnerabilities, including false positives, while Sonarqube has only discovered 11 vulnerabilities. At the end of the manual verification, we discovered that RIPS could find 36 zero-day vulnerabilities in the tested plugin, where 34 of them are XSS and the rest are LFI. Sonarqube was able to find 4 false positives SQL injections and 7 XSS zero days.

We have noticed that the nature of vulnerabilities discovered by both parties is not the same even for the false positives. This could be explained by the nature of the algorithm used by each tool.

However, the ZD and FP rates of both tools, give a better idea about the efficiency of each tool (Table 3 and Fig. 4):

Table 3. Tools result in rates

	RIPS	Sonarqube
Zero day rate ((ZD)x)rate	0,837209302	0,162790698
False positive rate ((FP)x)rate	0,006519119	0,000219515

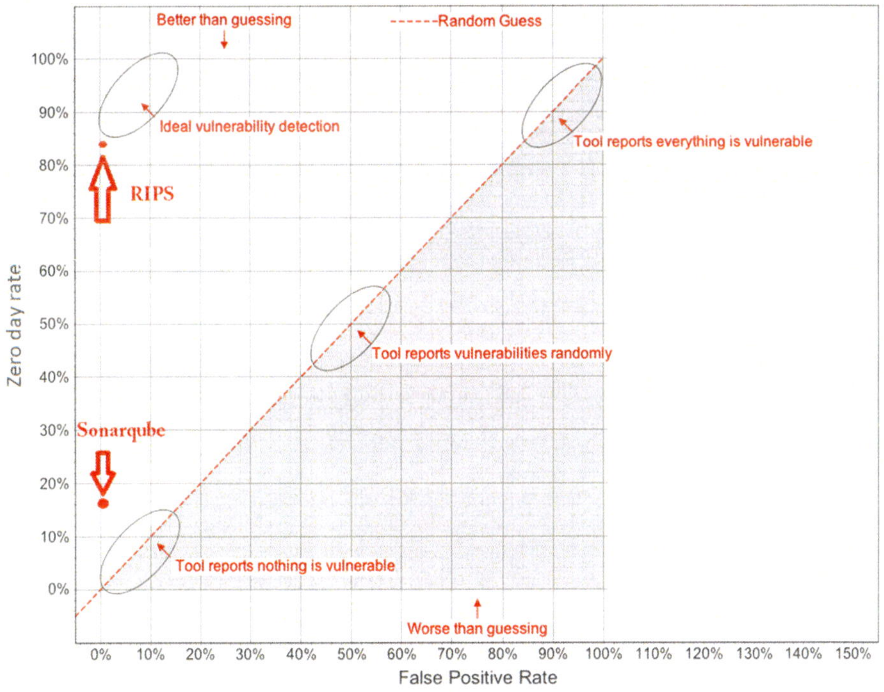

Fig. 4. Youden index for accuracy score

Two things could explain the number of false positives generated by RIPS or sonarqube:

- The scan was performed without considering the WordPress default core function, and a lot of them are used as filters for XSS vulnerabilities.
- Most filtering functions implemented by WordPress core, use Regess expressions.

By adding the WordPress filtering functions to the RIPS database, we reduced the number of false positives considerably.

6 Conclusion and Future Work

Static analysis is a powerful technique to find vulnerabilities in the source code, and its power comes from the fact that it can cover all the paths of a program. Unfortunately, this power comes at the cost of a high false positives rate due to the limitations of this technique, which was proved by experimentation of the two most popular Static vulnerability discovery tools in the industry.

The static analysis approach is still being used and recommended in the software development industry. The static analysis approaches can be used at any step of the Software Development Life Cycle (SDLC) as the access to the source code is always possible during development. This is not possible in the case of dynamic analysis, where it is required to have access to a running application, and this reduces the cost of correcting these vulnerabilities. In this paper, we have discussed some of the most relevant research in automated static vulnerability discovery. We have highlighted their power and their weaknesses. In future work, we will evaluate these static approaches to see which ones give better results according to each specific context.

References

1. Özkan, S.: Browse CVE vulnerabilities by date. https://www.cvedetails.com/browse-by-date.php. Accessed 18 Dec 2019
2. Liu, B., Shi, L., Cai, Z., Li, M.: Software vulnerability discovery techniques: a survey. In: 2012 Fourth International Conference on Multimedia Information Networking and Security, Nanjing, China, pp. 152–156 (2012)
3. Mokhov, S.A., Paquet, J., Debbabi, M.: MARFCAT: fast code analysis for defects and vulnerabilities. In: 2015 IEEE 1st International Workshop on Software Analytics (SWAN), Montreal, QC, Canada, pp. 35–38 (2015)
4. Kim, H., Choi, T.-H., Jung, S.-C., Lee, O., Doh, K.-G., Lee, S.-Y.: Rule-Based Source-Code Analysis for Detecting Security Vulnerabilities, p. 12 (2008)
5. Larochelle, D., Evans, D.: Statically detecting likely buffer overflow vulnerabilities. In: Proceedings of the 10th Conference on USENIX Security Symposium, USA, p. 13 (2001)
6. Jovanovic, N., Kruegel, C., Kirda, E.: Pixy: a static analysis tool for detecting web application vulnerabilities. In: IEEE Symposium on Security and Privacy, p. 6 (2006)
7. Chen, J., et al.: IoTFuzzer: discovering memory corruptions in IoT through app-based fuzzing. In: NDSS (2018)

8. Allen, E.E.: Control flow analysis. ACM SIGPLAN Not. **5**(7), 1–19 (1970). Proceedings of the Symposium Compiler Optimization
9. Kronjee, J., Hommersom, A., Vranken, H.: Discovering software vulnerabilities using data-flow analysis and machine learning. In: Proceedings of the 13th International Conference on Availability, Reliability and Security - ARES 2018, Hamburg, Germany, pp. 1–10 (2018)
10. Static Code Analysis Control—OWASP Foundation. https://owasp.org/www-community/controls/Static_Code_Analysis. Accessed 16 Oct 2020
11. Zhivich, M., Leek, T., Lippmann, R.: Dynamic buffer overflow detection. In: 2005 Workshop on the Evaluation of Software Defect Detection Tools, 12 June, Chicago, IL, p. 7 (2005)
12. Aggarwal, A., Jalote, P.: Integrating static and dynamic analysis for detecting vulnerabilities. In: 30th Annual International Computer Software and Applications Conference (COMPSAC 2006), Chicaco, IL, pp. 343–350 (2006). https://doi.org/10.1109/COMPSAC.2006.55
13. Dahse, J., Schwenk, J.: RIPS-A static source code analyser for vulnerabilities in PHP scripts. In: Seminar Work (Seminer Çalismasi). Horst Gortz Institute Ruhr-University Bochum (2010)
14. Fang, Y., Han, S., Huang, C., Wu, R.: TAP: a static analysis model for PHP vulnerabilities based on token and deep learning technology. PLoS ONE **14**(11), e0225196 (2019). https://doi.org/10.1371/journal.pone.0225196
15. SonarQube Documentation—SonarQube Docs. https://docs.sonarqube.org/latest/. Accessed 02 Jan 2021
16. Table of Contents—OWASP. https://owasp.org/www-project-top-ten/2017/. Accessed 02 Jan 2021

Machine Learning Techniques for Intrusion Detection in SDN: A Survey

Abass Adamou Djergou(✉), Yassine Maleh, and Soufyane Mounir

Laboratory of Sciences and Techniques for Engineering (LaSTI), ENSAK,
Sultan Moulay Slimane University (USMS), Beni Mellal, Morocco
abass.adamoudjergou@usms.ac.ma,
{y.maleh, s.moumir}@usms.ma

Abstract. The SDN (Software Defined Network) was developed to reduce the complexity of the traditional network by controlling and managing the entire network from a centralized location. This new model introduces the separation of the transmission plane and the control plane by providing a new independent plane called the SDN controller. Today, due to the various benefits of this new paradigm, many commercial and industrial enterprises are converging to this solution in their network environments, especially in data centers. Nevertheless, the emerging SDN technology can lead to many security vulnerabilities and threats such as man-in-the-middle (MITM) attacks, denial-of-service (DoS), overload saturation attacks. Therefore, deploying Intrusion Detection Systems (IDS) to monitor malicious activities is crucial for the SDN network architecture. This paper aims to review the different attacks and intrusions affecting the SDN environment through the various existing studies conducted in recent years by the research communities. Then, we will study the different approaches or machine learning techniques such as Naive Bayesian Classification, Support Vector Machine (SVM), etc., to classify and detect anomalies and intrusions in an SDN architecture to choose the most appropriate and efficient algorithm.

Keywords: SDN · Machine learning · Intrusion detection · Software-defined network

1 Introduction

In the traditional network architecture or conventional distributed network, the functionality of decision-making processes known as control plane and network traffic transmission (data plane) are implemented on all network assets (e.g., routers or switches). Thus, the configuration of traffic policies (e.g., routing, switching, QoS, ACL) is performed independently on each network device. In addition, deploying new protocols is a challenge because all devices must be updated or replaced [1]. In addition, device-level management tools must be used to update the configuration settings [2]. This makes the configuration both time-consuming and error prone.

A new concept has emerged to overcome the problems inherent in conventional distributed networks, the SDN. Its main objective is to make the network more flexible and straightforward to manage with efficient administration by decoupling the control

Y. Maleh et al. (Eds.): ICI2C 2021, LNNS 357, pp. 460–473, 2022.
https://doi.org/10.1007/978-3-030-91738-8_42

plane from the data plane of the network devices [3]. Thus, the new paradigm can control the whole system from a centralized remote device called an SDN controller. The centralized view of the entire network allows the SDN controller to provide a global view of the whole infrastructure. As a result, the cost of operation will be significantly reduced compared to the conventional network [3].

Despite the many benefits of SDN technology, it is susceptible to emerging security threats that attackers can exploit to perform various malicious tasks. Specifically, a centralized network controller is a valuable target for attackers for two reasons. First, the controller is centrally located between the application and the data planes. Second, a controller is software subject to vulnerabilities [4]. Some of these issues are inherited from the traditional network environment, while others are specific to the SDN architecture [5]. The security breach typically impairs the credibility, integrity, or availability of hardware, software, or an information resource. Attacking these components can cause considerable damage to the organization. The damage can be a monetary or reputational loss that can lead to the total or partial collapse of the organization. Therefore, an effective measure must be put in place to prevent damage. Many attacks are possible due to the centralized control introduced by the SDN architecture. The authors in [6] demonstrated the possibility of an attack to access the SDN controller. Once the controller is compromised, the attacker can modify the device rules and deny a legitimate user access to available resources (DoS attack). DoS attacks are not the only attacks for SDN, but among the common attacks, there are other attacks such as vulnerability scanning, man-in-the-middle (MITM) and side channel. Integrating an Intrusion Detection System (IDS) into the SDN architecture is potentially one of the best approaches to creating a secure SDN environment.

Intrusion detection systems (IDS) are designed to detect malicious activity, including viruses, worms, DDoS attacks. Thus, detection schemes detect attacks by finding malicious patterns in network traffic that do not conform to the normal expected behavior [7]. The critical success factors of the Intrusion Detection System (IDS) are speed of anomaly detection, accuracy and reliability [8]. Machine learning (ML) techniques are applied to develop IDS to improve these critical factors and achieve a low false alarm rate [9].

This paper will first review the various attacks targeting the SDN environment, then study the different machine learning approaches used for intrusion detection in SDN.

The rest of the paper is structured as follows. Section 2 will present the methodology used. Section 3 will present a brief description of the SDN architecture, network attacks, threat vectors, and attack surfaces in SDN. In Sect. 4, we will discuss intrusion detection methods while relying on machine learning-based IDS. Then Sect. 5 will mitigate attacks in SDN particularly on intrusion detection techniques using ML. The last part will present the paper's conclusion and perspectives.

2 Research Methodology

In this section, we present the methodology that we have undertaken to review the work done in recent years by the scientific community in the security of the SDN environment, including the development of intrusion detection systems based on Machine

Learning approaches. in this section, research questions are defined to meet the expectations of the SDN environment for intrusion detection. To select papers related to these research questions, we followed the recommendations described in [10, 11].

2.1 Research Questions

This section aims to identify the most relevant issues and challenges in IDS-ML for SDN environments, including detection accuracy, false-positive rate, resource consumption, real-time intrusion detection, performance, etc. the main research questions are presented in Table 1.

Table 1. Research questions

RQ#	Research questions	Explanation
RQ1	What are the latest studies on ML-based intrusion detection in SDNs?	Discuss recent studies on ML-based intrusion detection in SDN and their results
RQ2	What are the current challenges of using ML-based IDS in SDN?	Identify current issues and inherent limitations of existing IDS-based ML in the SDN environment

2.2 Search Terms

This step aims to select the most appropriate terms related to our research area i.e., intrusion detection in an SDN environment. Therefore, terms such as "intrusion detection system", "IDS", "anomaly detection", "signature detection", "misuse detection", "machine learning", "ML", "software defined networking", "SDN" were used to search for relevant information. The following search terms are used to select the main papers: (((machine AND learning) OR (ml)) AND ((intrusion AND detection) OR (anomaly AND detection) OR (signature AND detection)) AND ((software-defined AND networking) OR (software AND defined AND networking) OR SDN OR (software-defined AND network) OR (software AND defined AND network)).

2.3 Collection Sources

All papers selected in this study come from CNRST-IMIST electronic resources ("Eressources - IMIST," n.d.) as well as from other popular digital databases not partnered with IMIST or to which we do not have access from the IMIST platform (see Table 2). Journal articles and conference papers are considered for our study.

Table 2. Digital research databases

Online databases#	Online databases	URL
1	Web of science	https://apps.webofknowledge.com
2	SpringerLink	https://link.springer.com/
3	ScienceDirect	https://www.sciencedirect.com/

2.4 Inclusion and Exclusion Criteria

Table 3 presents the summary of the inclusion-exclusion criteria to obtain a unique and comparable set of papers.

Table 3. Inclusion and exclusion criteria

Criteria#	Criterion	Objective
Inclusion	A study that clearly defines how IDS - ML can be applied and supported in the SDN environment	This paper aims to make a study on IDS - ML in SDN. Thus, we want to know how IDS - ML affects the security of SDN. In particular, we need papers that directly propose IDS - ML in SDN environments
Exclusion	Papers not focusing on ML-based intrusion detection techniques in the SDN environment	This paper is aimed only at studies that present IDS-based ML exactly in SDN environments

3 SDN Architecture and Security Challenges

One of the key features of the SDN architecture is the separation of the control plane from the data plane. This will remove the control functionality of network devices thus becoming simple packet transmission elements [12]. In this section we will briefly introduce the description of the SDN architecture. This will make it easier for us to understand the security challenges surrounding the SDN environment.

3.1 SDN Architecture

According to the Open Network Foundation (ONF) [13] SDN is an architecture that separates the control plane from the data plane, and centralizes all network intelligence in a programmable entity called the "Controller", to manage multiple elements of the data plane (e.g., switches or routers, etc.) via APIs (Application Programming Inter face). OpenFlow is considered one of the first approaches to support SDN architecture. Figure 1 shows the SDN architecture. The top layer is the Application Layer that includes various SDN applications; the control plane is in the middle that serves as a communication bridge between the application layer and the infrastructure layer that consists primarily of transmission devices.

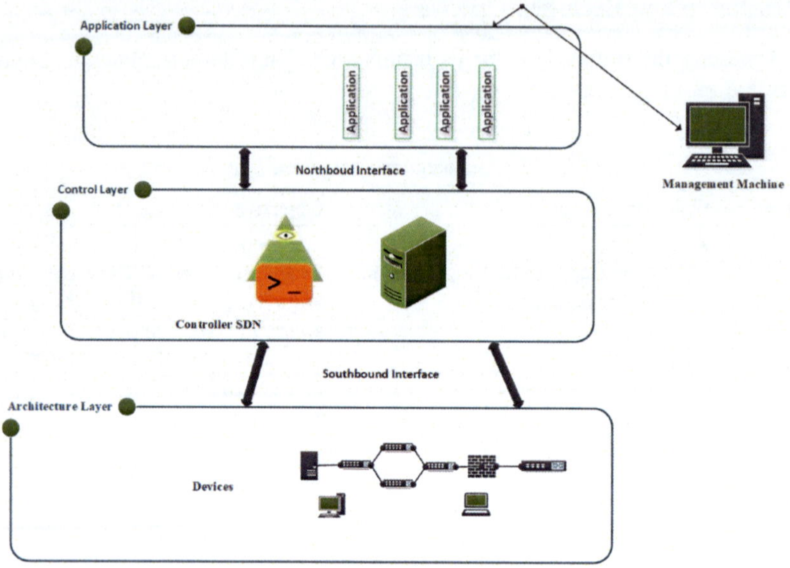

Fig. 1. SDN architecture

3.2 Security Challenges in SDN Architecture

Different security threats are detected in the network by separating control and data transmission. The consequences of these security threats can lead to data leakage, data corruption, access and denial of service (DoS).

In [5] the authors classified the SDN security issues by type and the SDN layer/interface affected by each issue/attack. The security issues identified in Table 4 are also mapped onto the SDN architecture in Fig. 2 to highlight the entity and interface impacted by the attack or vulnerability. Details of the attack scenarios in each of these categories are provided in the following subsections:

Unauthorized access: One of the original characteristics of SDN is described as the centralized control of the network, which allows SDN to provide an abstraction that allows applications to read/write the state of the network. If an attacker were to impersonate a controller/application, they could access network resources and manipulate the operation of the network.

Data leakage: In this category, malicious actors take information obtained from the network and use it to generate false flow requests leading to a DoS attack.

Compromised applications: Because the controller acts as a data plane abstraction for applications and SDN allows for the integration of third-party applications into the architecture, a malicious application could have as detrimental an effect on the network as a compromised controller.

Denial of Service: Due to the communication path between the controller and the network device, an attacker could flood the controller with packets requiring a flow rule decision and make it unavailable to legitimate users [5].

Table 4. Security issues at the environment by layer/interface affected

Security issue	Vulnerabilities/Attacks	SDN layer target				
		App layer	App-Ctrl interface	Ctrl layer	Ctrl-Data interface	Data layer
Unauthorized access	Unauthorized Controller Access / Controller Hijacking			✓	✓	✓
	Unauthorized / Unauthenticated Application	✓	✓	✓		
Data leakage	Flow Rule Discovery (Side-Channel Attack on Input Buffer)					✓
	Credential Management (Keys, Certificates for each Logical Network)					✓
	Forwarding Policy Discovery (Packet Processing Timing Analysis)			✓	✓	✓
Data modification	Flow Rule Modification to Modify Packets (MITM Attack)			✓	✓	✓
Compromised applications	Fraudulent Rule Insertion	✓	✓	✓		
Denial of service	Controller-Switch Communication Flood			✓	✓	✓
	Switch Flow Table Flooding					✓
Configuration issue	Lack of TLS Adoption	✓	✓	✓	✓	✓
	Policy Enforcement	✓	✓	✓		
	Lack of Secure Provisioning	✓	✓	✓	✓	✓

Configuration Issue: As highlighted in Table 4, misconfiguration or incorrect use of security features can impact all layers of the architecture [5].

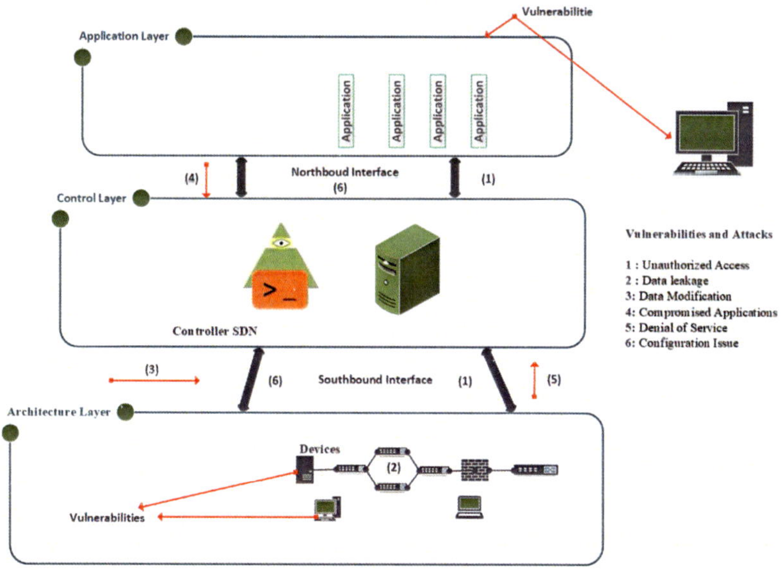

Fig. 2. Attack hot spots in SDN

A potential security vulnerability may also exist on the administrative machine used for the SDN controller. These endpoints are used to develop applications for the network devices' control logic and monitor activities in the network. Attackers could potentially exploit the vulnerability of this endpoint or its connection to the controller. Traditional network threats are always present in SDN technologies. It is possible to generate malicious activities in computer systems, such as denying services or exploiting vulnerabilities on servers, hosts, or network devices [14].

This section presents some of the open challenges and possible vulnerabilities in an SDN-enabled network. Therefore, it is essential to include protection in the architecture for the secure operation of the network infrastructure. To manage and deal with these intrusions and attacks on the computer system and networks, in this case the SDN network, many methods have been devised, one of which is a so-called intrusion detection system discussed in the next section.

4 Intrusion Detection System (IDS)

The terms attack and intrusion in networks have been defined in different ways. One of the frequently used definitions of intrusion is the external factor for a failure and defect in network performance [15].

The role of IDS is to examine network traffic and detect unwanted traffic for suspicious activity or patterns and alert the network administrator. Two different methods can be considered for intrusion detection: misuse detection or signature-based detection and anomaly detection. Figure 3 illustrates these intrusion detection methods.

Signature or misuse detection: The main idea behind misuse detection is to demonstrate attacks in patterns so that even changes in these attacks can be identified. The main drawback of this method is its inability to detect attacks whose traffic behavior cannot be predicted [16].

Anomaly detection: The objective of the anomaly detection method is to produce a statistical model to define normal traffic. In this case, any deviation from this model is considered an anomaly and is detected as an attack [16]. The ability to detect unknown and unfamiliar attacks is the main motivation and reason for developing new methods for detecting an anomaly.

This section made a small opening on intrusion detection mechanisms in networks and systems in general. In the following sections, we review existing intrusion detection techniques designed by the scientific community for the SDN environment and, in particular, those using machine learning to improve detection accuracy and obtain a low false alarm rate.

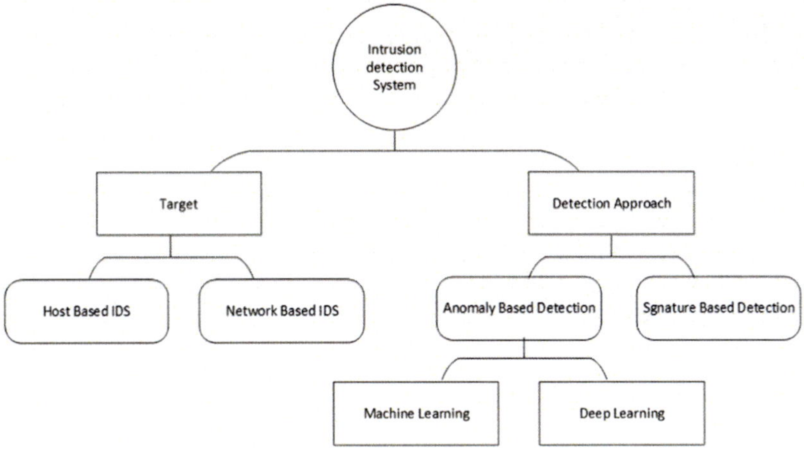

Fig. 3. Intrusion detection system

5 Attack Mitigation in SDN

Several research works have been conducted in intrusion detection using the ML approach in SDN environments. Studies have shown that the integration of ML with SDN has played an important role in intrusion identification over the past five years, which we discuss below (Table 5).

Table 5. Summary of IDS based machine learning approachs in SDN

Ref	Method	Targeted threat	Controller	Dataset	Accuracy
[17]	C4.5 DT	Botnet	Opendaylight	CTU-13, ISOT	97% 90%
[18]	SVM, Naive bayes, Random Forest	DDoS	Floodlight	Traffic generated in SDN	97%, 90%, 74%
[19]	ASVM	Flooding-based DDoS	OpenDaylight	SDNTrafficDS	97%
[20]	SVM	Probe, DoS, R2L, U2L, Data Attacks	Ryu	DARPA	84, 21%
[21]	SVM, NB, NC	–	POX	Traffic generated in SDN	92,3%; 96,79%; 91,02
[22]	SVM, Selective Logging, IP traceback	–	Ryu	NSL-KDD	95,98%
[22]	SVM, Selective Logging, IP traceback	–	Ryu	9 features of NSL-KDD	87,74%
[23]	SVC-RF	DDoS	Ryu	Traffic generated in SDN	98,8%
[24]	Decision Tree, Random Forest, XGBoos	DDoS, PROBE, R2L and U2R	–	NSL-KDD	95,95%
[25]	SVM	DDoS	POX	NSL-KDD	98,90%
[26]	Q-Learning	DoS	ONOS	Traffic generated in SDN	–

A study conducted by Tariq and al. [17], suggested the use of a flow-based approach that would detect botnets in the SDN environment by further using ML algorithms without the need to read the packet payload. This study aims to combine real-time flow trace data, coupled with historical context, allowing the author to extract an enriched feature set that further enables classification and improves detection and subsequently reduces the false positive rate. In addition, a supervised ML algorithm approach based on a C4.5 decision tree [27, p. 5], [28, p. 5] is applied for the botnet classification process. Botnet traces are derived from publicly available CTU-13 botnet datasets [29] and ISOT botnet datasets. Preliminary results show that the system can detect unknown botnets, with results achieved of 97% for identifying known botnets and 90% for unknown botnets.

In [18], the authors propose a model to detect DDoS attacks in SDN. Their work was divided into two steps, first, the model extracts several key features from the incoming packet traffic such as source address, destination address, source port, destination port and protocol type and then measures the distribution of each feature using entropy. In the second step, following the entropy of the resulting features, the model uses the Support Vector Machine (SVM) algorithm to predict whether the traffic is abnormal or not and therefore detect the DDoS attack. After evaluation, the experiments showed an accuracy rate of 97% with the SVM algorithm unlike the naive bayes and Random Forest algorithms where they obtained an accuracy rate of 90% and 74% respectively.

In [19], the authors proposed an advanced support vector machine (ASVM) technique as an improvement to an existing support vector machine (SVM) algorithm for detecting DDoS attacks. The ASVM technique is a multiclass classification method consisting of three classes. In order to evaluate their work, one hundred hosts, nine switches and in comparison, with many existing research on SDN security that used a single controller in their network configuration, three controllers are used in their architecture. Although one controller is down due to the attack, another controller can still be used. They were able to detect two types of flooding-based DDoS attacks successfully. The evaluations were based on the measures of false alarm rate, detection rate and accuracy. As a result, a detection accuracy rate of about 97% was achieved with the fastest training and testing time.

In [20], the authors proposed a technique that combines the advantages of flow-based IDS and packet-based IDS to provide a high detection rate without degrading network performance. The flow-based IDS uses an anomaly detection algorithm based on support vector machines (SVMs) trained with the DARPA intrusion detection dataset. This first line of defense detects any intrusion on the network. When an attack is detected, the malicious flow is mirrored to a packet-based IDS for further examination and action. Evaluation results show a detection rate of 84.21% and good performance with minimal additional overhead.

In [21], the authors proposed a technique based on three different supervised learning models, namely Support Vector Machine (SVM), nearest centroid and Naïve Bayes (NB). Their goal was to be able to classify data traffic based on applications in the SDN environment. Network traffic traces are captured and flow characteristics are generated, sent to the classifier for prediction. These authors mention having an accuracy of 92.3% with SVM, 96.79% with NB and 91.02% with Nearest Centroid.

In [22], the authors presented an SDN-based IDS using support vector machines (SVMs) along with selective logging for IP tracing. The evaluation achieved 95.98% accuracy using the entire NSL-KDD dataset and 87.74% accuracy on selected sub-features of the NSL-KDD dataset. These authors explain their choice of method because SDN provides a centralized architecture for detection analysis and SVM logging provides decent detection accuracy without much computation. Selective logging provides about 90–95% savings in terms of memory resources and IP tracing allows tracking the actual source of packets in case of an attack.

The authors in [23] studied a hybrid Support Vector Classifier (SVC) and Random Forest (RF) model (SVC-RF) that classifies traffic using the SVC results, which are again filtered by the random forest for distributed denial of service (DDoS) attack detection. It achieves an accuracy of 98.8%. With a sensitivity of 97.91%, high true positives are represented, which signifies the model's performance, an accuracy of 98.27% means that the model correctly predicts the class label in most cases. In [24], the authors demonstrated a method consisting of classical and advanced tree-based ML algorithms, Decision Tree, Random Forest and XGBoos for traffic monitoring to detect malicious behaviors on the network as part of IDS in the SDN controller. The NSL-KDD dataset is used to train and test the proposed method, considered a reference dataset for several state-of-the-art approaches in IDS. Several advanced preprocessing techniques are performed on the dataset to extract the best shape from the data. Using only five of the 41 features of NSL-KDD, a multi-class classification task is performed by detecting an attack and classifying the type of attack (DDoS, PROBE, R2L and U2R) accuracy of 95.95%.

In [25] the authors proposed a DDoS attack detection and mitigation model for an SDN system. This model is based on the SVM technique as a classifier. For better accuracy and to reduce the testing time, the KPCA (kernel principal component analysis) technique with GA (genetic algorithm) has been used in this model. The NSL-KDD dataset and another dataset specific to DDoS attacks were used for the test. According to the authors, the KPCA technique is used in this sense to extract the main features from the dataset and reduce their dimension from the DDoS vectors; GA is used to select the appropriate features in order to optimize the different SVM parameters. The experimental results show that on the DDoS dataset, KPCA works effectively and the accuracy of the proposed model is 98.907%.

The authors in [26] developed a machine learning based defense framework called Q-MIND. Their goal was to design a system capable of efficiently detecting and mitigating stealthy DoS attacks in SDNs. The said approach is based on Q-Learning based reinforcement learning to maximize its detection performance. Performance evaluation results show that their Q-MIND approaches, by applying the optimal Q-Learning agent policy, can detect and mitigate stealth DoS attacks more effectively than existing methods.

6 Discussion

Looking at all these studies, we can see that two main machine learning-based approaches are currently used to detect exploitable attacks on the SDN: simulation-based approaches and public dataset-based approaches.

6.1 ML-Based Simulation

The work presented in [18, 19, 21, 23, 26] uses the simulation-based approach for malicious traffic detection in SDN in their respective work. In these approaches, researchers establish the network topology with legitimate hosts to generate the traffics, and other hosts act as pots to create attack traffics. They use public tools such as Scapy or hping3, to simulate attacks such as the DoS attack. The characteristics, such as source IP address or source port, flow packets, etc. extracted from the collected traffic for normal and malicious data separately. Then all these samples are randomly shuffled into a.CSV file to create the data used in the training model. The training model can be used later to classify normal and intrusive packets within the SDN platform.

It is clear that these approaches are fast in the calculation, and simple to analyze. However, it has many limitations. First, the dataset created has a very small size and therefore, it is not enough to give accurate results (number of samples are a few hundred or a few thousand in most cases). Moreover, the generated data contains only a few types of attacks. These attacks are not realistic to represent the diversity of anomalies present on the Internet. Limiting the diversity of attack types helps the attacker know the detection mechanism's behavior, and he can easily fabricate an attack to reproduce this behavior. Second, the number of extracted features is insignificant, and the small number of features is not enough to cover the behavior of all attacks.

Furthermore, the features used for the training model were extracted mainly from the header packet without further inspection of the payload data. The header field can be easily modified to look like normal fields. In general, the attacker can easily implement malicious code inside the payload packets to fool detection systems such as Root to Local (R2L) attacks and worm infections. As a result, they return a low accuracy to detect the application-level attacks.

6.2 ML-Based Public Datasets

The work presented in [17, 20, 22, 24, 25] uses a public dataset to detect intrusions in the SDN environment. The selection of the appropriate dataset greatly impacts the evaluation of network intrusion detection systems (IDS). Unfortunately, most publicly available datasets are not realistic, and they lack the variety in attack type to cover all the trends found on the Internet today. One of the main reasons for this shortage is privacy and legal issues for service providers to publish their network data. As a result, most available data sets do not give acceptable accuracy when deployed with intrusion systems. Existing classical ML-based techniques have difficulties in detecting sophisticated attacks in large-scale networks. Several challenges can lead to misclassification of results due to irrelevant features and a lack of training data. These factors lead to difficulties in detecting attacks using classical ML methods.

7 Conclusion and Perspectives

In this article, we first described the SDN architecture, its contribution to the lack of manageability and programmability of the traditional or distributed architecture. With the separation of the control plane from the data plane that SDN brings, the network is more flexible than ever and simple to manage with efficient administration. We have also highlighted the insecurity that SDN-enabled networks face due to this famous separation of the control plane from the data plane. This has prompted the scientific community to develop a solution to mitigate the threats and intrusions that target the SDN environment. The intrusion detection system is the most effective mechanism in such an infrastructure to establish effective classification patterns to detect and distinguish normal behavior from abnormal behavior and examine network traffic and detect unwanted traffic for suspicious activity or patterns and alert the network administrator.

We have focused on machine learning-based IDS, view the critical success factors of an intrusion detection system (IDS) are the speed of anomaly detection, accuracy and reliability [8], as well as record machine learning (ML) techniques are known in improving these critical factors and in addition to obtaining a low false alarm rate.

We have provided a detailed study of various approaches based on ML techniques used in attack detection in SDN. We performed our experiment on all these studies and explain that there are two ML-based simulation approaches (using simulation data and using a public dataset) and explain why some traditional machine learning-based methods do not perform well.

In future works, we will propose a system based on one of the most efficient machine learning methods, taking into account the shortcomings that exist in some models and obtain a model that will have superior performance compared to state-of-the-art methods.

References

1. Farhady, H., Lee, H., Nakao, A.: Software-defined networking: a survey. Comput. Netw. **81**, 79–95 (2015). https://doi.org/10.1016/j.comnet.2015.02.014
2. Sudar, K.M., Deepalakshmi, P.: Comparative study on IDS using machine learning approaches for software defined networks (2020)
3. Elsayed, M., Le-Khac, Nhien-An., Jurcut, A.: InSDN: a novel SDN intrusion dataset. IEEE Access **8**, 165263–165284 (2020). https://doi.org/10.1109/ACCESS.2020.3022633
4. Dawoud, A., Shahristani, S., Raun, C.: Unsupervised deep learning for software defined networks anomalies detection. In: Nguyen, N.T., Kowalczyk, R., Xhafa, F. (eds.) Transactions on Computational Collective Intelligence XXXIII. LNCS, vol. 11610, pp. 167–178. Springer, Heidelberg (2019). https://doi.org/10.1007/978-3-662-59540-4_9
5. Scott-Hayward, S., Natarajan, S., Sezer, S.: A survey of security in software defined networks. IEEE Commun. Surv. Tutor. **18**(1), 623–654 (2016). https://doi.org/10.1109/COMST.2015.2453114
6. Benton, K., Camp, L.J., Small, C.: OpenFlow vulnerability assessment. In: Proceedings of the Second ACM SIGCOMM Workshop on Hot Topics in Software Defined Networking, New York, NY, USA, pp. 151–152 (2013). https://doi.org/10.1145/2491185.2491222

7. Chandola, V., Banerjee, A., Kumar, V.: Anomaly detection: a survey. ACM Comput. Surv. **41**(3), 1–58 (2009). https://doi.org/10.1145/1541880.1541882
8. Sultana, N., Chilamkurti, N., Peng, W., Alhadad, R.: Survey on SDN based network intrusion detection system using machine learning approaches. Peer-to-Peer Netw. Appl. **12** (2), 493–501 (2018). https://doi.org/10.1007/s12083-017-0630-0
9. García-Teodoro, P., Díaz-Verdejo, J., Maciá-Fernández, G., Vázquez, E.: Anomaly-based network intrusion detection: techniques, systems and challenges. Comput. Secur. **28**(1), 18–28 (2009). https://doi.org/10.1016/j.cose.2008.08.003
10. Kitchenham, B., Brereton, P.: A systematic review of systematic review process research in software engineering. Inf. Softw. Technol. **55**(12), 2049–2075 (2013). https://doi.org/10.1016/j.infsof.2013.07.010
11. Sadqi, Y., Maleh, Y.: A systematic review and taxonomy of web applications threats. Inf. Secur. J. Glob. Perspect. 1–27 (2020). https://doi.org/10.1080/19393555.2020.1853855
12. Kreutz, D., Ramos, F., Verissimo, P., Rothenberg, C., Azodolmolky, S., Uhlig, S.: Software-defined networking: a comprehensive survey. Proc. IEEE **103**(1), 14–76 (2015). https://doi.org/10.1109/JPROC.2014.2371999
13. Software-Defined Networking (SDN) Definition: Open Networking Foundation. https://opennetworking.org/sdn-definition. consulté le 09 mai 2021
14. Jankowski, D., Amanowicz, M.: Intrusion detection in software defined networks with self-organized maps. J. Telecommun. Inf. Technol. **2015**(4), 3–9 (2015)
15. par D. S. dans L. mise en réseau le 8 mai 2001 à minuit: Lock IT down: understand network intrusions and attacks. TechRepublic. https://www.techrepublic.com/article/lock-it-down-understand-network-intrusions-and-attacks/. consulté le 09 mai 2021
16. Altwaijry, H.: Bayesian based intrusion detection system. In: Kim, Haeng Kon, Ao, Sio-Iong., Rieger, Burghard B. (eds.) IAENG Transactions on Engineering Technologies, pp. 29–44. Springer, Dordrecht (2013). https://doi.org/10.1007/978-94-007-4786-9_3
17. Tariq, F., Baig, S.: Machine learning based botnet detection in software defined networks. Int. J. Secur. Appl. **11**(11), 1–12 (2017). https://doi.org/10.14257/ijsia.2017.11.11.01
18. D. Li, C. Yu, Q. Zhou, et J. Yu, « Using SVM to Detect DDoS Attack in SDN Network », 2018, vol. 466, n° 1. doi: https://doi.org/10.1088/1757-899X/466/1/012003
19. Oo, M., Kamolphiwong, S., Kamolphiwong, T., Vasupongayya, S.: Advanced support vector machine- (ASVM-) based detection for distributed denial of service (DDoS) attack on software defined networking (SDN). J. Comput. Netw. Commun. **2019**, 1–12 (2019). https://doi.org/10.1155/2019/8012568
20. Schueller, Q., Basu, K., Younas, M., Patel, M., Ball, F.: A hierarchical intrusion detection system using support vector machine for SDN network in cloud data center. In: présenté à 2018 28th International Telecommunication Networks and Applications Conference, ITNAC 2018 (2019). https://doi.org/10.1109/ATNAC.2018.8615255
21. Raikar, M., Meena, S.M., Mulla, M., Shetti, N., Karanandi, M.: Data traffic classification in software defined networks (SDN) using supervised-learning. Procedia Comput. Sci. **171**, 2750–2759 (2020). https://doi.org/10.1016/j.procs.2020.04.299
22. Hadem, P., Saikia, D., Moulik, S.: An SDN-based intrusion detection system using SVM with selective logging for IP traceback. Comput. Netw. **191**, 108015 (2021). https://doi.org/10.1016/j.comnet.2021.108015
23. Ahuja, N., Singal, G., Mukhopadhyay, D., Kumar, N.: Automated DDOS attack detection in software defined networking. J. Netw. Comput. Appl. **187**, 103108 (2021). https://doi.org/10.1016/j.jnca.2021.103108
24. Alzahrani, A., Alenazi, M.: Designing a network intrusion detection system based on machine learning for software defined networks. Future Internet **13**(5), 111 (2021). https://doi.org/10.3390/fi13050111

25. Sahoo, K.S., et al.: An evolutionary SVM model for DDOS attack detection in software defined networks. IEEE Access **8**, 132502–132513 (2020). https://doi.org/10.1109/ACCESS.2020.3009733
26. Phan, T.V., Gias, T.M.R., Islam, S.T., Huong, T.T., Thanh, N.H., Bauschert, T.: Q-MIND: defeating stealthy dos attacks in SDN with a machine-learning based defense framework. In: présenté à 2019 IEEE Global Communications Conference, GLOBECOM 2019 - Proceedings (2019). https://doi.org/10.1109/GLOBECOM38437.2019.9013585
27. C4.5 - 1st Edition. https://www.elsevier.com/books/c45/quinlan/978-0-08-050058-4. consulté le 12 juill 2021
28. Saha, S.: What is the C4.5 algorithm and how does it work? Medium (2018). https://towardsdatascience.com/what-is-the-c4-5-algorithm-and-how-does-it-work-2b971a9e7db0. consulté le 12 juill 2021
29. The CTU-13 Dataset. A Labeled Dataset with Botnet, Normal and Background traffic. Stratosphere IPS. https://www.stratosphereips.org/datasets-ctu13. consulté le 12 juill 2021

A Suricata and Machine Learning Based Hybrid Network Intrusion Detection System

Said Ouiazzane(✉), Malika Addou, and Fatimazahra Barramou

LaGeS Laboratory, ASYR Team, Hassania School of Public Works, Casablanca, Morocco

Abstract. The objective of this paper is to propose a hybrid model of Network Intrusion Detection System (NIDS) based on the use of two types of IDS: Signature-based NIDS (SNIDS) and Anomaly Detection-based NIDS (ADNIDS). Indeed, modern computer networks have become the backbone for all the most critical business sectors. In parallel with the evolution and expansion of computer networks, cyber threats keep improving day after day to become more and more sophisticated and capable of bypassing all security policies implemented by information security managers. Knowing that cyber-attacks can cause irreparable damage, costing the victim entity a lot of money, following a leak of critical and sensitive information. In addition, traditional prevention mechanisms such as network firewalls are no longer sufficient to counter cybercrime as they can only stop known attacks from the outside but not those coming from the inside or 0-day attacks. Therefore, intrusion detection systems are important devices to deploy in IT infrastructures to protect them from suspicious activities. However, SNIDS alone only provides detection of intrusions with known signatures but not unknown 0-day attacks. ADNIDS, on the other hand, can detect unknown intrusions but generate very high false alarm rates. Another approach is to use both types of NIDS to form a hybrid system and it is the most effective solution to counter any kind of attack, including unknown cyber threats. The use of both SNIDS and ADNIDS at the same time forms what is called a hybrid NIDS. Our hybrid NIDS model is based on Suricata as the SNIDS and ADNIDS based on the Machine Learning Decision Tree algorithm. The network baseline included the set of benign traffic patterns and was designed after balancing and optimizing the CICIDS2017 dataset. The classification of the benign traffic via Decision Tree yielded very conclusive results in accuracy, F-Measure, Recall, and precision.

Keyword: Intrusion Detection · Zero-day attacks · Machine Learning · Multi-agent systems · Cybersecurity

1 Introduction

Computer networks have become the backbone for all of the most critical business sectors [1]. As these networks evolve and expand, cyber threats continue to threaten the security of IT infrastructures. As a result, security threats are improving day by day, becoming more sophisticated and capable of circumventing all security policies implemented by government agencies and critical infrastructures.

Y. Maleh et al. (Eds.): ICI2C 2021, LNNS 357, pp. 474–485, 2022.
https://doi.org/10.1007/978-3-030-91738-8_43

Network intrusions can come from several sources of threats, including the presence of software bugs due to operating systems and software that are becoming larger and more feature-rich [2]. Cyberattacks can cause irreparable damage at a high cost to the victim entity by leaking critical and sensitive information. In addition, traditional firewalls and prevention mechanisms are no longer effective in protecting modern networks, as they only intercept attacks from the outside, but not those from the inside. Therefore, it is imperative to find alternatives to ensure the security of Local Area Networks (LAN) while detecting cyber threats that may target computer networks.

Network Intrusion Detection Systems (NIDS) are now very important appliances that must be deployed at the network infrastructure level to protect the computer networks from suspicious activities that may threaten network security [3]. We can distinguish between two types of NIDS, the first type is called Signature-based NIDS (SNIDS) and the second type is called Anomaly Detection-based NIDS (ADNIDS). On the one hand, the SNIDS is based on a signature database and can detect known intrusions with a very low false-positive rate. On the other hand, ADNIDS is based on detecting anomalies while identifying deviations from the normal operation of the network (baseline).

NIDS is the current trend and several researchers have addressed the problem of intrusion detection and have proposed very interesting and promising approaches. Nevertheless, most of the proposed works still suffer from inadequacies in zero-day attack detection, lack of modularity and scalability features, use of old or even obsolete training datasets, and lack of exploitation of open source NIDS supporting Multiple Threading.

This research work proposes a more efficient NIDS model for detecting intrusions targeting modern computer networks. Thus, we propose the NIDS based on two detection mechanisms to monitor networks against known and unknown (zero-day) attacks. On the one hand, the first mechanism aims to provide first filtering against known attacks while employing Suricata, an open-source NIDS having the Multiple Threading functionality. On the other hand, the second mechanism is dedicated to detecting unknown attacks that have deviated from the normal traffic baseline. Furthermore, our system is based on the multi-agent paradigm, which allows us to have a modular NIDS while supporting components to support the evolution of networks.

The rest of this paper is organized as follows: Sect. 2 discusses some related research works conducted by the scientific community. Section 3 highlights our hybrid NIDS model while presenting the proposed architecture and its operating principle. Section 4 deals with the experimentation and testing part, and finally, Sect. 5 concludes the article and updates our perspectives.

2 Related Work

In this section, we will highlight some of the work done by the scientific community to advance the field of intrusion detection.

Cybersecurity researchers have conducted several research works on intrusion detection systems. Notably, S. Smys et al. [4] proposed an IDS to detect intrusions in IoT networks based on a hybrid neural network model. According to the author, his system can give satisfactory results in high-dimensional IoT networks. Sydney et al. [5] proposed a Deep Learning-based method for developing a wireless IDS. The author used the Wrapper Based Feature Extraction Unit (WFEU) to reduce the dimension of UNSW-NB15 and AWID datasets. Hadeel Alazzam et al. [6] proposed a Pigeon Inspired Optimizer algorithm method to reduce the dataset dimension while keeping only the most relevant features. The evaluation of their approach was done on the datasets KDDCUP99, NLS-KDD, and UNSW-NB15. Ferrag et al. [7] proposed an IDS that he named Rules and Decision Tree-based Intrusion Detection System (RDTIDS) to detect intrusions that may target IoT networks. His approach combines various classification techniques: Decision tree, JRip algorithm, Forest PA, and REP Tree. Kumar et al. [8] proposed a signature-based IDS capable of detecting five categories of network traffic, namely: Exploit, Probe, Generic, DOS, and Normal. The proposed system behaves like a dog watcher capable of monitoring the network against suspicious activities. The author based on the UNSW-NB15 dataset and generated his test dataset bearing the Real-Time dataset at NIT Patna CSE lab (RTNITP18). S. Jin et al. [9] proposed a lightweight signature-based IDS capable of detecting intrusions targeting the In-Vehicle Controller Area Network Bus Network. The research confirms that this type of network still lacks protection mechanisms against cybercrime and requires the proposed IDS to be as lightweight as possible. These networks are very notorious for their limitation in computational resources. A. N. Cahyo et al. [10] confirmed that the hybrid NIDSs proposed by the scientific community are not yet mature and need to be improved to be usable in real networks. Thus, the researcher conducted a literature review of hybrid NIDS during the last five years while presenting the different architectures adopted in the various proposed approaches. H. Gajjar et al. [11] opted for analyzing and comparing different approaches which rely on the Openstack private cloud to detect intrusions based on either signatures or anomaly detection. Dinesh Goyal et al. [12] proposed a solution to reduce the number of false alarms generated from NIDS based on anomaly detection. The author designed an anomaly detection-based NIDS capable of detecting intrusions with a reduced false alarm rate. To achieve this, the researcher focused on three main components: the preprocessing component, the detection engine, and the module for alert processing. R. Samrin et al. [13] conducted a state-of-the-art study on the set of approaches for anomaly detection-based NIDS that used the KDDCUP99 dataset. Then, the researcher identified some shortcomings and weaknesses of the proposed methods in the literature while proposing improvement points to design a more effective anomaly-based NIDS. Asish Kumar Dalai et al. [14] conducted a state-of-the-art study on hybrid NIDS while opting for a comparative study between the various approaches. Said Ouiazzane et al. [15] proposed a NIDS to detect intrusions that could target UAV networks. The author used multi-agent systems and machine learning techniques to model the NIDS. Thus, a set of classification methods were used and evaluated to choose the most reliable one. The author of [16] proposed a mechanism for intrusion detection in networks generating big data. The researcher merged two classification techniques, namely: a Short-Term Long Memory (LSTM) and a Convolutional Neural Network (CNN). In addition, the author opted for balancing the CIDDS-001 dataset to limit the impact of the imbalance of the training dataset on the classification results. According to the researcher, the results were conclusive for both multi-class classification (99.83% accuracy) and binary classification (99.17% accuracy).

The scientific community has indeed started to focus on different types of NIDS and researchers are constantly proposing new approaches to improve the effectiveness of NIDS. Nevertheless, intrusion detection is not yet mature and requires more effort to succeed in developing more effective and efficient NIDS.

The rapid development of cybercrime via the appearance of new, more sophisticated, and targeted 0-day attacks is jeopardizing the security of networks in small and large companies. In addition, traditional security mechanisms can no longer protect networks against modern cyber attacks as they are mainly based on known attack signatures developed by security experts.

Most of the approaches proposed in the literature are unsatisfactory and deal with very limited problems. On the one hand, several researchers have just focused on signature-based detection mechanisms. However, classical SNIDS are no longer effective in detecting unknown 0-day attacks. In addition, SNIDS can no longer ensure the security of modern computer networks and require very frequent updates in signature development for newly emerging attacks. On the other hand, most research works have proposed approaches to improve the detection mechanisms adopted by ADNIDS by trying to propose highly efficient classification techniques based on Machine Learning and Deep Learning algorithms.

On the other hand, ADNIDS generates very high false-positive rates, which overlaps with the visibility of administrators on the security of their IT infrastructure. Furthermore, few research works have proposed to opt for hybrid approaches that combine both NIDS types (SNIDS and ADNIDS). Furthermore, hybrid detection-based approaches just deal with classification techniques and do not address the architectural aspect of the proposed NIDS, which hinders the proper understanding of the proposed models.

In our present research work, we will put forward a simple and efficient architecture to detect known and unknown attacks. Our model is based on the combination of two detection mechanisms: SNIDS and ADNIDS. The SNIDS of our model allows first filtering to detect known attacks whose signatures are already known by the experts. At the same time, the ADNIDS is implemented to detect deviations from the network baseline. The network baseline consists of developing a reference profile of the network during its normal operation in the absence of any trace of cyberattacks.

3 Proposed Approach

In this section, we will highlight our proposed NIDS model for detecting known and unknown attacks.

3.1 Proposed NIDS Model

We propose a hybrid NIDS based on a SNIDS and an ADNIDS. The SNIDS is used to detect intrusions whose signatures are already known by security experts. On the other hand, the second brick ADNIDS allows detecting deviations from the network baseline during its normal operation. Figure 1 shows the components of the proposed NIDS model. Indeed, our NIDS model is mainly based on two detection mechanisms:

- Signature-based NIDS (SNIDS): This is the first brick of our system and allows capturing network packets and checking the match against a signature base of known intrusions. This layer is capable of detecting known cyber-attacks;
- Anomaly Detection-based NIDS (ADNIDS): This detection layer can model normal based on machine learning algorithms. The ADNIDS component of our model is used to detect 0-day attacks whose signatures are not yet known by the security community. Indeed, this component can recognize normal behavior and considers any deviation from the baseline as an anomaly.

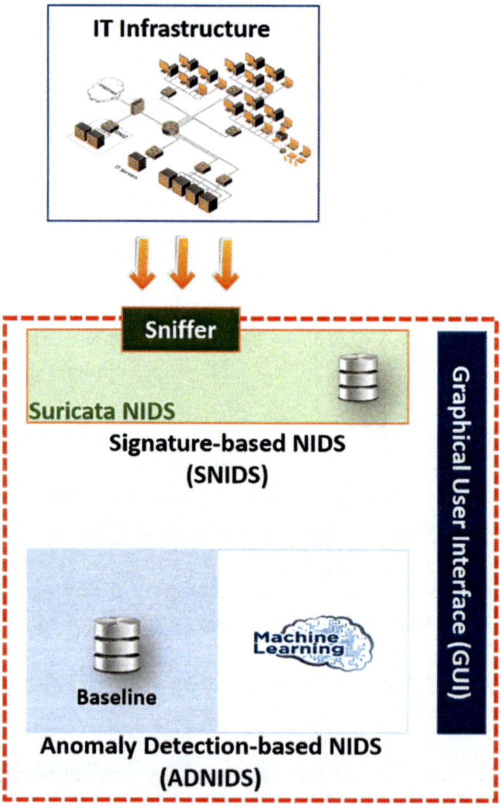

Fig. 1. The proposed model of NIDS

The diagram in Fig. 2 gives an overview of the workflow adopted by our NIDS system to detect known attacks and zero-day attacks. Indeed, our system is composed of a SNIDS and an ADNIDS that work together to detect all types of intrusions, including those whose signatures are not yet known. On the one hand, we have used Suricata as a SNIDS to perform network traffic capture and detection of known attacks corresponding to attack signatures. On the other hand, if the packet does not match a known intrusion by Suricata, the packet will be sent to ADNIDS to check if it matches the network baseline or not. Thus, ADNIDS is based on machine learning techniques to recognize normal and, therefore, detect any deviation from the baseline. Therefore, any network packet, whose characteristics do not match a known intrusion and which does not match the baseline patterns, is considered a zero-day attack if it is not a false positive or false negative.

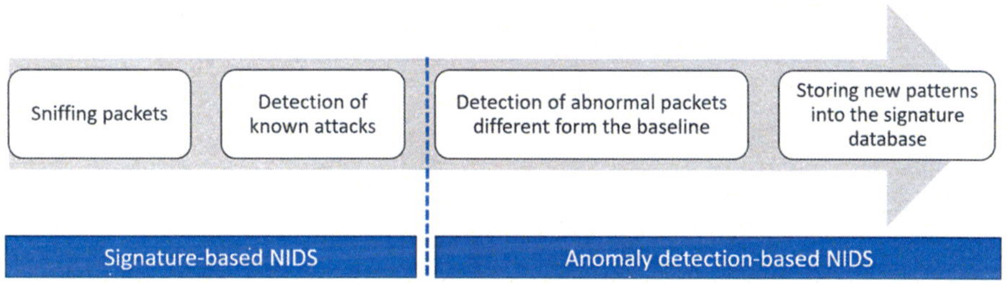

Fig. 2. The detection mechanism of known and zero-day attacks adopted by our proposed system.

3.2 Operating Principle of the Proposed Model

Figure 3 shows the intrusion detection mechanism adopted by our proposed system. Indeed, the system allows to check the matching of the network packet with two detection modules. The first module is a SNIDS ensuring the detection of attacks whose signatures are known. The second module is an ADNIDS, and it allows to verify the pattern matching of the packet (which does not correspond to a known attack) against the network baseline. Thus, our NIDS system can detect known intrusions using its first SNIDS module and also ensures the detection of deviations from the normal network profile (baseline) to detect zero-day intrusions whose signatures are not yet known or developed by security experts.

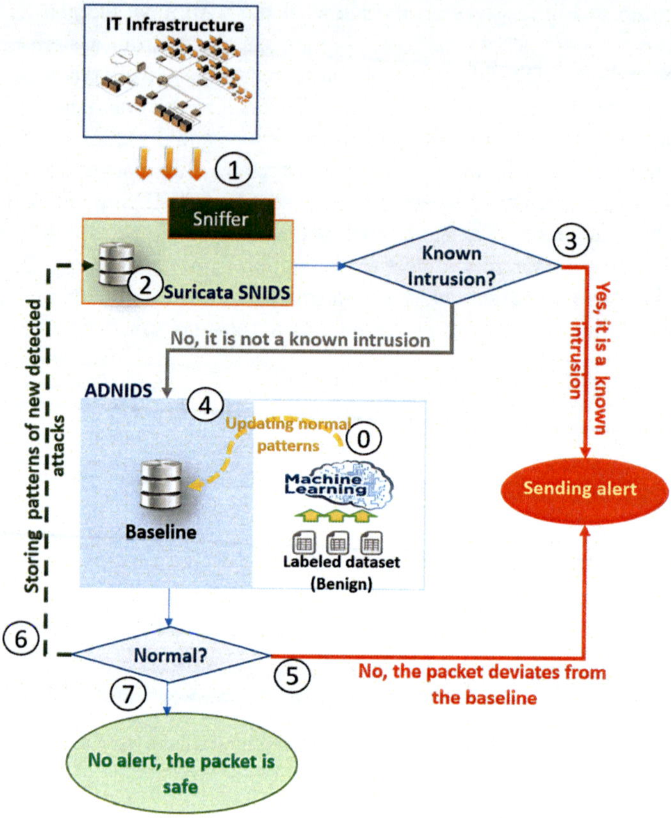

Fig. 3. The operating principle of the proposed NIDS

According to Fig. 3 above, the detection mechanism is done in 8 steps. Thus, the workflow adopted during the process of cyber attack detection by our NIDS model can be presented as follows:

- Step 0: Before counting the detection phases, it is wise to present what is ensured during phase 0. ADNIDS is used to model benign traffic that is free of any trace of the cyberattack. In Phase 0, ADNIDS is supposed to be trained on normal traffic, describing the normal operation of the network without the presence of a cyber attack. For this purpose, a set of machine learning techniques must be used and tested to select the most reliable accuracy and precision. In addition, annotated datasets can be used to model the normal behavior of network packets;
- Step 1: In this step, the proposed hybrid NIDS captures the network traffic via the Sniffer of the SNIDS component. To do this, the system must have a high-speed network card that must be connected to a SPAN port on the Federator Switch. This way, the Sniffer can collect a copy of the network traffic without causing network latency as opposed to the inline mode;

- Step 2: At this point, the Sniffer captured the network packets, and it is time to check their match against the SNIDS signature database. The latter checks if there is a possible match between the characteristics of the captured network packet and the known intrusion patterns stored in the SNIDS signature database. We can distinguish between two cases:
  - If the packet matches a pattern recognized by the SNIDS database, an alert is sent to the security administrator to get his attention regarding the detected intrusion;
  - If the packet does not match any known attack signature, the network packet in question must pass through the second component, which is the ADNIDS.
- Step 3: If the network packet matches a known intrusion, an alert is automatically sent to the security administrators to investigate and analyze the suspicious packets;
- Step 4: If the network packet is not recognized as a known intrusion, the ADNIDS intervenes to compare the characteristics of the packet to the network baseline;
- Step 5: If the characteristics of the network packet deviate from the baseline, an alert will be sent to the administrator to inform them that a suspicious packet is present in the network traffic;
- Step 6: When the packet deviates from the baseline, it is considered suspicious and its signature will be sent to the SNIDS so that the packet will be recognized as suspicious in the next detections;
- Step 7: If the packet corresponds to normal behavior, it is considered benign and does not represent a risk of intrusion on the IT assets.

3.3 Use Case of the Proposed Model

The proposed hybrid NIDS will be deployed at the network level in passive mode. The architecture diagram in Fig. 4 shows the mode of deployment of our NIDS system. Indeed, our NIDS will be connected to a Switch L3 on a port in SPAN mode, in this way, the NIDS does not intercept the network traffic and receives a copy of all the network traffic circulating in the L3 backbone switch. Thus, all traffic flowing through the different broadcast domains will be copied and sent to the NIDS to analyze and detect suspicious packets.

Our NIDS model is connected to the network on a SPAN port of the backbone switch. Thus, the system captures all traffic passing through the network so that it checks the match of the packets against the SNIDS signature database to detect known attacks. If the packet does not correspond to a known intrusion, the ADNIDS checks the packet 's match against the baseline of the network describing its normal operation. Suppose the packet deviates from the characteristics of benign traffic represented by the baseline. In that case, the packet in question is considered a zero-day attack that must be analyzed and investigated thoroughly to avoid falling into the trap of false positives. Thus, the new patterns are communicated to the signature database so that the system can now detect similar packets using SNIDS.

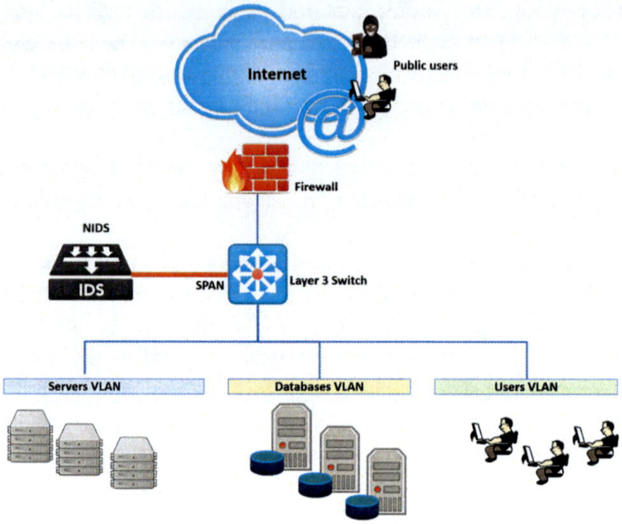

Fig. 4. Deployment mode of the proposed model

4 Experimentation

In this section, we will discuss the network baseline development mechanism at the ADNIDS brick level. For this purpose, we used the CICIDS2017 dataset after optimizing and balancing it. Then, we used the Decision Tree algorithm to classify the benign traffic of the optimized dataset.

4.1 CICIDS2017 Dataset Optimization

CICIDS2017 includes missing data and data with infinite values. This type of data could falsify the classification results using Machine Learning algorithms. Therefore, the CICIDS2017 dataset must be cleaned up to make the classification as reliable as possible.

Cleaning the CICIDS2017 Dataset. After analyzing the contents of the CICIDS2017 dataset, it was found to include missing values and infinite data. We used Pandas ' "isnull().any(1)" and dropna(axis = 0, how = 'any') functions to remove the missing values and then we removed the infinite values by using the function "replace([np.inf, -np.inf], np.nan, inplace = True)".

Balancing the CICIDS2017 Dataset. The CICIDS2017 dataset is unbalanced and consists of less abundant classes than others (see Table 1).

Table 1. The CICIDS2017 dataset is unbalanced

Class of network traffic	Frequency
Normal	**2272688**
Abnormal	556697

After balancing the CICIDS2017 dataset, the distribution of the normal and abnormal instances is presented in Table 2 below.

Table 2. The balanced CICIDS2017 dataset

Class of network traffic	Frequency
Normal	**250000**
Abnormal	582504

Feature Selection. We opted for feature selection to eliminate less relevant attributes that could falsify the classification results and cause huge learning delays. First, we removed attributes with constant or qua-constant values since they do not add value during the learning phase. Second, we removed duplicated features using the Python function duplicated(). Third and finally, we removed correlated attributes using the corr () function. Thus, the remaining relevant attributes are in Table 3.

Table 3. The retained attributes of the CICIDS2017 dataset

Selected features	Selected features	Selected features
Destination port	Flow IAT mean	FIN flag count
Flow duration	Flow IAT min	PSH flag count
Total Fwd packets	Fwd IAT min	ACK flag count
Total length of Fwd packets	Bwd IAT total	URG flag count
Fwd packet length max	Bwd IAT mean	Down/Up ratio
Fwd packet length min	Bwd IAT std	Init_Win_bytes_forward
Fwd packet length mean	Fwd PSH flags	Init_Win_bytes_backward
Bwd packet length max	Fwd Header length	Active mean
Bwd packet length min	Bwd Header length	Active std
Flow bytes/s	Bwd packets/s	Idle std
Flow packets/s	Min packet length	Label

4.2 Classification of Benign Traffic

After optimizing the CICIDS2017 dataset, we used the Decision Tree (DT) algorithm to classify the normal traffic and presented the results in Table 4. The DT was able to classify benign traffic with 99.9% of accuracy with a very low false alarm rate.

Table 4. Metrics to evaluate the use of Decision Tree algorithm

	TP	FP	TN	FN	Recall	Precision	Sensitivity	Specificity	F-Measure	Accuracy
Benign	74704	94	174779	135	99.8%	99.9%	99.8%	99.9%	99.8%	99.9%

5 Conclusion

In this work, we were able to propose a hybrid NIDS based on the use of both SNIDS and ADNIDS. The proposed model is capable of detecting any type of computer attack, including zero-day intrusions and exploits. It is based on the open-source Suricata system and machine learning techniques. The Decision Tree has performed well in classifying benign traffic while achieving 99.9% of accuracy. The work is not yet complete and we plan to perform the following tasks in our future publications:

- Using several Machine Learning algorithms and comparing them to choose the most reliable one to model the network baseline;
- Trying several datasets and comparing the results obtained;
- Testing the system in a real computer network.

References

1. Chiba, Z., Abghour, N., Moussaid, K., Omri, A.E., Rida, M.: Newest collaborative and hybrid network intrusion detection framework based on suricata and isolation forest algorithm. In: Proceedings of the 4th International Conference on Smart City Applications, pp. 1–11, October 2019
2. Maleh, Y., Ezzati, A., Qasmaoui, Y., Mbida, M.: A global hybrid intrusion detection system for wireless sensor networks. Procedia Comput. Sci. **52**, 1047–1052 (2015)
3. ElSayed, M.S., Le-Khac, N.A., Albahar, M.A., Jurcut, A.: A novel hybrid model for intrusion detection systems in SDNs based on CNN and a new regularization technique. J. Netw. Comput. Appl. **191**, 103160 (2021)
4. Maleh, Y., Sahid, A., Belaissaoui, M.: Optimized machine learning techniques for IoT 6LoWPAN cyber attacks detection. In: Ajith, A., et al. (eds.) Proceedings of the 12th International Conference on Soft Computing and Pattern Recognition (SoCPaR 2020), pp. 669–677. Springer International Publishing, Cham (2021). https://doi.org/10.1007/978-3-030-73689-7_64
5. Kasongo, S.M., Sun, Y.: A deep learning method with wrapper based feature extraction for wireless intrusion detection system. Comput. Secur. **92**, 101752 (2020)
6. Alazzam, H., Sharieh, A., Sabri, K.E.: A feature selection algorithm for intrusion detection system based on pigeon inspired optimizer. Expert Syst. Appl. **148**, 113249 (2020)
7. Ferrag, M.A., Maglaras, L., Ahmim, A., Derdour, M., Janicke, H.: Rdtids: rules and decision tree-based intrusion detection system for internet-of-things networks. Future Internet **12**(3), 44 (2020)
8. Kumar, V., Sinha, D., Das, A.K., Pandey, S.C., Goswami, R.T.: An integrated rule based intrusion detection system: analysis on UNSW-NB15 data set and the real time online dataset. Clust. Comput. **23**(2), 1397–1418 (2019). https://doi.org/10.1007/s10586-019-03008-x
9. Jin, S., Chung, J.G., Xu, Y.: Signature-Based Intrusion Detection System (IDS) for in-vehicle CAN bus network. In: 2021 IEEE International Symposium on Circuits and Systems (ISCAS), pp. 1–5. IEEE, May 2021
10. Cahyo, A.N., Sari, A.K., Riasetiawan, M.: Comparison of hybrid intrusion detection system. In: 2020 12th International Conference on Information Technology and Electrical Engineering (ICITEE), pp. 92–97. IEEE, October 2020

11. Gajjar, H., Malek, Z.: A survey of Intrusion Detection System (IDS) using OpenStack private cloud. In: 2020 Fourth World Conference on Smart Trends in Systems, Security and Sustainability (WorldS4), pp. 162–168. IEEE, July 2020
12. Bhatti, D.G., Virparia, P.V.: Soft computing-based intrusion detection system with reduced false positive rate. Des. Anal. Secur. Protoc. Commun. 109–139 (2020)
13. Samrin, R., Vasumathi, D.: Review on anomaly based network intrusion detection system. In: 2017 International Conference on Electrical, Electronics, Communication, Computer, and Optimization Techniques (ICEECCOT), pp. 141–147. IEEE, December 2017
14. Dalai, A.K., Jena, S.K.: Hybrid network intrusion detection systems: a decade's perspective. In: Lobiyal, D.K., Mohapatra, D.P., Nagar, A., Sahoo, M.N. (eds.) Proceedings of the International Conference on Signal, Networks, Computing, and Systems. LNEE, vol. 395, pp. 341–349. Springer, New Delhi (2017). https://doi.org/10.1007/978-81-322-3592-7_35
15. Ouiazzane, S., Barramou, F., Addou, M.: Towards a multi-agent based network intrusion detection system for a fleet of drones. Int. J. Adv. Comput. Sci. Appl. (IJACSA), **11**(10) (2020). https://doi.org/10.14569/IJACSA.2020.0111044
16. Al, S., Dener, M.: STL-HDL: a new hybrid network intrusion detection system for imbalanced dataset on big data environment. Comput. Secur. **110**, 102435 (2021)

Experimental Study on the Effectiveness of Machine Learning Methods in Web Intrusion Detection

Chakir Oumaima(✉), Rehaimi Abdeslam, Sadqi Yassine, and Farchane Abderrazek

Laboratory LIMATI, FPBM, USMS University, Beni Mellal, Morocco

Abstract. Web-based attacks have been more sophisticated and hard to detect in recent years, relying on just using traditional intrusion detection systems may not be enough. In this respect, intrusion detection and prevention systems using Machine learning methods have been important in recent literature. In this paper, we present an experimental study on the effectiveness of Machine learning methods in web intrusion detection. In which, We have investigated and compared four types of ML classifiers often used in the cybersecurity domain: KNN, Decision Tree, Multinomial, and Bernoulli Naive Bayes, SVM Linear, Sigmoid, and RBF. The experimental results based on ECML/PKDD 2007 and CSIC HTTP 2010 dataset showed that SVM RBF and Decision Tree classifiers achieved better performance in terms of Accuracy, Recall, Precision, F-value, FPR, and FNR than others.

Keywords: Web intrusion detection · Machine Learning (ML) · Web attacks · ECML/PKDD 2007 · CSIC HTTP 2010

1 Introduction

In recent years, web applications are becoming a lucrative and common target for attackers who purposefully exploit their vulnerabilities and launch attacks such as SQLI, XEE and XSS that could compromise the applications critical properties such as confidentiality, integrity or availability [1,2]. To prevent this kind of attacks, it is necessary to use a security system that can analyze, detect and prevent all malicious HTTP/HTTPS Requests/Responses, such as Web application firewalls (WAF) and web intrusion detection systems(IDS) [3].

There are two basic approaches to detect Web attacks, the signature-based and the anomaly-based, which are both inappropriate to ensure the security of web applications. Signature-based systems can effectively detect known attacks with a low false-positive rate, but they cannot detect unknown attacks such as Zero-day attacks. In contrast, anomaly-based systems perform better at detecting unknown attacks but have a higher false-positive rate.

According to recent researches, Machine Learning-based intrusion detection and prevention systems are the best solutions for securing web applications, the

Y. Maleh et al. (Eds.): ICI2C 2021, LNNS 357, pp. 486–494, 2022.
https://doi.org/10.1007/978-3-030-91738-8_44

experiments have been made proved that it can detect both known and unknown web attacks with a very low false-positive rate.

Generally, Machine learning is a technique that gives machines the ability to learn features from the dataset and then automatically assess whether an input x belongs to a benign or harmful class.

The main goal of this study is to evaluate the efficacity of machine learning algorithms such as KNN, Decision Tree (DT), SVM Linear (SVML), SVM RBF (SVMR), SVM Sigmoid (SVMS), Multinomial Naive Bayes (MNB) and Bernoulli Naive Bayes (BNB), in web intrusion detection using two open-source datasets generated for web attacks detection, ECML/PKDD 2007 and CSIC HTTP 2010.

This paper is organized as follows: Sect. 2 describes related works. The research methodology used is discussed in detail in Sect. 3. The experimental results are presented and discussed in Sect. 4. Section 5 is dedicated to the conclusion.

2 Related Work

In this paper, we focus on the experimental evaluation of the most popular ML algorithms used to mitigate web attacks. In particular, from 2011 onwards, web security techniques based on learning methods have received a remarkable increasing attention [4].

To the best of our knowledge, no previous work has been made to evaluate ML algorithms in the context of web attacks detection using both CSCI HTTP 2010 and ECML/PKDD 2007. These two datasets are the most used publicly available for the web attacks detection problems [5–10].

The authors in [11], evaluated the performance of three algorithms, namely J48, Adaboost and Naive Bayes. J48 exhibited a Detection Rate (True Positives) of 95.97% while NaiveBayes reached 88.89% and AdaBoost with 83.23%. Kozik et al., provided only the accuracy, which is not sufficient to prove the performance of any ML algorithm in web intrusion detection. In this paper, we consider other performance metrics like Precision, Recall, F1-score, FPR and FNR.

In [12] authors have demonstrated that SVM and Logistic regression performs well in detecting web attacks using CSIC HTTP 2010 dataset, they achieved an accuracy with a value of 95% and 97% respectively. To confirm the result obtained using CSIC 2010, in this work, we also tested the ML algorithms using ECML/PKDD 2007 dataset.

To detect XSS-based attacks, Khan et al. [13] evaluated the performance of four classifiers, KNN, Naive Bayes, J48, SVM using a customized dataset of 1924 instances including 1515 bengin and 409 of javascript codes as malicious instances. Their experimental results showed that the overall performances of all the four classifiers are good and efficient for XSS detection with a value of accuracy 97.14%, 95.06%, 99.22%, 94.55% respectively.

In [14], the authors have evaluated the performance of KNN, SVM with polynomial and Linear kernels and Random Forest classifiers in detecting XSS attacks. The experiments were carried out with the use of a customized training dataset including 2000 malicious and benign scripts, as well as the testing dataset containing 13,000 malicious and benign scripts. According to their results, these classifiers are efficient for detecting XSS-based attacks as they achieved a good performance in terms of accuracy, precision, Sensitivity and Specificity.

Kar et al. [15] presented SQLiGoT, a system for detecting injection attacks through SQL queries modeling such as token graph using SVM. The authors proposed alternative designs of the SVM classifier consisting of single and multiple SVMs. The experimental results of their work showed that SQLiGot achieved a good performance in terms of accuracy, precision, recall, F-value and FPR with a single SVM, Two SVM and Three SVM. All the three SVM classifiers achieved a value of 0.31% of FPR.

3 Methodology

This section discusses the research methodology used to evaluate the potential effectiveness of using machine learning classifiers to secure web applications. The parameters used in this experiment are presented in the following table (Table 1).

Table 1. The parameters used in this work.

Classifier	Parameters
KNN	n_neighbors = 5
SVM Linear	C = 1 kernel = 'linear' gamma = 20 decision_function_shape = 'ovr'
SVM RBF	kernel = 'rbf' random_state = 0 gamma = 20 decision_function_shape = 'ovr'
SVM Sigmoid	C = 1 kernel = 'sigmoid' gamma = 20 decision_function_shape = 'ovr'
DT	criterion = 'entropy' random_state = 0

3.1 Dataset Description

One of the biggest problems in web intrusion detection is the lack of a public dataset to evaluate the web intrusion detection systems. The current datasets are not appropriate for Web attack detection, because the majority of them do not target real Web applications. The ECML/PKDD 2007 and CSIC HTTP 2010 are two open-source datasets that were generated for web intrusion detection. In this work, we used two datasets ECML/PKDD 2007 and CSIC HTTP 2010, that are widely used publicly available datasets for the web attack detection problems.

ECML/PKDD 2007 is a dataset generated for the ECML/PKDD Discovery Challenge. This dataset contains 24504 and 10502 valid requests for training and testing, respectively, and 15110 malicious requests for testing [12,16,17]. Table 2 shows the size of the ECML/PKDD 2007 dataset used in this work.

Table 2. The ECML/PKDD 2007 dataset size used in this work.

	Train data	Test data
Normal	19649	8420
Attack	12510	2505

CSIC HTTP 2010 is a dataset that contains real HTTP requests generated to an e-commerce application. This dataset contains two types of requests, 36000 normal requests for training, 36000 normal for testing and more than 2500 requests classified as attacks for testing. It contains various types of web attacks such as SQL Injection, Cross-site scripting, CRLF Injection, and Buffer overflow [12,16,17].

The problem encountered during the use of this dataset is that it contains only valid requests for training. To solve this problem, we created our own dataset from the original CSIC 2010 dataset (Table 3).

Table 3. The CSIC HTTP 2010 dataset size used in this work.

Label	Train data	Test data
Normal	4000	1500
Attack	4000	1500

3.2 Training and Classification

As illustrated in Fig. 1, we started our experiment by extracting the data from ECML/PKKD 2007 and CSIC HTTP 2010 dataset using the urlparse library.

We mainly focus on extracting both legitimate and malicious data from GET, POST and PUT requests. We extracted the data that we would use for training and testing our algorithms after cleaning the data by removing all null and duplicated data. Since we are dealing with a binary classification problem and

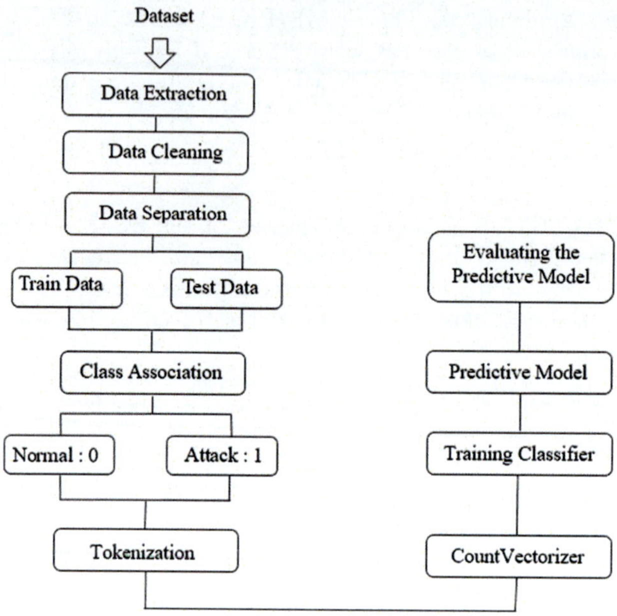

Fig. 1. Proposed methodology

we already know the classification of our data, we added a class for each sample of data. The class with a value of 0 for legitimate data and a value of 1 for malicious data. Then we created a tokenization algorithm to transform the input string data into a list of tokens through many steps, firstly we have decoded the input data using the same library urlparse to replace URL characters like "%xx" into its original character like '<', '>', ':' we have chosen to preserve those characters since several attacks, such as XSS attack, use them as payload construction. To keep the same format, we have converted each character to a lower case. Then, we split the data using special characters such as '&', 'white space', '=', '—'. Moving on, we have collected several keywords used to split data in [13], and our own keywords gathered by analyzing attack types and the content of the dataset to make our won bag of words that are used for tokenization. Since nearly all algorithms operate with numerical characteristics, we encoded each token to a vector using the CountVectorizer tool [14,18]. In order to categorize the data into malicious and legitimate, we supplied the training data to the classifiers. Finally, we evaluated the predictive models created by the classifiers with test data.

3.3 Evaluation Metrics

To evaluate the performance of our Machine Learning classifiers, we used the most common metrics, such as accuracy, recall, precision, F-value and False positive rate.

- Accuracy (A): Refers to the ratio between the number of correctly detected attacks and the total number of attacks.

$$A = \frac{TP + TN}{TP + TN + FP + FN} \quad (1)$$

- Recall (R): Calculate the number of correctly detected attacks.

$$R = \frac{TP}{TP + FN} \quad (2)$$

- Precision (P): This metric is used to calculate the number of correctly classified legitimate requests.

$$P = \frac{TP}{TP + FP} \quad (3)$$

- F-value (F1): The harmonic average of recall and precision.

$$F1 = 2.\frac{R.P}{R + P} \quad (4)$$

- False positive rate (FPR): Calculate the number of legitimate requests detected as an attack by the classifier.

$$FPR = \frac{FP}{FP + TN} \quad (5)$$

- False negative rate (FNR): Calculate the number of attacks detected as a legitimate requests by the classifier.

$$FNR = \frac{FN}{FN + TP} \quad (6)$$

3.4 Experiment Environment

The experiments presented in this paper were carried out on a PC equipped Intel(R) Core (TM) i5-3320M*64 based CPU running at 2.6 GHZ with 12G RAM installed and a 64-bit Windows 10 OS (Table 4).

Table 4. Tools used in this work.

Tool	Version
Ubuntu OS	18.04
Virtual box	6.1.12
Python	3.6.9
Scikit-learn	0.24.1
Jupyterlab	5.2.2

4 Results and Discussion

The experimental findings are presented and discussed in this section.

The performance evaluation of each classifier based on the ECML/PKDD 2007 and CSIC 2010 datasets is shown in Table 5 and Table 6.

The results shown in Table 5, show that SVM Linear, RBF, and DT classifiers achieved better performance compared to others with an accuracy value of 99.29%, 99.97% and 98.97% respectively. However, we can see that Bernoulli NB, KNN and SVM RBF gives better performance in terms of false-positive rate with a value of 0.01%, 0.09%, 0.03% respectively.

Table 5. The performance evaluation of each classifier based on the ECML/PKDD 2007 dataset.

Classifier	TP	TN	FP	FN	A	R	P	F1	FPR	FNR
MNB	2460	7862	558	45	94.48%	98.20%	81.51%	89.07%	6%	1%
BNB	2077	8419	1	428	96.07%	82.91%	99.95%	90.63%	0.01%	17%
KNN	1979	8412	8	526	95.11%	79%	99.59%	88.10%	0.09%	20%
SVML	2502	8346	74	3	99.29%	99.88%	97.12%	98.48%	0.879%	0.11%
SVMR	2505	8417	3	0	99.97%	100%	99.88%	99.93%	0.03%	0%
SVMS	1737	6972	1448	768	79.71%	69.34%	54.53%	61.04%	17%	30%
DT	2505	8308	112	0	98.97%	100%	95.72%	97.81%	1%	0%

In the intrusion detection domain, the false positive rate is very important as much as accuracy, because an intrusion detection system with a higher false-positive rate equates to poor precision, in this case, the defensive system will not just safeguard the web application instead it will affect its availability. According to [19], 1% increase in false-positive rate might cause an intrusion detection system to classify 1000 legitimate users incorrectly.

Compared to other classifiers, SVMR and DT classifiers performed better in terms of Recall and FNR with values of 100% and 0%, respectively. As a result, these classifiers are effective in detecting web attacks.

Table 6. The performance evaluation of each classifier based on the CSIC HTTP 2010 dataset.

Classifier	TP	TN	FP	FN	A	R	P	F1	FPR	FNR
MNB	1378	1436	64	122	93.8%	91.86%	95.56%	93.67%	4.2%	8.1%
BNB	1364	1429	71	136	93.1%	90.93%	95.05%	92.94%	4.7%	9%
KNN	1235	1424	76	265	88.63%	82.33%	94.20%	87.86%	5%	17%
SVML	1486	1406	94	14	96.40%	99.06%	94.05%	96.49%	6%	0.93%
SVMR	1500	851	649	0%	78.36%	100%	69.79%	82.20%	43%	0%
SVMS	962	721	779	538	56.1%	64.13%	55.25%	59.35%	51%	35%
DT	1500	1430	70	0	97.66%	100%	95.54%	97.71%	4%	0%

Likewise, the results based on the CSIC 2010 dataset demonstrate that SVML and DT classifiers outperformed other classifiers in terms of accuracy, recall, F1 and FNR.

As seen in Table 6, all of the classifiers had a greater false-positive rate. SVM Sigmoid achieved the highest false-positive rate with a value of 51%, while Multinomial NB achieved the lowest with a value of 4.2%.

5 Conclusion

This paper presents an experimental study on the effectiveness of ML algorithms in web intrusion detection using two open-source datasets: ECML/PKDD 2007 and CSIC HTTP 2010. KNN, Decision Tree, Multinomial and Bernoulli Naïve Bayes, SVM Linear, sigmoid and RBF were the four ML algorithms we tested.

The experimental results based on the ECML/PKDD 2007 dataset showed that SVM RBF achieved better performance in terms of Accuracy, Recall, Precision, F-value, FPR and FNR with values of 99.97%, 100%, 99.88%, 99.93%, 0.03% and 0% respectively. In contrast, the results based on CSIC HTTP 2010 dataset showed that the Decision Tree was the only classifier that achieved good performance with Accuracy, Recall, Precision, F-value, FPR and FNR values of 97.66%, 100%, 95.54%, 97.71%, 4%, 0% respectively.

References

1. Aburomman, A.A., Reaz, M.B.I.: A survey of intrusion detection systems based on ensemble and hybrid classifiers (2016). https://doi.org/10.1016/j.cose.2016.11.004
2. Valentini, G., Masulli, F.: Ensembles of Learning Machines. Springer, Heidelberg (2002). https://doi.org/10.1007/978-1-4419-9326-7
3. Zhang, M., Xu, B., Bai, S., Lu, S., Lin, Z.: A deep learning method to detect web attacks using a specially designed CNN. In: Liu, D., Xie, S., Li, Y., Zhao, D., El-Alfy, E.-S.M. (eds.) ICONIP 2017. LNCS, vol. 10638, pp. 828–836. Springer, Cham (2017). https://doi.org/10.1007/978-3-319-70139-4_84
4. Ghaffarian, S.M., Shahriari, H.R.: Software vulnerability analysis and discovery using machine-learning and data-mining techniques: a survey (2017). https://doi.org/10.1145/3092566
5. Vartouni, A.M., Teshnehlab, M., Kashi, S.S.: Leveraging deep neural networks for anomaly-based web application firewall. IET Inf. Secur. **13**, 352–361 (2019). https://doi.org/10.1049/iet-ifs.2018.5404
6. Tekerek, A., Gemci, C., Bay, O.F.: Design and implementation of a web-based intrusion prevention system: a new hybrid model. J. Faculty Eng. Arhitect. Gazi Univ. (2016). https://doi.org/10.17341/gummfd.63355
7. Torrano-Gimenez, C., Nguyen, H.T., Alvarez, G., Franke, K.: Combining expert knowledge with automatic feature extraction for reliable web attack detection (2015). https://doi.org/10.1002/sec.603
8. Tekerek, A.: A novel architecture for web-based attack detection using convolutional neural network. Comput. Secur. **100**, 102096 (2021). https://doi.org/10.1016/j.cose.2020.102096

9. Choraś, M., Kozik, R.: Machine learning techniques applied to detect cyber attacks on web applications. Logic J. IGPL (2015). https://doi.org/10.1093/jigpal/jzu038
10. Tian, Z., Luo, C., Qiu, J., Du, X., Guizani, M.: A distributed deep learning system for web attack detection on edge devices (2019). https://doi.org/10.1109/TII.2019.2938778
11. Kozik, R., Choraś, M., Renk, R., Hołubowicz, W.: A proposal of algorithm for web applications cyber attack detection. Comput. Inf. Syst. Ind. Manag. (2014). https://doi.org/10.1007/978-3-662-45237-0_61
12. Smitha, R., Hareesha, K.S., Kundapur, P.P.: A machine learning approach for web intrusion detection: MAMLS perspective. In: Wang, J., Reddy, G.R.M., Prasad, V.K., Reddy, V.S. (eds.) Soft Computing and Signal Processing. AISC, vol. 900, pp. 119–133. Springer, Singapore (2019). https://doi.org/10.1007/978-981-13-3600-3_12
13. Khan, N., Abdullah, J., Khan, A.S.: Defending malicious script attacks using machine learning classifiers. Wirel. Commun. Mob. Comput. (2017). https://doi.org/10.1155/2017/5360472
14. Mereani, F.A., Howe, J.M.: Detecting cross-site scripting attacks using machine learning. In: Hassanien, A.E., Tolba, M.F., Elhoseny, M., Mostafa, M. (eds.) AMLTA 2018. AISC, vol. 723, pp. 200–210. Springer, Cham (2018). https://doi.org/10.1007/978-3-319-74690-6_20
15. Kar, D., Panigrahi, S., Sundararajan, S.: SQLiGoT: detecting SQL injection attacks using graph of tokens and SVM. Comput. Secur. **60**, 206–225 (2016). https://doi.org/10.1016/j.cose.2016.04.005
16. GitLab. https://gitlab.fing.edu.uy/gsi/web-application-attacks-datasets. Accessed 27 June 2021
17. Betarte, G., Rodrigo, M., Pardo, A.: Web application attacks detection using machine learning techniques. IEEE (2018)
18. Scikit-learn/CountVectorizer. https://scikitlearn.org/stable/modules/generated/sklearn.feature_extraction.text.CountVectorizer.html. Accessed 27 June 2021
19. Pietraszek, T.: Using adaptive alert classification to reduce false positives in intrusion detection. In: Jonsson, E., Valdes, A., Almgren, M. (eds.) RAID 2004. LNCS, vol. 3224, pp. 102–124. Springer, Heidelberg (2004). https://doi.org/10.1007/978-3-540-30143-1_6

A Review of Indoor Positioning Systems (IPS) and Their Different Use-Cases

Bouchra Louafi[(✉)], Mustapha Hain, and Aissam Jadli

AICSE Laboratory, ENSAM Casablanca, University Hassan II, Casablanca, Morocco
bouchra.louafi-etu@etu.univh2c.ma

Abstract. As the world is actually moving faster like never before, the challenge of growing smart cities became the biggest concern of the 21st century, calling therefore different new information and communication technologies, human and environmental capital; allowing people to reach the best results ever with less energy and time, hence, it makes human's life easier. In this sense, Indoor positioning systems (IPS) are among these technologies which provides best services for humanity, specially in emergency cases, allowing them to easily move in indoor environments while the use of outdoor systems based on satellites is critical giving back wrong results. This paper presents a comparative analysis between existing Indoor Positioning Systems proposed in many research papers, based on different criteria.

Keywords: Indoor navigation · Indoor positioning systems · Tracking · Localization · ILONA · Inertial measurement unit · Received signal strength indicator · Radio frequency identification · Augmented reality

1 Introduction

For a target tracking (person, object) or a geographic guidance (department, library, block...) in an outdoor environment, there is a location based information system that receives data from GNSS (Global Navigation Satellite System) satellite, generating exact and accurate results.

However, GPS may cause large error when it's used indoor because of the complicated building blocks; that's why there is a persistent need for an alternative indoor positioning system, taking into consideration people with special needs and signal losses; these are called : IPS. They orient users inside buildings and can be used as tour guide in a museum, it also helps prevent the shortest path for blind people or lost children in malls. The concept of Indoor Navigation was first evoked in the robotic field, exactly in September 1973 when a NASA research paper [1] analyzes the behavior of a breadboard Mars robot while navigating inside the laboratory through a planning path, trying to avoid the obstacles faced. Since that time, many researches worked for automatic robots Indoor Navigation: they developed systems and sensors able to detect the location of mobile robots such as COPIS (1990) which aims to get omnidirectional

Y. Maleh et al. (Eds.): ICI2C 2021, LNNS 357, pp. 495–505, 2022.
https://doi.org/10.1007/978-3-030-91738-8_45

view information from one frame of a panoramic image; its configuration and its application is extensively detailed in the paper [2]. Since it is really an attractive research topic, IPSs have been improved parallely with the development of new technologies trying progressively to find limitations in order to enhance its performance and get adapted to people needs. Nowadays, we're talking about ILONA (Indoor LOcalization and NAvigation), emphasizing IPS systems that use integrated smartphones' sensors; this paper will focus on each one separately through the second part, making a general synthesis, and discussing which one will be more efficient and adapted to this century's needs in the third section.

2 State of the Art

The following section reviews previous experimental IPS' systems, based on recent researches and results, their pros and cons, and their added values.

2.1 RSSI Fingerprinting (Received Signal Strength Indicator)

A technology that allows distance's prediction between a receiver and a transmitter by calculating the power present in a signal being sent [3] following two phases:

* Data collection phase: When signals are transmitted, each location presents its distinctiveness (only one RSSI vector); this is the fingerprint. This phase corresponds to the'Offline Phase' in the scheme.
* Positioning phase: this phase is detailed in'localization server' and'online phase' on the scheme below. The matching algorithm used is KNN, executed on RSSI correlation coefficient.

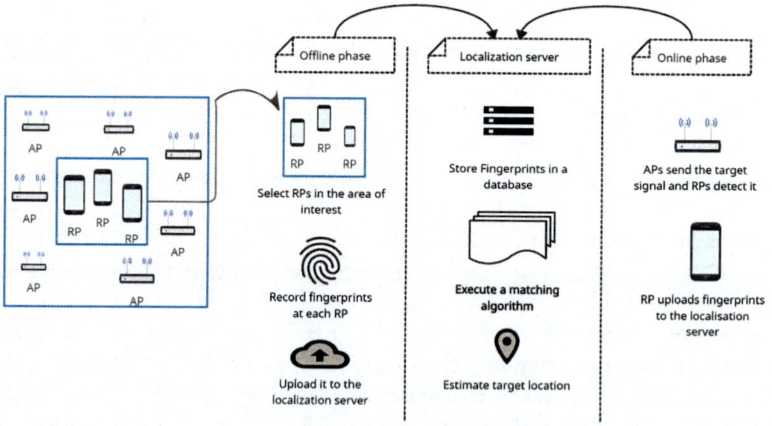

Fig. 1. RSSI fingerprinting process

Fingerprint positioning take building blocks into consideration. That's why it's seems suitable for indoor environments.But it still have some limitations, namely the time-consuming process of the radio map creation, that's why "Al Mamun, M.A. and Anaya, D.V. and Yuce, M.R" proposes **FaStER** [4]. That corrects errors generated from the Inertial Measurements Unit and due to the presence of human body, hence it ensures a high accuracy in terms of location estimation and RSSI acquisition. Experiment shows a high accuracy because the system has a FAst, and STable radio map, in addition to being Expendable, and Reliable.

2.2 Bluetooth Beacons

Specially designated for short-distance communication where the mobile device detects the radio frequency signals (identifiers) emitted by the Bluetooth beacons then calculates their distances, therefore the user's location is estimated.

The process starts by measuring and filtering RSSI for each Beacon. Then, based on the mean value of RSSI, the algorithm determines the distance from each beacon and estimates the current position of the user, then it assign it to the closest position stored in the database(the figure at right).

For best results, and shortest paths for directed graphs, the Dijkstra's algorithm calculate the navigational route [5].

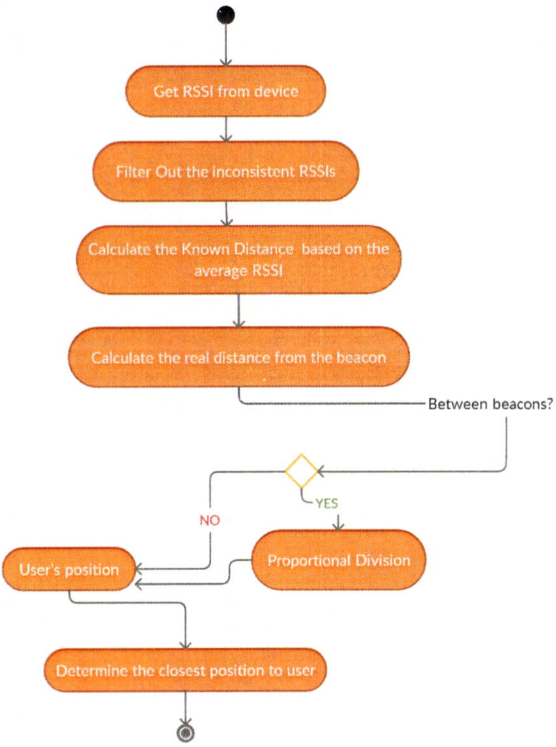

2.3 RFID (Radio Frequency Identification)

A technology based on small electronic devices used for Access control (e.g. Hotels' access), supply chain management... It includes an RFID tag and a reader, and a gateway to process the signal transmitted between them [6,7]; the reader uses an electromagnetic field by emitting radio waves that power the tag chip to retrieve data from it and identify the object to which it's attached. Tags can be active, passive, or semi-active. Active ones have embedded inbuilt battery in their circuitry so that they can operate in the UHF and SHF with a detection range of up to 100 m. Thus, they are useful for long-range localization and object tracking but they are expensive, doesn't give sub-meter accuracy, nor efficient tracking system. However, passive tags are cheapest and can detect targets in up to 10 m range [8], by measuring RSSI following the Friis equation:

$$\mathrm{P}_R = P_T \times G_T \times G_R \times \left(\frac{\lambda}{4\pi d}\right)^2$$

GT: transmitter's gain, GR: receiver's gain, λ: wave-length and d: the distance between the sender and the receiver [9]. For indoor Navigation, each tag is a location reference. When the user moves, the sensor attached to the reader records his movement and predicts his location [6].

RFID-based IPS starts early with **SpotON** and **LANDMARC** systems; the first investigates ad-hoc location sensing using an aggregation algorithm and the RSSI to estimate the distance between active tags [10]. The second is a location sensing prototype system serving for locating objects inside buildings [11].

A wise use of RFID is a blind navigation system proposed in [6], it includes: 1/ The track infrastructure: Each tag is embedded into a stone block along a pathway or at least at the junction of the footpath to store the ID and location; 2/ The navigation device: an embedded system with an RFID reader to get information from a tag and transfers it to the microprocessor unit; 3/ the navigation server: when the navigation starts, it receives request from the communication module and send a planned route as a feedback. In [7], some researches proposed a passive RFID based system able to track autonomous robots within a closed environment [7], to make the robot moving towards a target object without knowing its current location [9]. After experiments, it shows a good performance (error's average not exceeding 28cm within a navigation area of 7 m × 7 m, and a navigation duration between 176 s and 236 s).

2.4 NFC (Near Field Comm)

A combination of the contactless RFID and interconnection technologies that allows short-range wireless connection among mobile devices, PCs, and intelligent objects [12]; this is the case of: [13], which defines NFC as an innovative bidirectional short range wireless communication technology that can operate in card emulation, reader/writer, and peer-to-peer operating modes within few centimeters. The paper presents an NFC Internal enabled mobile phones, having an embedded navigation application allowing the user to indicate his destination by simply touching the NFC tag (that contains the URL of indoor map) in the entrance of the building in order to get the optimal route to it. As well as the user is navigating through the halls, he can touch other tags inside the building to make his mobile device receiving the coordinate data, hence, gathering his current position to get instructions for reorientation towards his destination [13].

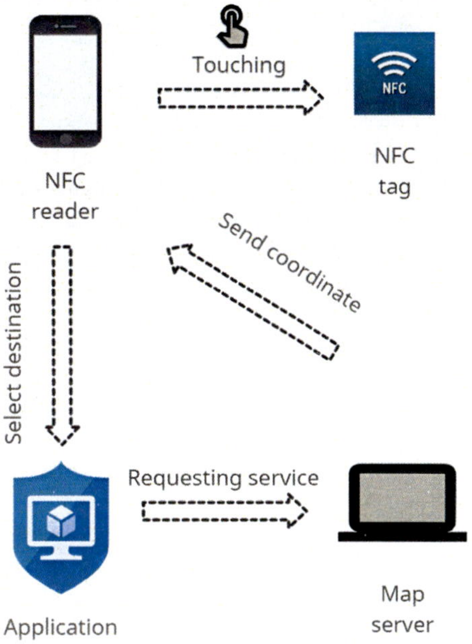

Many challenges are facilitated through NFC, such as touristic surfing [12] inside a museum or a historical monument, by touching the mobile phone to an NFC tag in a poster stuck at the entrance of the monument. Or for navigation inside sports hospitals as its complex layout is likely to lead to the loss of patients [14].

2.5 VLC (Visible Light Communication)

Relies on LIFI to transfer information extracted via image processing, between peers, through a line of sight (LOS), in order to track the device's position in the surroundings. VLC is considered one of the best IPSs, thanks to its ability to measure TOF, AOA or via vision analysis [15]; so, it gives back a good accuracy.

Since its appearance, several VLC-based IPS have been developed: **EPSILON** [16] proposed in 2014: it relies on LED beacons because they're more energy saving (2×) compared to CFL (less than 10% after 70,000 working hours). [16]. EPSILON measures the RSS from multiple bulbs and compute the distances to each one through an optical model; to estimate the receiver's location.

In the same year, 3 electrical engineers proposed **LUXAPOSE** [17] which relies on emerging LED equipped with on-off keying facilitating data transmission and getting many optical data with a single frame capture thanks to CMOS cameras.

LUXAPOSE gives decimeter accuracy and 3D positioning and orientation through the captured image uploaded in the cloud server [17].

To avoid the unnecessary high pulse rate and flickers created while using VLP systems that rely on intensity-based modulation, "Zhice Yang" and his colleagues substitute it by polarization-based modulation through a lightweight VLC: **PIXEL** (2018) [18] which reduces the computational complexity resulting from high-resolution images of CMOS image sensors. PIXEL is distinguished by:

- Having an accelerometer in client devices insuring accuracy by capturing only 2 location anchors instead of at least 3 ones required for other VLP.
- Making the goal possible not just for smartphones but also for wearables (google glasses) without incurring hardware modifications.

Since these VLP systems extract information via image processing, which requires high computational cost not suitable for embedded processors in smartphones, besides not supporting dimming, they are judged not enough for practice. That's why **high moving speed VLP** appeared in 2017 [15]; it aims to reduce the computational complexity for accurate LED signal based positioning and allows a high accuracy in the real-time without linking phone to some custom auxiliaries.

2.6 QR Codes

As existent IPS usually presents some deficiencies, researchers try always to make it better. In 2018, "Ying Zhuang" published a new idea to navigate, based on QR codes [19] because of its advantages to save money and to insure safety.

The authors have implemented two possible ways to get positions using the QR code: Either encoding the location information into QR codes as a unique URL ready to get scanned by the user to obtain the encoded information from the Net; OR providing decoders on users' devices to decode QR-code and extract information offline.[19] The whole process in shown in the Fig. 2.

Another use-case of QR-code based IPS appears in 2020 in [20] that consists of a hybrid system using both QR-codes and foot-mounted INS (Inertial Navigation System) for large-scale (even tens of kilometers) indoor patrol inspection. During a test in Wuhan University, the system proved its effectiveness, feasibility, robustness and resistance to environmental changes. Moreover, it's low-cost, user friendly, low computational complexity, and low power consumption for smartphones.

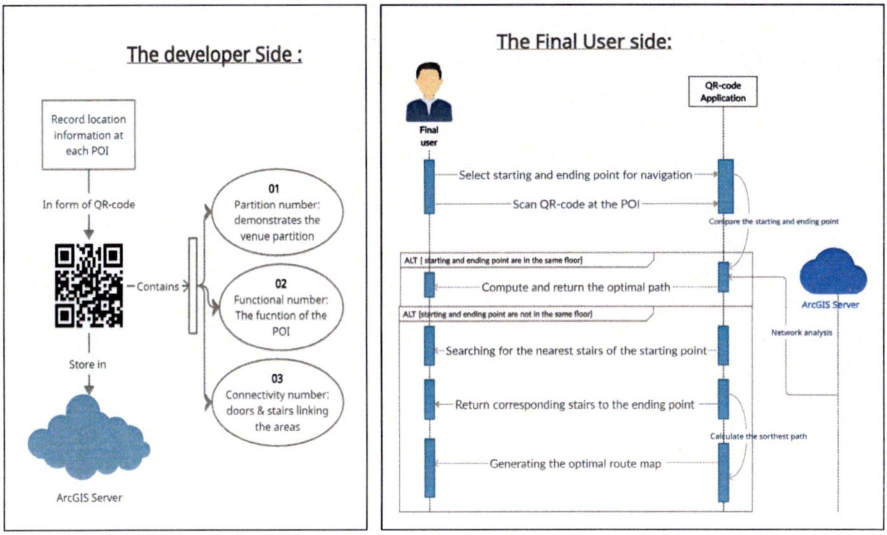

Fig. 2. The process of QR-Code based IPS

2.7 AR (Augmented Reality)

An interface between virtual data and the real world. AR indoor navigation refers to tracking the movement of a camera using computer vision. It's a part of SLAM (Simultaneous Localization and Mapping technology).

The idea was first introduced by "Michael Kalkusch" and other researchers in [21] (2002); through a hybrid system that guides the user by showing him directions and a floor map, using optical and inertial tracking of markers distributed in the environment from a single camera mounted on helmet worn by the user.

Later in 2020, "Gunwoo Lee and Hyun Kim" opted also for a hybrid solution: **HMIPS** [22], that uses both marker images and IMU data from smartphones for Pedestrian Tracking in Subway Stations; it comes as a solution for the low accuracy resulted in subway stations that are structured as an open space with high floor height, which involves variation of signal strength. It also solves the cost of building indoor maps, and the maintenance of infrastructure. While some researchers try to catch up different indoor environments, others focus on developing an useful IPS to insure autonomous mobility for people with special needs, this is the case of **ARIANNA** (pAth Recognition for Indoor Assisted Navigation with Augmented perception) [23]: a computer vision-based system that recognizes physical paths (painted lines on the floor, or tactile pavings). In 2020, the same team developed **ARIANNA+**, to eliminate the requirement of installing a physical line on the floor and substitute it by a virtual path, to guide the user through haptic, speech, and sound feedback [24]. After testing the ARIANNA+ in a laboratory of 32.4 m with a lot of separated turns, the team realized that the tracking performance is high on two conditions: 1/ setting the initial camera position perpendicularly to the floor, 2/ holding the smartphone horizontally while following the virtual path [24]. A year after (in 2021), an AR-based IPS that combines computer vision and QR code [25] was proposed to allow pedestrians to navigate through large indoor environments using only their smartphones, and to maximize the accuracy by tracking multiple phones' sensors continuously; relying on device orientation and image recognition.

3 Discussion

Since there are several IPS systems, we need to study the advantages and drawbacks of each one, based on some criteria.

Accuracy can be usually at the top of user's needs; therefore, we are starting here by comparing different IPS in the term of that main criteria (Accuracy): Fig. 3: The diagram shows a rapprochement between accuracy values, except for QR-codes which presents only 5% from laboratory experiences, due to the lack of a framework that can be able to locate POI, which means that this technology is too limited and the average of its errors is exceeding the right prediction values. On an other side, Augmented Reality has registered the biggest percentage among these systems. Specially by the use of HMIPS that shows 37.5% better positioning results compared to radio signals.

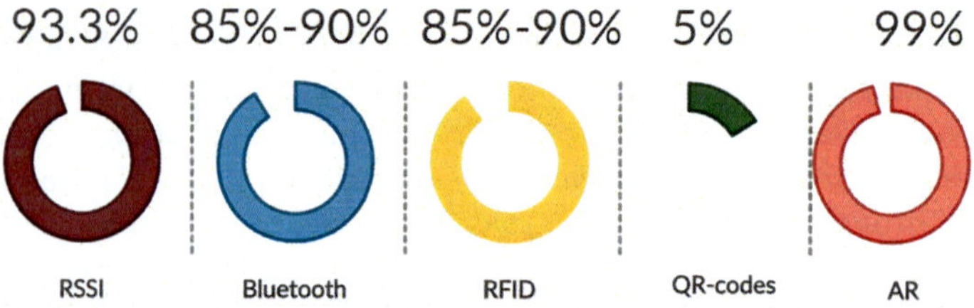

Fig. 3. Comparison of IPSs' accuracy

Since VLC proposed prototypes are significantly not similar in term of criteria, we are studying each one separately: the Fig. 4 below compare their accuracy.

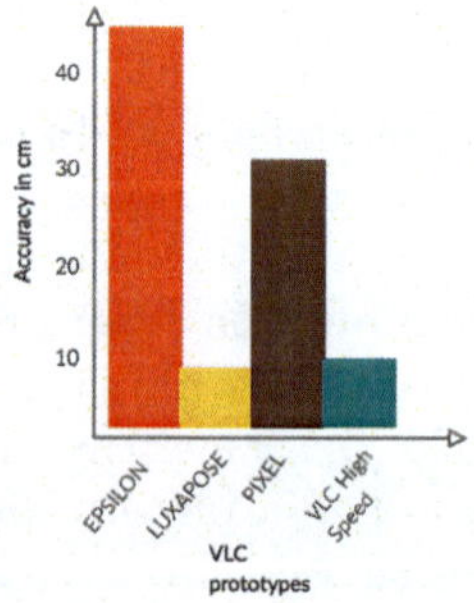

Fig. 4. Comparison of VLCs' accuracy

From the diagram, it's clearly observed that EPSILON won the first level of sub-meter accuracy, when compared to other VLC systems. In fact, experiment has shown that, for three typical offices it provides 0.4 m, 0.7 m and 0.8 m of accuracy for the 90th percentile. Even with a single light, the 90th percentile accuracy is 1.1 m [16]). The second place goes to PIXEL that provides: 1/Less than 10 cm of error in 90% of test cases, and a max positioning delay of 3 s; for devices with 60 × 80 pixel in camera resolution and a CPU frequency not high than 300 MHz. 2/Reliable communication over a long distance of 10 m using only an image/video capturing resolution of 120 × 160 pixel [18].

Table 1. Comparison of IPSs based on different criteria

IPS	Prototype	Cost	Flickering	Responding time
RSSI	[3,4]	High	No	large time delay
Bth	[5]	Very high	No	Not given
RFID	[1,6,11,15–17,21–23]	Very high	No	Between 176 s and 236 s
NFC	[12,14]	Medium	No	424 Kbit/s
VLC	EPSILON	High	Yes	2.1 s
	LUXAPOSE	Medium	Yes	9 s/33 MP images
	PIXEL	High	No	1200 ms/transmitting 17 bits
	VLC high speed	Medium	Mitigated	18 km/h (5 m/s)
AR	[5,7,9,24,25]	Medium	No	Real time

Besides all that, the choice of an IPS doesn't only rely on accuracy, but there still other criteria demonstrated in the Tables 1, 2. We can cite for example VLP High speed that gave a low accuracy, but it can reach a speed of up to 18 km/h, and computational time reduced to 22.7 ms for single-luminaire and to 35.7 ms for dual-luminaries. [15]; and it also supports dimming and avoids flickering. And of course, we can't neglect the importance of accessibility, an IPS that works by using the minimal phone configuration and doesn't require any other custom devices is better and cheaper than the one who isn't (The Table 2).

Table 2. Indoor positioning systems' accessibility

IPS	Prototype	Auxiliaries++	Mobile phone	Camera	Other access way
RSSI	[3,4]	APs+RP	Wifi integrated	Without	None
Bth	[5]	Beacons	Bth integrated	Without	None
RFID	[1,6,11,15–17,21–23]	RFID Tag	Any	Without	Phone/RFID-read
NFC	[12,14]	NFC tag	With NFC option	Not given	Phone/NFC-read
VLC	EPSILON	Light sensor	Galaxy S3	Custom receiver	None
	LUXAPOSE	No	Lumia 1020	33 MegaPixels	None
	PIXEL	A polarizer	Galaxy S2	8 MegaPixels	Google glasses
	High speed	No	OnePlus 2	5 megapixel	None
AR	[5,7,9,24,25]	No	Google PIXEL 2	60 fps	None

4 Conclusion and Perspectives

This paper cites the different indoor positioning systems actually used with an emphasis on the benefits, drawbacks and limitations of each one. It essentially aims to highlight the IPS which gives the best result in terms of quality, price, response time, distance, and other choice criteria previously mentioned. The results above are generated from a set of experiments done in laboratories with big surfaces and complicated turns to evaluate the efficiency of these IPS.

For future work, we are choosing AR based IPS since it presents the advantage of 99% of accuracy, in addition to a very high responding time (Real time), our work will focus on improving the way the AR do the matching between real location and its map information stored in a cloud server. Moreover, we are trying to include voice guidance and object detection to serve the visually-impaired people while moving inside buildings: The voice guidance will help them to know the location they are in, and to reorient them if they get lost; whereas the object detection will notify them if they touch something dangerous, or that can cause a harm. This will be implemented in the Immersal Mapper SDK.

References

1. Levine, M.D.: Scene analysis for a breadboard mars robot functioning in an indoor environment technical memorandum 33–645 scene analysis for a breadboard mars robot functioning in an indoor environment (1973)
2. Yagi, Y., Kawato, S.: Panorama scene analysis, 120(mm), 181–187
3. Yang, R., Zhang, H.: RSSI-Based Fingerprint Positioning System for Indoor Wireless Network. In: Li, K., Xue, Y., Cui, S., Niu, Q. (eds.) LSMS/ICSEE 2014. CCIS, vol. 463, pp. 313–319. Springer, Heidelberg (2014). https://doi.org/10.1007/978-3-662-45286-8_33
4. Al Mamun, M., Anaya, D., Yuce, M.: Faster: fast, stable, expendable and reliable radio map for indoor localization (2021). doi: 10.1109/INERTIAL51137.2021.9430464
5. Satan, A.: Bluetooth-based indoor navigation mobile system. In: Proceedings of the 2018 19th International Carpathian Control Conference, ICCC 2018, pp. 332–337. Institute of Electrical and Electronics Engineers Inc. (2018). doi: 10.1109/CarpathianCC.2018.8399651
6. Chumkamon, S., Tuvaphanthaphiphat, P., Keeratiwintakorn, P.: A blind navigation system using RFID for indoor environments. In: 5th International Conference on Electrical Engineering/ Electronics, Computer, Telecommunications and Information Technology, ECTI-CON 2008, vol. 2, pp. 765–768. IEEE (2008). doi: 10.1109/ECTICON.2008.4600543
7. Tesoriero, R., Tebar, R., Gallud, J.A., Lozano, M.D., Penichet, V.M.R.: Expert systems with applications improving location awareness in indoor spaces using RFID technology. Expert Syst. With Appl. **37**(1), 894–898 (2010). https://doi.org/10.1016/j.eswa.2009.05.062
8. Obeidat, H., Shuaieb, W., Obeidat, O., Abd-Alhameed, R.: A Review of Indoor Localization Techniques and Wireless Technologies. Wireless Personal Communications **119**(1), 289–327 (2021). https://doi.org/10.1007/s11277-021-08209-5

9. Teo, J.H., Loganathan, A., Goh, P., Ahmad, N.S.: Autonomous mobile robot navigation via rfid signal strength sensing, 9(8), 1140–1144 (2020). https://doi.org/10.18178/ijmerr.9.8.1140-1144
10. Choi, J.S., Lee, H., Elmasri, R., Engels, D.W.: Localization systems using passive UHF RFID. In: NCM 2009–5th International Joint Conference on INC, IMS, and IDC, pp. 1727–1732 (2009). doi: 10.1109/NCM.2009.198
11. Lionel, M.N., Liu, Y., Lau, Y.C., Patil, A.P.: LANDMARC: indoor location sensing using active RFID. Wirel. Netw. **10**(6), 701–710 (2004). https://doi.org/10.1023/B:WINE.0000044029.06344.dd
12. Borrego-Jaraba, F., Luque Ruiz, I., Gómez-Nieto, M.Á.: A NFC-based pervasive solution for city touristic surfing. Pers. Ubiquit. Comput. **15**(7), 731–742 (2011). https://doi.org/10.1007/s00779-010-0364-y
13. Ozdenizci, B., Ok, K., Coskun, V., Aydin, M.N.: Development of an indoor navigation system using NFC technology. In: Proceedings - 4th International Conference on Information and Computing, ICIC 2011, pp. 11–14 (2011). doi: 10.1109/ICIC.2011.53
14. Wang, X., Zhou, Z., Li, Y., Jin, Z.: An efficient sports navigation method based on nfc technology **3**, 850–855 (2020). https://doi.org/10.1109/ICITBS49701.2020.00187
15. Fang, J., et al.: High speed indoor navigation system based on visible light and mobile phone, 14(8) (2017). doi: 10.1109/JPHOT.2017.2687947
16. Li, L., Hu, P., Peng, C., Shen, G., Zhao, F.: Epsilon: a visible light based positioning system. In: Proceedings of the 11th USENIX Symposium on Networked Systems Design and Implementation, NSDI 2014, pp. 331–343 (2014)
17. Kuo, Y.S., Pannuto, P., Hsiao, K.J., Dutta, P., Arbor, A.: P447-Kuo. In: ACM MOBICOM, pp. 447–458 (2014)
18. Yang, Z., Wang, Z., Zhang, J., Huang, C.: Polarization-based visible light positioning 1233(LCD), 1–14 (2018). doi: 10.1109/TMC.2018.2838150
19. Zhuang, Y., Kang, Y., Huang, L., Fang, Z.: A geocoding framework for indoor navigation based on the QR code. In: Proceedings of 5th IEEE Conference on Ubiquitous Positioning, Indoor Navigation and Location-Based Services, UPINLBS 2018, pp. 1–4 (2018). doi: 10.1109/UPINLBS.2018.8559741
20. Liu, T., Kuang, J., Ge, W., Zhang, P., Niu, X.: A simple positioning system for large-scale indoor patrol inspection using foot-mounted INS, QR code control points, and smartphone. IEEE Sens. J. **21**(4), 4938–4948 (2021). https://doi.org/10.1109/JSEN.2020.3030934
21. Kr, M., Lidy, T., Knapp, M., Reitmay, G., Kaufmann, H., Sclunalstieg, D.: Structured visual markers for indoor pathfinding (2002)
22. Lee, G., Kim, H.: Applied sciences a hybrid marker-based indoor positioning system for pedestrian tracking in subway stations (2020). https://doi.org/10.3390/app10217421
23. Croce, D., Giarré, L., La Rosa, F.G., Montana, E., Tinnirello, I.: Enhancing tracking performance in a smartphone-based navigation system for visually impaired people. In: 24th Mediterranean Conference on Control and Automation, MED 2016, pp. 1355–1360 (2016). doi: 10.1109/MED.2016.7535871
24. Valvo, A.L., Croce, D., Garlisi, D., Giuliano, F., Tinnirello, I.: A navigation and augmented reality system for visually impaired people †, pp. 1–15 (2021)
25. Patil, M., Kadam, A., Choudhari, K., Prajapat, J., Kamble, J.: Augmented reality based indoor navigation system **6**(2), 383–385 (2021)

Towards Automatic Rule Conflict Detection in Snort

Michael Alicea(✉) and Izzat Alsmadi

Texas A&M University San Antonio, San Antonio, TX 78224, USA
malic01@jaguar.tamu.edu, Izzat.Alsmadi@tamusa.edu

Abstract. Firewalls have been one of the most important networks access control mechanisms for several decades. Their ability to filter traffic flowing through them has been instrumental to the task of keeping malicious activity from occurring on the network. This crucial role to the network means that properly configuring the firewall is of the utmost importance. Tools have been created to evaluate firewall rules to ensure that their policy is sound and does not contain errors that could either stop benign traffic or allow malicious traffic. In this scope, we proposed and developed a tool to automate the process of firewall role-conformance. Those tools can be used for Built-in-Testing (BiT) for firewalls to run frequently for high-level quality requirements such as: sanity checks, conformance, no roles-conflict or contradiction. Those can be important preventive measures to avoid serious vulnerabilities and attacks.

Keywords: Network firewalls · Access control · Firewall rules mis-configuration

1 Introduction

In recent years, the very idea of a network has been changing. With the introduction of the software defined network (SDN), networks have become programmable. This allows for networks to be much more dynamic and subject to change at a whim whereas in the past, a change to network infrastructure took careful planning and consideration. One of these considerations would be the effect it would have on network access controls.

Firewalls have served as the main method of access control to just about every network, usually sitting right at the entry point to filter anything going in or out. They started out simple, looking at packet headers and making decision based on them. These early firewalls only functioned at layer 2 and 3 in the OSI model. This only gave early firewalls the ability to filter traffic based on where packets came from, and where they are going. Other tools were also developed to function on other layers such as the application layer. Some of these additional tools were the first IDS and IPS, becoming an important part of network security often in addition to a firewall. Over time, it became increasingly apparent that businesses and organizations take cybersecurity more seriously. Tools

Y. Maleh et al. (Eds.): ICI2C 2021, LNNS 357, pp. 506–516, 2022.
https://doi.org/10.1007/978-3-030-91738-8_46

were developed to verify the integrity of a firewall's rule list. A mis-configured rule list can potentially nullify the firewalls own existence in the network. These tools allowed network admins to automatically check their rules for conflicts that could adversely affect the network's security.

This effort to make automatic policy correction tools was not equally divided amongst the different access control devices. While there are many tools to validate firewall rules, tools to examine the IDS and IPS never seemed to come to fruition the same way they did for firewalls. Tools like Snort give excellent detection and logging capabilities that can be of great importance to network admins and is still used today. As an important part of a network's security, it's rules should also be verified the same way the firewall would be. What we seek to do, is to develop a tool that can check IDS rules for rule conflicts the same way similar projects have done for firewalls.

Our main contributions in this paper can be summarized as the followings:

- We enumerate different categories of mis-configurations in firewall rules. We discuss the similarities and differences between the categories of misconfigurations. We noticed in the literature, while some of those categories are similar, they are interpreted differently.
- We propose and develop a tool that can automatically check for different categories of mis-configurations in firewall rules.
- We evaluated the tool using two different benchmark data-sets of rules and showed they value of building and using such tools.

2 Related Work

In the past, there have been similar tools created for firewalls. Most notably, [1] worked on a tool called 'Firewall Policy Advisor'. Back when this tool was created, proprietary solutions were the norm and they all had their own syntax to creating rules. To alleviate this issue, the tool allows you to input your rules into the program's GUI in a way that reflects how most modern GUI firewalls work today, with input fields for source and destination addresses and ports etc. A network administrator can input their rules and test for any mis-configurations between any given rules. The tool will take the first rule, then check it against every rule afterwards for any conflicts between any of the other rules. Each one of these conflicts is an anomaly caused by rule overlapping categorized as shadowing, correlation or redundancy. This process is repeated on every rule until every rule combination has been checked for conflicts. Firewall policy advisor is an older project that only worked on filtering firewalls, but it has inspired other tools since.

One of such tools inspired by Firewall Policy Advisor is the tool called PolicyVis created by [2]. PolicyVis is a firewall rule visualization tool. The primary creator of Firewall Policy Advisor also worked on this tool although not as the primary author this time. PolicyVis appears to be an extension of Firewall Policy Advisor. PolicyVis seeks to be more user friendly by using a 2D graph to display the firewall rules and any possible mis-configurations between rules. Every rule is shown as a rectangle on a graph that corresponds to their scope. Potential mis-configurations are displayed as two or more rectangles overlapping on the graph. This method of rule visualisation is quite helpful for a user that wants to quickly check for rule conflicts as a user can spot conflicting rules at a glance.

Another tool we have looked at is FIREMAN, a firewall modeling and analysis tool developed by [3]. FIREMAN takes the opposite approach as Firewall Policy Advisor does as it starts from the bottom of the rule list. It looks at the rules above it in the list to create a space of possible packets. If the rule being evaluated overlaps with this space, there is most likely some sort of mis-configuration. Instead of checking two rules at a time, it checks every other rule against one rule. This allows the tool to find when multiple rules together make up an anomaly. While multiple rules blocking another rule is multiple instances of anomalies, FIREMAN better groups the rules that conflict with any individual rule.

Some tools out there have the intended purpose of finding issues in networks with multiple access control components. The authors [4] focused their efforts on finding the same anomalies across multiple firewalls within the same network. All of the mis-configurations that can take place within a single firewall, can take place in a larger network with multiple firewalls. In this case, you can not just check a single firewall as if it were in a vacuum. That is what these authors have done. This paper focuses on inter-firewall anomalies that can occur throughout a network with multiple devices rather than just one firewall. The effects of inter-firewall anomalies only occur to the subnet behind the second firewall. In their tool, you input the policy of multiple devices along with how the network is structured with these devices. Anomaly detection works much the same as previous tools only using a rule list that combines the policies of multiple devices. This is a very thorough approach to detecting mis-configuration in the networks overall access control policy.

The work done in this subject did not end with the above tools. While some of the more recent work done in mis-configuration detection did not end in full fledged tools like the ones above, algorithms have been created to detect them. [5] created one of these algorithms using C++ which they called the PolicyAnomaly-Discovery Algorithm. Unlike the tools above, this algorithm exclusively works on shadowing and correlation anomalies as these are the two most dangerous anomalies to have in your policy. This algorithm works by building graphs that represents the firewall policy, with nodes being firewall rules and edges being the dependencies between rules. The algorithm works to remove the edges from the graph and creates new nodes if needed to fix mis-configurations in the rules. The algorithm was ran on both real and synthetic rule sets and was able to find the mis-configurations in these rule sets.

More work has also been done without resulting in a complete tool, at least not yet. [6] created an algorithm based on the Particle Swarm Optimization Algorithm (PSO). In short, PSO is a population based algorithm that flies particles representing potential solutions to a problem through a search space. In the author's implementation, each particle has seven attributes of each rule. These attributes are made up of the packet headers which are the rule id, protocol, source and destination addresses and ports, and the action associated with the rule. The algorithm then creates a second rule set. This second set starts blank and eventually gets filled in with an anomaly free ruleset as the algorithm is ran. The end result is a more optimized rule set than what was input to the system with no anomalies left over. This paper does not have any formal testing, however it is perhaps the most unique approach to the problem.

3 Access Control Mis-Configurations

One of the easiest ways to bypass access controls is if the access control rules were mis-configured in the first place. Mis-configurations can be just as bad as no access controls at all. It is important that your access controls work properly as intended. Whether it be conflicting rules or a wrongly implemented rule, these mis-configurations reduce security in your policies. Going forward, we will be looking into potential issues regarding the possible mis-configurations in both firewalls and IDS/IPS.

3.1 IDS/IPS Vs Firewalls

Firewalls and IDS/IPS actually suffer from a lot of the same types of misconfigurations. These two different types of access controls function differently so the conditions for a mis-configuration are a little different despite the end result being the same for any given mis-configuration. Firewalls work a little different to IDS/IPS as in most firewalls, the order in which rules are enforced is a little different. Firewalls typically check a list of rules one by one and enforce the first rule that matches the current traffic. Some newer ones allow setting a numbered priority and rules are checked in the order of their priority. In an IDS/IPS like Snort for example, rule enforcement is determined by the action given to a rule. Snort instead gives priority to actions in the order of activate, dynamic, pass, alert, log [7]. Two rules with the same action that also have the highest priority on a given piece of traffic will both trigger. For example, if two alert rules have an overlapping scope and a packet comes that matches both rules, Snort will trigger both alerts. On the other hand, if an alert and pass rule have an overlapping scope and traffic that matches both rules enters, the pass rule will be prioritized and the alert will not trigger.

3.2 Shadowing

The next concept of rule mis-configuration in access control is called shadowing. Shadowing is when a rule completely covers the scope of a second rule so that the second rule will never have a chance to trigger [8]. When this occurs, it is important to review the rules in question and consider either removing or editing one of them to fix the inconsistency in the rules. Shadowing can be harder to spot when you have a larger network. If a network contains multiple firewalls, a packet has to abide by all of the rules of each individual firewall it passes through [3].

When a packet crosses two firewalls with different actions, this is called an inter-firewall anomaly [9], which is conceptually similar to shadowing but over multiple firewalls. Different parts of the network could have their own access control policy and take a different action traffic that an earlier firewall [10]. Dealing with this issue is not a new concept. Back in 2005, [11] developed a program that could detect these issues not only within a single firewall, but across a network with multiple firewalls. Their program could even detect some of the issues that will be mentioned later in this paper.

The issue of shadowing is also present in Snort. In a firewall, you may have a rule that is intended to make an exception inside of another rule that either blocks or allows a range of ports/ip addresses. As discussed before, Snort sets priority of rules based on their associated action. This means that if you wanted to put an exception to an alert rule with a large scope by allowing a subset with a pass rule, this can be done easily as the pass will out prioritize the alert. The opposite is not true however. In Snort, you can not have an alert rule to make an exception in a pass rule with a large scope. The alert rule will always be shadowed due to the higher priority of pass. Snort will look at the two rules, determine that the pass is of higher importance and ignore the alert.

3.3 Correlation

Similar to shadowing, rule correlation may also potentially be an issue in your rules. Correlation is when a rule partially covers the scope of another rule and has a different action [2]. In simpler terms, correlation is essentially partial shadowing. Some of the desired range may have the wrong action due to two rules having an overlapping scope. This behavior may actually be desirable in some situations. For instance if a long port range is blocked, a rule that allows a subset of these ports may be made. Regardless of whether the correlation is intentional, checking that the correct rule has priority can ensure the exceptions work as intended.

In terms of an IDS/IPS, correlation also works essentially the same as shadowing. If two different rules have an overlapping scope and different actions, the rules covered in the overlap will be affected by the rule with the highest priority action. This can make the difference between correlation and shadowing. If a pass rule with a large scope exists at the same time as an alert rule with a subset of that scope, the alert will be shadowed. If an alert rule with a large scope exists with a pass rule with a subset of that scope, the rules will be correlated and the pass will create an exception to the alert rule for some traffic. Again, making an exception can be desirable but it is important to remember that an alert can not make an exception in a pass rule.

3.4 Redundancy

While having no impact on security, redundancy is still a mis-configuration. Redundancy is simply having two rule that perform the same action on the same packets [12]. Redundant rules are unnecessary and should be removed whenever found. Every rule added to a firewall or IDS makes inspecting packets slower as they have one more thing to check every packet. Having redundant rules therefore reduces the performance of the network with no benefit to it's security. Removing redundant rules is necessary to increase the performance of the network [13]. This performance gain is even greater in Snort. A firewall will look down the list until the first rule that applies, Snort on the other hand will look at all of the rules for one that applies. For example, if there are identical alert rules, both rules will trigger. Not only does this cause a slowdown in performance, but it also sends unneeded messages to the log file. If two alerts trigger on the same rule, both alerts will be printed to the log file. This can happen naturally, for instance a DOS attack that pings random ports pings a port with a different rule causing both the alert for that port and the DOS rule. In the case of two identical alert rules, this just wastes space and can cause confusion to the person reading the logs.

4 Snort Mis-Configuration Checker: Goals and Approaches

As a rule list gets longer, it gets more difficult to audit the rules. Especially when using community rules lists in combination with local rules, the audit process becomes longer and more error prone. A major change in the network infrastructure can trigger a need to re-review the rules on the network to ensure that the rules still function as originally intended. To help alleviate this issue, we have developed an automated tool that can detect many of these possible mis-configurations in a given Snort installation.

The goal of this tool is to find the mis-configurations that a machine can realistically find. This does not include items such as the flowbit mis-configuration because doing so would involve knowing exactly what a given rule is supposed to prevent along with knowing what the connection state is supposed to be. There are far to many assumptions to be made with some mis-configurations. This project will focus solely on the mis-configurations that are rule conflicts. This limits the tool to discovering the mis-configurations that occur when two rules overlap one another. This tool detects redundancy, correlation and shadowing. The tools from the literature review all have some sort of method of importing your rule list into the program and it runs independently. Our tool instead is run directly on the machine running Snort. The tool will read the Snort configuration and rules files directly to find rules. This program takes no user input and will automatically detect Snort's configuration file and rules.

4.1 Reading the Configuration File

To check the rules, the program must first find which rules are in use and where they are located. To do this, the configuration file, located at /etc./snort/snort.conf, is read for variables and rule files. In Snort, rule files are included in the configuration file with the keyword 'include' followed by a file path that ends with the extension '.rules'. To find all of the rules files, the configuration file is read and every line that starts with 'include' and ends with '.rules' is put into a list so that they can be iterated through later. Next, the IP variables within the configuration file need to be obtained. These are defined as 'IPVAR' followed by an address or another variable that holds an address. If a variable references another variable, the variable is recursively evaluated until the value is found. These variables are placed into a python dictionary to facilitate easy replacement inside of the rules later. Once all rules and variables have been extracted from the configuration file, the rules can be extracted.

4.2 Parsing Rules

The rules then have to be parsed from the rule files that are currently in use. The rules are parsed using a snort parser library [14] that extracts the rules into python dictionaries. Any rule that is commented, meaning the line begins with "#", is ignored. From there, the fields from these dictionaries are extracted to simpler rule object that holds the packet header fields and the snort option fields. The snort options, i.e. the part of the rule within the parenthesis that are not simple packet headers, are stored as a dictionary because options can be listed in any order unlike the headers. In python, two dictionaries with the same values in a different order are considered equivalent, so leaving options as dictionaries allows for easy evaluation regardless of order. For the purposes of evaluating rules, we also consider two sets of options to be equivalent if one is a subset of the other. This means that The rule objects are then put into a list that will be used to evaluate rules against one another.

5 Testing for Mis-Configurations

Once all the rules are correctly parsed, the program then begins checking for misconfigurations between any two rules. Using the list of rule objects, the rules are iterated through and each rule is tested against every subsequent rule in the list so that no two rules are tested twice. The mis-configuration checker looks for shadowing, correlation and redundancy in the rules based on the criteria from the mis-configurations section. Those results are printed to a CSV file which contains a table that displays every found mis-configuration between any two rules if one exists.

As a proof of concept, we created a set of rules that are designed to test every combination of conditions that could cause a mis-configuration. This list consists of rule pairs that deliberately cause the conditions that result in a mis-configuration. This rule set is not designed to be something that is realistically used but instead serves as a way to ensure that the classifications work correctly. The list of test cases the program can successfully classify the mis-configuration or lack thereof are as follows:

- Redundancy
  - Two identical rules
  - Overlapping rules with the same action
  - Identical rules with different options, one being a subset of the other
  - Overlapping rules with different options, one being a subset of the other
- Correlation
  - Source ports overlap, destination ports identical
  - Destination ports overlap, source ports identical
  - Source and destination ports overlap
  - Rule with lower priority action uses "any" keyword in ports
  - All test cases repeated with options where one is a subset of the other
- Shadowing
  - Rules are identical except for action
  - Shadowing rule's scope encompasses another rule
  - Shadowing rule uses "any" keyword for ports
- Non-Conflicting Rules
  - Different protocols
  - Rules with no overlap but the same protocol
  - Different IP addresses
  - Identical packet headers but with different options
  - Overlapping rules but with different option

Knowing that the tool can properly detect mis-configurations, we looked on Github for a set of rules to demonstrate its capabilities on a set of rules that could be used in a real environment. We decided on a repository that contained many rules for emerging threats that has been updated in March of 2021 [14]. These rules were generated based on real threats from this year. While representative of a set of rules an organization may actually have in place, it has one limitation in testing the tool. The rules are strictly for IDS mode, which means that there are only alert rules in place and therefore, no shadowing or correlation can occur. For a sample of rules to test, we chose three files that contained a total of 916 rules, which is a total of 419,070 unique rule pairs that need to be tested. The results show that of these unique rule pairs, 3,350 had at least some overlap triggering the redundancy message (Tables 1 and 2).

Table 1. Emerging threat rules results

Mis-Configurations		Test
Redundancy		3,350
No Conflict		834,790
	Protocol	151,616
	Address	217,272
	Port	26,970
	Options	438,932

6 Comparison

6.1 Past Authors

The majority of the tools can detect all of the different mis-configurations that we discussed. The one exception does not deal with redundancy. As stated earlier, redundancy has no impact on security and still results in a safe rule-set. The authors use rule-sets of various sizes in order to test their tools. Some of them use rather small rule-sets while one uses a large rule-set at 30,000. The exact numbers for each, if they were given, are in the following table

Table 2. Comparison chart

Author	Shadowing	Correlation	Redundancy	Rule-set size
[1]	×	×	×	Unspecified
[2]	×	×	×	45
[3]	×	×	×	10
[4]	×	×	×	70
[5]	×	×	×	30,000
[6]	×	×		Unspecified

6.2 Our Rule Mis-Configuration Detection Tool

Like the majority of the papers looked at, our tool can detect shadowing, correlation and redundancy. Some of them can even offer corrections, which ours does not do. Our tests were comprised of one synthetic test and a test with real rules. The synthetic test consisted of 16 test of rule pairs for 32 total rules. The real rule-set uses 916 rules of emerging threats for the year 2021. This real test however only tested for redundancy as a limitation of the rule-set not containing the drop action. The tool proved to have been able to find redundancy error and gave a reason as to why the rest of the rule pairs were not mis-configured.

Possibly the most unique point in our tool is how it approaches the task to begin with. Other tools require that the user inputs the rules into the program so that the programs stay as platform agnostic as possible. This is valuable, especially in the older papers when proprietary software splintered the syntax of how rules are written. Now that popular open source tools are available, we chose what is possibly the most popular and had our tool work on the device directly. Instead of reformatting and inputting the rules into our tool, we directly read the Snort configuration files without any input from the user.

7 Conclusion

The tool is a preliminary effort towards achieving automatic mis-configuration detection in Snort similar to that as the firewall tools from the results of our tests. We have created an application that is capable of taking in a list of Snort rules and detecting problems such as overlapping that can cause rules to conflict and poke holes into the access control policy. Networks in the modern age have become much more dynamic than in the past, thanks to software-defined networks. These programmable networks are responsible for the shift to dynamic networks. Any given change could require changing the access control policy on one or more devices, meaning that not only will control policy receive minor updates, but those updates also need to be verified. Our tool is a stepping stone into the automation of re-evaluating Snort rules.

8 Future Work

This tool is an initial effort to build a more comprehensive tool that can assess not only a Snort rule-set, but also the efficacy of the IDS itself. The tool so far can statically examine a rule-set in a given Snort installation and identify rule conflicts that can have adverse affects on the overall policy. Not all errors in rule enforcement comes from the set of rules alone. Sometimes, other errors in configuration of either Snort or the network can cause rules to not be enforced as expected. In the next phase of this project, we will create a routine in which the tool will take the existing rule list and send traffic with the intent of violating the rules. The Snort logs will be evaluated afterwards to ensure that the rules were correctly enforced. This will ideally allow for an admin to determine if an issue with their Snort installation is due to their rules, or some other configuration issue.

References

1. Al-Shaer, E.S., Hamed, H.H.: Firewall Policy Advisor for Anomaly Detection, Rules Editing and Translation (2000)
2. Tran, T., Al-Shaer, E.S., Boutaba, R.: Policyvis: Firewall security policy visualization and inspection. In: LISA, vol. 7, pp. 1–16 (2007)
3. Yuan, L., Chen, H., Mai, J., Chuah, C.-N., Su, Z., Mohapatra, P.: Fireman: a toolkit for firewall modeling and analysis. In: 2006 IEEE Symposium on Security and Privacy (S&P 2006), pp. 15-pp. IEEE (2006)
4. Alfaro, J.G., Boulahia-Cuppens, N., Cuppens, F.: Complete analysis of configuration rules to guarantee reliable network security policies. Int. J. Inf. Secur. **7**(2), 103–122 (2008)
5. Gobjuka, H., Ahmat, K.A.: Fast and scalable method for resolving anomalies in firewall policies. In: 2011 IEEE Conference on Computer Communications Workshops (INFOCOM WKSHPS), pp. 828–833. IEEE (2011)
6. Arthur, J.K., Kwadwo, E., Doh, R.F., Mantey, E.A.: Firewall rule anomaly detection and resolution using particle swarm optimization algorithm. Int. J. Comput. Appl. **975**, 8887 (2019)
7. Tran, T.: Misconfiguration analysis of network access control policies. Ph.D. dissertation, University of Waterloo (2008)

8. Oliveira, R.M., Lee, S., Kim, H.S.: Automatic detection of firewall misconfigurations using firewall and network routing policies. In: IEEE DSN Workshop on Proactive Failure Avoidance, Recovery, and Maintenance (PFARM). Citeseer (2009)
9. Taibah, M., Al-Shaer, E., Hamed, H.: Dynamic response in distributed firewall systems. DePaul CTI Technical Report, CTI-TR-05-002, Tech. Rep. (2004)
10. Ohri, S.: Distributed firewalls (2008)
11. Al-Shaer, E., Hamed, H., Boutaba, R., Hasan, M.: Conflict classification and analysis of distributed firewall policies. IEEE J. Sel. Areas Commun. **23**(10), 2069–2084 (2005)
12. Al-Shaer, E., Hamed, H.: Design and implementation of firewall policy advisor tools. DePaul University, CTI. Tech, Rep (2002)
13. Kadam, P., Bhusari, V.: Review on redundancy removal of rules for optimizing firewall. Int. J. Res. Eng. Technol. (2014)
14. Codecat007, l.: snort-rules (2021). https://github.com/codecat007/snort-rules

A Study of Connection Speeds in Transport Layer Security Version 1.3 (TLS 1.3) Using Different Handshake Modes

Mohamed Elnour Abdelhafez[1(✉)], Sureswaran Ramadass[1], Mustafa Abdelwahab[2], and B. G. D. A. Madhusanka[1]

[1] School of Science and Engineering, Malaysia University of Science and Technology (MUST), Petaling Jaya, Malaysia
mohamed.elnour@phd.must.edu.my, sures@nav6.org, bgmad@ou.ac.lk

[2] Internet Society Sudan Chapter, Khartoum, Sudan

Abstract. Reliable communication networks play a significant role in guaranteeing the optimum operation of several applications on the Internet, for instance, IoT and real-time applications, which are required to provide dynamic data flow over the network. A protocol with a faster connection and less latency is highly recommended to enhance the quality of service and satisfy the user experience in such applications. This paper examines the proposed enhancement in TLS version 1.3 handshakes, which is designed to lead to less latency by utilizing various operation modes in this protocol. Also, it evaluates the impact of 0-RTT operation mode on the protocol performance overall in the real setup and use.

Keywords: TLS · Latency · Handshake · Quality of service

1 Introduction

The demand to send confidential data over the Internet started in the 90s when the NETSCAPE company designed the first published protocol in 1995 known as Secure Socket Layer (SSL) version 2 [1]. Since the time, several versions have been published with new features and enhancements. In today's businesses, encryption has become essential to guarantee confidential connections over the public Internet for users to access resources and services; therefore, encryption on the Internet has risen dramatically to adhere to this demand.

According to the Google Transparency Report [2], encrypted web traffic has steadily increased on the Internet from around 50% in 2014 to between 80% and 90% in 2020. In addition, new security modules such as zero-trust will probably increase the corporate networks that utilize encryption in the internal data and public data. Conversely, the most recent version of (SSL) protocol, now known as Transport Layer Security version 3 (TLS 1.3), comes with significant enhancement in the protocol structure, security, and performance. All legacy algorithms have been removed and replaced with the Authenticated Encryption with Associated Data (AEAD) algorithms.

Y. Maleh et al. (Eds.): ICI2C 2021, LNNS 357, pp. 517–525, 2022.
https://doi.org/10.1007/978-3-030-91738-8_47

Also, a connection with zero round trip mode is introduced to reduce the protocol latency, a new feature that exists only in version 1.3. [9]

The protocols that utilize TLS can generate several types of traffic, for instance, data-based traffic like email or audio and video, which demands a real-time response. Thus, an effective method must be implemented to control the traffic and deliver adequate Quality of Service.

2 Latency and Protocol Performance

The traditional TCP/IP-based traffic can adjust according to changes in delay or throughput across a network and meet the application's needs; these applications either utilize TCP or UDP to transfer packets. It can use as much available capacity in UDP, the application can generate, while an end-to-end acceptable data rate in TCP is required. This type of traffic is classified as elastic traffic generated by various applications such as electronic mail using Simple Mail Transfer Protocol (SMTP) and File Transport Protocol (FTP). The inelastic traffic is sensitive to delay, and throughput as the nature of the traffic is real-time such as streaming protocols.

The basic requirements to indicate the state of the QoS in a network consist of Throughput, which relies on the adopted networking technology, packet loss impact varies based on the application type, latency (delay), and jitter (delay variation) are significantly critical factors for real-time applications.

The demand for a network with low latency is driven by the need to save time and resources, especially in a slow network such as a rural mobile network where it is not sufficient as it is desire. Also, there is an economic motivation to reduce latency as far as possible. For instance, in 2006, Amazon conducted a study by delaying their webpages delivery by 100 milliseconds (ms) resulted in a 1% reduction in revenue [4, 10], similar findings observed by Google and Akami in a study conducted in 2017 showing that a slight increase in latency increase the probability of clients to abandoning their connection [5].

3 Latency in TLS 1.3

A noticeable enhancement has occurred in the handshake of this protocol by introducing one round-trip time (1-RTT) mode to make the encrypted connection faster compared to the previous versions, which will positively impact the QoS and raise the application performance overall.

Two different operation modes are introduced to resume the connection in TLS 1.3, 1-RTT mode as full resumption handshake, and 0-RTT mode is also introduced to make the network overhead low by allowing the client to send a ClientHello (CH) message together with an encrypted application data which are known as early data or data in the first flight. This is considered a tremendous change in the protocol structure to adhere to the increasing demand for low latency protocol. Figure 1 and Table 1 illustrate the differences between the modes in regards to round-trips.

Table 1. Round-Trips and operation mode

Handshake	TLS 1.3 + TCP
Initial handshake	2-RTT
Full resumption handshake	2-RTT
0-RTT resumption handshake	1-RTT

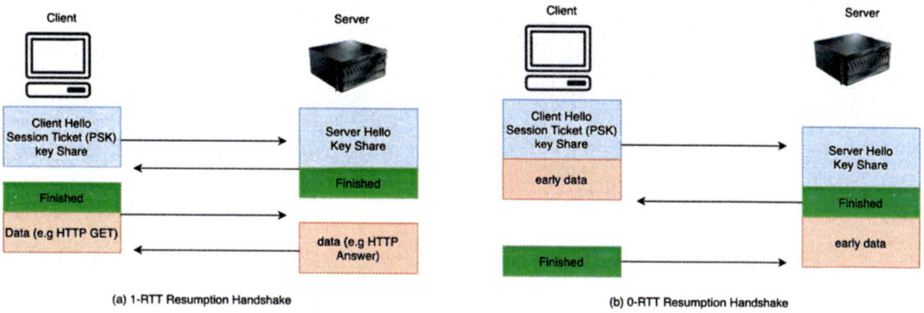

Fig. 1. Resumption HS modes and Round-Trip

On the other hand, a significant enhancement in the protocol encryption suite structure was introduced by eliminating the legacy and adopting modern algorithms. EdDSA (Edwards-curve Digital Signature Algorithm), ECDSA (Elliptic Curve Digital Signature Algorithm), and RSA (Rivest–Shamir–Adleman) are used as signature algorithms, while (elliptic curve) Diffie–Hellman ephemeral ((EC) DHE), PSK, and PSK with (EC) DHE) are used as key exchange algorithms [13].

To guarantee faster encryption and decryption, symmetric algorithms are utilized to encrypt the application data that uses TLS 1.3 [14]. The suite structure is shown in Fig. 2.

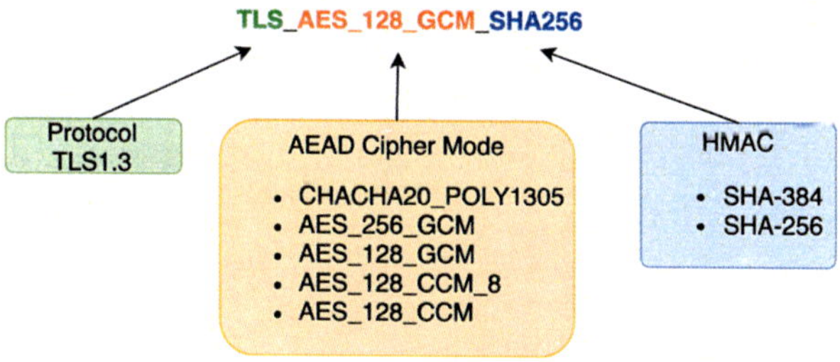

Fig. 2. Encryption suite structure

4 Experimental Testbed

The testbed builds in a real enterprise environment consist of 2 Virtual Machines (VM) located in a different geographical area connected through the Internet crossing four operators over IPv4. Figure 3 describes the connectivity used in the testbed.

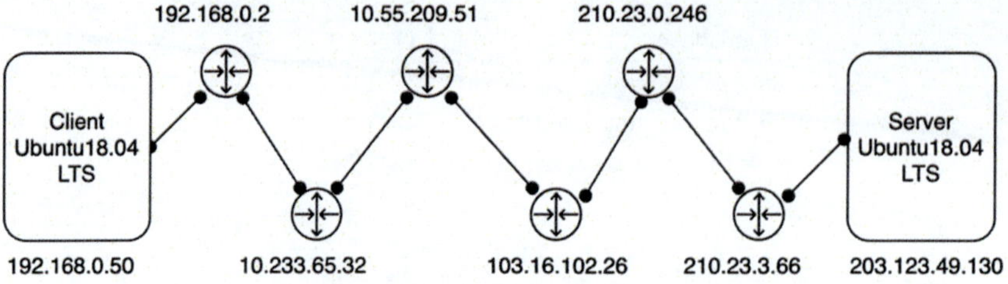

Fig. 3. Internet connection

A network performance measurement tool (iperf3) [6] showed the connectivity bandwidth between the server and the client is 49.1 Mbits/s and 48 Mbits/s for the sender and the receiver, respectively. Figure 4 shows the tool results.

```
[root@remote-access:~# iperf3 -c 203.123.49.130
Connecting to host 203.123.49.130, port 5201
[  4] local 192.168.0.50 port 47102 connected to 203.123.49.130 port 5201
[ ID] Interval           Transfer     Bandwidth       Retr  Cwnd
[  4]   0.00-1.00   sec  3.63 MBytes  30.4 Mbits/sec    1   90.0 KBytes
[  4]   1.00-2.00   sec  4.82 MBytes  40.4 Mbits/sec    0    122 KBytes
[  4]   2.00-3.00   sec  6.06 MBytes  50.8 Mbits/sec    0    155 KBytes
[  4]   3.00-4.00   sec  6.37 MBytes  53.4 Mbits/sec    0    183 KBytes
[  4]   4.00-5.00   sec  6.30 MBytes  52.9 Mbits/sec    0    207 KBytes
[  4]   5.00-6.00   sec  6.30 MBytes  52.9 Mbits/sec    0    229 KBytes
[  4]   6.00-7.00   sec  6.30 MBytes  52.9 Mbits/sec    0    249 KBytes
[  4]   7.00-8.00   sec  6.24 MBytes  52.4 Mbits/sec    0    267 KBytes
[  4]   8.00-9.00   sec  6.30 MBytes  52.9 Mbits/sec    0    284 KBytes
[  4]   9.00-10.00  sec  6.24 MBytes  52.4 Mbits/sec    0    301 KBytes
- - - - - - - - - - - - - - - - - - - - - - - - -
[ ID] Interval           Transfer     Bandwidth       Retr
[  4]   0.00-10.00  sec  58.6 MBytes  49.1 Mbits/sec    1             sender
[  4]   0.00-10.00  sec  57.2 MBytes  48.0 Mbits/sec                  receiver

iperf Done.
```

Fig. 4. Bandwidth measurement

The overall network latency average between the server and the client is 13.9 milliseconds (ms). Figure 5 illustrates the graph for 100 rounds of ICMP echo request and echo replay performed from the client host; the result showed no packet loss.

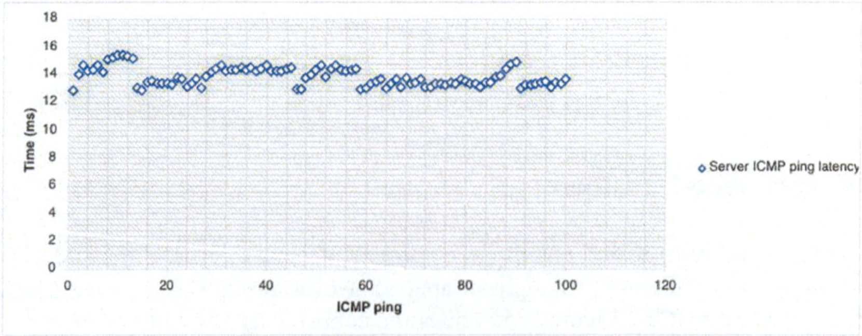

Fig. 5. Network latency

TLS implementation open-source software (OpenSSL) was used to lunch HTTP GET function in an HTTP server requested over TLS from the client-side, the request to be handled by the hosting server software (Nginx) that utilize the same (OpenSSL) version that has been used in the client-side. Two virtual host configurations linked to different testing domain names (https://13.pi.sd, https://131.pi.sd), operate 1-RTT and 0-RTT modes, respectively. The following Table 2 and 3 show the software versions used in the testbed, and server/client specifications respectively.

Table 2. Testbed components and software versions

Virtual Machine (VM)	Operating System	Network	CPU/RAM/SWAP	Application TLS software
Client VM	Ubuntu 18.04 LTS, Linux 4.15.0–147-generic	1 GB full duplex GigabitEthernet	Intel(R) Xeon(R) CPU D-1518 @ 2.20 GHz x2Core/2 Gb RAM/ 2 Gb SWAP	OpenSSL 1.1.1 g tcpdum
Server VM	Ubuntu 18.04 LTS, Linux 4.15.0–147-generic	1 GB full duplex GigabitEthernet	Intel(R) Xeon(R) CPU E5420 @ 2.50 GHz x2Core/2 Gb RAM/ 2 Gb SWAP	OpenSSL 1.1.1 gnginx/1.19.10

Table 3. Server and client specifications

Virtual Machine (VM)	Socket buffer size (read) (min, default, max) bytes	Socket buffer size (write) (min, default, max) bytes	CPU % idle	file system
Client VM	4096 131072 6291456	4096 16384 4194304	95.73%	ext4
Server VM	4096 131072 6291456	4096 16384 4194304	98.51%	ext4

Testing Domain 1. This domain (131.pi.sd) is designed to operate with 1-RTT mode only with customized session timeout by 3 seconds, which means the resumption HS can be performed immediately after 5secounds after the data follow stopped in the session.

Testing Domain 2. This domain (13.pi.sd) is designed to operate with 0-RTT mode, session timeout set to 5 s where the resumption HS can be performed after the data follow stopped in the session.

Both domains use digital certificates issued by an authorized certificate authority (Let's Encrypt certificate authority) to avoid using self-signed certificates. That is to have a testbed typically as a system in a real environment.

5 Methodology and Implementation

The usefulness of this methodology is to examine the impact of the latency enhancement on the protocol performance practicality in a real environment while the 0-RTT mode is enabled.

The resumption HS operation mode will be the only interchanging factor during the experiment's phases; all other factors remain without changes to focuses the study on this part of the protocol. The experiments consist of two scenarios explained as the following.

Scenario 1. The server uses 1-RTT mode with timeout set to 3 s; a script in the client-side designed to perform an initial handshake with an additional option to save the Ticket Session (TS) from being used later during the resumption HS phase (cost 1-RTT), the time interval between the initial HS and the resumption HS will be 5 s to guarantee the session is dead before the resumption is executed. The tcpdump packet capture tool will be running in the background to record and save the communication between the client and the server. The script will repeat the same scenario starting from the fresh initial HS up to 100 rounds. The phases are shown in Fig. 6.

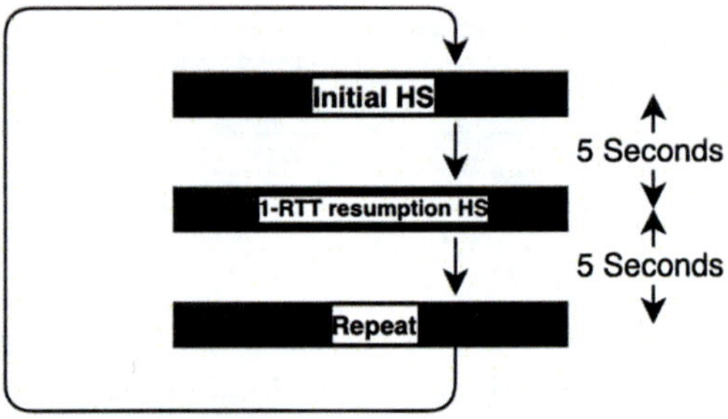

Fig. 6. 1-RTT test phases

Scenario 2. In this scenario, a 0-RTT mode will be enabled in the server with session timeout set to 3 s, a script in the client-side designed to perform an initial HS phase (cost 1-RTT), followed by a resumption HS with early data using the Session Ticket key copied from the initial HS, the time interval between the two HSs will be 5 s to guarantee the session is timeout before the resumption HS performed by the client. These test phases will be repeated 100 times while a tcpdump tool is in the background in the client capturing the communication packets. Figure 7 showing scenario 2 phases.

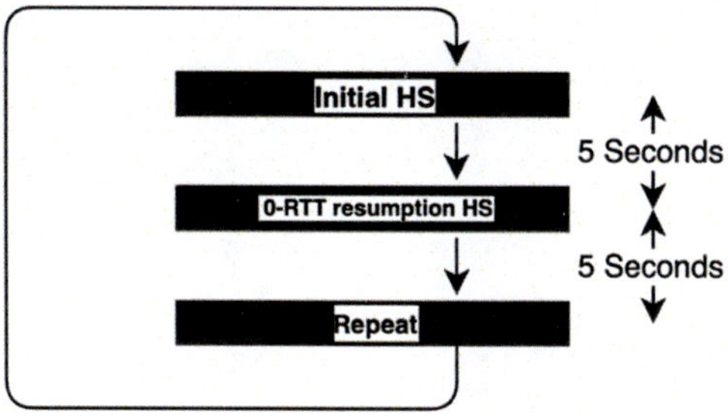

Fig. 7. 0-RTT test phases

Both scenarios are executed consecutively (not simultaneously) to avoid resource overloading in the testbed components and ensure the changes happen in the resumption HS modes only. To obtain the required time to resume the TLS connection (connection time), we calculated the time interval between the ClientHello message in the resumption HS until the data is received by the client from the server. The calculation uses the timestamps in the captured messages where the tcpdump tool is employed for this purpose.

The calculated connection time per round will be graphed to compare the needed time in each operation mode during the resumption HS. Figure 8 showing the graph for Latency between the server and the client, the required connection time in 0-RTT resumption HS, and the required connection time with 1-RTT resumption HS.

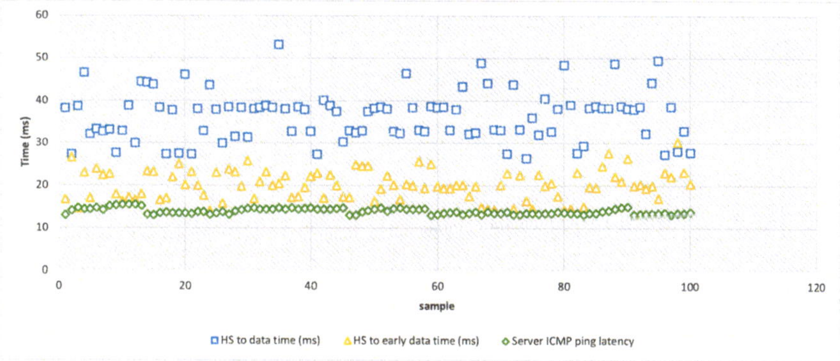

Fig. 8. Delay in HS, ICMP

Based on the average values, the 0-RTT mode reduced the required time for connection significantly by 44.7% (shown in Fig. 9), which will be reflected in the protocol performance overall and make it a suitable operation mode for real-time applications. However, the calculated reduction on time is half of the required time to perform the full resumption HS with miner delay on the processing of the messages [11, 7, 12].

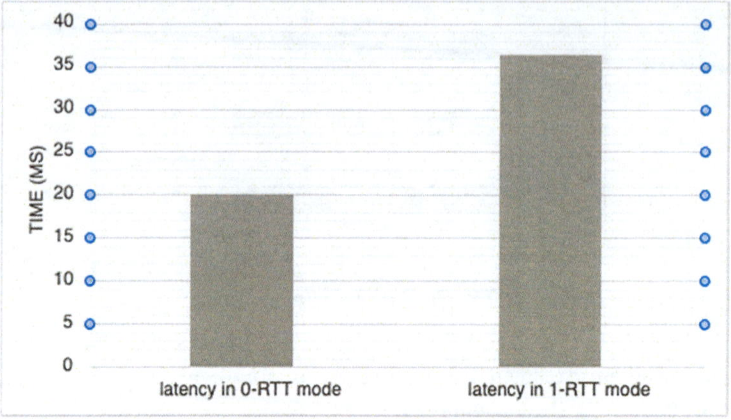

Fig. 9. Average latency in 0-RTT and 1-RTT mode

Some major security issues are raised with the use of the 0-RTT mode. For instance, the early data message is not a forward secret, which means if the private keys are leaked sometime before the HS, the attacker can decrypt the early data message from the recorded network traffic [7]. Another issue is when the attacker recorded the ClientHello message during the resumption HS and replay it again [8, 9]. However, since the message is encrypted with a correct key, the server could decrypt it and forward it to the application level without errors or interruption.

There are particular solutions has proposed, for instance, using Puncture Pseudorandom Function (PPRF) to allow the 0-RTT message to be decrypted once at the server-side [14, 15], Single-Use Ticket, Client-Hello Recording, Freshness checks, Application Profile, and Separate API [9].

6 Conclusion

The significant enhancements in the protocol, specifically the adoption of the 0-RTT operation mode, make the protocol suitable for real-time applications such as audio, video, self-driving cars, and some IoT application where time is a sensitive factor, testing the protocol in a real environment showed that 0-RTT reduced the connection time by 44.7%, which means the impact of 0-RTT operation mode on the protocol enhances the performance compared to the full resumption handshake. However, countermeasures and procedures need to be employed to eliminate the potential security threats due to the usage of the 0-RTT mode.

Acknowledgments. We would like to thank NLTVCEDU SDN BHD - Malaysia for providing the necessary resources needed in the testbed.

References

1. Hickman, K., Elgamal, T.: The SSL protocol (1995)
2. Farley, E., et al.: Google transparency report. Know Your Neighborhood (2014)
3. Stallings, W.: Data and computer communications. Pearson Education India (2007)
4. Yoav, E.: Amazon Found Every 100ms of Latency Cost them 1% in Sales. Gigaspaces, 20 January 2019. https://www.gigaspaces.com/blog/amazon-found-every-100ms-of-latency-cost-them-1-in-sales
5. Tian, X., Han, R., Wang, L., Lu, G., Zhan, J.: Latency critical big data computing in finance. J. Finan. Data Sci. **1**(1), 33–41 (2015)
6. Mortimer, M.: iperf3 Documentation (2018)
7. Aviram, N., Gellert, K., Jager, T.: Session resumption protocols and efficient forward security for TLS 1.3 0-RTT. J. Cryptology **34**(3), 1–57 (2021)
8. Van Tilborg, H.C., Jajodia, S. (Eds.): Encyclopedia of cryptography and security. Springer Science & Business Media, Boston (2014). https://doi.org/10.1007/0-387-23483-7
9. Rescorla, E., Dierks, T.: The transport layer security (TLS) protocol version 1.3 (2018)
10. Chen, S., Jero, S., Jagielski, M., Boldyreva, A., Nita-Rotaru, C.: Secure communication channel establishment: TLS 1.3 (over TCP fast open) versus QUIC. J. Cryptol. **34**(3), 1–41 (2021). https://doi.org/10.1007/s00145-021-09389-w
11. Cremers, C., Horvat, M., Scott, S., van der Merwe, T.: Automated analysis and verification of TLS 1.3: 0-RTT, resumption and delayed authentication. In: 2016 IEEE Symposium on Security and Privacy (SP), pp. 470–485. IEEE, May 2016
12. Rescorla, E.: TLS 1.3 (2015). http://web.stanford.edu/class/ee380/Abstracts/151118-slides.pdf
13. Sikeridis, D., Kampanakis, P., Devetsikiotis, M.: Assessing the overhead of post-quantum cryptography in TLS 1.3 and SSH. In: Proceedings of the 16th International Conference on Emerging Networking Experiments and Technologies, pp. 149–156, November 2020
14. Dowling, B., Fischlin, M., Günther, F., Stebila, D.: A cryptographic analysis of the TLS 1.3 handshake protocol. J. Cryptology **34**(4), 1–69 (2021). https://doi.org/10.1007/s00145-021-09384-1
15. Dowling, B., Fischlin, M., Günther, F., Stebila, D.: A cryptographic analysis of the TLS 1.3 handshake protocol candidates. In: Proceedings of the 22nd ACM SIGSAC Conference on Computer and Communications Security, pp. 1197–1210, October 2015

Robust Secret Share to Reinforce the Security of IBE's Master Key

Ahmed Bentajer[1(✉)], Youssef Said[1], Zakaria Igarramen[2], and Mustapha Hedabou[3]

[1] SIGL LAB., ENSA of Tetouan, University Abdelmalek Essaadi Tetouan, Tetouan, Morocco
{abentajer,youssef.said}@uae.ac.ma
[2] MTI LAB., ENSA Safi, Safi, Morocco
[3] UM6P-CS LAB., UM6P, Benguerir, Morocco
mustapha.hedabou@um6p.ma

Abstract. Organizations are deploying different cryptosystems to protect their sensitive data, especially those outsourced to a cloud provider. Key Management System is in the heart of those solutions, its failure may lead to vulnerabilities and sensitive data exposure. Bentajer et al. proposed a system based on ID-based cryptography, which encrypt data before outsourcing it to a third party cloud storage provider. The solution enhances data confidentiality and enables secure deletion since data are stored in an encrypted manner. However the security of the private key generator related to ID-based cryptography was not well addressed. In this paper, we present an enhancement to the proposed system by strengthening the security of its key management system. For this purpose, we use Shamir secret share to securely store and manage the secret key. Furthermore, we implemented a prototype of the proposed system to study its practicability. The conducted analysis has demonstrated that it offers a significant value-added of security with a negligible overhead amount of computing time.

Keywords: Shamir secret share · IBE · Data confidentiality · Key management security · Multi-cloud

1 Introduction

Cloud services offers an abstraction of infinite capabilities on a metered way and helps reduce investments on data centers. Cloud storage (STaaS) is amongst the most adopted solution especially for the general public. In the model of StaaS, the user get access to an abstraction of infinite storage resource in a pay-as-you-go manner. It also reduces the burden of data management, data storage and makes data sharing and collaborating over documents more convenient and easy[1, 2].

[1] https://docs.microsoft.com/en-us/.
[2] https://docs.google.com/.

Y. Maleh et al. (Eds.): ICI2C 2021, LNNS 357, pp. 526–534, 2022.
https://doi.org/10.1007/978-3-030-91738-8_48

The service may be provided directly by a Cloud Service Provider (CSP) or provided by a Service Provider that uses the infrastructure of a third party Cloud Provider (CP).

Therefore, outsourcing data to, a non-trusted, third-party cloud provider raises many security issues. Data confidentiality in the cloud is not as easy to manage, since once data are outsourced the customer have no control over them and the multi-tenancy feature of the cloud makes it easier for data to be compromised. In the model of the StaaS, users are often wondering how data are stored and protected, especially replicated copies of data [1]. Lot of efforts have been done by CSPs, researchers and experts in the fields to provide users with the necessary assurances regarding data confidentiality and data integrity.

Researchers in the field agreed that encryption is a prominent solution to reduce the risk related to data security. For this purpose, many client side encryption has been proposed [9–13]. Amongst those solutions there is ADS-IBE [7] which is a flexible and secure deletion scheme built upon the use of identity-based encryption (IBE). By taking advantage of IBE features, the proposed scheme leverage the burden of key management. Using IBE leverage the issue of managing decryption keys since the Private Key Generator (PKG) compute the corresponding private key (d_{ID}) each time the user needs it for decryption purpose.

In this paper, we describe a protocol based on Shamir secret share (SSS) for improving the security of IBE's PKG and its Master Secret Key (*msk*). The proposed design splits the *msk* on n shares distributed to users, interacting with the system, and reconstruct it each time the PKG needs to compute the private corresponding key.

The remainder of this paper is as follow. Section 2 presents a Background of IBE, SSS and ADS-IBE. The third sections presents our proposed scheme and its implementation. The fourth section discuss the security analysis and performances. Finally, we come-up with our conclusions and assumptions.

2 Background

2.1 ID-Based Encryption

ID-based encryption is a public key cryptosystem that was first introduced by Adi Shamir in 1984 [2]. The main idea is to simplify key management on public key cryptosystem by eliminating Certificates and Certificate Authority (CA) deployment and relying only on a unique identity (ID) of the receiver/sender to make cryptographic operations. This involves a new trusted entity called the Private Key Generator (PKG) that computes the corresponding private key to user.

In 2001, Boneh and Franklin proposed the first provable secure and practical system which implements IBE using bilinear pairing in elliptic curve groups [3]. Also in 2001, Cocks proposed an, inefficient, IBE scheme based on the quadratic residuosity problem [4]. Other schemes were also proposed after [5,6].

Basically, an ID-based cryptosystem implements four algorithms:

- **Setup**: Run once by the PKG. It takes a security parameter as input and generates the whole system parameters including the master public key (*mpk*) that is shared with all the users that will use the system, and the master secret key (*msk*) which is kept secret.
- **KeyGeneration**: The algorithm is also run by the PKG and used to compute the private key (d_{ID}) corresponding to the user's ID
- **Encrypt**: Run by users interacting the with PKG. Takes as input the secret to be encrypted *S*, the *mpk* and the receiver's *ID* and output the cipher *C*.
- **Decrypt**: Run by the user willing to decrypt the cipher. It takes as input the *C* with d_{ID} and output the *S*.

Actually, the most attractive area of research related to IBE is the construction of an efficient scheme based on lattice problems, because it offers a lower asymptotic computational complexity than the schemes based on pairings or quadratic residues, and may resist attacks from quantum computers.

2.2 ADS-IBE Solution

The ADS-IBE solution enables a cloud user to encrypt its data before sending them to a cloud storage [7]. The data are associated with the users' ID. ADS-IBE ensure also secure deletion of data, since data are outsourced in an encrypted manner it will be hard for a malicious user or curious CSP to learn something without the corresponding d_{ID}. The components of the proposed design are:

- **Data owner**: Entity that own the data
- **Cloud storage provider**: a non-trusted third party storage provider
- **PKG**: Server hosting the IBE (PKG) cryptostem that execute the *SetUp* and *KeyGeneration* algorithms.

In the upload process (Fig. 1), the system takes as input the file F, the policy P (in our case the ID of the user), and the IBE master public key *mpk*. For policy P associated with file F, the data owner generates a random AES key K and uploads the encrypted file $\{F\}_K$ as well as P and $\{K\}_P$ to the cloud storage. The encryption of K with the public key P is performed using IBE [7].

For the download process (Fig. 2) the system retrieves $\{F\}_K$, P, and $\{K\}_P$ from cloud storage and requests the key manager for the private key d_p corresponding to the control key P in order to decrypt data [7].

ADS-IBE as it was proposed, rely to the operating system security mechanisms and policies to ensure the confidentiality of the *PKG* which make the whole system vulnerable if an attacker breaks the server and get it. For this purpose, we propose a new enhancement for the system in order to protect the *msk* by using SSS that will be described in the next section.

2.3 Secret Sharing

It was formulated by Adi Shamir [8] and is one the most used cryptography calculation for sharing a secret in a distributed way, which in our case resolve

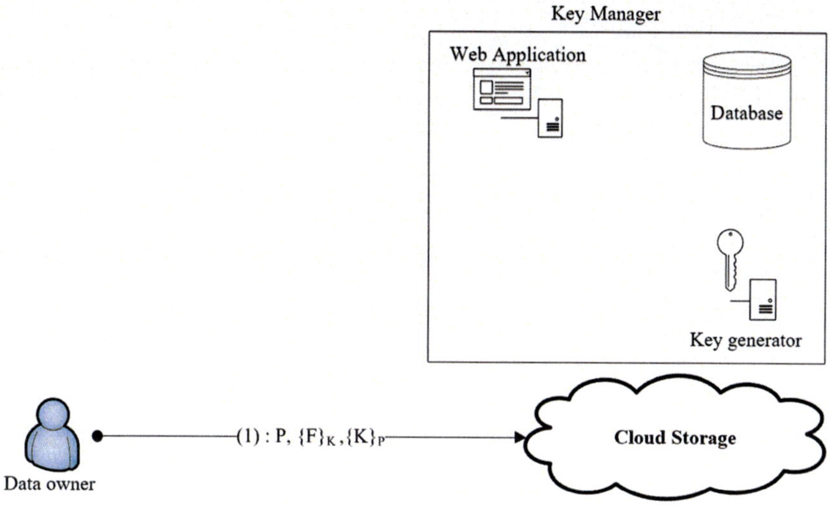

Fig. 1. ADS-IBE file upload.

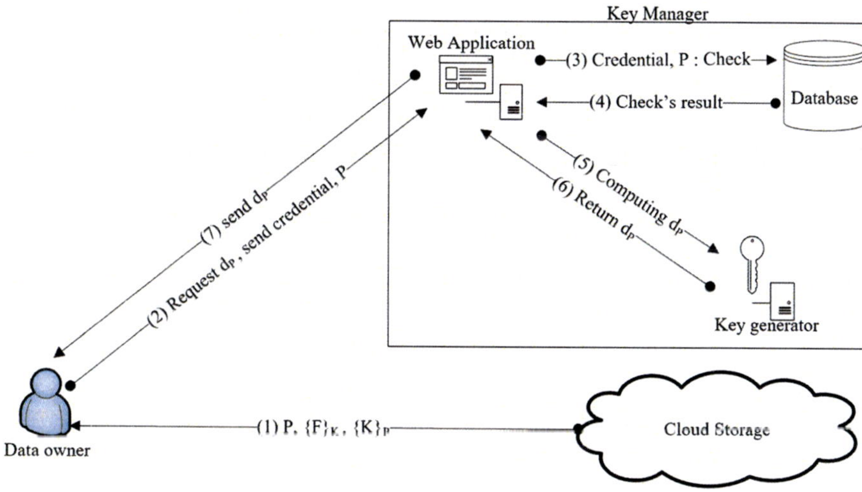

Fig. 2. ADS-IBE file download.

the problem of multi-cloud. The user can share a secret s amongst n parties (called also the shares) such as with any k out of n shares can reconstruct the original s. However, for less than **k** shares it is impossible to compute the secret. This is refereed to as (k, n)-threshold schema. SSS is built based on polynomial interpolation over finite fields.

Secret Sharing found its use on different domain. For example, in [10], the author proposed a multi-signature scheme in conjunction with Trusted Platform Module (TPM) for using BLS signature scheme with TPM in cryptocurrency wallet.

3 Proposed Scheme

In this section we present our proposed scheme to strengthen to security of the PKG's *msk*.

The proposed scheme deals with scenarios where the PKG is already deployed and the *SetUP* algorithm was executed to generate system parameters including the *msk*. The secret **s**, which is in our case is the *msk*, will be split into multiple parts (shares) that will be distributed to users who interact with ADS-IBE.

The proposed scheme can be modeled as follow:

- **Secure Storage**: A new server that will be used to store one of the $\boldsymbol{n}$ shares.
- **System users**: The **n** users using ADS-IBE who will receive the SSS computed shares of **s**.

The new protocol will be deployed with the PKG. The techniques used for establishing secure channels and for authenticating stakeholders, including SSL, SSH, DH key exchange are assumed secure in standard models.

3.1 Basic Design

We will here give an informal description of our proposed scheme that implement Cloud-KMS based on SSS. The protocol consists of three main phases. The **SetUP** algorithm that will responsible for computing the shares of the *msk* and shares them with the involved entities (users and servers) using the **Share** algorithm. The final phase is where the system conducts secure computation to reconstruct the *msk* through the **Reconstruction** algorithm. We note also that two computed shares are stored on a storage server on the cloud and one with the Cloud SSS-KMS.

Figure 3 presents the architecture of our proposed solution. The Cloud SSS-KMS is deployed with ADS-IBE. We note that phase 1 has already been executed and was set up.

Phase 1: Once the Setup function of ADS-IBE is executed and parameters are generated, including the *msk*, the SetUP function of Cloud SSS-KMS is executed in order to compute system parameters and the shares that are distributed to the entities concerned. The storage server is also deployed on cloud side, it may store some shares.

Phase 2: Is where a user wants to compute its d_{ID}. At the first stage the Reconstruction algorithm gets the shares through Sharing and reconstruct the *msk* and sends it to PKG.

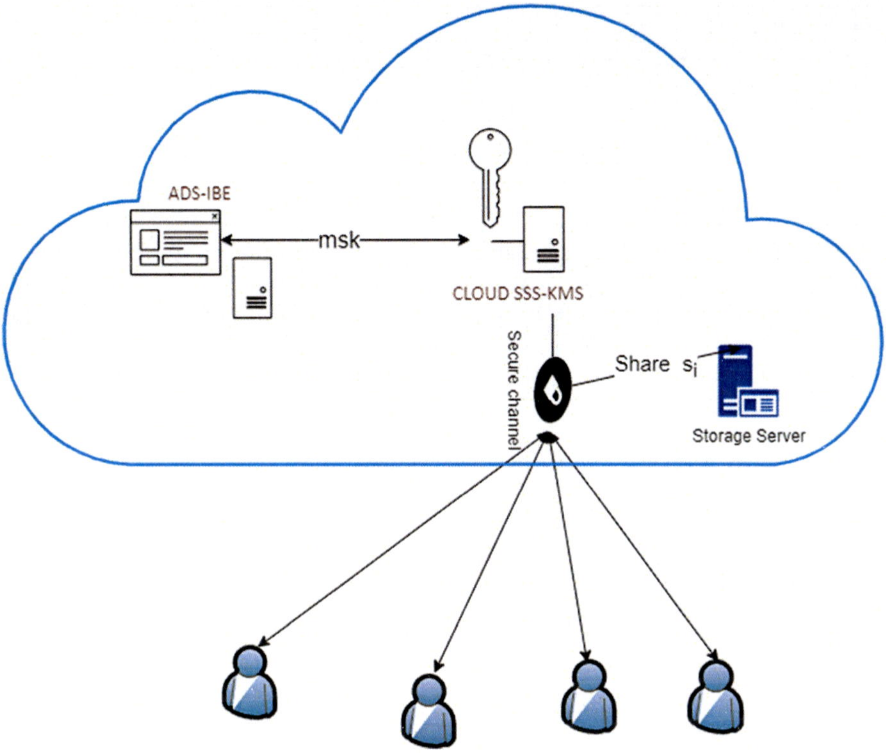

Fig. 3. IBE Cloud KMS architecture

SetUP Algorithm. This Algorithm 1 is used to compute and to initiate the process of *msk* shares. The algorithm outputs the finite field F_p, the public shares $[s]_i$ of *msk* along with its MAC γ and the public shares of MAC γ_i' for $i = 1, \cdots, n$.

Sharing Algorithm. This program (Algorithm 2) is invoked by the SetUP and Reconstruction algorithms. It takes the computed shares $[s]_i$ and $[\gamma']_i$ along with the value of MAC γ. The algorithm upload/download the shares to concerned entities and store the value of MAC γ.

Algorithm 1. SetUP

```
1: function SETUP(msk)
2:     Compute γ ← MAC(msk)
3:     Compute [s]_i for i = 1, ··· , n
4:     Compute γ'_i for i = 1, ··· , n
5:     Invoke the Share Algorithm for γ and ([s]_i, γ'_i) for i = 1, ··· , n
```

Algorithm 2. *Sharing*

```
1: function SHARING(γ, [γ']_i, [s]_i)
2:     Store the MAC γ
3:     Send/retrieve to/from each entity E_i: ([s]_i, γ'_i) for i = 1, ··· , n
```

Reconstruction Algorithm. This program is called when a user wants, the PKG, to generate its d_{ID}. The algorithm call the Sharing protocol in order to get $[\gamma']_i$ and $[s]_i$. We mention that the algorithm compute the MAC value from its sharing $[\gamma']_i$ and compare it with the stored value of MAC γ prior to compute the *msk*.

Algorithm 3. Sub protocol *Reconstruction*

```
1: function POST([γ']_i, [s]_i)
2:     number_share Integer ← 1
3:     Compute L_i ← ∏_{j≠i} (x_j / (x_j − x_i))
4:     Compute γ'' = p(0) ← Σ_{i=1}^{i−t} L_j[γ']_j
5:     if γ == q(0) then
6:         Compute msk ← q(0) ← Σ_{i=1}^{i=t} L_i[s]_i
7:     else
8:         number_share ← 0
9:     if number_share == 0 then
10:        Print : A share has been compromised
```

4 Discussion

We implemented our proposed scheme with Java 1.8 using a 64 bits Windows operating system with Intel Xeon Gold 5217 (3.0 GHz) processor and 256Go installed RAM. The Sharing protocol has been developed using JCraft library which is a pure Java implementation of SSH2 that is known to have more defensive mechanisms to avoid vulnerabilities. The t and n were $t = 3$ and $n = 10$.

We measured the performance of the proposed design with our developed prototype. Analysis proves that the additional security operations, namely SSS computation and reconstruction, do not heavily penalize the performance and does not imply a remarkable overload time to ADS-IBE (computation and reconstruction of the *msk* is ≈ 0.0002) and the exchange between the Share algorithm and users is a dominant factor ($\approx$ 15 s in upload operation and $\approx$ 5 s when downloading the shares).

The main contribution of our work is in its straightforward management infrastructure. The choice of the Shamir Secret Sharing reinforce the security of IBE's *msk* and add a new security layer to ADS-IBE. Instead of storing the *msk* in a centralized way, we computed an encrypted pieces of it and share them with, an un-trusted, users in a zero-trust network. This reduce the risks of targeting a centralized storage and make it impossible for each user to get the secret individually.

The originality of our model is double benefits. First, it increases the confidentiality of the *msk*, since it remains "encrypted" when it is not used by the PKG. Second, if a share is compromised it does not affect the whole system and the user with affected hash can not access the *msk* until the issue is remedied by getting a new share. Besides the prototype can work atop any cloud storage solution since no engineering changes are needed on the CSP side.

5 Conclusion

In this paper we have proposed an enhancement to ADS-IBE system that reinforce the security of IBE's *msk*. In our model, we apply shamir secret sharing to the *msk* and send each user a piece of share of it. This enable each stakeholder to hold a piece of the *msk* and make it available to any entity that needs to compute its d_{ID}. In the future, we plan to explore the implementation with a Trusted Platform Module (TPM) to explore the ability to isolate the execution of a piece of code from other application hosted on the same machine, this turned to be practicable for preventing. We also aim to extend the security analysis of the proposed scheme for a more active adversary in the presence of quantum computing abilities.

References

1. Google Cloud Whitepaper, Data deletion on Google Cloud Platform June 2021. https://cloud.google.com/security/deletion/. Accessed 20 June 2021
2. Shamir, A.: Identity-based cryptosystems and signature schemes. In: Blakley, G.R., Chaum, D. (eds.) CRYPTO 1984. LNCS, vol. 196, pp. 47–53. Springer, Heidelberg (1985). https://doi.org/10.1007/3-540-39568-7_5
3. Boneh, D., Franklin, M.: Identity-based encryption from the Weil pairing. In: Kilian, J. (ed.) CRYPTO 2001. LNCS, vol. 2139, pp. 213–229. Springer, Heidelberg (2001). https://doi.org/10.1007/3-540-44647-8_13
4. Cocks, C.: An identity based encryption scheme based on quadratic residues. In: Honary, B. (ed.) Cryptography and Coding 2001. LNCS, vol. 2260, pp. 360–363. Springer, Heidelberg (2001). https://doi.org/10.1007/3-540-45325-3_32
5. Choon, J.C., Hee Cheon, J.: An identity-based signature from gap Diffie-Hellman groups. In: Desmedt, Y.G. (ed.) PKC 2003. LNCS, vol. 2567, pp. 18–30. Springer, Heidelberg (2003). https://doi.org/10.1007/3-540-36288-6_2
6. Hess, F.: Efficient identity based signature schemes based on pairings. In: Nyberg, K., Heys, H. (eds.) SAC 2002. LNCS, vol. 2595, pp. 310–324. Springer, Heidelberg (2003). https://doi.org/10.1007/3-540-36492-7_20
7. Bentajer, A., Hedabou, M., Abouelmehdi, K., Igarramen, Z., El Fezazi, S.: An IBE-based design for assured deletion in cloud storage. Cryptologia **43**(3), 254–265 (2019). https://doi.org/10.1080/01611194.2018.1549123
8. Shamir, A.: How to share a secret. Commun. ACM **22**, 612–613 (1979). https://doi.org/10.1145/359168.359176

9. Igarramen, Z., Bentajer, A., Hedabou, M.: TPM based schema for reinforcing security in IBE's key manager. In: Attiogbé, C., Ferrarotti, F., Maabout, S. (eds.) MEDI 2019. CCIS, vol. 1085, pp. 146–153. Springer, Cham (2019). https://doi.org/10.1007/978-3-030-32213-7_11
10. Hedabou, M., Abdulsalam, Y.S.: Efficient and secure implementation of BLS multisignature scheme on TPM. In: IEEE International Conference on Intelligence and Security Informatics (ISI), pp. 1–6 (2020). https://doi.org/10.1109/ISI49825.2020.9280511
11. Bentajer, A., Hedabou, M., Abouelmehdi, K., ELFezazi, S.: CS-IBE: a data confidentiality system in public cloud storage system. In: The 9th International Conference on Emerging Ubiquitous Systems and Pervasive Networks (EUSPN-2018)/The 8th International Conference on Current and Future Trends of Information and Communication Technologies in Healthcare (ICTH-2018)/Affiliated Workshops Procedia Computer Science, pp. 559–564 (2018). https://doi.org/10.1016/j.procs.2018.10.126
12. Geambasu, R., Kohno, T., Levy, A.A., Levy, H.M.: Vanish: increasing data privacy with self-destructing data. In: Proceedings of the 18th Conference on USENIX security symposium (SSYM 2009), USENIX Association, Montreal, Canada, pp. 299–316 (2009)
13. Tang, Y., Lee, P.P.C., Lui, J.C.S., Perlman, R.: Secure overlay cloud storage with access control and assured deletion. IEEE Trans. Dependable Secure Comput. **9**(6), 903–916 (2012). https://doi.org/10.1109/TDSC.2012.49

Single Sign-On Revocation Access

Yousra Belfaik(✉), Abdelhak Lmouhandiz, Yassine Sadqi, and Said Safi

Laboratory LIMATI, FPBM, USMS University, Beni Mellal, Morocco
belfaik.yousra@usms.ac.ma, y.sadqi@usms.ma

Abstract. Single Sign-On (SSO) is a mechanism that allows users to seamlessly access multiple services using a single authentication action with an Identity Provider (IdP). OpenID Connect (OIDC) is an authentication and authorization protocol widely used in internet-based environments, and it presents the new standard for SSO. Due to its favorable usability and security, SSO has been widely used for online authentication services. Nevertheless, this technology pose a huge security risk in the case of a compromised IdP account. If an attacker gets to hijack the IdP account, he can easily access and control all the Internet-based services associated with the victim's hijacked account. Furthermore, SSO technology does not introduce efficient mechanisms to notify services that the centralized IdP account has been compromised, and/or revoke access from all the services in the same login session with it. Thus, it is mandatory to have a mechanism to protect user's accounts and revoke access in the case of a hijacked IdP account. In this paper, we discuss the authentication and authorization process of SSO, the protocol OIDC, and the IdP accounts compromise threat. Next, we perform a comparative analysis of four OIDC mechanisms that can be used as authentication revocation' solutions for SSO systems: OIDC Session Management, OIDC Front Channel Logout, OIDC Back-channel Logout, and Single Sign-Off.

Keywords: Single sign-on · OpenID connect · Revocation access · OIDC session management · OIDC back-channel logout · OIDC · Front-channel logout · Single sign-off

1 Introduction

Now day, all internet services require an authentication process to accept or reject a user's request to access their features. The existing user authentication schemes are commonly based on something you know, such as passwords, something you have, such as hardware tokens, or something you are, such as biometrics. The password-based authentication is the most deployed authentication scheme in computer systems, despite its security and usability problems that have been studied for decades [1,2]. One of the biggest challenges of password based-authentication is the limitations of human memory. Users can't remember passwords for many accounts, which make them choose easilyguessable passwords, write them down, share them with friends, or use a small collection of

Y. Maleh et al. (Eds.): ICI2C 2021, LNNS 357, pp. 535–544, 2022.
https://doi.org/10.1007/978-3-030-91738-8_49

passwords to manage their multiple accounts (e.g., a user can use 5 to 6 passwords to manage his 30 accounts).

Recently, a technique called Single Sign-On (SSO) has been proposed as an answer to the complexity and fragility of authentication with passwords. SSO is a centralized identity management system that allows users to authenticate once for the duration of a session, regardless of services' number that require authentication, only by using one credential on an identity provider. This system is based on three parties: Identity Provider (IdP) which is the entity that manages users' identities, Relying party (RP)(also called client) which is a public available service offers access to resources for registered end-users, and the End-User (EU) [8,9]. Moreover, OpenID Connect (OIDC) protocol has been defined as the new standard of SSO, as it enables billions of users to log in at millions of websites [6] easily, safely, and more securely compared with previous solutions. OIDC is an authentication protocol based on building a simple identity layer on top of the authorization protocol OAuth2.0 [10]. It allows internet-based clients to authenticate an end-user by obtaining his basic profile information (name, family_name, given_name, email, picture, etc.) from an IdP.

Due to the favorable usability and security of SSO, it has known a big proliferation in the Web environment. In 2018, Ghasemisharif et al. [7] analyzed a list of the top 1M websites according to Alexa, and they found that 57 555 (6.30%) websites of this list support SSO. In 2019 and according to Zhang and al. [5], more than one million websites support SSO via OpenID Connect. This wide use of SSO has made this system a double-edged sword. Despite the good benefits this technology offer, we can not ignore the huge security risk that presents in the case of a hijacked IdP account. If the centralized IdP account is compromised, the attacker can gain control of all the victim's accounts in the associated services to it. Hence, doing unwanted actions such as checking and sending messages, changing personal information, sharing posts in the name of the victim, etc. Furthermore, SSO does not have an efficient tool to revoke access to all services created or accessed from a compromised IdP account. The main contributions of this work are as follows.

- Describe the authorization and authentication workflow of Single Sign-On system, and the OpenID Connect protocol.
- Describe the IdP account hijacking threat of Single Sign-on.
- Describe in details the four present authentication revocation solutions to mitigate this threat: OIDC Session Management, OIDC Front-channel Logout, OIDC Back-channel Logout, and Single Sign-off.
- Perform a depth-comparative Analysis between these approaches.

2 Related Works

Single Sign-On systems are widely used and have been studied for years, but logging out of a service using SSO did not receive the attention it deserves even though it is considered one of the big limitations of SSO.

In 2005, Linden et al. [19] studies the usability of logout in Single Sign-On systems by examining the users' satisfaction with the current implementation of SSO. More precisely, this study was carried out in a university using Pubcookie SSO in its intranet services. This study uses two research methods to get both qualitative and quantitative data. The qualitative data was collected with two focus groups and, based on the results, the quantitative data was gathered by tests using a structured inquiry. The result of this study proves that users expect logging out of all the services when they log out of one of them (Single Logout), users need to know if they are not anonymous or identified in a service, and users do not resist the idea of having services provided by other organisations in the Single Sign-On System.

In 2011, Mustafic et al. [20] describe a persistent Single Sign-On system for ubiquitous home environments that involve the capabilities of Behavioral Biometrics to check if the user is still the same as that authenticated at the beginning of the session. If the system notices that the user has changed, it initiates a Single Logout which is supported by SAML 2.0.

In 2014, Suoranta et al. [21] describe challenges related to logout and its implementation in services that use Shibboleth SSO and analyzes logout behavior of different SSO technologies, such as SAML, Shibboleth, OAuth and OpenID logout. Based on the analysis, they give guidelines for improving the security and usability of logout in the federated SSO, and describe a polling-based solution for creating a system-wide logout mechanism that only requires minor changes to the existing code and does not burden the identity provider excessively. This solution gives users the option to choose between logging out of one service (local logout) and logging out of all the services at the same time (single logout). A usability test was conducted with 18 users to check the understanding of both logout options from the users' perspective. The results show that the users do not log out of the services before closing the web browser window, and they do not understand the wording used by security professionals for different logout options.

3 IdP Account Hijacking Threat

There is a wide range of attacks that can be used to compromise the user's IdP account. We can separate these attacks into two categories: Compromised IdP password attacks and Session hijacking attacks. These two categories of attacks are possible with physical access, network access and/or web access, and are caused by the implementation errors committed by the developers and the nature of the IdP.

- **Compromised IdP password attacks.** The attacker can gain the IdP account's password by using different techniques, such as installing malicious software on the victim's device, using sniffing attack (especially easy if the victim is navigating through an open wireless network), or by using social engineering attacks, etc. [12]. Phishing attack is the most common social engineering technique used to obtain users' IdP passwords, even in major

IdPs. Once an attacker obtains the IdP password, he would have the same level of access and control as the victim over all the RPs associated to this IdP. Furthermore, the adversary can take advantage of the IdP password to preemptively create accounts on other RPs to which the victim does not yet have access, in order to misuse information and accounts 'functionality once the victim joined these RPs. To determine the level of access and control an attacker can gain after the victim joins the RP, Ghasemisharif et al. [7] have manually tested this attack on 95 RPs, and they have found that a preemptive account hijacking attack is indistinguishable from the normal scenario of an RP account hijacking attack when it comes to the information and account functionality that the attacker can access.

- **Session hijacking attacks.** In this case, the adversary extracts HTTP cookies from the IdP, which allows him to attack the user's accounts on all RPs associated with this IdP, even if he never has the original password. The attacker can hijack session cookies either directly from the browser, or by using a sniffing attack, or by using malware, or by relying on some kind of XSS attack, etc. This category of attacks is less powerful than the first one because the attacker need to be within WiFi range, and can be thwarted by correct deployment of HTTPS. The fact of not having the IdP password not pose a huge problem since the majority of RPs do not require the IdP password to be re-entered. To determine thc level of risk this category present, Ghasemisharif et al. [7] evaluated 29 websites out of the Alexa top 500 and 66 popular iOS apps that support Facebook SSO and the Android version for a subset of these apps. They have found that in the majority of cases, the attacker has the same level of access as the user when using hijacked session cookies, and only three services record the user's re-authentication. Therefore, this category of attacks is successful as long as the IdP and/or RP does not ask the user to re-authenticate by re-entering their IdP password.

4 Access Revocation Solutions

In this section, we present four logout mechanisms that can be used as solutions to automatically logged out from all services in the same login session with a compromised IdP.

4.1 OIDC Session Management

OIDC Session management is a logout mechanism defines how to monitor the end-user's login status at the IdP (supporting OIDC) on a continuous basis so that the RP can automatically log out a user who has logged out of the IdP [14]. That means the RP would permanently ask the IdP if the user is still logged in or not. If not, the RP would clear session data related to that user and log him out. Furthermore, his solution uses front-channel communication, which communicate logout requests from the IdP to RPs via a user-agent (generally a browser). OIDC Session Management uses two iframes called OP iframe and RP iframe:

- **RP iframe.** is an invisible iframe that the RP loads from itself. This iframe must know the ID of the OP iframe in order to be able to send a postMessage to the OP iframe, and the origin URL of the OP iframe, so that it can ensure that messages are sent and processed only when they originate from the OP.
- **OP iframe.** is an invisible iframe that the RP loads into itself from the OP's check_session_iframe. This iframe must only accept postMessages from its parent iframe, the RP iframe, and must reject postMessage requests from any other source origin to prevent cross-site scripting attacks.

OIDC Session management logout flow starts with loading the two iframes from the RP. Next, the RP iframe polls the OP iframe with postMessage at an interval suitable for the RP application. The postMessage from the RP iframe provides the following concatenation as data: Client_id + " " + session_state. Next, the OP iframe accesses the OP browser state cookie and calculates the new session_state parameter and compares it with the one sent by the RP, and then sends a postMessage containing the session state as changed, unchanged or error. If the response is unchanged, the RP iframe does nothing. If the response is changed, the RP sends a passive request for re-authentication with prompt=none to obtain the current session state at the OP. Last, if the response is error, the RP must not perform re-authentication with prompt=none since it would generate a lot of unnecessary network traffic.

4.2 OIDC Front-channel Logout

OIDC Front-channel Logout mechanism uses a front-channel communication via the user-agent between the OP and RPs being logged out that does not need an OP iframe on RP pages, as OIDC Session management does [17]. The RPs supporting this logout mechanism must register a logout URI called frontchannel_logout_uri, which is added during the registration process. This logout URI must be accessible by the user's user-agent. Upon receiving a logout request (Fig. 1, 1), the IdP provides a response page with iframes referencing the logout URI for each RP has an active session with the IdP as a source to trigger the logout actions by these RPs (Fig. 1, 2). Next, The user-agent sends a logout request to each RP having an active session with the user, containing the frontchannel_logout_uri indicated in the source parameter of the iframe with the parameters "iss" (issuer identifier for the OP issuing the front-channel logout) and "sid" (session identifier) (Fig. 1, 3). Upon receiving a logout request, the RP validates it and then clears state associated with the connected session, removing all associated cookies and data stored locally within the user's browser. If the end-user is already logged out from the RP when the logout request is received, the logout is considered successful.

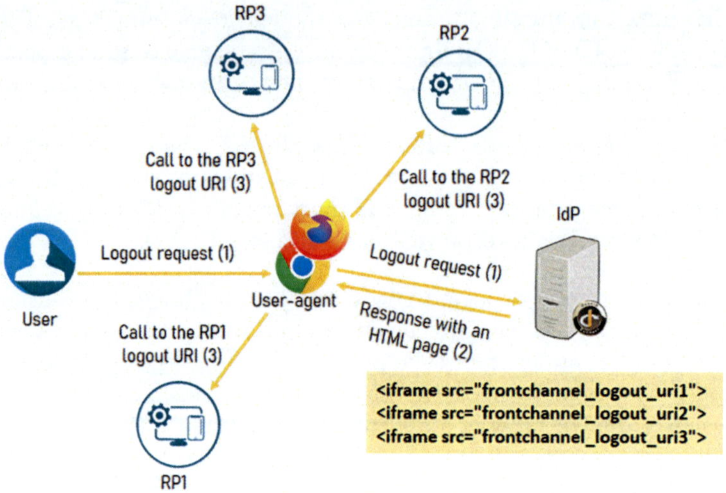

Fig. 1. OIDC front-channel logout flow.

4.3 OIDC Back-channel Logout

OIDC Back-channel logout mechanism uses direct back-channel communication between the IdP and RPs being logged out [16]. That means, RPs in the same login session with a user can be logged out by directly communicating the logout requests between the IdP and these RPs, without relying on a user-agent; which differs from the front-channel logout mechanisms mentioned above (OIDC Front-channel Logout and OIDC Session management). Furthermore, RPs supporting back-channel logout should register a back-channel logout URI with the IdP as part of their client registration. In the other hand, the IdPs supporting this logout mechanism need to monitor the group of the logged-in RPs so that they know what RPs to contact at their back-channel logout URIs to log out them.

The back-channel logout flow (Fig. 2) start with a logout request that a user initiated it from an RP on the IdP (1) (2). Next, The IdP will first generate a logout token, which is a JWT similar to the ID token, and then send an HTTP post request to the registered back-channel logout URI for each RP, including the generated logout token (3). Upon receiving a valid logout token, each RP will clear any state associated with the identified sessions (4) (5).

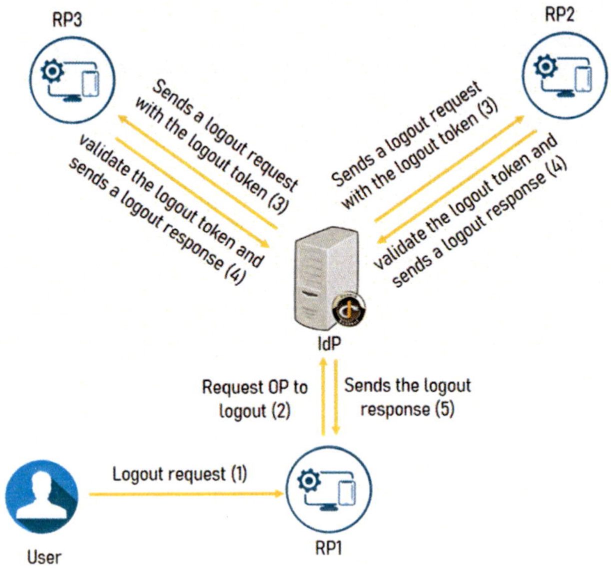

Fig. 2. OIDC connect back-channel logout flow.

4.4 Single Sign-Off

Single Sign-off is a logout protocol that aims to universally revoke access to all RP accounts created or accessed from a compromised IdP account. This solution relies on an extension compatible with OpenID Connect that support the authentication revocation feature. During the OIDC client registration, this extension adds a single authentication revocation URI to the list of redirection callbacks URI defined on the client metadata, which the IdP uses to notify the RP that a user's authentication has been revoked, and user sessions should be expired.

Single Sign-off authentication revocation flow (Fig. 3) start when a user realizes that his IdP account has been compromised. He will first log in to his IdP account, change her password, and then complete a two-factor authentication step. This will invalidate all active sessions of the IdP on all connected devices. Next, the IdP revokes access permission to all RPs associated with that account over SSO, and sends an authentication revocation request to each one of them, containing a JWT-type revocation token. Upon receiving a valid authentication revocation request, the RP logs that user out of all active sessions.

5 Comparative Analysis

OIDC session management and OIDC front-channel logout mechanisms are both based on a front-channel communication, which communicate logout requests between the IdP and RPs via a user-agent. This means that the user-agent is able

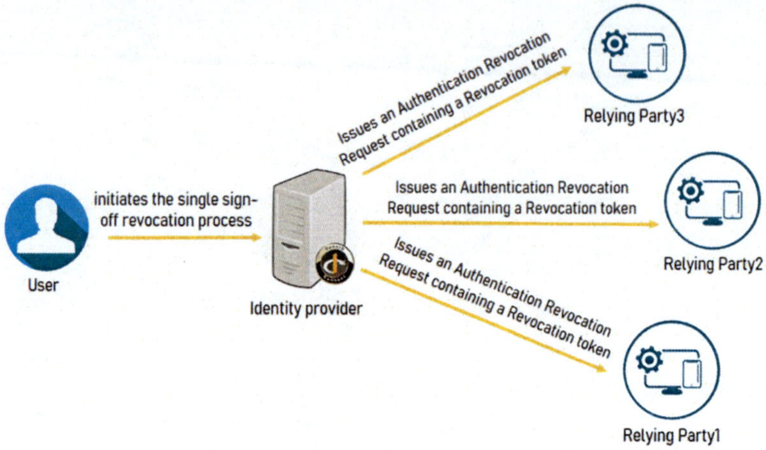

Fig. 3. Single Sign-Off authentication revocation flow.

to collect information transferring between these entities, which can pose security and privacy threats if the communication is not well signed and encrypted. Furthermore, this type of communication obligates the RP's browser session to be active to have a successful communication. In order to defend against clickjacking attacks, websites can use the X-Frame-Options or Content Security Policy (frame-ancestors) HTTP headers to prevent the browser from loading the page in an iframe. They can also use the "SameSite" cookie attribute to prevent session cookies from being included when the page is loaded in an iframe [18]. Thus, session management and front-channel logout mechanisms are not useful in this case.

In contrast, Single Sign-Off and back-channel logout mechanisms use a direct back-channel communication between the IdP and Rps for being logged out. This type of communication is considered more reliable than communication through a browser. In addition, it relies on TLS (Transport Layer Security) protocol to transfer messages securely between RPs and the IdP. Back-channel communication presents a downside since the session state maintained between the IdP and RP is not available and must be communicated explicitly. This means, that RPs must implement a specific method to terminate RP sessions with the IdP when receiving back-channel logout requests; which makes this mechanism more complicated than using a front-channel-based mechanism. Another significant limitation of back-channel logout is that the RP's back-channel logout URI must be reachable from all the IdPs used, so they should not be behind a firewall or NAT when used with public IdPs. Moreover, Back-channel logout specification have an open issue related to the refresh tokens, since they should not be revoked if they are issued with the offline_access scope value. Thus, it should define an additional optional parameter in the logout token, to explicitly revoked offline_access refresh tokens (which increase more the complexity).

If such a parameter is not defined, attackers can easily maintain access to the user's accounts through these tokens.

Single Sign-off is very similar to the back-channel logout solution with some key differences, since they both send logout requests from the IdP to the RPs containing a JWT-type revocation token. The complicated nature of OIDC back-channel logout specification, make things difficult for developers to understand the full implications of security mechanisms, and to implement it correctly. Moreover, Back-channel logout is more flexible than single sign-off, since it allows the IdP to specify which user sessions at the RP are to be terminated. In contrast, Single Sign-Off states that all active sessions on all devices must be terminated. This flexibility can be beneficial and useful in normal operations, but it increases the implementation complexity and the possibility of an improper deployment. The simple design of single sign-off minimizes implementation complexity and avoid adding optional features that can lead to an incorrect implementation. But, it still have a big limitation, that is the attacker can start the revocation process and exclude the user of all RPs. Although the back-channel logout and single sign-off approaches differ in terms of flexibility and implementation complexity, they are considered correct to mitigate this threat of SSO system.

6 Conclusion

SSO systems does not provide a functionality for IdPs to universally revoke access to all RP accounts created or accessed from a compromised IdP account. In this paper, we provided a depth-comparative analysis of the four different approaches to mitigate this significant threat of SSO. As a result, we have found that both back-channel logout and single sign-off approaches are correct to mitigate this threat. We hope of this work to highlight the big limitation of SSO schemes within the security community and to kick-start efforts to find an efficient tool, flexible and simple in terms of design.

References

1. Golla, M.: On the usability and security of password-based user authentication, p. 264 (2020)
2. Bonneau, J., Preibusch, S.: The password thicket: technical and market failures in human authentication on the web, p. 48 (2010)
3. Password management strategies for online accounts—Proceedings of the second symposium on Usable privacy and security. https://dl.acm.org/doi/10.1145/1143120.1143127
4. Florencio, D., Herley, C.: A large-scale study of web password habits. In: Proceedings of the 16th International Conference on World Wide Web - WWW 2007, Banff, Alberta, Canada, p. 657 (2007). https://doi.org/10.1145/1242572.1242661
5. Zhang, Z., Król, M., Sonnino, A., Zhang, L., Rivière, E.: EL PASSO: privacy-preserving, asynchronous single sign-on. ArXiv200210289 Cs, June 2020. http://arxiv.org/abs/2002.10289. Accessed 14 Aug 2020

6. OpenID Market Share and Web Usage Statistics, SimilarTech. https://www.similartech.com/technologies/openid
7. Ghasemisharif, M., Ramesh, A., Checkoway, S., Kanich, C., Polakis, J.: O Single sign-off, where art thou? An empirical analysis of single sign-on account hijacking and session management on the web, p. 19
8. Mladenov, V.: On the security of single sign-on
9. Mainka, C., Mladenov, V., Schwenk, J., Wich, T.: SoK: single sign-on security—an evaluation of openID connect. In: 2017 IEEE European Symposium on Security and Privacy (EuroSP), pp. 251–266 (2017)
10. Final: OpenID Connect Core 1.0 incorporating errata set 1. https://openid.net/specs/openid-connect-core-1_0.html
11. Hardt, D.: dick.hardt@gmail.com, The OAuth 2.0 Authorization Framework. https://tools.ietf.org/html/rfc6749
12. Navas, J., Beltrán, M.: Understanding and mitigating OpenID connect threats. Comput. Secur. **84**, 1–16 (2019). https://doi.org/10.1016/j.cose.2019.03.003
13. Fett, D., Kuesters, R., Schmitz, G.: The web SSO standard OpenID connect: in-depth formal security analysis and security guidelines. ArXiv170408539 Cs, avr. http://arxiv.org/abs/1704.08539
14. Draft: OpenID Connect Session Management 1.0 - draft 17. https://openid.net/specs/openid-connect-session-1_0-17.html
15. Silva, A.D.: OpenID Connect—Session Management, Medium, 27 August 2018. https://medium.com/@desilvaashen.15/openid-connect-session-management-d0c8e7dc252b
16. Jones, M.B., Bradley, J.: OpenID connect back-channel logout 1.0 - draft 06 (2020). https://openid.net/specs/openid-connect-backchannel-1_0.html
17. Jones, M.: OpenID connect front-channel logout 1.0 - draft 04, 7 August 2020. https://openid.net/specs/openid-connect-frontchannel-1_0.html
18. Clickjacking Defense - OWASP Cheat Sheet Series. https://cheatsheetseries.owasp.org/cheatsheets/Clickjacking_Defense_Cheat_Sheet.htmldefending-with-x-frame-options-response-headers
19. Linden, M., Vilpola, I.: An empirical study on the usability of logout in a single sign-on system. In: Deng, R.H., Bao, F., Pang, H.H., Zhou, J. (eds.) ISPEC 2005. LNCS, vol. 3439, pp. 243–254. Springer, Heidelberg (2005). https://doi.org/10.1007/978-3-540-31979-5_21
20. Behavioral biometrics for persistent single sign-on. In: Proceedings of the 7th ACM Workshop on Digital Identity Management. https://dl.acm.org/doi/10.1145/2046642.2046658
21. Suoranta, S., Manzoor, K., Tontti, A., Ruuskanen, J., Aura, T.: Logout in single sign-on systems: problems and solutions. J. Inf. Secur. Appl. **19**(1), 61–77 (2014). https://doi.org/10.1016/j.jisa.2014.03.005

New Way to Enumerate Large Quadratic Residue Codes Based on Hash and Automorphism Group

Hamza Faham[1](✉), Saïd Nouh[1], My Seddiq El Kasmi Alaoui[2], Mounir Sadiq[1], and Mohamed Azzouazi[1]

[1] LTIM Lab, Ben M'sick Sciences Faculty, Hassan II University of Casablanca, Casablanca, Morocco

[2] LIS Lab, Aïn Chock Sciences Faculty, Hassan II University of Casablanca, Casablanca, Morocco

Abstract. Quadratic Residue (QR) codes are cyclic binary linear block codes. They are known for their remarkable high error correction capacity. However, it is very hard to enumerate and analyze them. A very interesting method called Hash_MIM_Weights_Enumerators (HMWE) was developed previously to determine the weights enumerators of QR codes having lengths in the form 8m-1. Nevertheless, these enumerators were found just for lengths less than or equal to 223. In this paper, we propose an important improvement to HMWE method by new exploitation of Automorphism Group. Instead of the list of codewords collected by cyclic permutations, a very large list of all codewords obtained by using Automorphism group will be generated. This important modification on HMWE has significantly reduced both the run time complexity and the spatial memory complexity. Consequently, the likelihood weights enumerator of the QR (239, 120) is determined for the weights 31, 32, 34 and 35. For the weights 39 and 40, approximated values are computed with confidence interval.

Keywords: Quadratic residue codes · Error-correcting codes · Weights enumerators · Automorphism group

1 Introduction

Error detection and correction works are various, such as decoding algorithms [1–6], finding codes minimum distance [7, 8] and the linear codes weights enumeration [9, 10].

To assess the error correction performance of a C code during an information exchange, it is highly recommended to know its analytical characteristic, which requires prior study on its weight enumerator. The weight enumerator provides valuable information about the analytical performance of the C code in terms of error correction [11, 12].

In this work, we suggest improving a weights enumeration method called (HMWE). Our significant modification of this method is the following: We will generate a larger list using the Automorphism group instead of the codewords list gathered by cyclic permutations. This new idea has enabled us to minimize both the run time and the spatial memory complexities remarkably.

Y. Maleh et al. (Eds.): ICI2C 2021, LNNS 357, pp. 545–556, 2022.
https://doi.org/10.1007/978-3-030-91738-8_50

The rest of the work is arranged in this way. In the following section, we expose some related works. In the third section, we describe the proposed method. In the fourth section, we show our results and we make a discussion. Finally, we give a conclusion through the fifth section.

2 Related Works

Given the difficulty of the weights enumeration of linear codes, several methods were developed to find A_w (the codewords number of weight w in the C code). Among these methods, we cite an applicable one to linear codes of small dimensions that explore all the codewords space exhaustively and enumerate them one by one. This method is tough and requires the enumeration of 2^k codewords, making it inapplicable for long and medium codes.

Another method of Gaborit et al. [13] is applicable to the $\frac{1}{2}$ rate linear codes. It requires the enumeration of $2\sum_{i=1}^{\frac{w}{2}-1}\binom{k}{i}+\binom{k}{w/2}$ codewords. A simplified version is presented by Tjhai et al. [14] for dual circulating codes that are formally self-dual. This version requires only the enumeration of $2\sum_{i=1}^{\frac{w}{2}}\binom{k}{i}$ words. As a result of this method, the weights enumerators of the quadratic residual codes of lengths 76, 124 and 138 are determined.

We cite also a method that uses Chen's theorem [15] for cyclic codes. Indeed, Chen has shown that if c is a codeword of weight w, of a cyclic code C, it exists a cyclic permutation of c whose information part weight is equal to $r = \left\lfloor \frac{kw}{n} \right\rfloor$. Therefore, to find the codewords number of weight equal to w, it is sufficient to code all the information vectors v of weight equal to r and if the weight of the codeword c associated with v is equal to w, then we count all its cyclic permutations. Another method applicable to quadratic residual codes is presented in [16] where the authors have combined the theorems of Assmus-Mattson and Gleason to determine the weight enumerator of the binary extended quadratic residual code of length 168.

In [17–20], the authors have constructed and determined the weight enumerators of some classes of linear codes with two or three non-zero weights. In [12], the form of the weight enumerator of some irreducible cyclic codes, of length n over a finite field $GF(q)$, is determined under the condition of the existence of a prime integer p that divides n and $q-1$. Other algorithms use a Monte Carlo method to approximate the weight enumerator of linear codes [21, 22].

The authors of [23, 24] have developed two algorithms exploiting the hashing, the Monte Carlo estimation method and the stability of linear codes by their automorphism group to approach the weight enumerators of linear codes. They have applied their algorithms to determine at 99% the partial weight enumerators and the analytical performances of several BCH codes of lengths 255. The same authors have presented in [25] another method based on hashing techniques, MacWilliams and Pless identities to determine the weight enumerators of the quadratic residue codes of lengths of the $8m$-1 form. The proposed method determines the weight enumerators and the analytical performances of the quadratic residue codes of lengths 191, 199 and 223.

3 Description of the Proposed Method

Let the Hash table have a size N, a positive integer, and let L be a list of several weight w (just the information part). All L elements belong to QR (n). To precipitate the search of an element in L, this latest is separated on N sub-sets: $L[0]$, $L[1]$,…, $L[N-1]$. Each sub-set L includes the words having the same hash value given by the Hash function introduced in the algorithm A1 below. The list L is implemented as a three dimensions table.

```
Function hash (word, N)
Pos←0
For i=1 to the dimension k of the code
   If word [i] =1 then
      Pos←Pos + i;
   End If
End For
Return (Pos modulo N)
End Function
```

Algorithm A1: The used hash function

To examine if a codeword (just the information bits) is in the list L, the Fast_Search_By_Hash function (algorithm A2) is used. Algorithm A3 shows our modification on «Number of CodeWords Without Cyclic Copies» function presented in [25].

```
Function Fast_Search_By_Hash (L,e)
L: a set of binary vectors of length k, divided into N sub-sets numbered
from 0 to N-1.
e : a binary vector of length k
Outputs:
True if e in L and False otherwise.
Begin Function
h←hash(e,N)
   If e in L[h] then
         Return True
   Else
         Return False
   End If
End Function
```

Algorithm A2: Fast Search of a vector in a list

```
Function NumberOfCodeWords (w, GEN, Part_Aut, N)
 Inputs:
      -  The weight w
      -  The generator matrix GEN of QR(n)
      -  Part_Aut a sub-group of Automorphisms of QR(n)
      -  SizePA : the size of the set Part_Aut
Begin Fun ;lction
      -  L;:kàl)
   Repeat
      - Find a codeword c of weight w by the MIM method
      - For each element σ from Part_Aut do:
  If there not exists any permutation φ in Part_Aut:
  Fast_Search_By_Hash(information_part(φ(σ(c))),L)=True
   then
                  - h←hash(information part of (σ(c)),N)
                  - Add σ(c) in the sub list number h of L
                  - Aw← Aw+size of {φ(σ(c)): φ∈ Part_Aut}
  End
  Number_of_iterations ← Number_of_iterations +1
  End
   Until there not exists any element to add in L
Outputs: The number N of elements in L
End Function
```

Algorithm A3: Find the number of codewords of a given weight in a given QR(*n*) code

The A4 Algorithm presents the improvement proposed for the HMWE method [25]:

Inputs:

- The length n of the quadratic residue code QR(n).
- The generator matrix GE(n) of EQR(n) and GEN(n) of QR(n).

Outputs: The weight enumerators E(n) and A(n) of EQR(n) and QR(n) codes, respectively.

1. Find the binary weights enumerator B(n) of EQR(n) using GE(n) and the MIM method
2. Create the system S(n) by using B(n) and the MacWilliams identity
3. Solve S(n) to obtain the form of E(n) and that of A(n) and find the list R(n) of the residual unknowns of the form R_{4j} which are sufficient to determine A(n) and E(n)
4. Find A(n) as follows:

For each element R_{4j} in R(n) do

Find the number Γ_{4j-1} of codewords of weight 4j-1 in the QR(n) code without cyclic copies: A_{4j-1}←NumberOfCodeWords(4j-1,GEN,Part_Aut)

End For

5. Determine the weights enumerator E(n) by using the Pless identity and A(n).

Algorithm A4: HMWE steps

The proposed improvement concerns the use of permutations defined by the Automorphism group instead of cyclic permutations. The idea proposed in this paper has allowed us, on the one hand, to reduce the memory space required to store codewords used as representatives of Automorphism classes (AutClass). On the other hand, it has enabled us to reduce the time complexity. Indeed, from one codeword, we can extract up to «SizePA» other codewords, knowing that in the old version the maximum number of codewords that can be extracted from a codeword was equal to the length n of the studied code.

In this work, we propose replacing the cyclic class with an Automorphism class (AutClass) as shown in Fig. 1. The codewords of QR(n) are divided in many AutClasses: two codewords c and c' are in the same class if it exists an integer u such that $c' = \varphi_u(c)$.

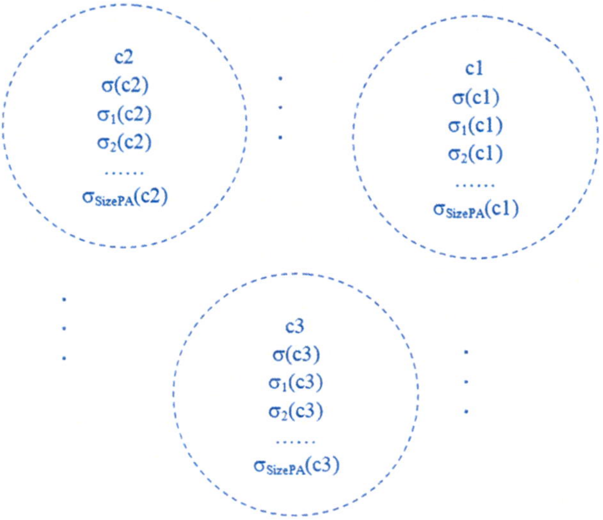

Fig. 1. Automorphism class

4 Results and Discussion

To validate our proposed improvement of HMWE method, we have computed its results for quadratic residue codes of lengths 167 and 199. Obtained results are the same compared to the available known enumerators of these two codes.

Table 1. Comparison of run time

	Run time (in hour)		
	167		199
	weight = 23	weight = 27	weight = 31
Algorithm A3 (old version), with cyclic permutation	0.006944	2.008333	26.80166
Algorithm A3 (new version), with Automorphism group	0.004722	1.138428	7.999104
Rate of reduction of run time	31.99%	43.31%	70.15%

To visualize the reduction of run time by using part of the Automorphism group (Algorithm A3: new version) instead of the cyclic permutations (Algorithm A3: old version), we propose comparing the run time of both versions of this algorithm. Table 1 presents the results of applying the two algorithms on the RQ (167) and RQ (199) codes.

From Table 1 we notice that by the use of a part of the Automorphism group, the execution time decreases in a very remarkable way. For example, the run time of algorithm A3 (b) requires 7.99 h against 26.80 h for A3 (a) which allows us to reduce the reduction time by 70.15%.

The found result is quite logical because from a given codeword and by using cyclic permutations it is possible to extract exactly n-1 other codewords while using the group of Automorphism we can extract SizePA-1 other codewords, such that «SizePA» far exceeds *n*. For example, for the code RQ (167) the size of the used Automorphism sub-group is 13861, which means that from a one codeword and with the use of the Automorphism group we can extract up to 13861 other codewords against 166 codewords deduced by a cyclic permutation. This clearly justifies the rapidity of the proposed version of HMWE method.

By application of A4 algorithm, we give respectively in Table 2 and 3 the form of E (239) and A(239).

Table 2. The form of E (239)

E[32]=15x
E[36]=20z
E[40]=6y
E[44]=9826364199840+458160x-240y-1020z
E[48]=2763033644875595-7367295x+4680y-76160z
E[52]=563183085819550560+45174480x-59280y+1788360z
E[56]=77261037323961831840-211734480x+548340y-18213120z
E[60]=7322909683422267896416+2887694160x-3948048y+101078600z
E[64]=487483179585402441457185-38820361515x+23030280y-199658880z
E[68]=23124096173038326785561760+327419611440x-111861360y-1779111180z
E[72]=791383455611960799658986240-1875782163840x+461428110y+21398087680z
E[76]=19750373698021058193293512320+7775368386240x-1640633280y-132944108220z
E[80]=362781601859740311736058315640-23803035878085x+5085963168y+601491878400z
E[84]=4943822284212046405243976027520+51916075625280x-13870808640y-2171777877280z
E[88]=50328269435601995149122652732800-64195613926080x+33521120880y+6522539197440z
E[92]=384989002876779970003077984401280-49952422572480x-72199337280y-16681854212640z
E[96]=2224039989507921879532824380655560+538240970911590x+139241579040y+36884173150720z
E[100]=9743513111032809569936392963683456-1767326895297600x-241352070336y-71234628471000z
E[104]=32483734621357894577193583667662080+4051617391497600x+377112609900y+121056231905280z
E[108]=82640392751525515503632846749283520-7351805034764640x-532394272800y-181977826241400z
E[112]=160772023328997238294213807312836435+11041134037414650x+680281570800y+242886434630400z
E[116]=239533136767497410553085899185966400-14005420639236000x-787694450400y-288550223010000z
E[120]=273553025761929117200239161941471680+15146808533056800x+827079172920y+305557758528000z

Table 3. The form of A (239)

A[31]=2x
A[32]=13x
A[35]=3z
A[36]=17z
A[39]=y
A[40]=5y
A[43]=1801500103304+83996x-44y-187z
A[44]=8024864096536+374164x-196y-833z
A[47]=552606728975119-1473459x+936y-15232z
A[48]=2210426915900476-5893836x+3744y-60928z
A[51]=122023001927569288+9787804x-12844y+387478z
A[52]=441160083891981272+35386676x-46436y+1400882z
A[55]=18027575375591094096-49404712x+127946y-4249728z
A[56]=59233461948370737744-162329768x+420394y-13963392z
A[59]=1830727420855566974104+721923540x-987012y+25269650z
A[60]=5492182262566700922312+2165770620x-2961036y+75808950z
A[63]=129995514556107317721916-10352096404x+6141408y-53242368z
A[64]=357487665029295123735269-28468265111x+16888872y-146416512z
A[67]=6551827249027525922575832+92768889908x-31694052y-504081501z
A[68]=16572268924010800862985928+234650721532x-80167308y-1275029679z
A[71]=237415036683588239897695872-562734649152x+138428433y+6419426304z
A[72]=553968418928372559761290368-1313047514688x+322999677y+14978661376z
A[75]=6254285004373335094542945568+2462199988976x-519533872y-42098967603z
A[76]=13496088693647723098750566752+5313168397264x-1121099408y-90845140617z
A[79]=120927200619913437245352771880-7934345292695x+1695321056y+200497292800z
A[80]=241854401239826874490705543760-15868690585390x+3390642112y+400994585600z
A[83]=1730337799474216241835391609632+18170626468848x-4854783024y-760122257048z
A[84]=3213484484737830163408584417888+33745449156432x-9016025616y-1411655620232z
A[87]=18453698793054064888011639335360-23538391772896x+12291077656y+2391597705728z
A[88]=31874570642547930261111013397440-40657222153184x+21230043224y+4130941491712z
A[91]=147579117769432321834513227353824-19148428652784x-27676412624y-6394710781512z
A[92]=237409885107347648168564757047456-30803993919696x-44522924656y-10287143431128z
A[95]=889615995803168751813129752262224+215296388364636x+55696631616y+14753669260288z
A[96]=1334423993704753127719694628393336+322944582546954x+83544947424y+22130503890432z
A[99]=4059797129597003987473497068201440-736386206374000x-100563362640y-29681095196250z
A[100]=5683715981435805582462895895482016-1030940688923600x-140788707696y-41553533274750z
A[103]=14076285002588420983450552922653568+1755700869648960x+163415464290y+52457700492288z
A[104]=18407449618769473593743030745008512+2295916521848640x+213697145610y+68598531412992z
A[107]=37188176738186481976634781037177584-3308312265644088x-239577422760y-81890021808630z
A[108]=45452216013339033526998065712105936-4043492769120552x-292816850040y-100087804432770z
A[111]=75026944220198711203966443412657003+5152529217460170x+317464733040y+113347002827520z
A[112]=85745079108798527090247363900179432+5888604819954480x+362816837760y+129539431802880z
A[115]=115774349437623748433991517939883760-6769286642297400x-380718984360y-139465941121500z
A[116]=123758787329873662119094381246082640-7236133996938600x-406975466040y-149084281888500z
A[119]=136776512880964558600119580970735840+7573404266528400x+413539586460y+152778879264000z

From these two tables, we notice that the list of unknowns R(239) contains three unknown variables R(239) = {R32 = E [32] = 15x, R36 = E [36] = 20z, R40 = E [40] = 6y}. By executing algorithm A4 for weights 31, 35 and 39 we can find the values of A[31] = 2x, A[35] = 3z and A[39] = y.

Table 4 summarizes the found results, noting that the value of A[39] was obtained by a Monte Carlo approximation on a sample of 500 000 000 codewords after 100 computing repetitions. So we have found A[39]$\simeq$1817857311 with the confidence interval [1702772771, 1949625796]. Therefore, E[40]$\simeq$10907143866. From this table we deduce the values of the three unknowns (x, y, z). By solving the systems presented in Tables 2 and 3, we find the likelihood weights enumerators of A (239) and E (239) shown in Tables 7 and 8.

Table 5 presents the Mykkeltveit method results. To verify the validity of these results we apply the Chinese remainder theorem and present the found results in Table 5.

Table 4. Results of HMWE method for EQR (239)

QR(239)	Weight w	Size of Automorphism	A_{w-1}	E_w
	32	28 441	113 764	853 230
	36	28 441	7 081 809	47 212 060

Table 5. The Mykkeltveit method results for EQR (239)

	H2	G_4^0	G_4^1	S3	S5	S7	S17	S239
Dimension	60	31	30	40	24	18	8	1
E [32]	150	6	0	0	0	0	0	0
E [36]	2 340	30	0	160	0	0	0	0

Table 6. Congruence of the number of codewords of weights 32 and 36 in EQR (239)

n	|G|	w	E_w	$E_w$mod|G|
239	6 825 840	32	853 230	853 230
239	6 825 840	36	47 212 060	6 257 020

From this table, the values of E32 and E36 are well verified and then the likelihood weight enumerators of A (239) and E (239) are presented respectively in Table 7 and 8.

Table 7. The likelihood weights enumerator of A (239)

A[31]=113764
A[32]=739466
A[35]=7081809
A[36]=40130251
A[39]=1817857311
A[40]=9089286555
A[43]=1725850809331
A[44]=7687880877929
A[47]=554188473418481
A[48]=2216753893673924
A[51]=122001124799863166
A[52]=441080989661043754
A[55]=18027795121003113334
A[56]=59234183969010229526
A[59]=1830725727324653130602
A[60]=5492177181973959391806
A[63]=129995525005778709075572
A[64]=357487693765891449957823
A[67]=6551827195499205471403413
A[68]=16572268788615637368843927
A[71]=237415036918375623555456783
A[72]=553968419476209788296065827
A[75]=6254285003469572557567995599
A[76]=13496088691697498676857253661
A[79]=120927200623017262103431093706
A[80]=241854401246034524206862187412
A[83]=1730337799464630173716341253160
A[84]=3213484484720027465473205184440
A[87]=18453698793080715015306776442088
A[88]=31874570642593962299166250218152
A[91]=147579117769365825491108839028536
A[92]=237409885107240675790044654089384
A[95]=889615995803317074387190772465416
A[96]=1334423993704975611580786158698124
A[99]=4059797129596709225226948680733650
A[100]=5683715981435392915317728153027110
A[103]=14076285002588941749029065527250142
A[104]=18407449618770154594884162612557878
A[107]=37188176738185664965815820348961718
A[108]=45452216013338034958219335982064322
A[111]=75026944220199848962994248061496943
A[112]=85745079108799827386279140641710792
A[115]=115774349437622342066920566222256500
A[116]=123758787329872158761190950099653500
A[119]=136776512880966101796742431102985700

Table 8. The likelihood weights enumerator of E (239)

E[32]=853230
E[36]=47212060
E[40]=10907143866
E[44]=9413731687260
E[48]=2770942367092405
E[52]=563082114460906920
E[56]=77261979090013342860
E[60]=7322902909298612522408
E[64]=487483218771670159033395
E[68]=23124095984114842840247340
E[72]=791383456394585411851522610
E[76]=19750373695167071234425249260
E[80]=362781601869051786310293281118
E[84]=4943822284184657639189546437600
E[88]=50328269435674677314473026660240
E[92]=384989002876606501281153493117920
E[96]=2224039989508292685967976931163540
E[100]=9743513111032102140544676833760760
E[104]=32483734621359096343913228139808020
E[108]=82640392751523699924035156331026040
E[112]=160772023328999676349273388703207735
E[116]=239533136767494500828111516321910000
E[120]=273553025761932203593484862205971400

5 Conclusion and Perspectives

In this paper, we have presented a new efficient version of the HMWE method for finding the weights enumerators for binary quadratic residue codes with lengths of 8m-1. These codes are very old, but they are very useful. In consequence of the new results found in this paper, approximated analytical performances of the length 239 are now available. The use of the Automorphism sub-group has yield to a height reduction of the run time complexity. For example, the reduction rate of run time passes 70% for weight 31 in the QR (199) code. In the perspective, we are envisaging a new technique that will enable us to gain run time and spatial complexities more and more.

References

1. Nouh, S., Aylaj, B.: Efficient serial concatenation of symbol by symbol and word by word decoders. Int. J. Innovative Comput. Inf. Control, **14** (2018)
2. Alaoui, M.S.E.K., Nouh, S., Marzak, A.: High speed soft decision decoding of linear codes based on hash and syndrome decoding. Int. J. Intell. Eng. Syst. **12**(1) (2019)
3. El Kasmi Alaoui, M.S., Nouh, S., Marzak, A.: Two new fast and efficient hard decision decoders based on hash techniques for real time communication systems. In: Mizera-Pietraszko, J., Pichappan, P., Mohamed, L. (eds.) RTIS 2017. AISC, vol. 756, pp. 448–459. Springer, Cham (2019). https://doi.org/10.1007/978-3-319-91337-7_40
4. Alaoui, M.S.E.K., Nouh, S., Marzak, A.: A low complexity soft decision decoder for linear block codes. In: Proceedings of the First International Conference on Intelligent Computing in Data Sciences (2017)

5. Faham, H., El Kasmi Alaoui, M.S., Nouh, S., Azzouazi, M.: An efficient combination between Berlekamp-Massey and Hartmann Rudolph algorithms to decode BCH codes. Periodicals Eng. Nat. Sci. **6**(2), 365–372 (2018)
6. Faham, H., El Kasmi Alaoui, M.S., Nouh, S., Azzouazi, M.: High performance decoding by combination of the Hartmann Rudolph decoder and soft decision decoding by hash techniques. In: Motahhir, S., Bossoufi, B. (eds.) ICDTA 2021. LNNS, vol. 211, pp. 781–790. Springer, Cham (2021). https://doi.org/10.1007/978-3-030-73882-2_71
7. Joundan, I.A., Nouh, S., Namir, A.: A new way to find the minimum distance of BCH and quadratic double circulant codes. In: Smart Application and Data Analysis for Smart Cities, 9 May 2018
8. Joundan, I.A., Nouh, S., Azouazi, M., Namir, A.: A new efficient way based on special stabilizer multiplier permutations to attack the hardness of the minimum weight search problem for large BCH codes. Int. J. Electr. Comput. Eng. **9**(2), 1232 (2019)
9. Yang, S., Kong, X., Shi, X.: Complete weight enumerators of a class of linear codes over finite fields. Adv. Math. Commun. **15**(1), 99 (2021)
10. Zheng, D., Zhao, Q., Wang, X., Zhang, Y.: A class of two or three weights linear codes and their complete weight enumerators. Discrete Math. **344**(6), 112355 (2021)
11. Clarck, G.C., Cain, J.B.: Error-Correction Coding for Digital Communication, New York Plenum (1981)
12. Brochero Martínez, F.E., Giraldo Vergara, C.R.: Weight enumerator of some irreducible cyclic codes. Des. Codes Crypt. **78**(3), 703–712 (2014). https://doi.org/10.1007/s10623-014-0026-6
13. Gaborit, P., Nedeloaia, C.-S., Wassermann, A.: On the weight enumerators of Duadic and quadratic residue codes. IEEE Trans. Inf. Theor. **51**, 402–407 (2005)
14. Tjhai, C., Tomlinson, M., Horan, R., Ahmed, M., Ambroze, M.: Some results on the weight distributions of the binary double-circulant codes based on primes. In: IEEE International Conference on Communications Systems, 1–30 October November 2006
15. Chen, C.: Computer results on the minimum distance of some binary cyclic codes. IEEE Trans. Inf. Theor. **16**, 359–360 (1970)
16. Su, W., Lee, C., Lin, T., Truong, T., Chang, Y.: On determination of the weight distribution of binary (168, 84, 24) extended quadratic residue code. In: Proceedings of the ISIT 2008, Toronto, Canada (2008)
17. Yang, S., Yao, Z.A.: Complete weight enumerators of a family of three-weight linear codes. Des. Codes Cryptogr. **82**, 663 (2017). https://doi.org/10.1007/s10623016-0191-x
18. Yang, S., Yao, Z.-A., Zhao, C.-A.: The weight enumerator of the duals of a class of cyclic codes with three zeros. Appl. Algebra Eng. Commun. Comput. **26**(4), 347–367 (2015). https://doi.org/10.1007/s00200-015-0255-6
19. Wang, X., Gao, J., Fu, F.-W.: Complete weight enumerators of two classes of linear codes. Cryptogr. Commun. **9**(5), 545–562 (2016). https://doi.org/10.1007/s12095-016-0198-1
20. Liu, H., Liao, Q., Wang, X.: Complete weight enumerator for a class of linear codes from defining sets and their applications. J. Syst. Sci. Complexity **32**(3), 947–969 (2019). https://doi.org/10.1007/s11424-018-7414-3
21. Nouh, S., Belkasmi, M.: A genetic algorithm for finding the weight enumerator of binary linear block codes. Int. J. Appl. Res. Inf. Technol. Comput. **2**, December 2011
22. Nouh, S., Aylaj, B., Belkasmi, M.: A method to determine partial weight Enumerator for linear block codes. Comput. Eng. Intell. Syst. **3**, October 2012
23. Alaoui, M.S.E.K, Nouh, S., Marzak, A.: Determination of partial weight enumerators of BCH codes by Hash methods. In: 4th International Conference on Wireless Technologies, Embedded and Intelligent Systems (WITS 2017), Fez-Morocco, 19–20 April 2017

24. Alaoui, M.S.E.K., Nouh, S., Marzak, A.: A fast method to estimate partial weights enumerators by hash techniques and Automorphism group. Int. J. Adv. Comput. Sci. Appl. (IJACSA), **8**(9) (2017)
25. Nouh, S., Alaoui, M.S.E.K., Marzak, A.: New efficient and fast method to compute the weights enumerators of large doubly even self-dual quadratic residue codes. Int. J. Intell. Eng. Syst. **11**(6) (2018)

Two-Stage Pipelining Implementation of the Secure Hash Algorithm SHA-3 on Virtex-5 and Virtex-6 FPGAs

F. Assad[1(✉)], M. Fettach[1], F. Elotmani[1], and A. Tragha[2]

[1] Information Processing Laboratory, Hassan II Casablanca University, Casablanca, Morocco
{fatimazahra.assad-etu, fadwa.alotmani-etu}@etu.univh2c.ma, mohammed.fettac@univh2c.ma

[2] Information Treatment and Modeling, Laboratory Hassan II Casablanca University, Casablanca, Morocco
abderrahin.tragha@univh2c.ma

Abstract. In recent years, many hardware implementations of the SHA-3 have been proposed to improve throughput, frequency, efficiency, or area. However, the increasing demand for hashing larger data requires other improved implementations of this function in terms of throughput. In this work, two different implementations of the Secure Hash Algorithm SHA-3 are proposed, the designs are based on the pipeline technique. Our main focus was to maximize the throughput of this function, also to analyze the impact of the use of the pipelining technique on this function The proposed designs are coded using VHDL language and implemented on two different devices Virtex 5 and 6 FPGA. The designs reach a maximum Throughput almost double 16.24 Gbps in the 2 stages external pipelining, and 17.01 Gbps in the internal pipelining. In contrast, it's worth 8.12 Gbps on Virtex 5 for 512 bits output length in the case of one stage. However, this increase in the throughput will negatively affect the Area needed to implement the design.

Keywords: SHA-3 · High throughput · Two-stage pipelining · KECCAK · Cryptographic hash functions · Hardware implementation · FPGA · FSM

1 Introduction

Cryptographic hash functions aim to generate message digest or hash value based on messages of arbitrary size. Cryptographic hash functions intersect with many important areas such as IOT [1, 2], WSN [3] Medicine [4], and many more.

Due to the successive attacks on hash algorithms like MD5, SHA-0, SHA-1, and SHA-2, The National Institute of Standard and Technology (NIST) has organized a competition to develop a more robust and secure new standard. The selection of the candidates was based on several Evaluation criteria, security, performance on hardware devices, simplicity, and flexibility. The competition was finalized in October 2012 with the KECCAK algorithm being the winner [5].

Y. Maleh et al. (Eds.): ICI2C 2021, LNNS 357, pp. 557–567, 2022.
https://doi.org/10.1007/978-3-030-91738-8_51

Differently, from the already existing hash functions MD5 [6], SHA-1 [7], and SHA-2 [8], Keccak or SHA-3 is based on the sponge function. The SHA-3 cryptographic hash algorithm is standardized in FIPS 202.

Generally, hardware implementations of hash functions are more secure than their software realizations. Furthermore, hardware implementations offer the flexibility to be optimized for high throughput or low hardware area cost [9].

During and after the NIST's competition, several implementations of SHA-3 (Keccak) are proposed [10–18]. Several optimization techniques are used to ameliorate the throughput, such as loop unrolling, pipelining, and re-timing.

The pipelining technique has been used in two different manners: internal pipelining, where the registers are inserted between the sub-function, in [13] Anashtasiou et al. introduces a pipeline stage between the π and the χ sub-rounds of the algorithm. In contrast, Mestiri et al. inserted two registers in the round transformation: the first register is inserted between the Pi operation and Chi operation, and the second register is implemented at the end of the KECCAK round [18]. The second method to exploit the pipelining is external pipelining, where each stage contains the entire compression function as the implementation of Michail et al. [19].

Summarizing, the mains contributions of this work are recapitulated as follows:

- We designed the pipelined architectures for the Keccak algorithm, considering all different output lengths (128, 256, 384, and 512), which provides optimized performance on FPGA devices.
- We provided two different approaches. The first one is a Sub-pipelining with two pipeline registers. The first register is inserted between the Pi and Chi steps, and the second is at the end of the round. The second approach is an external pipelining where the register is inserted between the two cores. Therefore, we conducted an extensive analysis of the advantages and disadvantages of each type of pipeline.
- We have checked the proper functioning of the proposed architecture through Post-Place and Route simulation via the ModelSim simulator.
- We conducted an extensive analysis and evaluation to compare the area (Slices), frequency (MHz), throughput (Gbps), and efficiency (Mbps/Slices) of the proposed designs with previous works.

The rest of this paper is organized as follows: The second section briefly introduces the SHA-3 specifications. Section 3 presents the proposed designs of the SHA-3 hash function. Section 4 presents the FPGA synthesis results and comparison, while the last section gives the paper conclusions.

2 SHA-3 Specifications

First of all, we should introduce the concept of state used in the rest of this work.

The 1600-bit state matrix of KECCAK is composed of 5×5 matrixes of 64-bit words, as appeared in Fig. 1.

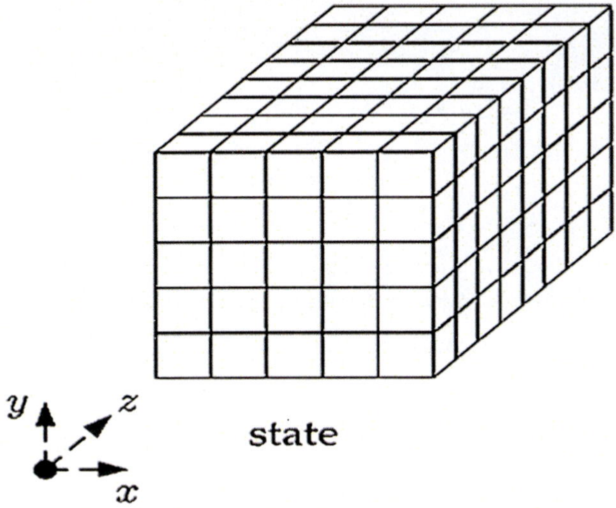

Fig. 1. The state of SHA-3

The SHA-3 algorithm is a cryptographic hash function designed by G. Bertoni et al. [20]. It is based on the sponge construction that comprises two phases as shown in Fig. 2. The main unit of SHA-3 (KECCAK-f permutation) consists of five separate steps: θ (Theta), ϱ (Rho), π (Pi), χ (Chi) and ι (Iota).

These are the Pseudo Codes of these sub-functions, including the mathematical formulas:

Pseudo Code: Theta step

```
1) for 0 ≤ x ≤ 4 do
        for 0≤ z ≤ 63
                C(x, z) = Theta-in(x, 0, z) xor Theta-in(x,1, z) xor Theta-in(x, 2, z)
                                        xor Theta-in(x, 3, z) xor Theta-in(x, 4, z)
        end for
   end for
2) for 0 ≤ x ≤ 4 do
        for 0 ≤ z ≤ 63 do
                D(x, z) = C[(x–1)mod 5 , z)] _ C[(x + 1)mod 5, (z–1)mod 64]
        end for
   end for
3) for 0 ≤ x ≤ 4 do
        for 0 ≤y ≤4 do
                for 0 ≤z ≤ 63 do
                        Theta-out(x, y, z) = Theta-in(x, y, z) xor D(x, z)
                end for
        end for
   end for
4) Return Theta-out
```

```
Pseudo Code: Rho step
1) for 0 ≤ z ≤63 do
        Rho-out (0, 0, z) = Rho-in (0, 0, z)
   end for

2) Let (x, y) = (1, 0)

3) for 0 ≤ t ≤ 23 do
a- for 0 ≤z ≤63 do
        Rho-out (x, y, z) = Rho-in [x, y, (z □ (t + 1)(t + 2)=2)mod 64)]
   end for
b- (x,y) = (y, (2x + 3y)mod 5)
   end for

4)Return Rho-out
```

```
Pseudo Code: PI step
1) for 0 _ x _ 4 do
        for 0 ≤ y ≤4 do
                for 0 ≤ z ≤ 63 do
                        Pi-out (x, y, z) = Pi-in [(x + 3y) mod 5, x, z]
                end for
        end for
   end for

2) Return Pi-out
```

```
Pseudo Code: Chi step
1) for 0≤ x ≤ 4 do
        for 0 ≤ y ≤ 4 do
                for 0≤ z ≤63 do
                        Chi-out(x, y, z) = Chi-in(x, y, z) xor not [Chi-in((x + 1)mod 5, y,z)]
                                and Chi-in((x + 2)mod 5, y, z)
                end for
        end for
   end for

2) Return Chi-out
```

```
Pseudo Code :Iota step
1)Let Iota-out (x, y, z) = Iota-in (x, y, z)

2) for 0≤ z ≤63 do
        Iota-out (0; 0; z) = Iota-in (0, 0, z) xor RC(z)
   end for

3) Return Iota-out
```

RC is a 64-bit value prefixed for each round.

The output hash value is truncated from the first r-bit [21]. It should be noted that the same function of compression is used during the 2 phases as shown in Fig. 2.

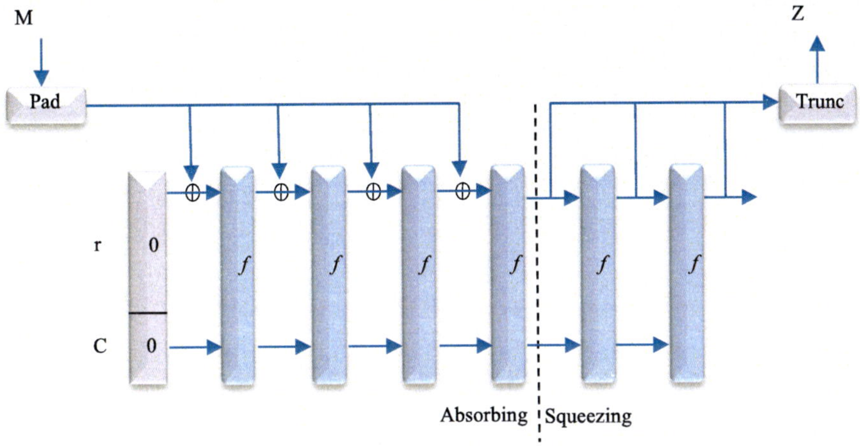

Fig. 2. Absorbing and squeezing phases of the sponge construction

3 The Proposed Designs

In this section, three different architectures are presented, the 1st one is basic architecture, the 2nd and 3rd designs are based on the pipelining technique. The basic proposed architecture has five inputs and two outputs, as shown in Fig. 3. This design is able to apply 4 types of padding [r, c] = [1152, 448], [1088, 832], [832, 768], as well as the Keccak [1024, 576], and generate 4 different types of outputs with different lengths.

3.1 The Basic Design

The message should be padded by adding a determined number of bits according to the "pad10*1" rule. After this, the message of multiple or equal size r must be cut into equal blocks. The below formula is used to get 1600 bits of data, that will be converted subsequently to a state, data = (Block xor r) ||C.

The last pre-treatment step is the conversion of the data to the state of 5×5 matrixes of 64-bit words where the state(x,y,z) = f(z + 64*(5*y + x)).

The main part of the design is the compression function which consists of 5 sub-functions; the mathematical formulas of these sub-functions are given in Sect. 2.

A Mux 2: 1 and a complementary DMux 1: 2 are used to apply the compression function 24 times on the data since Keccak is based on 24 rounds. These 2 components are controlled by 2 signals coming from the control unit S1 and S2, A register is needed to store the result of each round. Plus, a four to one Multiplexer selects the desired output size and the truncating unit to produce the final hash value with the desired length.

We opted for a ROM to store the 24 constants needed to calculate the Iota sub-function for the 24 rounds. The appropriate round constants (RC) are selected using a 5-bit counter Modulo 24 generated by the control unit which this counter S3 is the address-bus of the ROM.

The control unit is used to control and synchronize the tasks of the different units of the design.

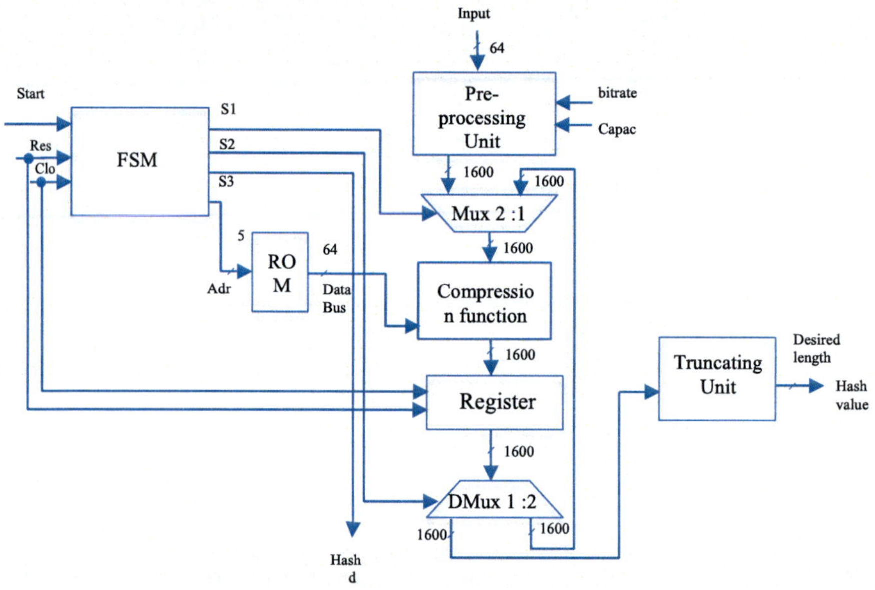

Fig. 3. Basic architecture

Based on the basic design presented in Fig. 3, two different architectures are derived using the pipelined structure. The first design is based on sub-pipelining, where the first register is inserted between the Pi and Chi sub-functions. In contrast, the second register is inserted after the IOTA sub-function at the end of the round, as depicted in Fig. 4. The second one is based on external pipelining. In this case, the register is inserted outside the sub-functions as depicted in Fig. 5.

3.2 The External Pipelined Design

The design allows outputting the first block after exactly 24 clock cycles and can process a new block every 12 clock cycles. This allowed processing multiple messages for hashing simultaneously. The output register of the first stage is used to feed the input of the second stage and two 4-bit counters are now used. These 4-bit counters steer the twelve iterations at each core and the respective round constants. In this case, two pipeline cores are used and according to that the size of the constant ROM is adjusted as shown in Fig. 4.

The first stage will receive the first block of the message while the 2nd stage is empty, then the block processed by the first stage will be sent to the 2nd stage to continue the remaining 12 rounds and the 1st stage will receive a new block and so on until the end of processing of all message blocks.

At the last cycle, the output is fed to the truncating unit in order to produce the final hash value.

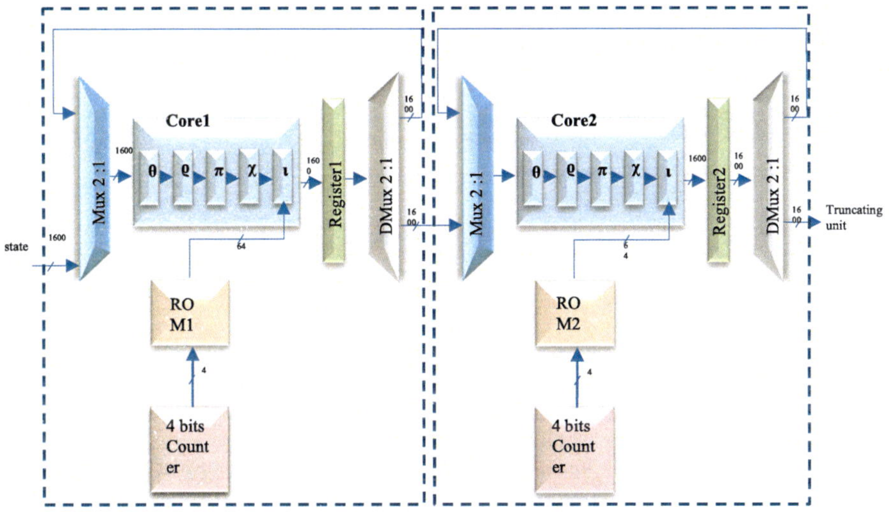

Fig. 4. External two-stages pipelining

3.3 The Internal Pipelined Design

Figure 5 represents the proposed design with the internal pipeline. At each pipeline stage, it is necessary to insert a register capable of storing the information processed in the previous iteration. This register stores the intermediate state processed in the previous cycle.

In the first round, the pipeline is filled with the original blocks of data. After that, the processed blocks are reused in the new round.

The principal disadvantage of this type of pipelining is that the transformations of the next block cannot start before the output of the previous block becomes available.

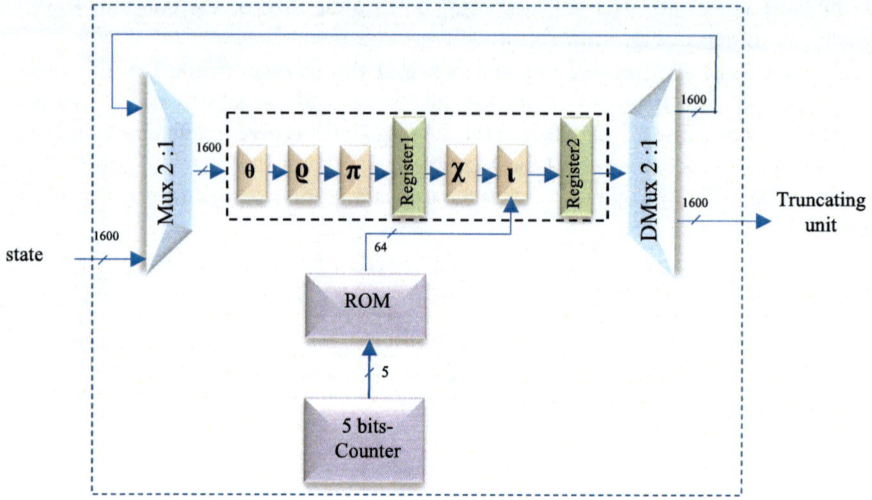

Fig. 5. Internal two-stages pipelining

4 Implementation Results

This section's basic objective is to analyze the results obtained, and the comparison between our implementation results and the results of the previous implementations already done for SHA-3 to examine their efficiency.

Two FPGA boards were used: a Xilinx Virtex 5 XC5VLX110-1ff1760, and a Xilinx Virtex 6 XC6VLX760-1ff1760. The complete implementation results are given in Table 1 for Virtex-5 and Virtex-6 devices, respectively. Moreover, Table 2 presents a comparison of Results for 512 bits output length. It should be noted that these designs are implemented without the use of any dedicated resource.

The measurements were focused on the consumed FPGA area resources, frequency, design output throughput, and efficiency.

The throughput is obtained by using the following Formula:

$$Throughput = \frac{block\,size(r) \times frequency}{Latency} \quad (1)$$

Where Latency presents the number of clock cycles needed to compute the final hash value.

The efficiency is calculated by using the formula below:

$$Efficiency = \frac{Throughput}{Area} \quad (2)$$

The hardware implementation of the SHA-3 in the case of external pipelining takes 2366 slices for 338.40 MHz frequency, this design achieves a throughput of 16.24 Gbps and 6.86 Mbps/slices for efficiency in Virtex 5. In the case of the internal

Table 1. Implementation results on virtex 5 and virtex 6 for 512 bits output length

FPGA devices	Architecture	Maximum frequency (MHz)	Area (slices)	Throughput (Gbps)	Efficiency (Mbps/Slices)
Virtex5	Basic design	338.409	935	8.12	8.68
	external 2 stages pipelining	338.40	2366	16.24	6.86
	2 stages sub-pipelining	354.50	2115	17.01	8.04
Virtex6	Basic design	376.081	1019	9.02	8.85
	external 2 stages pipelining	393.10	2499	18.87	7.55
	2 stages sub-pipelining	384.30	1813	18.44	10.17

pipelining, the implementation takes 2115 slices for 354.50 MHz frequency, where it achieves a throughput of 17.01 Gbps and 8.04 Mbps/slices for efficiency in Virtex 5.

In the case of the external pipelining 2 stages, the throughput is almost double 16.24 Gbps, and 17.01 Gbps in the case of the internal pipelining. Whereas it's worth 8.12 Gbps on Virtex 5 for 512 bits output length in the case of one stage.

By comparing the performance of the basic implementation and the implementations based on both types of pipelining, we notice that the throughput increases noticeably, this is justified by the fact that the number of clock cycles is the dominant factor of the throughput equation as shown in Eq. 1. On the downside, the occupied area increases noticeably, which affects also the TPS which decreases in the case of 2 stages as shown in Table 1.

The external pipeline technique brings the advantages of combined data throughput enhancement in multi-message hashing, where the function processes more than one message concurrently.

The internal pipeline technique also offers a data throughput improvement. The principal disadvantage of this type of pipelining lies in the fact that the transformations of the next block cannot start before the output of the previous block becomes available. In other words, the design can support single-block messages and multi-block messages. However, it cannot operate on multi messages cases.

In Table 2 the implementation results of the proposed designs are reported and the comparisons with some previous implementations.

Based on Table 2 [14] excels in efficiency with a value equal to 30.1 Mbps/Slices in V6. This is due to the reduced value of the area (Slices) occupied by the implementation 240 slices since this implementation belongs to the lightweight implementation category.

Table 2. Comparison of Results on Virtex 5 and Virtex 6 for 512 bits output length

References	Area (slices)		Frequency (MHz)		Throughput (Gbps)		Efficiency (Mbps/slices)	
	V5	V6	V5	V6	V5	V6	V5	V6
[18]	4793	–	317.11	–	12.68	–	2.71	–
[14]	240	–	301.02	–	7.224	–	30.1	–
[22]	1388	1167	287.39	394.01	11.50	15.76	8.28	13.50
[15]	1647	1181	137.01	251.7	2.39	4.39	1.45	3.72
[16]	1192	–	223	–	5.35	–	4.49	–
[23]	2573	–	285	–	5.70	–	2.21	–
2 stages sub-pipelining	2115	1813	354.50	384.30	17.01	18.44	8.04	10.17
2 stages external-pipelining	2366	2499	338.40	393.10	16.24	18.87	6.86	7.55

5 Conclusion

In this paper, Two pipelining designs of the SHA-3 algorithm were proposed, The implementation results are provided for 2 different FPGA devices, Virtex 5 and Virtex 6. Four metrics are set to make the comparison between the different implementations Area, Maximum frequency, throughput. As proved, the pipelining method has a positive effect on SHA'3 throughput. Still, the penalty is an increase in the area since the barriers of the stage add a significant area to the final hardware.

References

1. Rao, M., Newe, T., Grout, I.: Secure Hash Algorithm-3(SHA-3) implementation on Xilinx FPGAs, suitable for IoT applications. Int. J. Smart Sens. Intell. Syst. **7**(5), 1–6 (2020). https://doi.org/10.21307/ijssis-2019-018
2. Velmurugadass, P., Dhanasekaran, S., Anand, S.S., Vasudevan, V.: Enhancing Blockchain security in cloud computing with IoT environment using ECIES and cryptography hash algorithm. Materials Today: Proceedings, p. S2214785320364014 (2020). https://doi.org/10.1016/j.matpr.2020.08.519
3. Yang, Y., He, D., Kumar, N., Zeadally, S.: Compact hardware implementation of a SHA-3 core for wireless body sensor networks. IEEE Access **6**, 40128–40136 (2018). https://doi.org/10.1109/ACCESS.2018.2855408
4. Li, X., Ibrahim, M.H., Kumari, S., Sangaiah, A.K., Gupta, V., Choo, K.-K.R.: Anonymous mutual authentication and key agreement scheme for wearable sensors in wireless body area networks. Comput. Netw. **129**(2), 429–443 (2017). https://doi.org/10.1016/j.comnet.2017.03.013
5. Dworkin, M.J.: SHA-3 standard: permutation-based hash and extendable-output functions. Nat. Inst. Stand. Technol. NIST FIPS 202 (2015). https://doi.org/10.6028/NIST.FIPS.202
6. Rivest, R.: The MD5 Message-Digest Algorithm (1991). http://altronic-srl.com.ar/md5%20algoritmo.pdf

7. SHA-1 Standard, National Institute of Standards and Technology (NIST), Secure Hash Standard, FIPS PUB 180–1 (1995)
8. NIST: Federal information processing standard 180–2, secure hash standard. https://csrc.nist.gov/csrc/media/publications/fips/180/2/archive/2002-08-01/documents/fips180-2.pdf
9. Wong, M.M., Haj-Yahya, J., Sau, S., Chattopadhyay, A.: A new high throughput and area efficient SHA-3 implementation. In: 2018 IEEE International Symposium on Circuits and Systems (ISCAS), Florence, May 2018, pp. 1–5 (2018). https://doi.org/10.1109/ISCAS.2018.8351649
10. Neelima, S., Brindha, R.: 512 bit-SHA3 design approach and implementation on field programmable gate arrays. IJRES **8**(3), 169 (2019). https://doi.org/10.11591/ijres.v8.i3.pp169-174
11. Sideris, A., Sanida, T., Dasygenis, M.: High throughput implementation of the keccak hash function using the Nios-II processor. Technologies **8**(1), 15 (2020). https://doi.org/10.3390/technologies8010015
12. Mestiri, H., Kahri, F., Bouallegue, B., Machhout, M.: An efficient hardware implementation of KECCAK algorithm. **3**(12), 5 (2017)
13. Athanasiou, G.S., Makkas, G.-P., Theodoridis, G.: High throughput pipelined FPGA implementation of the new SHA-3 cryptographic hash algorithm. In: 2014 6th International Symposium on Communications, Control and Signal Processing (ISCCSP), Athens, Greece, May 2014, pp. 538–541 (2014). https://doi.org/10.1109/ISCCSP.2014.6877931
14. Arshad, A., Kundi, D.-S., Aziz, A.: Compact implementation of SHA3–512 on FPGA. In: 2014 Conference on Information Assurance and Cyber Security (CIACS), Rawalpindi, Pakistan, June 2014, pp. 29–33 (2014). https://doi.org/10.1109/CIACS.2014.6861327
15. Honda, T., Guntur, H., Satoh, A.: FPGA implementation of new standard hash function Keccak. In: 2014 IEEE 3rd Global Conference on Consumer Electronics (GCCE), Tokyo, Japan, October 2014, pp. 275–279 (2014). https://doi.org/10.1109/GCCE.2014.7031105
16. Sundal, M., Chaves, R.: efficient FPGA implementation of the SHA-3 hash function. In: 2017 IEEE Computer Society Annual Symposium on VLSI (ISVLSI), Bochum, Germany, July 2017, pp. 86–91 (2017).https://doi.org/10.1109/ISVLSI.2017.24
17. Rao, M., Newe, T., Grout, I., Mathur, A.: High speed implementation of a SHA-3 core on Virtex-5 and Virtex-6 FPGAs. J. Circ. Syst. Comput. **25**(07), 1650069 (2016). https://doi.org/10.1142/S0218126616500699
18. Mestiri, H., Kahri, F., Bedoui, M., Bouallegue, B., Machhout, M.: High throughput pipelined hardware implementation of the keccak hash function. In: 2016 International Symposium on Signal, Image, Video and Communications (ISIVC), pp. 282–286 (2016). https://doi.org/10.1109/ISIVC.2016.7894001
19. Michail, H.E., Ioannou, L., Voyiatzis, A.G.: Pipelined SHA-3 implementations on FPGA: architecture and Performance analysis. In: Proceedings of the Second Workshop on Cryptography and Security in Computing Systems - CS2 2015, Amsterdam, Netherlands, pp. 13–18 (2015). https://doi.org/10.1145/2694805.2694808
20. Keccak specifications.pdf
21. Paar, C., Pelzl, J.: Understanding Cryptography: A Textbook for Students and Practitioners. Springer, Berlin, Heidelberg (2010). https://doi.org/10.1007/978-3-642-04101-3
22. Kahri, F., Mestiri, H., Bouallegue, B., Machhout, M.: High speed FPGA implementation of cryptographic KECCAK hash function crypto-processor. J. Circ. Syst. Comput. **25**(04), 1650026 (2016). https://doi.org/10.1142/S0218126616500262
23. Provelengios, G., Kitsos, P., Sklavos, N., Koulamas, C.: FPGA-based design approaches of keccak hash function. In: 2012 15th Euromicro Conference on Digital System Design, Cesme, Izmir, Turkey, September 2012, pp. 648–653 (2012). https://doi.org/10.1109/DSD.2012.63

Securing MQTT Architecture Using a Blockchain

Raouya Akninᵃ and Youssef Bentaleb

ENSA, UIT, Kenitra, Morocco
raouya.aknin@uit.ac.ma

Abstract. Since MQTT (Message Queuing Telemetry Transport) protocol is designed for resource-constrained environments, it includes only elementary security measures, which are mainly sufficient for transmitting non-critical data. However, for applications that transmit critical data, it is recommended to add some security measures. Many articles have used blockchain to improve security in the IoT environment; however, most of them neglect that many IoT systems are already functional. It is tough to rebuild the architecture from scratch. This article aims to propose a secured MQTT architecture using blockchain and smart contracts without affecting the overall operation of the MQTT protocol. Instead, it adds some parameters to the standard MQTT messages. The new architecture is based on blockchain and smart contract to control and store the transactions between clients and brokers. It is composed of two phases: The registration phase, where a client asks for a token to use for publishing or receiving data, and the Publishing phase, where the brokers network verifies the validity of the token and the integrity of the data before allowing the client publication. By comparing the current architecture and the improved one, we can conclude that the architecture proposed in this article will improve the security of MQTT protocol by using tokens for authentication, topic key for ensuring data confidentiality, brokers network to avoid a single point of failure and data hash to ensure the integrity of data. Also, the overall operation of the protocol was not affected.

Keywords: Internet of things · MQTT · Blockchain · Smart contract

1 Introduction

According to oracle [1] "IOT (internet of things) is a network of physical objects "things"—that are embedded with sensors, software, and other technologies to connect and exchange data with other devices and systems over the internet". The protocols used in IoT are lightweight to meet the requirements of constrained environments in terms of resource and energy. Hence, they include just elementary security measures.

There are some attempts to improve the security of IOT protocols by using the internet security measures such as the TLS (Transport layer security) protocol. However, they are still heavy and inadequate for this environment. Because IoT systems transit from infancy step to a mature one, and to avoid security breaches, the security measures must be taken seriously. Blockchain is one of the technologies used to ensure

Y. Maleh et al. (Eds.): ICI2C 2021, LNNS 357, pp. 568–578, 2022.
https://doi.org/10.1007/978-3-030-91738-8_52

the confidentiality, integrity, and authenticity of the data. It facilitates recording transactions and tracking assets through a shared, immutable ledger [2].

Many articles have proposed secured architectures using a blockchain [3–6]. However, these architectures must be rebuilt from scratch. The challenge in the IoT environment is to find a lightweight protocol that is robust in terms of security without changing the overall operation.

To meet these needs, this paper's contribution consists of proposing an enhancement of MQTT architecture, as one of the most used protocols in the IoT environment, by using blockchain and smart contract and adding some parameters to the standard MQTT messages without affecting the overall operation. The architecture will use a token for authentication, a Topic secret key to ensure confidentiality, a broker's network to avoid a single point of failure, and a data hash to ensure integrity. The remainder of this article is structured as follows: First, it exposes the current MQTT architecture, its strengths, and its challenges. Next, it briefly defines the technologies used to secure the new architecture, namely Blockchain and smart contract. Then, it discusses some related work. After that, it proposes an improvement of MQTT architecture, and finally, it summarizes the main ideas that have been discussed in this article.

2 The Current MQTT Architecture: Strengths and Challenges

2.1 MQTT Components

As depicted in Fig. 1, The MQTT architecture is composed of three components: Publisher who sends the data related to a specific topic to a broker. Subscriber: who registers in a particular topic in the broker to receive a notification when this topic is updated. The broker gets the data from different publishers and sends them to subscribers already registered in this topic.

2.2 MQTT Architecture: Strengths and Challenges

MQTT protocol presents many advantages, making it the most used protocol in the IoT field [7]. Indeed, it is a lightweight protocol, and consequently, it can be used in constrained environments. It prevents data loss because it uses a TCP as an underlying transport protocol, and it uses an acknowledgment for every message. In addition to that, the MQTT protocol avoids subscribers requests and, consequently, bandwidth congestion thanks to its publish-subscribe model. On the other hand, In the current MQTT architecture, hackers can exploit some vulnerabilities and, therefore, can cause harm to confidentiality, integrity, and data availability [8]. Indeed, When MQTT protocol doesn't use the Transport layer security (TLS) protocol, the communication between the clients and the broker is in plain text format.

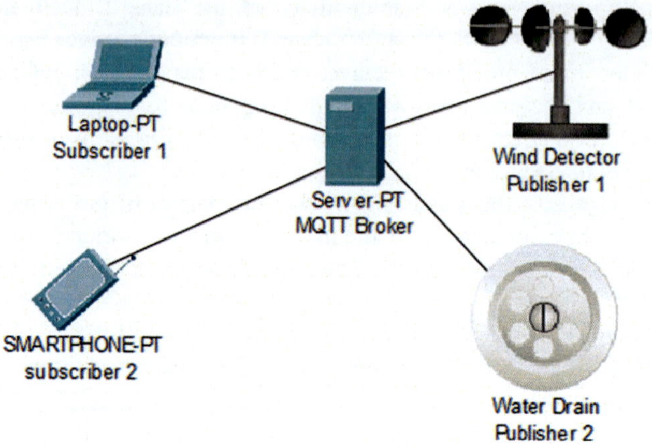

Fig. 1. The current MQTT architecture

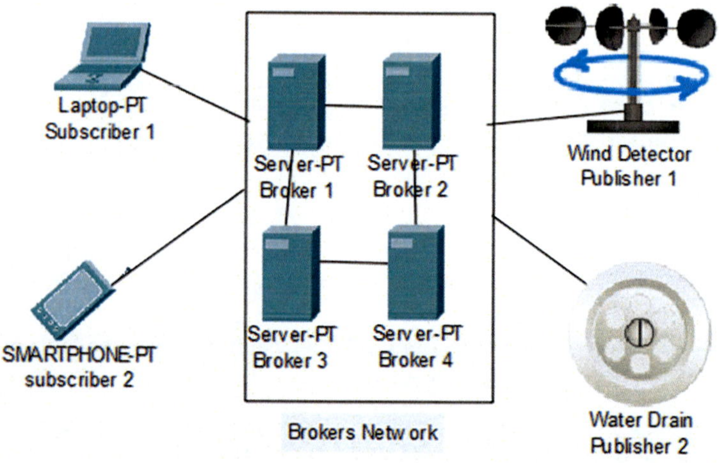

Fig. 2. The enhanced MQTT architecture

Hence, any sniffer can read the packets transmitted in the network (Man in the Middle). It can also steal the credentials and use them to publish or subscribe to the broker (Identity spoofing). It can publish a large amount of data to saturate the brokers or subscribers' memory or saturate the bandwidth (Denial of service).

In addition to that, Credentials are permanent; therefore, if a sniffer gets this critical information, he will use it until the client changes it. In the MQTT protocol, the broker also presents a single point of failure. Indeed, if a broker has broken down, the communication between publisher and subscriber is no longer possible. To avoid sending the credentials every time, the broker stores the session information in its internal storage. Thus can provide storage saturation at the broker level.

3 Blockchain and Smart Contract

3.1 Blockchain

A blockchain is distributed database composed of a list of ordered blocks Fig. '3'. Each block is composed of a header and body. A body block contains transactions. A header is composed of information related to a block, namely Block version, Timestamp, Merkle tree root hash, Parent Block hash. Each block is connected to the previous one using the Parent block hash. The first block in the blockchain is called a genesis block, and its field Parent block hash is set to zero. Because all the users in the network can verify and validate a blockchain without an intermediary, a consensus algorithm is required. A consensus algorithm [9] is a procedure through which all the Blockchain network peers reach a common agreement. In other words, consensus algorithm establishes trust between unknown peers in a distributed computing environment. The most common consensus algorithm are: Proof of work (POW), and Proof of stake (POS).

3.2 Smart Contract

According to the definition of IBM [10] "Smart contract is a program stored on a blockchain and executed when predetermined conditions are met. It typically is used to automate the execution of an agreement so that all participants can be immediately certain of the outcome, without any intermediary's involvement or time loss".

4 Related Work

Several works have proposed security improvement of the IoT architecture. Indeed, The article [8] proposed a token for authentication to an MQTT broker. Although this architecture resolves the problem of sending credentials in plain text format and consequently avoids passive sniffing, the problem of a single point of failure persists since a single server issue the token. In addition to that, the delivered token is stored in the local device's storage which can cause problems in a constrained environment. The articles [3–6] have proposed a new blockchain architecture to improve security in an IoT environment. These architectures can be applied to new IoT systems; however, for IOT functional systems, it will be challenging to rebuild them from scratch. The article [11] has proposed an architecture that combines MQTT protocol and blockchain to improve IoT data security. The architecture has used brokers network to avoid a single point of failure and blockchain to record all transactions history. However, the article doesn't mention the communication between different architecture components, the authentication method between the IoT devices and the gateway, and the MQTT broker. This article will propose an improvement of the MQTT protocol as the most used one in the IoT environment without affecting the overall operation of its architecture [12].

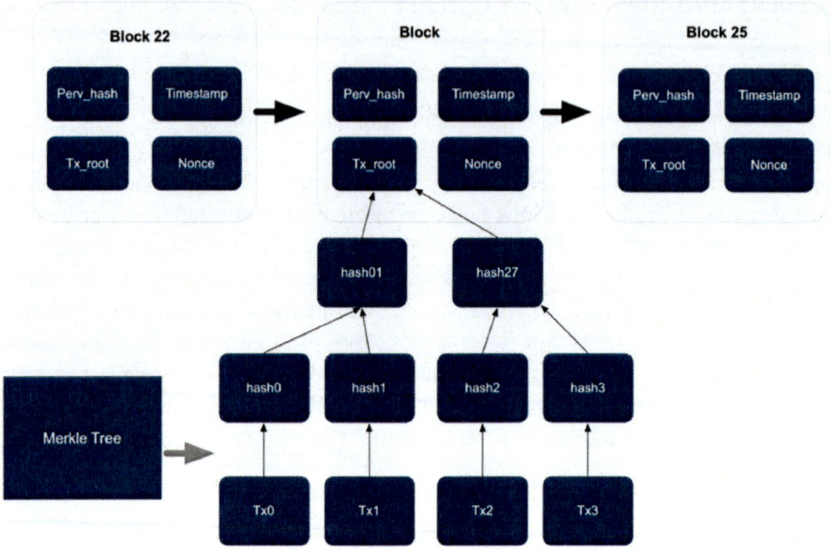

Fig. 3. Blockchain structure

It is based on adding parameters to standard MQTT messages and using brokers network instead of one to avoid a single point of failure.

5 An Improvement of MQTT Architecture

This section proposes an improvement of MQTT architecture (Fig. 2) using blockchain and smart contract for controlling and storing all the transactions between clients and brokers network. The new architecture comprises two main components: Clients (Publisher/Subscriber) and Brokers network (Each broker executes smart contract) instead of one to avoid a single point of failure. It is based on two communication phases: Registration and publishing. And it defines three types of transactions: Creating a new topic, Token Request and publishing data.

5.1 Messages

Table 1 below establishes a comparison between the messages used by MQTT [7] and enhanced MQTT. The new architecture has introduced a new type of message calling "Request for publishing," a Publish message containing the following information: id_client, topic_Name and Hash of data. Connect Message in the enhanced MQTT must include a list of topics that the client will publish or subscribe on.

Table 1. Comparison between MQTT and enhanced MQTT messages

Messages	Connect	Connect ACK	Publish	Publish ACK	Subscribe	Subscribe ACK	Request for Publishing
MQTT	X	X	X	X	X	X	–
Enhanced MQTT	X (It also includes a list of topics that the client must publish or subscribe on)	X	X	X	–	–	X(A special type of publish message which contains id_client, topic_Name and Hash of data)

5.2 Transactions Types

The proposed architecture uses three types of the transaction generated by brokers network: the first transaction is the creation of a new topic: The result of this transaction is the generation of secret topic key for a new topic. The second transaction is creating a token: at the end of this transaction, a winner broker generates a token containing a client identity, a list of topics, the rights on publishing or subscribing, and the expiration date. The last transaction is publishing data, and the result of this transaction consists of sending a Topic secret key after verifying a client's token. The activity diagram depicted in Fig. 4 shows all conditions that brokers network must verify before creating a token, a new topic, or publishing data.

5.3 Token

The new architecture requires a token to publish in or subscribe on a specific topic. The token is stored in the blockchain and it is composed of the following elements: ID_-client, list of topics that contains couples of name of the topic and the client right and expiration date of the token.

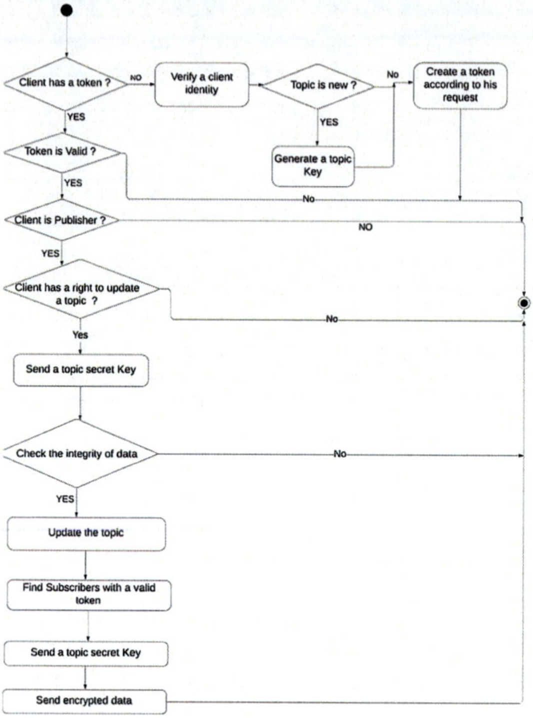

Fig. 4. Smart contract conditions

5.4 Registration Phase

As illustrated in the sequence diagram 7.a and algorithm 5, during this phase, a client sends to network brokers all the topics that will publish or subscribe on, in its turn, the brokers' network verifies the client identity, creates a new topic if it doesn't exist and generates a token for a client (Figs. 5 and 7).

```
Algorithm Name:Registration_pbase
Declaration of variables:
Connect connect_Message;Booleen certificate; int i,j,N, Length; Topic
Blockchainl;String Winner broker,Topic_Name,Key_topic ;Date exp_date; Token
Blockchain2[];
Struct Topicrights {
Topic_Name   String
Topic_rights  char};
Struct Connect {
Id_client String
List_of_topics[]  Topicrights };
Struct Topic{
Topic_Name String
Key_Topic int };
Struct Token{
Identification     Connect
Exp_date       Date };

Start
Read(connect_Message)
If (certificate=0)
Then break;
Else
For (i=0;i< connect_Message . List_of_topics .length;i++)
 For(j=0;j<Blockchainl.length;j++)//Search the topic in the blockchain
   If (Blockchainl [j]Name==list_of_topics.Name)
   Then
   Length=j;
   break;
   else
   Length=j;
    end if ;
 end for;
 if(Length == list_of_topics.length)//if the topic doesn't exist create one
 then
   Winner_broker←Mining (N)//function for Mining with 'N 'the numbers of the participating brokers
   Topic_Name← list_of  _tocics[i].Name;
   Key_Topic←Random(2048 bits);
   Blockchainl ← (Topic_Name, key_Topic);//Add the new topiv and its key to a blockchain
end if
end for
Winner_broker← Mining(N);
Exp_date←current_date()+60j//function which returns the current date
Blockthain2←(connect_Message .Id_client ,connect_Message. List_of
_Topics,exp_date)//client Token
Write  (connect_ACK)//MQTT Message connect_ACK
end if
end
```

Fig. 5. Registration phase algorithm

5.5 Publishing Phase

As illustrated in sequence diagram 7.b and algorithm 6, Before publishing in a specific topic, the client sends a request for a publication. When receiving this message, the brokers' network verifies the client's token, and then a winner broker sends a secret topic key. When a client sends encrypted data, the winner broker verifies its integrity, finds subscribers, and sends them the key and the encrypted data (Figs. 6, 7).

```
Algorithm Name:Publishing_phase
Declaration of variables : Req_for_Pub  Request_Pub; Pub_Transaction  Blockchain3[];String
Key_Secret_Topic=0;int i,j,l,k=0;
Struct Req_for_Pub{
Id_client    string
Topic_Name   string
Hash         string};
Struct Pub_Transaction{
Id_client     string
Topic_Name    string
Winner_broker string};

Start
  Read(Request_Pub);
  While (i< Blockcahin2.length|| Key_Secret_Topic!=0)
       If (Blockchain2[i].Id_client== Request_Pub. Id_client&& Blockchain2[i].
exp_date<date_actuelle) //Verify the client identity and validity of the Token
             Then
                 While(j< Blockchain2[i].  list_Topics.length || Key_Secret_Topic!=0)
                     If (Blockchain2[i].list_Topics[j].name== Request_Pub .Topic_Name&&
                      Blockchain2[i].list_Topics[j].rights==w)//Verify the right of the writing in this topic
                      Then
                      Winner_broker← Minning (N);
                      Blockchain3← Request_Pub .id_client, Request_Pub .Topic_Name,
                      Winner_broker ;
                      While (k< Blockchain1.length)//Find the Topic secret Key
                        If(Blockchain1[k].Topic_Name== Request_Pub .Topic_Name)
                        Then
                          Key_Secret_Topic← Blockchain1.Key_Topic
                          Write(Key_Secret_Topic);//Send the Topic secret Key
                          Break;//sortie de la boucle while
                         else  k++;
                         end if
                         end while
                       else j++;
                       end if

                 end while
         else i++;
         end if

  end while
if(Key_Secret_Topic!=0)
Then
Read(Encypt PubMessage);
PlaintextMessage← Decrypt(Encypt PubMessage);
Empreinte← Hash(PlaintextMessage);
   If (Empreinte== Request_Pub .Hash)//Verify the data integrity
   Then
     Write (Publish_ACK);
     For(i=0;i++;i< Blockchain2.length)//Find the subscribers in the topic
      If (Blockchain2[i].exp_date>current_date)//Verify the validity of the token
      Then
         For(j=0;j++;j< list_Topics.length)
           If (Blockchain2[i]. list_Topics[j].name==Topic_Name &&
              Blockchain2[i].List_Topics[j].rights==r)//Verify the rights of reading
           Then
             Write( Key_Secret_Topic);
             write (encrypted_data);
             Break();

             End if
          End for
          End if
End  for
End if
End if
End
```

Fig. 6. Publishing phase algorithm

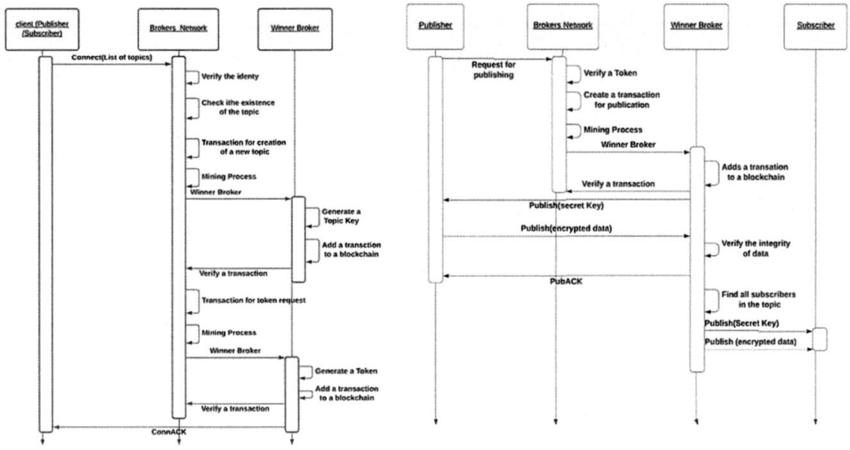

Fig. 7. (a) Registration phase; (b) Publishing Phase

5.6 The Contribution of the Article

By comparing the proposed architecture and the current one, we can conclude that the proposed architecture in this article will improve the security of the MQTT protocol. Indeed, all transactions history is stored in a shared, immutable ledger. In addition to that, the new architecture used a token for authentication instead of sending credentials in plain text format or using TLS protocol in the resources constraint environment. For each topic, a secret key is generated and stored in the blockchain to ensure data confidentiality. The new architecture relies on a Brokers network instead of one broker to avoid a single point of failure. The brokers' network also verifies the data integrity thanks to a data hash. Publishing and registration phases are implemented in smart contract to keep all the history of the communication. The advantage of this architecture also lies in the fact that it does not affect the overall operation of the MQTT protocol; Instead, it just adds parameters to the standards MQTT messages.

6 Conclusion

In summary, it is important to note that the proposed architecture in this article is dedicated to the IoT applications that transmit critical data. For others, the current MQTT architecture is very satisfying. The new architecture uses a blockchain and smart contract to enhance the security of the MQTT protocol. The purpose of the proposed architecture is to improve the security of MQTT architecture without affecting the overall operation. The improved architecture uses a token for authentication. It also ensures data confidentiality by using a secret topic key, data integrity with hashes, and the high availability by using brokers network. Although this architecture provides advantages for enhancing security parameters, implementing this architecture in a constraint IOT environment may introduce some problems in terms of delay, power consumption, or network congestion. It is also important to note that this article focuses just on the application layer security without considering other layers. Thus, in future work, we focus on the test and the improvement of this architecture.

References

1. What is the Internet of Things (IoT)? https://www.oracle.com/internet-of-things/what-is-iot/. Accessed 19 Sep 2021
2. What is Blockchain Technology? - IBM Blockchain. https://www.ibm.com/topics/what-is-blockchain. Accessed 19 Sep 2021
3. Singh, M., Singh, A., Kim, S.: Blockchain: a game changer for securing iot data. In: 2018 IEEE 4th World Forum on Internet of Things (WF-IoT), pp. 51–55 (2018). https://doi.org/10.1109/WF-IoT.2018.8355182
4. Alphand, O., et al.: IoTChain: a blockchain security architecture for the internet of things. In: 2018 IEEE Wireless Communications and Networking Conference (WCNC), pp. 1–6 (2018). https://doi.org/10.1109/WCNC.2018.8377385
5. Fakhri, D., Mutijarsa, K.: Secure iot communication using blockchain technology. In: 2018 International Symposium on Electronics and Smart Devices (ISESD), Bandung, pp. 1–6 (2018). https://doi.org/10.1109/ISESD.2018.8605485
6. Truong, H.T.T., Almeida, M., Karame, G., Soriente, C.: Towards secure and decentralized sharing of iot data. In: 2019 IEEE International Conference on Blockchain (Blockchain), pp. 176–183 (2018). https://doi.org/10.1109/Blockchain.2019.00031
7. Steve: How MQTT Works -Beginners Guide, 04 June 2018. http://www.steves-internet-guide.com/mqtt-works/. Accessed 15 Mar 2021
8. Bhawiyuga, A., Data, M., Warda, A.: Architectural design of token based authentication of mqtt protocol in constrained iot device. In: 2017 11th International Conference on Telecommunication Systems Services and Applications (TSSA), pp. 1–4 (2017). https://doi.org/10.1109/TSSA.2017.8272933
9. Consensus Algorithms in Blockchain - GeeksforGeeks. https://www.geeksforgeeks.org/consensus-algorithms-in-blockchain/. Accessed 15 Mar 2021
10. What are smart contracts on blockchain | IBM. https://www.ibm.com/topics/smart-contracts. Accessed 29 June 2021
11. Katende, M.M.: Combining MQTT and Blockchain to improve IoT Data Security, IoTBDS (2020). https://www.academia.edu/42677166/Combining_MQTT_and_Blockchain_to_improve_IoT_Data_Security. Accessed 15 Mar 2021
12. Zheng, Z., Xie, S., Dai, H., Chen, X., Wang, H.: An overview of blockchain technology: architecture, consensus, and future trends. In: 2017 IEEE International Congress on Big Data (BigData Congress), pp. 557–564. Honolulu, HI, USA (2017). https://doi.org/10.1109/BigDataCongress.2017.85

The Human Factor Capabilities in Security Operation Center (SOC)

Samir Achraf Chamkar(✉), Yassine Maleh, and Noreddine Gherabi

LaSTI, ENSAK, Sultan Moulay Slimane University, Beni Mellal, Morocco
yassine.maleh@ieee.org, n.gherabi@usms.ma

Abstract. The need for Cybersecurity Operation Center to protect the information system against cyberattacks has become increasingly common and a must to face the new sophisticated and advanced threat. The socio-technical system of the SOC contains three building blocks Human resources, technologies, processes. The literature review, announcements, and publications of the cybersecurity industry reveal that the human factor is considered the weakest link in cybersecurity and inside the Security Operation Centers (SOC). It represents the most important factor to guarantee the best performance. Many researchers shed light on the human factor challenges inside the SOC environment and how these challenges could be addressed, reduced or mitigated. However, these research papers do not consider the complexity, unpredictability and interdependent and evolving nature of the SOC systems.

This study aims to explore the human challenges inside the complex and interdependent socio-technical SOC system. To this end, we employed survey-based questionaries with 40 SOC analysts and 5 experts alongside the daily observation of SOC analysts and interviews with SOC experts. The finding of this study will help SOC managers and SOC designers better understand the challenges faced by the SOC analysts and take into account the interdependent and evolving nature of the Security Operation Centers.

Keywords: Human factors · Security Operation Center · Weakness · Challenges

1 Introduction

Maintaining the availability, confidentiality and authenticity of the high value information systems becomes a real challenge for security professionals under the increasing number of sophisticated attacks and the continuous improvement of evasion techniques using advanced and various attack vectors to gain access and conduct a full system compromise.

Many papers and threat reports from the leading security suppliers reveals that 10.18% of internet user computers experienced at least one malware-class attack targeting several industries [1]. Also, these attacks become complex and nearly impossible to avoid, given the openness of today's network. And the most of the leading security suppliers' stats are that the human is the weakest link in the security chain.

Y. Maleh et al. (Eds.): ICI2C 2021, LNNS 357, pp. 579–590, 2022.
https://doi.org/10.1007/978-3-030-91738-8_53

Companies need mature processes to prevent human errors and respond to threats with speed and agility [2, 3].

In response, the practice of cybersecurity focuses on preventing business damage or data loss by adopting new technologies to face the targeted and sophisticated threats using multilayered and defense in depth strategies using next-generation firewalls, antivirus solutions, hybrid intrusion detection and prevention systems, SIEM, EDR … etc.

Moreover, they opt to create SOC with effective, highly focused resources to detect and respond to cyber-attacks and security breaches by monitoring all security feeds in real-time and respond immediately in case of abnormal behavior. However, organizations fail to get the full benefits from their investment in SOC due to many reasons related to organizational, technological and human capabilities and issues [4].

Cybersecurity professionals and security researchers report the importance of frequently measuring the SOC performances, capabilities, and effectiveness toward facing treats and responding to cyber-attacks to maintain continuous improvements. Anything that is not measured cannot be improved [5]. However, this is not a simple task due to the interdependent and evolving nature of the Security Operation Centers as a socio-technical system [6].

In this article, we shed light on the human factor capabilities as the most important component and the weakest link simultaneously with the care of the specification of the nature of the SOC sociotechnical system. Hence, we ask the following questions in that respect:

- What are the most frequently faced challenges by the SOC analysts from a narrative perspective?
- How can these interconnected challenges impact the SOC the overall SOC capabilities?
- What are the weaknesses of the actual performance metrics used by SOC managers and decision-makers to assess the SOC analysts' capabilities?

The remainder of this paper is structured as follows. First, we provide a general overview of the SOC components and background of the SOC. In the third section, we present the research methodology. In the fourth section, we present the finding and result discussions based on the information gathered in the second phase in conjunction with the evolving and interdependent SOC. Finally, we provide a conclusion in the last section.

2 Research Methodology

The research methodology in this study consists of two main phases. In the first one, a literature review on security operation centers' challenges related to human factors has been conducted. In the second phase, we use the determined challenges during phase one on the survey-based questionaries and the interviews to determine the real challenges from multiple sources and multiples perspectives.

In this research, quantitative and qualitative analyses are conducted with both survey-based questionaries and interviews. The role of a qualitative study is to explore information from participants to generate the said case study rather than just list numeric data. Thus, this could also yield enough information to answer the research questions by exchanging questions between interviewer and interviewees in two-way communication during the interview.

According to [7] simple size of qualitative search is usually a range (4–50) due to the large volume of data collected and it's selected based on the appropriateness of participants as the best representatives with the required knowledge of the research topic. For the research questionaries, we used the announced challenges from:

- SOC managers identified problems.
- SOC Analysts reported challenges.
- Security professional and read team experts.
- Publications of cyber security industry related to SOC.
- Research papers on the field.

The table below describes the questionaries used on the questionary based survey (Table 1):

Table 1. Survey questions

Survey questions
Too many alerts that we can't look into (lack of correlation between alerts)
Lack of visibility into the IT security infrastructure
Performances of hardware and software resources (SIEM, Firewalls, IPS, IDS, EDR … etc).
Lack of Cyber Awareness, Knowledge and Skills
Lack of automation and orchestration.
Lack of Global Communication & Collaboration
Not following procedures and best practice.
Lack of process or playbooks.
Insufficient local innovation in cybersecurity.
Not following guidelines and best practices.
Not sharing responsibility
Lack of accountability.
Issues between the organization's IT operations and SOC
Lack of Metrics: Metrics should be Meaningful, Measurable, Support Improvement.

We posed other direct questions for all the SOC human resources, such as:

- Are there any other problems not mentioned in the list above?
- What are the most sought-after skills when hiring a SOC analyst?
- What are the three most common reasons that cause tourn over of SOC talents?
- Do you have any suggestions to improve SOC service performance?

The simple included 40 participants for the quantitative analysis and the qualitative study was conducted with 5 SOC experts.

The study was conducted with 45 participants currently working in several internal SOC and SOC service providers in their various roles and level of experience to make a more balanced view for this study.

3 Data Analysis and Results

The result of the quantitative or the questionary-based survey are shown in Fig. 1 below:

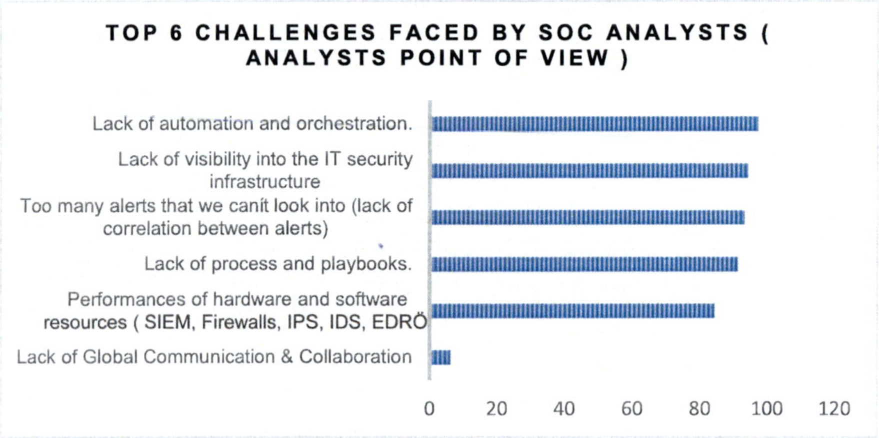

Fig. 1. Statistics of the challenges faced by SOC Analysts

Depending on the SOC analyst's answers, we have the following challenges:

- Lack of automation and orchestration comes in first place with 97%;
- Lack of visibility into the IT security infrastructure with 95%;
- The large volume of alerts means analysts often struggle to find accurate attacks among the mist of many alerts 93%;
- Lack of process or playbooks with 92%;
- Lack of SOC training and attack simulations;
- Lack of business knowledge and associated risks to adapt detection on customer risks;
- Lack of practical security knowledge;
- Lack of adequate metrics for SOC analyst's evaluation;
- Lack of treat hunting and red teaming inside the SOC.

3.1 Lack of Automation

Automation is the machine implementation of low-level security actions that are small parts of a more significant task to reduce human intervention in security operations. The purpose is to ensure better harmony and collaboration between the SOC analysts and the automation and not replace SOC analysts better at pattern recognition, improvisation, and decision-making with incomplete information.

Automation within the SOC is generally referred to as Security Automation and Orchestration (SAO) or Security Automation, Orchestration, and Response (SOAR), as shown in Fig. 2.

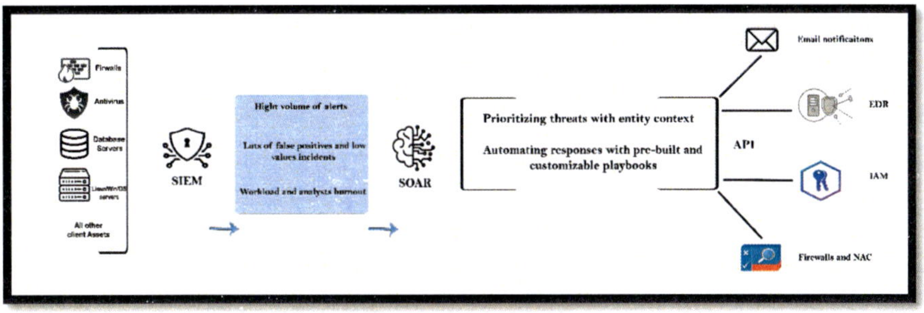

Fig. 2. Automation in the SOC

Many SOC Analysts are worried about the other effects of automation and whether it will take their roles and jobs in the future. But on the contrary, automation should not be seen as a limitation to the analyst's career, instead of a springboard that can help you become a better analyst.

The SOC analyst is the most valuable resource inside the SOC. He can perform tasks where the machines cannot, such as discovering new compromise and attacks patterns and taking charge of the automation task.

Also, automation can be considered a force multiplier when it comes to assisting analysts in handling the number of events and responding to the alerts with specific details. For example, after a malicious file detection and deletion by the antivirus, many events need to be addressed, and appropriate notes need to be written on the ticketing platform to close the ticket. This operation is performed manually by the SOC analysts, who spend what seems like hours a day copying and pasting the same notes and making the same actions. Moreover, the automation reduces the analyst's fatigue and overall stress. It makes the SOC a fun and challenging place to work while allocating more time to perfume more advanced and deeper event analysis. And senior analysts will spend more time training junior analysts and developing documentation and skills.

The automation can depend on many internal factors such as the complexity and the number of events gathered by the network and whether the tasks are possible to be automated or always need a human verification and intervention due to the changes and complexity of the pattern. It requires time, planning, coordination and maintenance to ensure the automated tasks are performing well.

For mature SOCs, we can see a dedicated profile alongside the SOC analysts for automating tasks to ensure consistency of some incident handling and responses, reducing the information bias and using higher fidelity data sources.

3.2 Lack of Visibility into IT Infrastructure

SOC needs to achieve and maintain clear visibility on the network, and any blind spot can increase the probability of a successful breach and reduce the SOC effectiveness. The absence of data and logs of a partition of the information system means that the SOC ensures no monitoring at that point.

It is straightforward to get flooded with data logs from different log sources using Syslog, agents, and API system call. The SOC needs to take concrete, requirements-based, value-driven approach to decide whether the logs must be collected, processed, used, and stored. The most frequent question, in this case, is what log data should be used and gathered by the SOC.

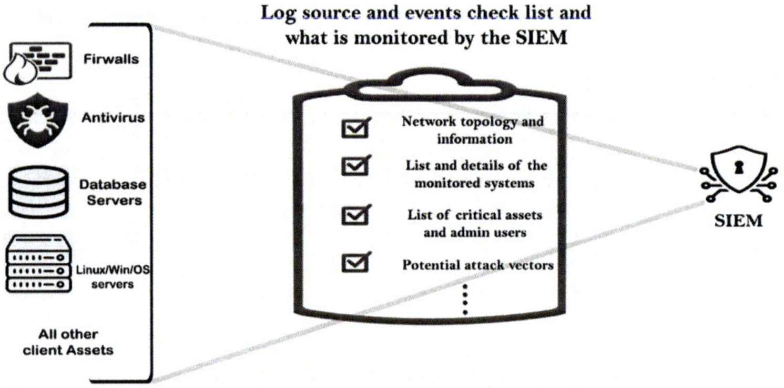

Fig. 3. Event logs collection

Data logs received from different sources are not equal. They contain different types of information (connections with external domains by proxy servers, IP addresses from firewalls, malicious file detections from antivirus servers, audit logs from servers and database audit from database servers, etc.), as shown in Fig. 3. and the SOC must prioritize which data to aggregate, process, and store, according to what it expects to get out of that data.

The answer to this question will always vary depending on the target environment and it requires an operation driven approach to prioritize which system we collect from and what data from the same itself by answering questions related to the following factors [8]:

- Mission performed by the monitored system;
- Mission criticality;
- Clarity and visibility of logs produced by the log source;

- Position of the log source in the potential attack vectors;
- Availability of fields of interest for correlation or forensics purposes;
- The overall volume of logs being generated.

3.3 False Positives

False-positive remains a significant challenge for SOC analysts and the whole SOC capabilities for any monitored system. It consumes analysts' time dealing with false security incidents and occasionally responding to real attacks requiring deep analysis, which impacts both the SOC capabilities and the experience of the SOC analysts inside, leaving them vulnerable to severe attacks.

Alarms do not always fire when a real security incident occurs and does not lead to immediate blocking hosts and IP addresses and isolate machine to remediate and reduce the potential impact.

Security tools do not always alert when something real attack happens and these attack alerts fall into the following four categories, as shown in Fig. 4:

- **True positives:** A real and correct detection of malicious activity.
- **True negatives:** The activity is benign, and no alert has been generated.
- **False positives:** A fake detection, the system alerts when a benign activity occurred
- **False negatives:** A real attack without being detected by the security measures.

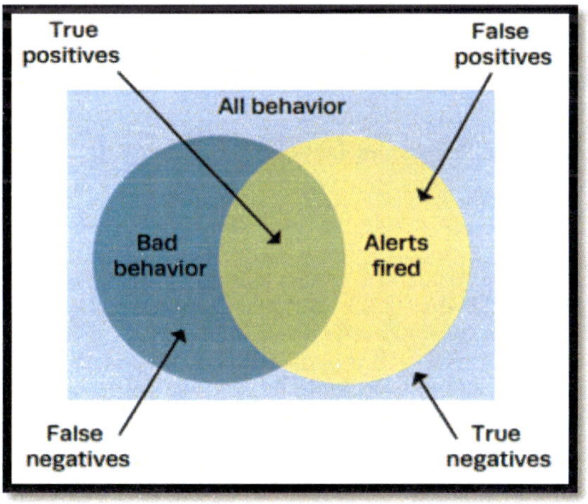

Fig. 4. The four categories of activities

The security solution could fire false positives due to weak defined signatures and patterns, leading to incorrect matching of the expected behavior, which compels the security analysts to spend inordinate amounts of time investigating and digging into legitimate alarms. Suppose the number of false positives bypasses the accepted threshold that the SOC analysts can handle. In that case, they ignore the full alerts related to the log sources that expose the systems to unmonitored attacks.

Achieving a high detection rate involves using the correct data feeds in coordination with SIEM engineers and rules implementers alongside the continuous tuning of the noisy rules. Others suggest the automation to reduce the number of false positives [9] requiring receiving the correct data with the coordination of SIEM engineers and rules/use cases implementers alongside the continuous fine-tuning of the noisy rules and use of automation reduce the number of false positives. Businesses can also put more effort into addressing misconfiguration issues to minimize the burden on analysts. Tuning of policies and excluding false-positive forms part of the work of an analyst [6].

3.4 Lack of Processes and Playbooks

Cybercriminals are sophisticated and in continuous improvement to bypass the up-to-date security solutions and at the same time, they know what exactly works for them. They are business savvy and try to target the weakest links in the security chain, starting with human errors and skills shortage to follow events from inception to resolution. That great responsibility comes with tremendous pressure for SOC analysts.

Playbooks help address the skill shortage and human error and ensure consistency by providing a clear insight into the most common types of investigation taken by the SOCs to drive mean time to resolution. Also, playbooks are considered as skills improvement factor for new analysts as it defines the SOC source of truth and common security knowledge.

The playbooks provide a clear insight into responding to cyber incidents and step-by-step actions and guidance during a cyber incident response in coordination with all SOC resources. The aim is to face security incidents using the most appropriate incident response, engaging the right resources.

Playbooks serve as a development factor toward sharpening the incident response plan, processes and procedures to detect, investigate, eradicate, and recover from a cyber incident and ensure the mapping, coordination, and communication with the stockholders.

3.5 Lack of SOC Training and Attack Simulations

The challenges of continuous monitoring, detecting and responding to a cyber incident inside the SOC are located in the intersection between technology, human, and processes. The initial detection of malicious behaviors and activities inside the SOC relay on human actors in their various roles in the first place and then to technological measures such as firewalls, IDS, antivirus solutions that warn human operators and forward information for further processing.

Responding to cybersecurity incidents requires necessary security background; moreover, an offensive security background can be much more effective due to the prior knowledge of cyber-attacks behavior and potential patterns generated by security solutions.

SOC analysts need to know what they are monitoring and the potential attack vectors for what they have. And what are the potential log sources, event logs among the significant number of collected data that need to be tackled and processed first for fast and effective detection, investigation, eradication, and recovery from a security incident.

Depending on the 2020 Devo SOC Performance Report, Fig. 5 below shows the most OSC-identified attacks in 2020 and 2019 [10].

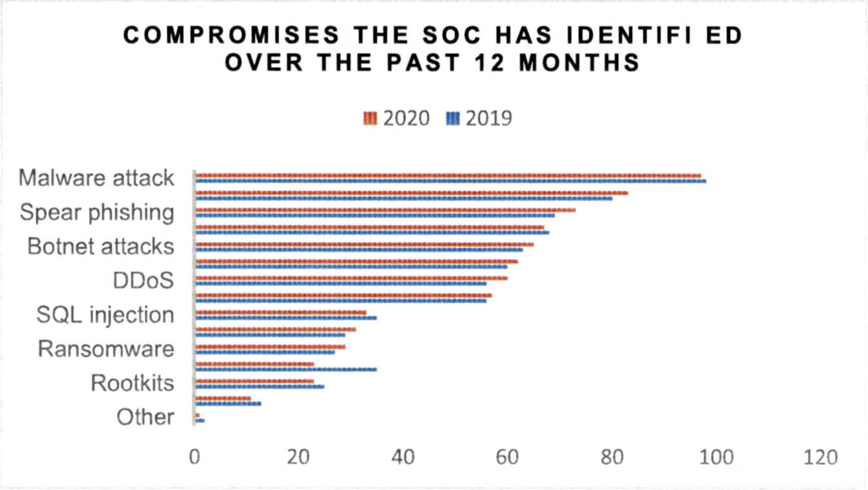

Fig. 5. Compromises identified by the SOC during 2020

Detecting and responding to each attack requires the basic knowledge of the attack lifecycle in coordination with generated patterns from firewall systems, Intrusion Prevention Systems, Web Application Firewall, Database Firewall, Intrusion Detection System, Anomaly Detection System, Web and database vulnerability scanners, anti-virus system other system administrators to identify the expected behavior of the monitored system and any deviation or suspicious pattern detection could be flagged and investigated. Figure 6 shows the attack lifecycle.

Fig. 6. Attack lifecycle

SOC analyst's security capabilities are considered the combination of security proficiencies knowledge and the practical skills for the analyst to behave correctly to ensure effective security of operational systems [11]. Security knowledge defines the thoracal information acquired by the SOC analysts about cyber threats, attacks life cycle, vulnerabilities, attack patterns and impact on the target. Skills outline the technical ability to detect and respond immediately in case of a cybersecurity incident.

Security knowledge and technical skills need to be measured frequently and improved. The literature review reveals an attacks simulation where SOC members conduct a simulated attack against a constituency segment to assess the target's resiliency to an actual attack. To get a clear insight into what can be detected using the implemented measures and evade being detected.

Attack's simulation can improve knowledge and technical skills alongside identifying the monitored system's blind spots and potential attack vectors. This simulation improves analysts' performances and provides new patterns to develop new use cases to protect against successful attacks during the simulation.

3.6 Performance of Hardware and Software Resources

We noticed that when security analysts talk about hardware and software performance during our survey, they usually mean the SIEM, including event log management and correlation of events gathered from different logs or security sources and event notification.

The implementation of SIEM requires deep analysis and previous planning and it involves a lot of time to achieve defined security goals with the SIEM system implementation.

As much as the deployment of the SIEM seems to be simple at first, it is necessary to know that it is only initial implementation that involves locating and configuring devices to send logs to the SIEM and achieving a higher level of security in the SIEM entire network never stops. Correlation rules adjustment is an iterative process and by analyzing the results of the created correlations, the methods for decreasing the number of false-positive/negative results are constantly revealed. For example, rules thresholds can be set too high resulting in unnoticed actual attacks. In contrast, on the other side, too low thresholds can give a large number of false positives and should be continuously fine-tuned [12].

3.7 Lack of Adequate Metrics for SOC Analyst's Evaluation

SOC managers use several methods to assess the performance and capabilities of SOC analysts. However, Agyepong et al. [13] suggest a lack of adequate metrics and measures for evaluating the performance of analysts. And SOC analysts are not satisfied with how their efforts are measured.

SOC managers focus on quantitative more than qualitative metrics to assess the performance of the SOC analysts. Among these metrics, we have the list below:

- The average time to detect a security incident is the time frame between the security incident and its detection by the SOC analysts;
- The average time to respond which the tame taken to respond to the detected incident properly involve applying the technical countermeasures and closing the case;

- Number of processed and analyzed alerts during the analyst's shift;
- The number of resolved security issues or closed tickets by shift;
- Number of incidents detected within the shift or a specific timeframe;
- Time spent on operations such as analyzing, responding, delivering details, etc.;
- Time spent on each security incident;
- A measure of the competency and experience of the analyst.

The literature that analysts are not satisfied with this time-based measurement as it does not consider several aspects of their work pointed out how many of their efforts are not considered under existing metrics.

Such metrics do not provide a clear insight on analyst's performance and can't be used to differentiate analyst's due to the following reasons:

- Time-based does not consider the complexity and the severity of the raised and closed incidents and deal with all tickets the same way.
- Deal with all analysts as if they were equal and with the same capabilities and skills.
- Does not take into consideration the other operational tasks.
- Does not consider the whole SOC system and only focuses on analyst-security incident interaction statistics.

4 Conclusion

Assuming the decomposability of the Security operation center and then focusing attention solely on one SOC analyst as the weakest link is unrealistic, given the complex and evolving nature of SOC systems. Moreover, measuring analysts independently as part of the whole system is unrealistic and can't differentiate and give the exact rating for two analysts in different environments. It becomes clear that more research is needed to improve existing metrics and measures to capture the overall efforts of analysts, given the inter-dependent of the SOC components.

The finding of our studies will provide a clear insight to the SOC managers, SOC analysts, and decision-makers about the challenges faced by the SOC analysts and the actual measurement metrics. However, more research is needed on the relationship of the three factors human, technology and processes inside the SOC sociotechnical system to figure out all the challenges that impact the analyst's performances and how to face these challenges using a complete model based on the analysts and SOC managers perspective and how to address these challenges using the same model.

References

1. Kaspersky: Kaspersky Security Bulletin 2020 Statistics, p. 26 (2020)
2. F. M. Services and S. Report: HR-Trends 2020. HR-Trends 2020 (2019). https://doi.org/10.34157/9783648132616
3. Crowdstrike: CrowdStrike 2020 Global Threat Report. Crowdstrike, pp. 1–32 (2020)
4. Mansfield-Devine, S.: Creating security operations centres that work. Netw. Secur. **2016**(5), 15–18 (2016). https://doi.org/10.1016/S1353-4858(16)30049-6

5. Forte, D.: An inside look at security operation centres. Netw. Secur. (2003). https://doi.org/10.1016/S1353-4858(03)00509-9
6. Zimmermann, V., Renaud, K.: Moving from a "human-as-problem" to a "human-as-solution" cybersecurity mindset. Int. J. Hum. Comput. Stud. **131**, 169–187 (2019)
7. Quick, J., Hall, S.: Part two: qualitative research. J. Perioper. Pract. **25**(7–8), 129–133 (2015). https://doi.org/10.1177/1750458915025007-803
8. Zimmerman, C.: Cybersecurity Operations Center (2014)
9. Wall, T., Rodrick, J.: Jump-Start Your SOC Analyst Career (2021)
10. Socs, T.: 2020 Devo SOC Performance Report TM (2020)
11. Ani, U.D., He, H., Tiwari, A.: Human factor security: evaluating the cybersecurity capacity of the industrial workforce. J. Syst. Inf. Technol. **21**(1), 2–35 (2019). https://doi.org/10.1108/JSIT-02-2018-0028
12. Vilendečić, B., Dejanović, R., Ćurić, P.: The impact of human factors in the implementation of SIEM systems. J. Electr. Eng. **5**(4), 196–203 (2017). https://doi.org/10.17265/2328-2223/2017.04.004
13. Agyepong, E., Cherdantseva, Y., Reinecke, P., Burnap, P.: Challenges and performance metrics for security operations center analysts: a systematic review. J. Cyber Secur. Technol. **4**(3), 125–152 (2020). https://doi.org/10.1080/23742917.2019.1698178

A Systematic Review on Software Defined Networks Security: Threats and Mitigations

Yassine Maleh(✉), Imade Fahd Eddine Fatani, and Khalid El Gholami

TCI Team, ENSAK, University Sultan Moulay Slimane, Beni Mellal, Morocco
{y.maleh,i.fatani,k.elgholami}@usms.ma

Abstract. The centralized architecture of the SDN is proving to be a double-edged sword. On the one hand, it gives the network administrator a global view through the controller to facilitate network management. On the other hand, this centralization is a vulnerability for SDN networks, taking over a controller is taking over the overall network. This paper aims to provide a comprehensive review of state of art, accompanied by categorizing the research literature into a taxonomy that highlights each proposal's main characteristics and contributions to the SDN's different layers. Based on the analysis of existing work, we also highlight key research gaps that could support future research in this area.

Keywords: Software defined network · SDN threats · SDN attacks · SDN mitigations

1 Introduction

Today, the Internet, or so-called computer networks, is a key component of businesses, schools, and homes. The Internet connects firms, knowledge and families on a global scale. However, over the last 20 years of traditional computer networks, we have seen the emergence of limitations such as dependence on unchanged protocols, distributed control, hardware that is difficult to reprogram [1], complex and unwieldy networks [2]. The need for new devices with networking capabilities and the Internet of Things (sensors, cameras, alarms, among others) brings challenges to network management. To solve traditional networks' problems, the Software-Defined Network technology emerged, which brings a new paradigm for network management. It separates the control plane from the data plane, centralizing the former in a device, the controller. All decisions to forward a packet with SDN go through this centralized software outside the switches. The switches forward, when so defined, packets based on decisions made in that software. Each new requirement or behavior with SDN does not require a change in hardware, either changing or upgrading, so only the control software needs to be updated or modified. SDN lowers operating costs since it only requires switches to communicate using OpenFlow protocol.

Despite its advantages, the SDN faces several threats, such as the denial of service (DoS) attack, considered the primary threat to these networks. Because of the centralized management in the SDN architecture, this type of attack can easily overload the switches' SDN controller and switch tables (TCAMs), resulting in a critical degradation of network performance. Research topics based on traditional security techniques

Y. Maleh et al. (Eds.): ICI2C 2021, LNNS 357, pp. 591–606, 2022.
https://doi.org/10.1007/978-3-030-91738-8_54

are numerous to protect networks from different types of attacks. Still, techniques using the SDN paradigm to improve security are new, promising and widely discussed in the literature. Hence, it has become necessary to analyze the recent SDN security threats and propose an up-to-date and detailed classification, which considers current and future threats related to the three layers of SDN [3]. We believe that this contribution would help to identify future research directions in this domain. This paper mainly aims to make the following contributions:

- It defines various requirements of a good threat and mitigations classification;
- It analyses the SDN security threats;
- It studies and analyses in-depth the SDN security mitigations and solutions for the different SDN layers.

The structure of this paper is as follows. Section 2 describes the research methodology. Section 2 presents the survey methodology. Section 3 reviews the related SDN threats and vulnerabilities. Section 4 presents the most important SDN mitigation techniques for each SDN Layer classification. Finally, Section 5 concludes the paper.

2 Research Methodology

This research investigates the following hypothesis: It is possible to balance security and performance in the control plane communication of an SDN environment, between controllers and switches, with a security level. It does not compromise the network's security. To support the hypothesis, the following research questions will be answered: 1. What are the main vulnerabilities, attacks, and existing solutions to provide security in SDN environments? This question aims to understand the current security scenario in SDN to provide a framework of knowledge on the subject to define the requirements to support the hypothesis. Although several works, which aim to organize the research on the topic, have already been developed. None of the existing studies in the literature on security in SDN used a well-defined methodology to cover the entire scope of works involving the topic.

To assist in developing the research proposal, a systematic mapping study (SMS - Systematic Mapping Study) [4] was conducted on SDN's current state of security. An SMS is a common methodology to help conduct research, such as in the medical area. Since 2007, its applicability in the Computing area has grown substantially, especially in Software Engineering [5]. According to Petersen [4], an SMS provides a categorization structure, offering a visual summary, mapping, and results. This requires less effort, while providing an overview with a better selection of information.

Initially, SMS was geared towards software engineering and was usually recommended for research areas with a range of relevant and high-quality early-stage studies. As security in SDN is early and high-quality studies are emerging, SMS can contribute to categorizing these studies. The SMS was first organized by defining the research questions. For this, it was necessary to think about the opportunities that security in SDN produces. The primary categories related to the topic were defined: vulnerabilities, attacks, and defense mechanisms. To highlight the necessity of creating an up-to-date and detailed survey of SDN threats and mitigations, we propose two research questions that we will try to address in this paper:

- RQ1: What are the main threats and vulnerabilities in SDN?
- RQ2: What are the main approaches to provide security in SDN?

To select the papers relating to these research questions, we rely on a three-stage selection procedure: (1) identifying search terms, (2) selection of sources, and (3) identifying and applying inclusion/exclusion criteria for selected papers:

Search Terms: This stage aims to select the most appropriate keywords in terms of providing our review. Also, synonyms, alternative spellings of the main elements of the field of threats, and mitigations in the SDN context are included to find relevant papers. Therefore, keywords such as "SDN threats", "SDN vulnerabilities", "SDN attacks", "SDN Security", "SDN security issues", "SDN mitigations", "SDN security solutions" were used for finding relevant papers.

To define the search string, the Boolean operations "OR" was used to select optional words and synonyms and "AND" to choose the terms related to population, intervention, and effect, resulting in the search string (("software-defined network" OR "software-defined networking" OR "SDN") AND ("security" OR "secure" OR "attack" OR "vulnerability" OR "threats" OR "integrity" OR "mitigations")).

Selection of Sources: The articles from journals as well as conference proceedings are considered for our review. Some references for downloaded articles were looked upon, and publications related to interest were also included. Other databases were not used due to either the complexity (or impossibility) of instantiating the search string, presenting a high index of repeated articles with the IEEE and ScienceDirect databases, or performing only one indexing (like Google Scholar).

Inclusion/Exclusion Criteria: Table 1 shows the summary of the inclusion-exclusion criteria for the SMS.

Table 1. Summary of the inclusion-exclusion criteria for selection papers

Criterion	Rational
Inclusion 1. The early-stage study should propose a solution for security in SDN	The SDN is our reference field. Thus, we need papers that directly proposed threat classification for SDN
Inclusion 2. The early phase study must describe attacks or vulnerabilities in SDN	Both academic and industrial taxonomies are essential to this study
Inclusion 3: Publication period	Papers published between 2010 and 2020
Exclusion 1. A study that does not focus on SDN security	The emphasis of this paper is only on taxonomies that discuss web security issues
Exclusion 2. Non-English papers	The papers that did not publish in English are excluded
Exclusion 3. The study has not been published between 2010 and 2020	The papers will be excluded after a full-text reading. If the focus is not on SDN threats and mitigations

3 SDN Security Threats and Attacks

3.1 SDN Security Threats and Attacks Classification

The SDN has many advantages over traditional networks due to the separation between the control plane and the data plane, making the networks programmable, flexible, and scalable. However, this has led to other security issues unique to the SDN. Although the idea seems to have convinced many cloud and network service providers of its many benefits, it cannot be widely adopted if its security challenges are not addressed.

Targeted attacks can disrupt the SDN software controller, and since the management subsystem is placed on the outside, it becomes the most vulnerable point of the entire system. Once every packet was routed separately, and there was no single control center that could be easily attacked, a controller failure would disrupt the whole network. Such a center is an attractive target for attackers. However, if the central control fails, the switches are likely to go offline in route calculation mode and start working with the old technology - albeit slowly, without breaking the connection. Therefore, device designers are unlikely to abandon proven technologies completely and will initially use "pure" SDN/OpenFlow.

The objects for a targeted attack in SDN have become more than for a classic IP network: the controller, the communication channels between the controller and the switches, the switches themselves, and the communication channels for data transmission. IP protocol is focused on the fact that routers monitor the load and performance of the device's channels. In SDN, the controller monitors the network's state, which receives data on channels' failure and load "from outside". That means there is a delay in receiving information and making decisions, which does not contribute to reliability.

Also, the network management protocol can be dangerous because it does not assume "intelligence" in the switch, so sending multiple commands to devices that already support OpenFlow could disrupt the network or a specific application. It is possible that by using this protocol, attackers will be able to insert devices that they control into the data flow route, allowing them to intercept network traffic and conduct other types of attacks. Incidentally, hardware manufacturers have already begun to implement support for OpenFlow in their products, and users who had not planned to build their SDN may be vulnerable to the unauthorized use of this technology, so companies will have to monitor the presence of the protocol OpenFlow in their network.

Another SDN issue is applications running on the controller and reconfiguring the network to meet an individual application or virtualization management system's requirements. At this point, it is an open question as to how these applications will compete for network resources in real-time. Developing such network applications is complex, and there is doubt whether cloud users will do so optimally. This means that it is highly likely that at some point, some applications will monopolize network resources, effectively shutting down all other applications. Thus, there is another point at which application developers can go wrong and implement the network optimization algorithm incorrectly. Finding this error is problematic because it only appears under certain conditions and high loads. Theoretically, configuration errors should be detected by the software that manages the network, but this requires SDN technology to be supported by the manufacturers of such software. In any case, the data center operator must monitor user applications and restrict them from monopolizing network resources.

These challenges need to be highlighted so that appropriate security measures can be taken proactively. Security analysis by several authors [6, 8] in the literature has shown that the SDN suffers from many security threats. As shown in Fig. 1., there are several threats and vulnerabilities on the different layers and interfaces of the SDN network architecture. In this section, we will list threats to all layers and interfaces of the SDN.

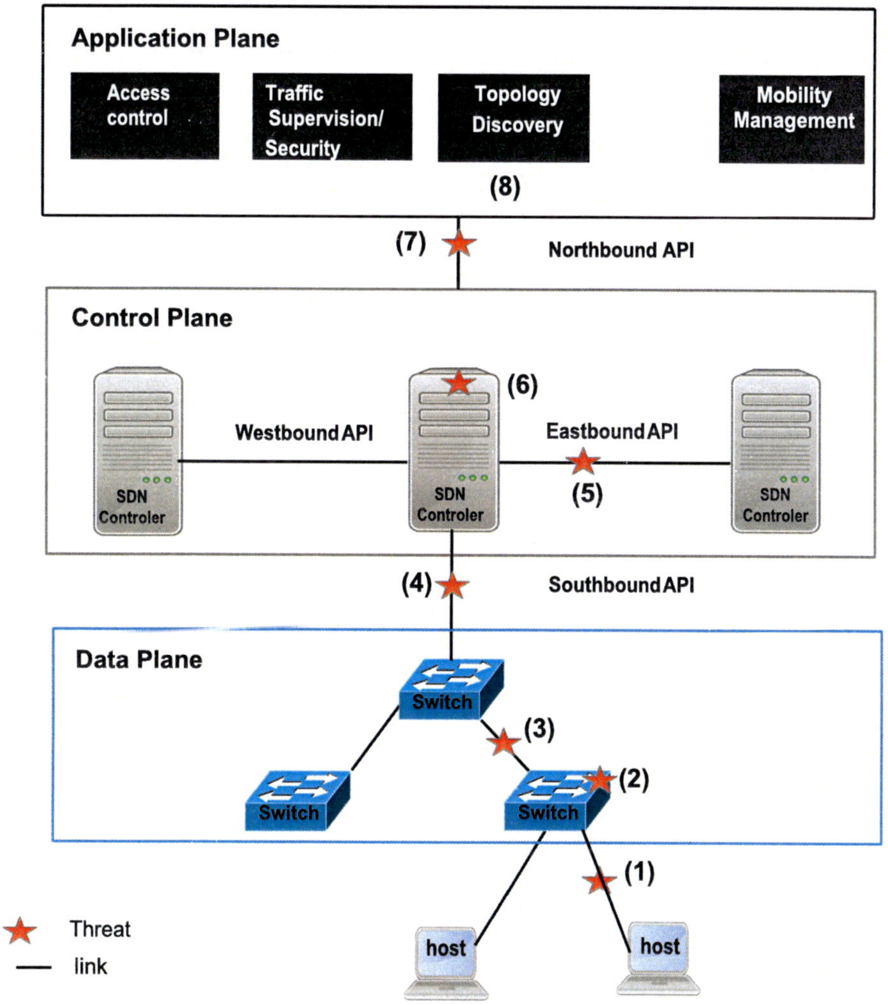

Fig. 1. Threats analysis in SDN

The centralized management of a network through a controller is fundamental for future networks, especially the Internet of Things. However, it also opens up new security challenges. The challenges are numerous, from the risk of a critical point in the SDN controller to interface vulnerabilities and the need for authentication and authorization of network applications. Moreover, the list of these security challenges is expected to increase as the SDN is progressively deployed. Table 2 below summarizes the main security issues that exist in the three layers of the SDN.

Table 2. Major Security threats in SDN

SDN plane/Layer	Proposal	Threat type	Description
Application plane	[9–13]	Lack of authentication and authorization	No robust authentication and authorization mechanism for applications
	[14, 15]	Fraudulent insertion of flow rules	Malicious or compromised applications can generate false flow rules, and it is complicated to verify if it comes from a compromised application
	[10, 16]	Lack of access control and traceability	Difficulty in implementing access control and traceability on third-party applications
Control plane	[17, 18]	Denial of Service attacks (DoS)	The centralization of network intelligence on a controller and the controller's limited processing resources are the main reasons for DoS attacks
	[19, 20]	Unauthorized access to the controller	No powerful mechanism to implement access control on an application
	[21]	Interception	Centralizing intelligence in a single entity can pose scalability and availability issues
Data plane	[22, 23]	Misconfiguration	The flow tables of OpenFlow switches can store a limited number of flow rules
	[24, 25]	Controller compromise	The data plan depends on the control plan that ensures its security. The security of the data plan depends on the security of the controller
	[26, 27]	Man-in-the-middle attack	Due to the optional use and the complexity of TLS protocol

DoS is within the vulnerabilities category because the selected studies focused on its vulnerabilities, although it is an attack. In it were grouped the papers involving DoS and DDoS attacks. The selected papers focus on exploiting the Controller's difficulty or a switch running with resource exhaustion (CPU and memory usage). The ease of increasing network latency with packet flooding and the difficulty of identifying the real devices that exist in the network, expanding the network's size with fake devices in a way that makes network management unfeasible.

The vulnerabilities related to the Access Control category exploit the possibility of accessing SDN components based on access permission policies and delegations problems. The Authentication category presents vulnerabilities to prospect the problem of impersonation: the difficulty of identifying whether an entity in the network is itself (be it a Controller device, switch, or even network administrator). The Confidentiality Breach category organizes studies in which the focus is on accessing sensitive information. Flow tables and flow rules are examples of information hijacked due to vulnerabilities in this category. The Misconfiguration category refers to vulnerabilities caused by network administrators, who intentionally or unknowingly inject misconfigurations into the network. The Integrity Violation category deals with the possibility of inserting spoofed information into the SDN network. Studies in this category focus on modifying the original data, such as packet data and flow rules. In the Controller Failure category, problems caused by Controller malfunction are related. This group of vulnerabilities studies the causes and consequences of this malfunction. The Network Update category focuses on the time lapse between network configuration changes, such as flow rules, firmware, etc. If a device depends on the device's configuration being updated during this period, conflicts can occur and open security holes. Finally, the lack of TLS category studies identifies the impacts of not using TLS to secure communication between SDN devices [28]. Although the TLS protocol is a possible solution, concerns are raised about how network devices manufacturers have implemented TLS. For example, the Openflow protocol specification is weak in this regard [29]. It only suggests using TLS in communication between devices without providing more details about how this use should occur.

Attacks directly related to SDN are rare, due to the early stage of research related to security in SDN. However, some attacks were found in the literature and are represented in Table 2. The identified attacks were Control Plane Saturation, which occurs when an attacker continuously injects requests for flows, to flood the communication between the Controller and the switch, Interception, which occurs when the attacker captures the control plane traffic, to divert the traffic, inspect the data or modify it.

In addition to these attacks, Poisoning attacks were identified, where an attacker injects packets on the network, to flood the data plane, resulting in a DoS; the Flow Table Overflow attack, where this category of attacks aims to produce new flows, in such an amount that exceeds the capacity of the flow tables, and thereby cause a DoS on switches. This can occur via a controller or a Men-in-the-Middle (MitM) attack [30]. Finally, the Fingerprinting attack, where this type of attack aims to obtain information about the Controller, routes, or key devices on the network, such as servers, for example, from variations in network metrics, such as through the time difference when traversing packets over the network.

4 SDN Security Mitigations

This section discusses some of the protection methods suggested for these particular threats and offers some general solutions to ensure the security of both the control level and the level of data transmission.

As a result, centralized network management is available to the network controller and the applications integrated. The SDN architecture provides an opportunity both to improve existing and create fundamentally new means of information protection. A wide range of research examines methods of protecting OpenFlow networks from specific, specific to Open-Flow threats to information security. In addition, means to counter traditional network threats, which can be prevented by using the features of the OpenFlow protocol, are being improved.

A practical attack prevention method for OpenFlow networks is pre-detection of network configuration inconsistencies, which can be achieved by applying verification models. Al-Shaer et al. [31] described FlowChecker, which uses binary detection diagrams to divide network resources into layers. Khurshid et al. [32] propose to check traffic flow processing policies using modular theory. The VeriFlow project implements a real-time flow rule checking policy. A utility has been developed that intercepts flow rules sent by a network controller to an OpenFlow switch to analyze those flows further. C. Schlesinger et al. [33] propose a technology to verify the isolation of traffic of a particular application, dividing the network infrastructure into layers to achieve confidentiality and data integrity.

It is important to note that SDN features can improve the traditional approaches to control traffic flows. X. Liu et al. [34] propose an original approach to network management in which traffic flows are allowed or denied based on source and destination security levels. G. Yao et al. [35] propose a network architecture with mandatory sender address verification for each data source. The network controller performs the verification in reactive mode for each packet transmitted to the controller due to non-compliance with any flow table rule.

J. Jafarian et al. [36] present an implementation of Moving Target Defense (MTD) for networks based on the OpenFlow protocol. MTD is implemented by hiding real internal host addresses for external sources. The controller assigns virtual IP addresses to network hosts and performs translation while forwarding. Due to the unpredictable distribution of virtual addresses, attacking specific infrastructure elements becomes much more labor-intensive.

Several papers investigate authentication and authorization mechanisms at the application level. Mechanisms adapted for differentiation of access of different services and network services are proposed. The FortNox framework resolves conflicts related to conflicting rules from different sources. It is an extension for the NOX controller, which checks flow rules before applying them to network devices. Shin et al. [37] have developed a framework for OpenFlow network security control (FRESCO) that allows information security-related applications to be integrated into OpenFlow networks. Wen et al. [38] present the PermOF control system to give OpenFlow applications the minimum rights necessary to prevent abuse.

Since the most probable area of SDN implementation in RF is cloud technologies, the most serious security threat from the commercial point of view is "denial of service" (DoS-, DDoS-attacks), this topic deserves special attention. It is fundamentally important to note that OpenFlow technologically allows you to use the SDN network as a sensor, because information about the flows can be obtained from all critical points.

The AVANT-GUARD controller extension [39] optimizes network resources by reducing the amount of control traffic sent between the controller and switches to mitigate DDoS attacks. One of the most effective protection methods against DoS attacks in networks of traditional architecture is intrusion prevention systems (IDS). Several researchers have suggested architectures or implemented experimental IDS based on the principles of OpenFlow networks. Hu et al. [40] developed an IDS that analyzes events in an OpenFlow network with multiple controllers; the architecture of the developed system allows the analysis of a large number of traffic flows, based on intrusion detection rules configured by the network engineer. NICE [41] is a mechanism for detecting DoS attacks in OpenFlow networks, based on an analytical model of attack graphs.

Moazenni et al. [42] propose a Reliable Distributed SDN (RDSDN) method to improve the fault tolerance of SDN controllers. An architecture of multiple distributed controllers is used, where each controller is responsible for one subnetwork as a master, and defined as a slave for the other subnets. A new formula is proposed to calculate the reliability rate of each subnet. Thus, the calculated reliability rates are shared among the controllers, to select the controller with the highest reliability value, chosen as the coordinator for the network. In RDSDN, the failure of the controllers is detected by the coordinator, who decides which other controller is the most appropriate to take over the subnetwork whose controller has failed. Each of the above methods of protection against attacks was evaluated in terms of its strengths and weaknesses. The solutions were analyzed to identify the gaps that had not yet been filled from the point of view of SDN security. To summarize the methods analyzed above, we created a table showing each of the solutions, the type of attack they are trying to protect against, the logical level of SDN at which they are working, and the security aspects they are trying to maintain.

As SDN vulnerabilities are still in the discovery and solution phase, several works that answer RQ1 (main vulnerabilities) do not yet have a defense mechanism. Several works that answer RQ2 (main defense mechanisms) exploit SDN's benefits to solve security problems in traditional networks, thus not directly correlated with the vulnerabilities categorized in this paper.

Table 3 examines possible attacks on all software and hardware network levels and analyzes protection systems. The structure of these systems, their advantages and limitations were analyzed, a comparative characteristic of protection methods was carried out. It was shown that the general problem with the proposed protection solutions is that they are designed to protect against only one particular type of attack and do not provide for general protection in network settings. To ensure a protected software configuration network, it would be worth considering parallel implementation of combinations of the proposed schemes. This can lead to a large number of processing costs and the creation of conflicting configurations. These problems lead to the possible emergence of additional variability and obtaining compromising solutions when scanning the network for attacks. However, it is clear that by adding additional protection mechanisms to the controller, the productivity of the network in terms of its throughput capacity can be negatively changed. To improve safety in SDN networks, it would be important to balance security and productivity by making a critical, network-wide safety mode.

Therefore, developing a comprehensive system that would simultaneously protect against several types of attacks is necessary.

Table 3. Security mitigations for SDN Layers (Planes)

SDN layer (Plane)	Security projects	Targeted threat	Main contribution	Mitigation type	Limitations
Control Plane	HyperFlow [43]	Controller availability	Provides a scalable logically centralized but physically distributed control platform based on event propagation	Detection framework	The additional delay when a controller converges or reaches network synchronization increases the response time
	McNettle [44]	Controller scalability	A multiple processor core SDN controller developed for large networks supporting resource intensive control algorithms	Control event processing throughput scales	Control modeling greatly impacts the network scalability
	AVANT-GUARD [39]	DDoS	Reduced communication overhead between data and control plane, fast response to changing flow dynamics at the data layer	Control plane saturation attack	The problem is that Avant-Guard can only handle TCP-based attacks and introduces latency for legal SYN packets
	DISCO [45]	Controller scalability	Provides security functions at the control layer for distributed heterogeneous networks	Distributed controller architecture	The main challenge is the overall consistency of views across all controllers
	Floodguard [46]	DoS	Provides a scalable, lightweight security solution for SDN networks using two modules: Proactive Flow Analyzer and Packet Migration	Proactive flow rule analyzer	The buffer saturation attack between control-data plane
	SoftGuard [47]	Low-rate TCP attack	Installs flow rules on switches to detect low-rate TCP attacks, and performs mitigation on ingress switchs	Detection framework	Processing load on the controller
	Enhanced Solid-Flow ([48]	Malicious Flow Rules	Propose a new security mechanism based on hashing algorithms and authorization to handle flow rules insertion	Flow rules database integrity	Effective only against flow rules integrity

(*continued*)

Table 3. (*continued*)

SDN layer (Plane)	Security projects	Targeted threat	Main contribution	Mitigation type	Limitations
Data Plane	FlowChecker [31]	Faulty flow rules	Configuration Analyzer of OpenFlow switches	Configuration verification tool	Bad effects on performance - Does not check entire network
	Monitoring Function Openflow [49]	Data plane fault recovery	Monitoring function on OpenFlow switches	Place a general message generator and processing function	Processing load on the controller
	VeriFlow [32]	Faulty flow rules	Flow rules checker	Use prefix tree (an ordered tree data structure)	Bad effects on performance
	FortNOX [50]	Flow rules contradictions	Provides an authorization mechanism based on digital signatures and security constraint thanks to an extension on a NOX controller	Controller framework	FortNOX records rule relations in alias sets, which are unable to track network traffic flows accurately
	Packet-In Filtering [51]	Packet-In messages	Filtering the mechanism on OpenFlow switches	Extend the OpenFlow specification	CPU loads in the switches/
	ISAVA [52]	DoS	An SDN-based integrated IP source address validation architecture	Packet signature signing and verification protocol between AS allianc	It complicates system implementation as it has to modify client's host-stack
	TMPTCP [53]	Multipath transmission control protocol (MPTCP) vulnerabilities	Combines the SAV technology and the conventional MPTCP	Source Address Validation (SAV)	Switch computing overhead
Application Plane	SE-Floodlight [50]	Applications authorization	Confict resolution, authorization, audit system	Secure controller architecture & secure AppCtrl API	The design presentation also does not discuss the administrative responsibilities in configuring and operating a secure control layer in an operational deployment scenario

(*continued*)

Table 3. (*continued*)

SDN layer (Plane)	Security projects	Targeted threat	Main contribution	Mitigation type	Limitations
	FRESCO [37]	Threats within/from applications	Using the FRESCO scripting language, applications can define constraint-based actions for flows such as drop, forward, redirect, mirror, network separation, etc	Security applications development framework	Leads to many issues of reliability and scalability
	PERM-GUARD [15]	Hijacked/Rogue Controller	Cross-layer protection framework	Flow rules	Controllers might potentially become a bottleneck for the network operations
	PermOF [38]	Access control	Resource isolation and access control	Minimize SDN application privileges	Leads to many issues of reliability and scalability
	FLOVER [54]	Security policy violation	Flow rules checker	Utilize the satisfiability modulo theories	VeriFlow boasts low-latency of the checking process, it cannot handle multiple controllers
	SDNSOC [55]	Flow rules Conflict Detection	Proposed an object oriented programming framework - SDN Security Operation Center (SDNSOC), which handles policy composition	Policy Conflict Detection	Does not consider indirect security policy violation detection
	SAIDE [56]	Application Interference	Combine the action and refactor fields of policies to detect the application interferences	SDN Application Interference Detection and Elimination	Cannot handle multiple controllers

5 Conclusion

Since the beginning of computer network development, several new protocols, specifications, methodologies, techniques, approaches, and technologies have emerged for the network to solve society's growing demands. However, the increased complexity of these solutions and the accelerated growth of new solutions made networks ossified, increasing the challenge of proposing innovative solutions due to the need to keep compatibility with what has already been implemented in the past, which impacts the resistance to disruption. As an alternative to the status quo, the SDN paradigm was proposed, enabling the development of solutions more decoupled through the separation of the network plans (management, control, and data). The break of the vertical integration and applying the separation of responsibilities principle make some new functionalities available, such as the network logical centralization and the network programmability. This allows the development of new solutions and functionalities for the network to emerge without compromising its operation. As the paradigm evolved, security started to be a critical point. Several vulnerabilities were discovered and exploited to provide several types of attacks, thus requiring new defense mechanisms. This paper categorized the main vulnerabilities, attack types, and mitigations mechanisms involving SDN described in the literature.

References

1. Son, J., Buyya, R.: A taxonomy of software-defined networking (SDN)-enabled cloud computing. ACM Comput. Surv. **51**(3), 59 (2018)
2. Singh, J., Behal, S.: Detection and mitigation of DDoS attacks in SDN: a comprehensive review, research challenges and future directions. Comput. Sci. Rev. **37**, 100279 (2020). https://doi.org/10.1016/j.cosrev.2020.100279
3. Banse, C., Schuette, J.: A taxonomy-based approach for security in software-defined networking. In: 2017 IEEE International Conference on Communications (ICC), pp. 1–6 (2017). https://doi.org/10.1109/ICC.2017.7997245
4. Petersen, K., Feldt, M., Mujtaba, S., Mattsson, M.: Systematic Mapping Studies in Software Engineering. In: 12th International Conference on Evaluation and Assessment in Software Engineering, vol. 12, pp. 1–10 (2008)
5. Bailey, J., Budgen, D., Turner, M., Kitchenham, B., Brereton, P., Linkman, S.: Evidence relating to object-oriented software design: a survey. In: First International Symposium on Empirical Software Engineering and Measurement (ESEM 2007), pp. 482–484 (2007). https://doi.org/10.1109/ESEM.2007.58
6. Ahmad, I., Namal, S., Ylianttila, M., Gurtov, A.: Security in software defined networks: a survey. IEEE Commun. Surv. Tutor. **17**(4), 2317–2346 (2015). https://doi.org/10.1109/COMST.2015.2474118
7. Shaghaghi, A., Kaafar, M.A., Buyya, R., Jha, S.: Software-defined network (SDN) data plane security: issues, solutions and future directions, arXiv:1804.00262 (2018)
8. Dargahi, T., Caponi, A., Ambrosin, M., Bianchi, G., Conti, M.: A survey on the security of stateful SDN data planes. IEEE Commun. Surv. Tutorials **19**, 1701–1725 (2017)
9. Namal, S., Ahmad, I., Gurtov, A., Ylianttila, M.: Enabling secure mobility with OpenFlow. In: 2013 IEEE SDN for Future Networks and Services (SDN4FNS), pp. 1–5 (2013). https://doi.org/10.1109/SDN4FNS.2013.6702540
10. Shuangyu, H., Jianwei, L., Jian, M., Jie, C.: Hierarchical solution for access control and authentication in software defined networks. In: Au, M.H., Carminati, B., Kuo, C.-C. (eds.) NSS 2014. LNCS, vol. 8792, pp. 70–81. Springer, Cham (2014). https://doi.org/10.1007/978-3-319-11698-3_6
11. Wang, H.: Authentic and confidential policy distribution in software defined wireless network. In: 2014 International Wireless Communications and Mobile Computing Conference (IWCMC), pp. 1167–1171 (2014). https://doi.org/10.1109/IWCMC.2014.6906520
12. Cui, H., Chen, Z., Yu, L., Xie, K., Xia, Z.: Authentication mechanism for network applications in SDN environments. In: 2017 20th International Symposium on Wireless Personal Multimedia Communications (WPMC), pp. 1–5 (2017). https://doi.org/10.1109/WPMC.2017.8301788

13. Nife, F., Kotulski, Z.: New SDN-oriented authentication and access control mechanism. In: Gaj, P., Sawicki, M., Suchacka, G., Kwiecień, A. (eds.) Computer Networks, pp. 74–88. Springer, Cham (2018). https://doi.org/10.1007/978-3-319-92459-5_7
14. Mekky, H., Hao, F., Mukherjee, S., Zhang, Z.-L., Lakshman, T.V.: Application-aware data plane processing in SDN. In: Proceedings of the Third Workshop on Hot Topics in Software Defined Networking, pp. 13–18 (2014). https://doi.org/10.1145/2620728.2620735
15. Wang, M., Liu, J., Chen, J., Liu, X., Mao, J.: PERM-GUARD: authenticating the validity of flow rules in software defined networking. In: Proceedings of 2nd IEEE International Conference Cyber Security Cloud Computing CSCloud 2015 - IEEE Int. Symposium Smart Cloud, IEEE SSC 2015, no. 37, pp. 127–132 (2016). https://doi.org/10.1109/CSCloud.2015.89
16. Klaedtke, F., Karame, G.O., Bifulco, R., Cui, H.: Access control for SDN controllers. In: Proceedings of the Third Workshop on Hot Topics in Software Defined Networking, pp. 219–220 (2014). https://doi.org/10.1145/2620728.2620773
17. Alcorn, J.A., Chow, C.E.: A framework for large-scale modeling and simulation of attacks on an OpenFlow network. In: 2014 23rd International Conference on Computer Communication and Networks (ICCCN), pp. 1–6 (2014). https://doi.org/10.1109/ICCCN.2014.6911848
18. Chin, T., Mountrouidou, X., Li, X., Xiong, K.: Selective packet inspection to detect DoS flooding using software defined networking (SDN). In: 2015 IEEE 35th International Conference on Distributed Computing Systems Workshops, pp. 95–99 (2015). https://doi.org/10.1109/ICDCSW.2015.27
19. Feghali, A., Kilany, R., Chamoun, M.: SDN security problems and solutions analysis. In: 2015 International Conference on Protocol Engineering (ICPE) and International Conference on New Technologies of Distributed Systems (NTDS), pp. 1–5 (2015). https://doi.org/10.1109/NOTERE.2015.7293514
20. Klaedtke, F., Karame, G.O., Bifulco, R., Cui, H.: Towards an access control scheme for accessing flows in SDN. In: Proceedings of the 2015 1st IEEE Conference on Network Softwarization (NetSoft), pp. 1–6 (2015). https://doi.org/10.1109/NETSOFT.2015.7116185
21. Zhang, L., Wang, Z., Gu, K., Miao, F., Guo, Y.: Transparent synchronization based port mutation scheme in SDN network. In: 2016 5th International Conference on Computer Science and Network Technology (ICCSNT), pp. 581–585 (2016). https://doi.org/10.1109/ICCSNT.2016.8070225
22. Matsumoto, S., Hitz, S., Perrig, A.: Fleet: defending SDNs from malicious administrators. In: Proceedings of the Third Workshop on Hot Topics in Software Defined Networking, pp. 103–108 (2014). https://doi.org/10.1145/2620728.2620750
23. Lévai, T., Pelle, I., Németh, F., Gulyás, A.: EPOXIDE: a modular prototype for SDN troubleshooting. SIGCOMM Comput. Commun. Rev. **45**(4), 359–360 (2015). https://doi.org/10.1145/2829988.2790027
24. Brooks, M., Yang, B.: A man-in-the-middle attack against OpenDayLight SDN controller. In: Proceedings of the 4th Annual ACM Conference on Research in Information Technology, pp. 45–49 (2015). https://doi.org/10.1145/2808062.2808073
25. Chi, P.-W., Kuo, C.-T., Guo, J.-W., Lei, C.-L.: How to detect a compromised SDN switch. In: Proceedings of the 2015 1st IEEE Conference on Network Softwarization (NetSoft), pp. 1–6 (2015). https://doi.org/10.1109/NETSOFT.2015.7116184
26. Schehlmann, L., Abt, S., Baier, H.: Blessing or curse? Revisiting security aspects of software-defined networking. In: 10th International Conference on Network and Service Management (CNSM) and Workshop, pp. 382–387 (2014). https://doi.org/10.1109/CNSM.2014.7014199

27. Mihai-Gabriel, I., Victor-Valeriu, P.: Achieving DDoS resiliency in a software defined network by intelligent risk assessment based on neural networks and danger theory. In: 2014 IEEE 15th International Symposium on Computational Intelligence and Informatics (CINTI), pp. 319–324 (2014). https://doi.org/10.1109/CINTI.2014.7028696
28. Ranjbar, A., Komu, M., Salmela, P., Aura, T.: An SDN-based approach to enhance the end-to-end security: SSL/TLS case study. In: NOMS 2016 - 2016 IEEE/IFIP Network Operations and Management Symposium, pp. 281–288 (2016). https://doi.org/10.1109/NOMS.2016.7502823
29. Midha, S., Triptahi, K.: Extended TLS security and defensive algorithm in OpenFlow SDN. In: 2019 9th International Conference on Cloud Computing, Data Science & Engineering (Confluence), pp. 141–146 (2019). https://doi.org/10.1109/CONFLUENCE.2019.8776607
30. Zhang, K., Qiu, X.: CMD: a convincing mechanism for MITM detection in SDN. In: 2018 IEEE International Conference on Consumer Electronics (ICCE), pp. 1–6 (2018). https://doi.org/10.1109/ICCE.2018.8326334
31. Al-Shaer, E., Al-Haj, S.: FlowChecker: configuration analysis and verification of federated Openflow infrastructures. In: Proceedings of the 3rd ACM Workshop on Assurable and Usable Security Configuration, pp. 37–44 (2010). https://doi.org/10.1145/1866898.1866905
32. Khurshid, A., Zou, X., Zhou, W., Caesar, M., Godfrey, P.B.: VeriFlow: verifying network-wide invariants in real time. In: 10th {USENIX} Symposium on Networked Systems Design and Implementation ({NSDI} 13), pp. 15–27 (2013). https://www.usenix.org/conference/nsdi13/technical-sessions/presentation/khurshid
33. Schlesinger, C., Story, A., Gutz, S., Foster, N., Walker, D.: Splendid isolation: Language-based security for software-defined networks. In: Proceedings of First Working Hot Topics Software Defined Networks, pp. 79–84. ACM (2012)
34. Liu, B., Bi, J., Zhou, Y.: Source address validation in software defined networks. In: Proceedings of the 2016 ACM SIGCOMM Conference, pp. 595–596 (2016). https://doi.org/10.1145/2934872.2960425
35. Yao, G., Bi, J., Xiao, P.: Source address validation solution with OpenFlow/NOX architecture. In: 2011 19th IEEE International Conference on Network Protocols, pp. 7–12 (2011). https://doi.org/10.1109/ICNP.2011.6089085
36. Jafarian, J.H., Al-Shaer, E., Duan, Q.: Formal approach for route agility against persistent attackers. In: Crampton, J., Jajodia, S., Mayes, K. (eds.) ESORICS 2013. LNCS, vol. 8134, pp. 237–254. Springer, Heidelberg (2013). https://doi.org/10.1007/978-3-642-40203-6_14
37. Shin, S., Porras, P., Yegneswaran, V., Gu, G.: FRESCO: modular composable security services for software-defined networks. In: Annual Network & Distributed System Security Symposium, NDSS, vol. 1, no. 1, pp. 1–16 (2013)
38. Wen, X., Chen, Y., Hu, C., Shi, C., Wang, Y.: Towards a secure controller platform for openflow applications (2013)
39. Shin, S., Yegneswaran, V., Porras, P., Gu, G.: Avant-guard: scalable and vigilant switch flow management in software-defined networks. In: Proceedings of the 2013 ACM SIGSAC Conference on Computer & Communications Security, pp. 413–424 (2013)
40. Hu, Y., Su, W., Wu, L., Huang, Y., Kuo, S.: Design of event-based intrusion detection system on OpenFlow network. In: 2013 43rd Annual IEEE/IFIP International Conference on Dependable Systems and Networks (DSN), pp. 1–2 (2013). https://doi.org/10.1109/DSN.2013.6575335
41. Chung, C., Khatkar, P., Xing, T.: NICE: network intrusion detection and countermeasure. IEEE Trans. Dependable Secur. Comput. **10**(4), 1–14 (2013). http://dblp.uni-trier.de/db/journals/tdsc/tdsc10.html#ChungKXLH13

42. Moazzeni, S., Khayyambashi, M.R., Movahhedinia, N., Callegati, F.: On reliability improvement of software-defined networks. Comput. Netw. **133**, 195–211 (2018). https://doi.org/10.1016/j.comnet.2018.01.023
43. Tootoonchian, A., Ganjali, Y.: HyperFlow: a distributed control plane for OpenFlow (2010)
44. Voellmy, A., Wang, J.: Scalable software defined network controllers. In: Proceedings of the ACM SIGCOMM 2012 Conference on Applications, Technologies, Architectures, and Protocols for Computer Communication, pp. 289–290 (2012). https://doi.org/10.1145/2342356.2342414
45. Phemius, K., Bouet, M., Leguay, J.: DISCO: distributed multi-domain SDN controllers. In: 2014 IEEE Network Operations and Management Symposium (NOMS), pp. 1–4 (2014). https://doi.org/10.1109/NOMS.2014.6838330
46. Wang, H., Xu, L., Gu, G.: FloodGuard: a DoS attack prevention extension in software-defined networks. In: 2015 45th Annual IEEE/IFIP International Conference on Dependable Systems and Networks, pp. 239–250 (2015). https://doi.org/10.1109/DSN.2015.27
47. Xie, R., Xu, M., Cao, J., Li, Q.: SoftGuard: defend against the low-rate TCP attack in SDN. In: ICC 2019 - 2019 IEEE International Conference on Communications (ICC), pp. 1–6 (2019). https://doi.org/10.1109/ICC.2019.8761806
48. Qasmaoui, Y., Malehh, A., Haqiq: Enhanced solid-flow: an enhanced flow rules security mechanism for SDN. IAENG Int. J. Comput. Sci. **47**(3) (20200
49. Kempf, J., Bellagamba, E., Kern, A., Jocha, D., Takacs, A., Sköldström, P.: Scalable fault management for OpenFlow. In: 2012 IEEE International Conference on Communications (ICC), pp. 6606–6610 (2012). https://doi.org/10.1109/ICC.2012.6364688
50. Porras, P., Cheung, S., Fong, M., Skinner, K., Yegneswaran, V.: Securing the software defined network control layer. In: Proceedings of the 2015 Network and Distributed System Security Symposium (NDSS), pp. 8–11 (2015). https://doi.org/10.14722/ndss.2015.23222
51. Kotani, D., Okabe, Y.: A packet-in message filtering mechanism for protection of control plane in OpenFlow switches. IEICE Trans. Inf. Syst. **E99.D**(3), 695–707 (2016). https://doi.org/10.1587/transinf.2015EDP7256
52. Zhang, C., et al.: Towards a SDN-based integrated architecture for mitigating IP spoofing attack. IEEE Access **6**, 22764–22777 (2018). https://doi.org/10.1109/ACCESS.2017.2785236
53. Wu, B., Li, H., Wu, Q., Jiang, Z., Liu, J.: TMPTCP: a lightweight trust extension for multipath-TCP. In: 2020 International Conference on Networking and Network Applications (NaNA), pp. 342–347 (2020). https://doi.org/10.1109/NaNA51271.2020.00065
54. Son, S., Shin, S., Yegneswaran, V., Porras, P., Gu, G.: Model checking invariant security properties in OpenFlow. (2013)
55. Chowdhary, A., Huang, D., Ahn, G.-J., Kang, M., Kim, A., Velazquez, A.: SDNSOC: object oriented SDN framework. In: Proceedings of the ACM International Workshop on Security in Software Defined Networks & Network Function Virtualization, pp. 7–12 (2019). https://doi.org/10.1145/3309194.3309196
56. Hu, T., Yi, P., Hu, Y., Lan, J., Zhang, Z., Li, Z.: SAIDE: efficient application interference detection and elimination in SDN. Comput. Netw. **183**, 107619 (2020). https://doi.org/10.1016/j.comnet.2020.107619

Author Index

Y. Maleh et al. (Eds.): ICI2C 2021, LNNS 357, pp. 607–609, 2022.
https://doi.org/10.1007/978-3-030-91738-8